Charles Seale-Hayne Library
University of Plymouth
(01752) 588 588
LibraryandITenquiries@plymouth.ac.uk

Muhammad Sahimi

Flow and Transport in Porous Media and Fractured Rock

© VCH Verlagsgesellschaft mbH, D-69451 Weinheim (Federal Republic of Germany), 1995

Distribution:

VCH, P. O. Box 10 1161, D-69451 Weinheim (Federal Republic of Germany)

Switzerland: VCH, P. O. Box, CH-4020 Basel (Switzerland)

United Kingdom and Ireland: VCH (UK) Ltd., 8 Wellington Court,
    Cambridge CB1 1HZ (United Kingdom)

USA and Canada: VCH, 220 East 23rd Street, New York, NY 10010–4606 (USA)

Japan: VCH, Eikow Building, 10-9 Hongo 1-chome, Bunkyo-ku, Tokyo 113 (Japan)

ISBN 3-527-29260-8

Muhammad Sahimi

Flow and Transport in Porous Media and Fractured Rock

From Classical Methods to Modern Approaches

VCH

Weinheim · New York
Basel · Cambridge · Tokyo

Dr. Muhammad Sahimi
Department of Chemical Engineering
University of Southern California
Los Angeles
CA 90089-1211
USA

Published by
VCH Verlagsgesellschaft mbH, Weinheim (Federal Republic of Germany)

Production Manager: Dipl.-Wirt.-Ing. (FH) Hans-Jochen Schmitt

Cover design: The boxes show, from left to right, (1) a schematic of the types of fractures found in heterogeneous rock, (2) a sample-spanning cluster (orange) superimposed on its backbone (red) and (3) miscible displacement patterns. The background shows a fractal front produced by a simulated system of particles diffusing on a 256 by 256 pixel screen (from B. Sapoval, M. Rosso, J. F. Gouyet and J. F. Colonna, 1986, *Solid State Ionics,* volume 18/19, pages 21–30; reproduced by permission of Elsevier Science B. V.).

Library of Congress Card No: applied for

A catalogue record for this book is available from the British Library.

Die Deutsche Bibliothek – CIP-Einheitsaufnahme
Sahimi, Muhammad:
Flow and transport in porous media and fractured rock : from classical methods to modern approaches / Muhammad Sahimi.
– Weinheim ; New York ; Basel ; Cambridge ; Tokyo : VCH, 1995
  ISBN 3-527-29260-8

© VCH Verlagsgesellschaft mbH, D-69451 Weinheim (Federal Republic of Germany), 1995

Printed on acid-free and chlorine-free paper

Printing: Druckhaus Diesbach, D-69469 Weinheim
Bookbinding: Großbuchbinderei J. Schäffer, D-67269 Grünstadt
Printed in the Federal Republic of Germany

Dedicated to my mother
Fatemeh Fakour Rashid

Preface

Disordered porous media are encountered in many different branches of science and technology, ranging from agricultural, ceramic, chemical, civil and petroleum engineering, to food and soil sciences, and powder technology. Over the past many decades, porous media have been studied both experimentally and theoretically. With the advent of precise instruments and new experimental techniques, it has become possible to measure a wide variety of physical properties of porous media and flow and transport processes therein. New computational methods and technologies have also allowed us to model and simulate various phenomena in porous media, and thus a deep understanding of them has been gained.

Whether we like it or not, we have to accept the fact that many natural porous systems are fractured, understanding of which requires new methodologies and ways of thinking. In the past two decades understanding fractured rock has taken on new urgency since, in addition to oil reservoirs, many groundwater resources are also fractured. Thus, flow in fractured rock has attracted the attention of scientists, engineers and *politicians* as a result of growing concerns about pollution and water quality. Highly intense exploitation of groundwater, and the increase in solute concentrations in aquifers due to leaking repositories and use of fertilizers, have made flow in fractured rock a main topic of research.

I have been working on such problems for 15 years, and during this period I have realized that there are two distinct approaches to modelling flow phenomena in porous media and fractured rock. Some of these approaches belong to a class of models that I call the *continuum models*. Largely based on the classical equations of flow and transport, the continuum models have been very popular with engineers. Although not as widely used as the continuum models, the second approach, which is based on the *discrete models* that represent a porous system by a discrete set of elements and use large-scale Monte Carlo simulations and various statistical methods to analyze flow phenomena in porous media and fractured rock, has also attracted wide attention. Many new ideas and concepts have been developed as the result of using this class of models, and new results have emerged that have helped us gain a much better understanding of porous systems. In addition, such ideas and concepts as percolation processes, universal scaling laws and fractals, the basic tools of the discrete models, have gradually found their rightful positions in the porous systems literature. Currently, such concepts are even taught in graduate courses on flow through porous media, and in courses on computer simulation of disordered media and statistical mechanical systems.

Realizing these facts, and given that there was no book that discusses and compares *both* approaches, I decided to write this book. Even a glance at the immense literature on these subjects reveals that it is impossible to discuss every issue and present an in-depth analysis of it in just one book. Percolation theory, fractals, Monte Carlo simulations, and similar topics have been by themselves the subjects of several books and monographs. Unless one explains the most important concepts and then provides references where the reader can find more material to read, such a book can easily contain over a thousand pages. Based on this realization and limitation, I selected the topics that are discussed in this book. Largely based on this limitation, I also had to ignore several important topics, such as flow of non-Newtonian fluids in porous media, filtration, and dissolution of rock by an acid which creates large fractures in the rock. In spite of such limitations, this book represents, in my opinion, a comprehensive review and discussion of the most important experimental and theoretical approaches to flow phenomena in porous media and fractured rock, and as such it can be used both as a reference book and a text for graduate courses on the subjects that it discusses. Considering the fact that this book discusses experimental measurement of the most important morphological and transport properties of porous systems, and the fact that many topics, especially those of single-phase flow and transport, are discussed in great detail, I believe that roughly half of the book can also be used in a senior-level undergraduate course on porous media problems that is taught in many chemical, petroleum and civil engineering and geological science departments.

As the famous song by John Lennon and Paul McCartney goes, "I get by with a little help from my friends," except that in my case my friends and colleagues have given me *a lot* of help. Many people have contributed to my understanding of the topics discussed in this book, a complete list of whom would be too long to be given here. However, I would like to mention a few of them who have had great influence on my way of thinking about these problems. I would like to thank Professors H. Ted Davis and L. E. (Skip) Scriven of the University of Minnesota who introduced me to various porous media problems and percolation theory, and taught me the fundamental concepts when I was their doctoral student. For over a decade, Dietrich Stauffer has greatly influenced my way of thinking about percolation, disordered media, and critical phenomena. I am deeply grateful to him. I would like to thank all of my past collaborators with whom I have published many papers on flow in porous media, especially Adel A. Heiba and Barry D. Hughes. Three other persons helped me write and finish this book. Michael Poulson, my publisher at VCH Publishers, was very patient and helpful. Drs. Sherry Caine and Dalia Goldschmidt helped me organize my thoughts, focus on writing this book, and have a more positive outlook of life.

My debts of gratitude to them, and to many more who taught me and influenced me, and thus made writing this book possible.

Muhammad Sahimi
Los Angeles,
August 1994

Contents

Chapter 1

Continuum versus Discrete Models

1.0 Introduction

Transport and flow phenomena in porous media and fractured rock, or industrial synthetic porous matrices, arise in many diverse fields of science and engineering, ranging from agricultural, biomedical, construction, ceramic, chemical, and petroleum engineering to food and soil sciences, and powder technology. Fifty percent or more of the original oil-in-place is left in a typical oil reservoir by traditional (primary and secondary) recovery techniques. This unrecovered oil is the main target for enhanced or tertiary oil recovery methods now being developed. However, oil recovery processes constitute only a small fraction of an enormous, and still rapidly growing, literature on porous media. In addition to oil recovery processes, the closely related areas of soil science and hydrology are perhaps the best-established topics. The study of groundwater flow and the restoration of aquifers that have been contaminated by various pollutants are important current areas of research on porous media problems. Classical research areas of chemical engineers that deal with porous media include filtration, centrifuging, drying, multiphase flow in packed columns, and diffusion and reaction in porous catalysts. Lesser known, but equally important, phenomena involving porous media are also numerous. For example, for the construction industry, transmission of water by building materials (bricks or concretes) is an important problem to consider when designing a new building. Various properties of wood, an interesting and unusual porous medium, have been studied for a long time. Some of the phenomena involving wood include drying and impregnation by preservatives. Civil engineers have long studied asphalts as water-resistant binders for aggregates, protection of various types of porous materials from frost heave, and the properties of road beds and dams with respect to water retention. Some porous media whose pore space morphology and wetting behavior are of physiological interest are skin, hair, feathers, teeth and lungs. Other types of porous media that are widely used are ceramics, pharmaceuticals, contact lenses, explosives, and various kinds of membranes. In any phenomenon that involves a porous system one has to deal with the complex pore structure of the porous medium and how it affects the distribution, flow, displacement of one or more fluids, or dispersion (i.e., mixing) of one fluid in another. Each process can, by itself, be very complex. For example, displacement of one fluid by another can be carried out by many different mechanisms, which may involve heat and mass transfer, thermodynamic phase change, and the interplay of various forces such as viscous, buoyancy, and capillary forces. If the solid matrix of the porous medium is deformable, its porous structure may change during flow or some other transport phenomenon. If the fluid is reactive, or if it carries solid particles of various shapes, sizes, and electrical charges, the pore structure of the medium may change due to the reaction of the fluid with the pore surface, or the physicochemical interaction between the particles and the pore surface.

In this book, we study and discuss various experimental, theoretical, and computer simulation approaches to flow, dispersion, and displacement processes in porous media and fractured rock. Most of our discussions regarding porous media are equally applicable to a wide variety of systems, ranging from oil reservoirs to catalysts, woods, and porous composite materials. We discuss flow phenomena only in a *static* medium, i.e., a medium whose morphology does not change during a given process. Thus deformable media, as well as those whose morphology changes due to a chemical reaction, or because of physicochemical interactions between the pore surface and a fluid and its contents, are not discussed here, except when we discuss in Chapter 4 diagenetic processes that form the present natural rock. The interested reader is referred to Sahimi *et al.* (1990) for a comprehensive discussion of transport and reaction in porous media and the resulting changes in their structure.

1.1 A hierarchy of heterogeneities and length scales

The outcome of any given phenomenon in a porous system depends on several length scales over which the system may or may not be homogeneous. By homogeneous we mean a system whose properties are *independent* of its linear size. When there are inhomogeneities in the system that persist at different length scales, the overall behavior of the system is dependent on transport processes such as diffusion, conduction, and convection, the way the fluids distribute themselves in the medium, and the morphology of the system. Often, the morphology of the system plays a role that is more important than that of other influencing factors. Thus, it is important to discuss the effect of various length scales at which the system may be considered homogeneous.

Consider, as an example, a subsurface natural reservoir, perhaps the most common heterogeneous porous medium. In principle, the reservoir is completely deterministic, in that it has potentially measurable properties and features at various length scales, and it could have been easy for us to obtain a rather complete description of the reservoir if only we could excavate each and every part of it. In practice, however, this is not possible, and therefore a description of any reservoir, or any natural porous medium for that matter, is a combination of the deterministic components–the information that can be measured–and indirect inferences which by necessity have stochastic or random elements in them. Over the past two decades, statistical physics of disordered media has played a fundamental role in developing the stochastic component of description of porous structures. There are several reasons for this development. One is that the information and data regarding the structure and various properties of many porous structures are still vastly incomplete. Another reason is that any property that we ascribe to a medium represents an average over some *suitably selected volume* of the medium. However, the relationship between property values and the volume of the system over which the averages are taken remains unknown. The issue of a suitably selected volume reminds us that any proper description of a porous medium or fractured rock has to have a length scale associated with it. In general, the heterogeneities of a natural porous structure can be described at mainly four different length scales which are as follows (Haldorsen and Lake, 1984).

(i) The *microscopic* heterogeneities are at the level of the pores or the grains, and are discernable only through scanning electron microscopy or thin sections.

(ii) The *macroscopic* heterogeneities are at the level of core plugs, and are routinely collected in fields and analyzed. Such heterogeneities are found in every well with property values varying widely from core to core. However, in most theoretical studies, cores are assumed homogeneous and average effective properties are assigned to them, notwithstanding their microscopic heterogeneities.

(iii) The *megascopic* heterogeneities are at the level of the entire reservoirs which may have large fractures and faults. They can be modelled as a collection of thousands, perhaps millions, of cores, oriented and organized in some fashion.

(iv) The *gigascopic* heterogeneities are encountered in landscapes that may contain many such reservoirs as described in (iii), along with mountains, rivers, etc.

Of course, not all of the above heterogeneities are important to all porous media. For example, porous catalysts usually contain only microscopic heterogeneities, and packed beds can be heterogeneous both at the microscopic and macroscopic levels. In this book, we consider only the first three classes of heterogeneities and their associated length scales.

1.2 Long-range correlations, fractals and percolation

At the early stages of studying flow phenomena in porous media and fractured rock, most researchers almost invariably assumed that the heterogeneities in one region or segment of the system were random and uncorrelated with those of other regions and sections. Moreover, it was routinely assumed that such heterogeneities occur at length scales much smaller than the overall linear size of the system. These assumptions were partly due to the fact that it was very difficult to model the system in a more realistic way, because of the computational limitation and lack of precise experimental techniques for collecting the required information. At the same time, these simple conceptual models did help us gain a better understanding of some of the issues. However, increasing evidence suggests that natural rock and soils do not conform to such simplistic assumptions. They exhibit *correlations* in their properties, and such correlations are often present at *all* scales. The existence of such correlations has necessitated the introduction of *fractal geometry*, which tells us how property values of various regions of a system depend on the length (or even time) scale of the observations, how they are correlated with one another, and how one can model such correlations realistically. Such concepts and modelling techniques are discussed in this book.

Once we accept that natural porous media and fractured rock are heterogeneous at many length scales, we also have to live with its consequences. As a simple, yet very important, example, consider the permeability of a porous medium, which is a measure of how easily a fluid can flow through the medium. In a natural porous medium and at large length scales (of the order of a few hundred meters or more) the permeabilities of various regions of the medium exhibit a broad distribution. That is, while parts of the medium may be highly permeable, other parts of it can be practically impermeable. If we consider a natural reservoir, then the low permeability regions can be construed as the impermeable zones, since they contribute little or nothing to the overall permeability, while the permeable zones provide the paths through which a fluid can flow. Thus, the impermeable zones divide the reservoir into compartments according to their permeabilities. This implies that the permeable regions may or may not be connected to one another, and that there is disorder in the interconnectivity of various regions of the reservoir. Thus, if we are to develop a realistic description of a natural porous structure, the interconnectivity of various regions of the system has to be taken into account.

The language for taking into account the effect of the interconnectivity of various regions of a pore space is *percolation theory*. Similar to fractal concepts, percolation has its roots in the mathematics and physics literature, although it was first used by chemists for describing polymerization and gelation phenomena. Percolation tells us how the interconnectivity of various regions of a given system affects its overall properties. It also tells us that if the volume fraction of the permeable regions is below some critical value, the pore space is not permeable, and the overall permeability of the system is zero. In the classical percolation that was studied over 35 years ago, it was assumed that the permeable and impermeable regions are distributed randomly and independently of each other throughout the system. Since then, more refined and realistic percolation models have been developed for taking into account the effect of correlations and many other influencing factors. These ideas and concepts are developed and used throughout this book for describing various flow phenomena in porous media and fractured rock.

1.3 Continuum versus discrete models

Now that we know what kinds of heterogeneities we deal with in this book, it is also necessary to consider the types of models that have been developed over the past several decades for describing flow and transport phenomena in porous media and fractured rock. The analysis of flow, dispersion, and displacement processes in such systems has a long history in connection with the production of oil from underground reservoirs. However, it is only in the past fifteen years that this analysis has been extended to include detailed structural properties of the media. These studies are quite diverse in the physical phenomenon that they consider. In this book we classify the models for flow, dispersion, and displacement processes in systems as *continuum models* and *discrete models*. Continuum models represent the classical engineering approach to describing materials of complex and irregular geometry, characterized by several length scales. The physical laws that govern flow and transport at the microscopic level are well understood, with the possible exception of ultramicroporous structures. Leaving aside that case, one could in principle write down differential equations for momentum, energy, and mass and the associated initial and boundary conditions at the fluid-solid interface. However, as the interface in typical porous structures is very irregular, practical and computationally and economically feasible techniques are not yet available for solving such boundary-value problems–even in the unlikely event that one knows the detailed morphology of the medium. Determination of the precise solid-fluid boundary in anything but the simplest porous media is, and will probably remain, a very difficult (if not impossible) task; the boundary (even if known) within which one would have to solve the equations of change would be so tortuous as to render the problem mathematically intractable. Moreover, even if the solution of the problem could be obtained in such great detail, it would contain much more information than would be useful in any practical sense. Thus, it becomes essential to adopt a macroscopic description at a length scale much larger than the dimension of individual pores or fractures.

Effective properties of a porous medium are defined as averages of the corresponding microscopic quantities. The averages must be taken over a volume that is small enough compared with the volume of the system, but large enough for the equation of change to hold when applied to that volume. At every point in the medium one uses the smallest such volume, and thereby generates macroscopic field variables obeying equations such as Darcy's law of flow or Fick's law of diffusion. The reasons for choosing the smallest suitable volume for averaging are to allow in the theory suprapore variations of the porous medium and to generate a theory capable of treating the usual macroscopic variations of the effective properties. In this book we encounter several situations where the conditions for the validity of such an averaging are not satisfied. Even when the averaging is theoretically sound, the prediction of the effective properties is often difficult because of the complex structure of the system. In any case, with empirical, approximate, or exact formulae for the transport coefficients and other effective properties, the consequences of a given phenomenon in a porous medium can be analyzed with such a theory. As mentioned above, many of the past theoretical attempts to derive effective transport coefficients of porous media from their microstructure entailed a simplified representation of the pore space, often as a bundle of capillary tubes. In this model, the capillaries were initially treated as parallel, and then later, as randomly oriented. These models are relatively simple, easy to use, and sufficiently accurate, provided that the relevant parameters are determined experimentally and the interconnectivity of the pore space does not play a major role. Having derived the effective governing equations and suitable transport properties, one has the classical description of the system as a continuum. We shall therefore refer to various models associated with this classical description as *continuum models*. These models have been widely used because of their convenience and familiarity to the engineer. They do have some limitations, one of which was noted above in the discussion concerning scales and averaging. They are also not well-suited for describing those phenomena in which the interconnectivity of the pore space or the fracture network, or that of a fluid phase, plays a major role. Such models also break down if there are correlations in the system whose extent is comparable with the linear size of the system.

The second class of models, the *discrete models*, are free of these limitations. These models have been advanced to describe phenomena at the microscopic level and have been extended in the last few years to describe various phenomena at the macroscopic and even megascopic levels. Their main shortcoming, from a practical point of view, is the large computational effort required for a realistic discrete treatment of the system. They are particularly useful when the effect of the pore space or fracture network connectivity, or long-range correlations, is strong. The discrete models that we consider in this book are mostly based on a network representation of the rock. The original idea of network representation of a pore space is rather old and goes back to the early 1950s, but it was only in the early 1980s that systematic and rigorous procedures were developed to map, in principle, any disordered porous medium onto an equivalent network. Once this mapping is complete, one can study a given phenomenon in porous media in great detail.

However, only in the past fifteen years have ideas from the statistical physics of disordered media been applied to flow, dispersion, and displacement processes in porous media and fractured rock. These concepts include: percolation theory, the natural language for describing interconnectivity effects; diffusion-limited aggregation processes, which describe fundamentally non-equilibrium growth processes and have been found to have close connections with various flow instability phenomena in porous media; fractal concepts, which are the main tool for describing the scale-dependence of the effective properties of disordered media and how long-range correlations affect them; and universal scaling laws, which describe how and under what conditions the effective macroscopic properties of a system may be independent of its microscopic details. What we intend to do in this book is to review the relevant literature on the subject, define and discuss relevant ideas and techniques from the statistical physics of disordered media and their applications to the processes of interest in this book, and discuss the progress that has been made as a result of such applications. In particular, we emphasize the important effect of the interconnectivity of the pores or fractures of a system on the phenomena of interest, and point out how scaling, percolation, and fractal concepts provide powerful tools for describing flow, dispersion, and displacement processes in porous media and fractured rock. We also treat the subject of characterization of fractured rock and flow and transport in the fracture network of heterogeneous rock in great detail.

1.4 Summary

Summarizing this Chapter, we study models of porous media and fractured rock, explain various experimental techniques that are used for characterizing the morphology of these systems and flow and transport therein, discuss continuum models of flow and transport in such systems, and compare them with discrete models of porous media and fractured rock that are based on a network representation of the system, and employ statistical physics of disordered media and Monte Carlo simulations. In all cases, we contrast the classical models and techniques with the modern approaches. Thus, for example, when we study characterization of porous media and fractured rock (Chapter 5), we discuss both mercury porosimetry and adsorption-desorption methods (the classical techniques), and nuclear magnetic resonance, small-angle scattering, and fractal characterization (the modern approaches). As such, we believe that our book is unique. We hope that this book can give the reader a clear view of where we stand in the middle of 1994.

Chapter 2

The Equations of Change

2.0 Introduction

The purpose of this Chapter is to give the equations of change, the continuity and momentum equations, and the continuity equation for individual chemical species. These equations will be used in the rest of this book, and therefore it may be useful if we summarize them here. We do not derive these equations, as their derivations can be found in any standard book on transport phenomena. Thus we summarize only the differential forms of the equations of change, since their macroscopic or integral forms are not used in this book. We assume that the reader is familiar with basic vector and tensor calculus. We follow the classical book of Bird *et al.* (1960) that this author has used over the past two decades, both as a student and as a teacher. Other readable accounts of these equations are given by Aris (1962) and Batchelor (1967).

2.1 The equation of continuity

If ρ is the density of a flowing fluid and $\mathbf{v}$ is its velocity vector at time t, then the continuity equation is given by

$$\frac{\partial \rho}{\partial t} + \nabla \cdot (\rho \mathbf{v}) = 0 , \tag{2.1}$$

where ∇ is the "del" operator, and $\nabla \cdot (\rho \mathbf{v})$ is the *divergence* of $\rho \mathbf{v}$. For example, in Cartesian coordinates

$$\nabla = \mathbf{e_x} \frac{\partial}{\partial x} + \mathbf{e_y} \frac{\partial}{\partial y} + \mathbf{e_z} \frac{\partial}{\partial z} , \tag{2.2}$$

where $\mathbf{e_x}$, $\mathbf{e_y}$ and $\mathbf{e_z}$ are unit vectors in the x-, y- and z-directions, respectively. In cylindrical coordinates we have

$$\nabla = \mathbf{e_r} \frac{\partial}{\partial r} + \mathbf{e_\theta} \frac{1}{r} \frac{\partial}{\partial \theta} + \mathbf{e_z} \frac{\partial}{\partial z} , \tag{2.3}$$

where $\mathbf{e_r}$, $\mathbf{e_\theta}$, and $\mathbf{e_z}$ are the corresponding unit vectors, and in spherical coordinates the del operator is given by

$$\nabla = \mathbf{e_r} \frac{\partial}{\partial r} + \mathbf{e_\theta} \frac{1}{r} \frac{\partial}{\partial \theta} + \mathbf{e_\Phi} \frac{1}{r \sin \theta} \frac{\partial}{\partial \Phi} . \tag{2.4}$$

If we use the *substantial derivative* operator

$$\frac{D}{Dt} = \frac{\partial}{\partial t} + \mathbf{v} \cdot \nabla \ , \tag{2.5}$$

then the equation of continuity can be rewritten as

$$\frac{D\rho}{Dt} + \rho \nabla \cdot \mathbf{v} = 0 \ . \tag{2.6}$$

For an incompressible fluid ρ is constant, and Eq. (6) reduces to

$$\nabla \cdot \mathbf{v} = 0 \ . \tag{2.7}$$

2.2 The momentum equation

If P is the pressure, $\boldsymbol{\tau}$ is the stress tensor, and $\mathbf{g}$ is the gravitational acceleration vector, then the momentum equation is given by

$$\frac{\partial(\rho \mathbf{v})}{\partial t} = -\nabla \cdot (\rho \mathbf{v}\mathbf{v}) - \nabla \cdot \boldsymbol{\tau} - \nabla P + \rho \mathbf{g} \ . \tag{2.8}$$

Observe that the first term on the right-hand side of Eq. (8) is the rate of convective momentum, while the second term is the rate of viscous momentum, both per unit volume of the system. The term $\mathbf{v}\mathbf{v}$ is a *dyadic product* of two vectors which, similar to $\boldsymbol{\tau}$, is a second-rank tensor. $\nabla \cdot \boldsymbol{\tau}$ and $\nabla \cdot (\rho \mathbf{v}\mathbf{v})$ are thus not simple divergences. Moreover, the stress tensor is symmetric. What distinguishes a Newtonian fluid from a non-Newtonian one is the form of the stress tensor $\boldsymbol{\tau}$. For a Newtonian fluid and in Cartesian coordinates the components of the stress tensor are given by

$$\tau_{\zeta\zeta} = -2\eta \frac{\partial v_\zeta}{\partial \zeta} + \frac{2}{3}\eta(\nabla \cdot \mathbf{v}) \ , \quad \zeta = x, y, z \ , \tag{2.9}$$

$$\tau_{\zeta\xi} = -\eta \left(\frac{\partial v_\zeta}{\partial \xi} + \frac{\partial v_\xi}{\partial \zeta} \right) \ , \quad (\zeta, \xi) = (x, y), \ (y, z), \ (z, x) \ , \tag{2.10}$$

where η is the viscosity of the fluid. Similar expressions for the components of $\boldsymbol{\tau}$ in cylindrical and spherical coordinates are given by Bird *et al.* (1960). In terms of the substantial derivative, the equation of motion can be written as

$$\rho \frac{D\mathbf{v}}{Dt} = -\nabla \cdot \boldsymbol{\tau} - \nabla P + \rho \mathbf{g} \ . \tag{2.11}$$

If the density and viscosity of the fluid are constant, then

$$\rho \frac{D\mathbf{v}}{Dt} = \eta \nabla^2 \mathbf{v} - \nabla P + \rho \mathbf{g} \ , \tag{2.12}$$

which is the well-known *Navier-Stokes equation.* If the viscous effects are negligible, then we obtain the *Euler equation*

$$\rho \frac{D\mathbf{v}}{Dt} = -\nabla P + \rho \mathbf{g} \ , \tag{2.13}$$

which is used for describing flow of an inviscid fluid. If the inertial effects, represented by $\rho D\mathbf{v}/Dt$, are negligible, which is the case for the flow phenomena considered in this book (for which the Reynolds number is very small), we obtain the *Stokes' equation*

$$-\nabla P + \eta \nabla^2 \mathbf{v} + \rho \mathbf{g} = \mathbf{0} \ , \tag{2.14}$$

which is used heavily throughout this book.

2.3 The diffusion and convective-diffusion equations

We consider a binary mixture of two miscible fluids, one of which is the solvent, while the other one is the solute. Suppose that the concentration of the solute is C, and that the molecular diffusivity of the solute in the solvent is D_m. Then the continuity equation for the solute is given by

$$\frac{\partial C}{\partial t} + \nabla \cdot \mathbf{J} = R_A \ , \tag{2.15}$$

where R_A is the molar rate of reaction (if there is any) per unit volume, and $\mathbf{J}$ is the total flux of the solute given by

$$\mathbf{J} = C\mathbf{v} - D_m \nabla C \ . \tag{2.16}$$

Therefore, the equation of continuity for the solute is given by

$$\frac{\partial C}{\partial t} + \nabla \cdot (C\mathbf{v}) = \nabla \cdot (D_m \nabla C) + R_A \ . \tag{2.17}$$

A similar equation can be written down for the solvent. If both $\mathbf{v}$ and D_m are constant and if there is no reaction in the system, then

$$\frac{\partial C}{\partial t} + \mathbf{v} \cdot \nabla C = D_m \nabla^2 C \ , \tag{2.18}$$

which is the well-known *convective-diffusion equation*, heavily used in Chapters 9-11. Here ∇^2 is the Laplacian operator, which in Cartesian coordinates is given by

$$\nabla^2 = \frac{\partial^2}{\partial x^2} + \frac{\partial^2}{\partial y^2} + \frac{\partial^2}{\partial z^2} \ , \tag{2.19}$$

while in cylindrical coordinates we have

$$\nabla^2 = \frac{1}{r} \frac{\partial}{\partial r} \left(r \frac{\partial}{\partial r} \right) + \frac{1}{r^2} \frac{\partial^2}{\partial \theta^2} + \frac{\partial^2}{\partial z^2} \ , \tag{2.20}$$

and in spherical coordinates one has

$$\nabla^2 = \frac{1}{r^2}\frac{\partial}{\partial r}\left(r^2\frac{\partial}{\partial r}\right) + \frac{1}{r^2\sin\theta}\frac{\partial}{\partial\theta}\left(\sin\theta\frac{\partial}{\partial\theta}\right) + \frac{1}{r^2\sin^2\theta}\frac{\partial^2}{\partial\Phi^2} \ . \tag{2.21}$$

If $R_A = 0$ and the fluids are stagnant, then we obtain the well-known diffusion equation

$$\frac{\partial C}{\partial t} = \nabla \cdot (D_m \nabla C) \ . \tag{2.22}$$

More generally, instead of a single-valued diffusivity one may have an *effective diffusivity tensor*. For example, if a porous medium is anisotropic, then each principal direction of the system is characterized by a distinct diffusivity. Even if the medium is not structurally anisotropic, the overall behavior of transport of the solute in the solvent may be characterized by an effective diffusivity tensor. Dispersion phenomena that are studied in Chapters 9 and 10 (and are also important to miscible displacements that are studied in Chapter 11), provide an example of a system in which there is a *flow-induced* dynamical anisotropy, and thus one needs more than one effective diffusivity to characterize the phenomena.

These equations have to be supplemented by additional correlations by which we can calculate physical properties of the fluids, such as their viscosities, densities and diffusivities. Moreover, for flow through porous media we need to relate macroscopic and measurable quantities, such as the average fluid velocity, to the morphological properties of the media. For example, Darcy's law relates the fluid velocity to the permeability K of a porous medium

$$\mathbf{v} = -\frac{K}{\eta}(\nabla P - \rho\mathbf{g}) \ . \tag{2.23}$$

The permeability K depends on the porosity ϕ of the medium, i.e., the volume fraction of its pore space, and a major task is to predict K for a given porous medium. These matters are discussed in Chapter 8. Moreover, when these equations are used for a porous medium, one has to take into account the fact that only a fraction ϕ of the medium is actually used for flow, diffusion and dispersion. Thus, for example, Eq. (17) should be rewritten as

$$\frac{\partial(\phi C)}{\partial t} + \nabla \cdot (\phi C \mathbf{v}) = \nabla \cdot (\phi D_m \nabla C) + R_A \ . \tag{2.24}$$

In most cases studied in this book, the boundary conditions are either of *Dirichlet type*, in which the value of the unknown on the boundaries or a portion of them is specified, or are of *Neumann type* in which the flux of the quantity in the direction normal to the external surface of the system is specified. The effect of the length scale for macroscopic homogeneity of the system, and the correlations between various regions of the system are also important. How they affect the form of the continuum equation of motion for describing a flow phenomenon in a porous medium is an important subject in this book.

Chapter 3

Fractal Concepts and Percolation Theory

3.0 Introduction

In recent years, the complex behavior of a wide variety of phenomena that are of interest to chemists, physicists, and engineers has been quantitatively characterized by using the ideas of fractal and multifractal distributions, which correspond in a unique way to the geometrical shape and dynamical properties of the systems under study. Beginning with the pioneering book of Mandelbrot (1982), many books and review articles have appeared that discuss various aspects of fractal phenomena and their applications to various branches of science and technology; see for example, Feder (1988), Avnir (1990), Bunde and Havlin (1991), Family and Vicsek (1991), Sahimi (1992a), and Vicsek (1992). A key to this remarkable progress has been the observation that many objects with disordered structure possess scale symmetry, such that they look similar on many different length scales. Such structures are thus called *self-similar*, and throughout this book we refer to such system as *geometrical fractals*. Another important observation has been the fact that many *dynamical* properties of disordered materials and media are related to the process time by *power laws* with *non-integer exponents*. We refer to such systems as *dynamical fractals*. Examples include the moments of the distribution of the times that fluid particles spend in a porous medium to travel through it. As we show later in this book, fractal concepts play a fundamental role in characterizing morphology of porous media and fractured rock, and flow and transport processes therein.

Another important tool of characterizing flow and transport phenomena in porous systems is percolation theory. As we already discussed in Chapter 1, percolation is an invaluable tool for studying flow phenomena in porous media and fractured rock, and has given us a much deeper understanding of such phenomena. Moreover, as we discuss in Chapter 5, percolation concepts are essential to the correct interpretation of experimental data obtained with the traditional methods of characterizing porous structures, such as mercury porosimetry and sorption isotherms. Depending on the state of a percolation system and how far it is from its critical point, a percolation process can give rise to fractal objects, so that there is a close connection between percolation and fractal structures and phenomena. As we shall see in Chapters 11 and 12, various flow instability phenomena can also give rise to fractal structures, and therefore it is essential to understand fractal phenomena before discussing various problems in porous media and fractured rock. Thus, the purpose of this Chapter is to review and discuss the essential concepts and ideas of fractals and percolation processes.

We begin by reviewing the general properties of fractal systems, after which percolation phenomena are discussed.

3.1 Box-counting method and self-similar fractals

Although it is possible to give a formal mathematical definition of a fractal system or set, an intuitive definition of fractal systems is probably more useful: In a geometrical fractal *the part is reminiscent of the whole.* This implies the self-similarity and scale-invariance that were already mentioned above. It is important to remember that a system can be fractal above, below, or in between upper and lower cutoff length scales, and be Euclidean otherwise. Non-fractal objects, such as a straight line, a square, or a sphere, are Euclidean and their effective dimensionality is d, the dimensionality of the space in which they are embedded.

The simplest characteristic of a geometrical fractal is its fractal dimension D_f, which is defined as follows. The fractal system is covered by non-overlapping d-dimensional hyperspheres of Euclidean radius r, and the number of such spheres $N(r)$ that is required for the coverage is counted. For a fractal system one has

$$N(r) \sim r^{-D_f} , \qquad (3.1)$$

where $\sim$ means an asymptotic proportionality. Equation (1) can be rewritten as

$$D_f = \frac{\ln N}{\ln(1/r)} . \qquad (3.2)$$

Thus, Euclidean objects such as straight lines, squares, or spheres follow $N(r) \sim r^{-1}$, r^{-2}, and r^{-3}, respectively. Another way of defining the fractal dimension is as follows. One cuts out a segment of a fractal system with linear dimension L and studies the volume $V(L)$ of the system as L is varied. If $V(L)$ is calculated by covering the system by hyperspheres of radius unity, then $V(L) = N(L)$, where N is the number of such spheres required to cover the system. For a fractal system one finds that

$$N(L) \sim L^{D_f} , \qquad (3.3)$$

where D_f is the same as in Eq. (1). Note that with this definition we implicitly assumed a lower and an upper cutoff for the fractal behavior of the system, namely, the radius of the hyperspheres (the lower cutoff) and the linear size of the system L (the upper cutoff). However, we can also define a dimensionless quantity, $\delta = r/L$, and rewrite Eq. (1) in terms of δ. Thus, we may write $N(\delta) \sim \delta^{-D_f}$, where $\delta << 1$, and $N(\delta)$ is the number of hyperspheres of radius δL that are needed for covering the system. The advantage of doing this is that we can use it for *any* fractal system, and also compare various fractal objects since all measurable quantities are now written in dimensionless form. This method of calculating D_f is called the *box-counting method*.

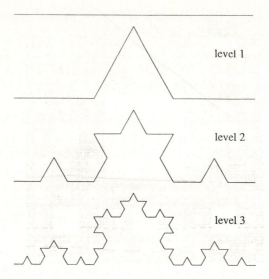

Figure 3.1: Construction of the Koch curve.

Let us provide a few examples of how this definition is actually used for calculating the fractal dimension of a system. Consider, for example, the Koch curve shown in Fig. 3.1. This curve is obtained from a straight line segment by replacing the middle third by the two sides of an equilateral triangle, and successively ad infinitum. Note that, after n iterations, the system is covered by $N = 4^n$ straight line segments, each having a length $r = 3^{-n}$. Therefore, if we write $N = r^{-D_f}$, we obtain

$$D_f = \frac{\ln 4}{\ln 3} \simeq 1.263 , \tag{3.4}$$

so that $D_f > d = 1$. As another example, consider the so-called Sierpinski carpet shown in Fig. 3.2. It is not difficult to see that $D_f = \ln 8/\ln 3 \simeq 1.89$, and thus $D_f < d = 2$

The Koch curve and the Sierpinski carpet are examples of what we call *exact* fractals, because their self-similarity is exact at every stage of their construction. Another type of geometrical fractals is what we call *statistically self-similar* fractals. These systems are self-similar and fractal only in an average sense. Two of the most important examples of such fractals are *percolation clusters* that are discussed below, and *diffusion-limited aggregates* that are considered in Chapter 11. Statistical self-similarity means that if we consider only one snapshot of a disordered system, it does not look self-similar at various scales. However, if we look at many realizations of it and superimpose them on top of one another (i.e., if we find the *average* of all realizations), then the system looks self-similar. Moreover, the *average* number of spheres $N(L)$ required to cover the system obeys Eq. (3).

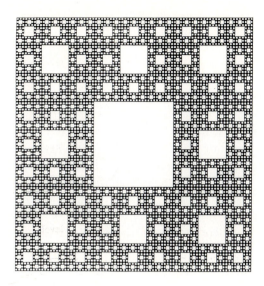

Figure 3.2: The Sierpinski carpet.

Because percolation clusters and many other disordered systems are only statistically self-similar, it may be better to use the term scale-invariance for describing them. Since these systems are disordered, a check of their self-similarity based on visualization alone is also impossible. A powerful method for testing the self-similarity of disordered systems is to construct an *autocorrelation function* $C(\mathbf{r})$ defined by

$$C(\mathbf{r}) = \frac{1}{V} \sum_{\mathbf{r}'} s(\mathbf{r}')s(\mathbf{r} + \mathbf{r}') , \qquad (3.5)$$

where $s(\mathbf{r})$ is a function such that $s(\mathbf{r}) = 1$ if a point at $\mathbf{r}$ belongs to the system, and $s(\mathbf{r}) = 0$ otherwise. If a system is fractal, then its correlation function defined above should remain the same, up to a constant factor, if all the length scales of the system are rescaled by a constant factor b. Thus, one must have $C(br) \sim b^{-x}C(r)$, where $r = |\mathbf{r}|$. It is not difficult to see that one must have $x = d - D_f$. Therefore, for a fractal object the autocorrelation function $C(r)$ *decays* as

$$C(r) \sim r^{D_f - d} . \qquad (3.6)$$

Note that for a Euclidean system $D_f = d$ and the autocorrelation function is a *constant*. This power-law decay of $C(r)$ not only provides a test of self-similarity or fractality of a disordered system, it also gives us a means of estimating its fractal dimension since, according to Eq. (6), if one prepares a logarithmic plot of $C(r)$ versus r, then for a fractal system one should obtain a straight line with a slope $D_f - d$. Experimental methods of measuring fractal properties of porous media and fractured rock are discussed in Chapter 5, where we show that the pore space of many types of porous media, and fracture network of rock have fractal properties.

3.2 Self-affine fractals

The self-similarity of a fractal system implies that its structure is invariant under an isotropic rescaling of lengths, i.e., if all lengths in all directions are rescaled by the same scale factor. However, there are many fractals that preserve their scale-invariance only if lengths in different directions are rescaled by factors that are direction dependent. In other words, the scale-invariance of such systems is preserved only if lengths in x, y, and z directions are scaled by scale factors b_x, b_y, and b_z, where in general these scale factors are not equal. This scale-invariance under a direction-dependent rescaling implies that the fractal system is *anisotropic*. Such fractal systems are called *self-affine*, a term that was used first by Mandelbrot (1985). If a fractal system is self-affine, it can no longer be described by a single fractal dimension. Instead, its *local* or small scale properties can be described by an *effective* fractal dimension, whereas its *macroscopic* behavior is characterized by another *integer* dimension which is, however, *less* than d. An example of a self-affine fractal is the *fractional Brownian motion*, to be discussed below, which has found wide applications. A well-known example of a process that gives rise to a self-affine fractal is a marginally stable growth of an interface. For example, if water displaces oil in a porous medium, the interface between water and oil is a self-affine fractal. This is discussed in Chapter 12.

3.3 Multifractal systems

Exact self-similar fractals usually require only their fractal dimension to be completely characterized. In many cases, especially with statistically self-similar fractals, this characterization of the systems by only their fractal dimension is inadequate, and many more parameters may be needed. Such systems possess *multifractal* properties and require an *infinite* family of parameters to be completely characterized, the simplest of which is D_f. Consider a partition of a set or an object, which may not necessarily be fractal, into N subsets or cells of size r. We take a measure, e.g., in terms of the probability $p_i(r)$ that cell i of the partition has a certain property. We then define the following moments

$$M(r,q) = \sum_{i=1}^{N} p_i^q(r) . \tag{3.7}$$

Two special cases are immediately recognizable. For $q = 0$ we have

$$M(r,0) = N \sim r^{-D_f} , \tag{3.8}$$

which is nothing but the box-counting method discussed above, whereas for $q = 1$ we have

$$M(r, 1) = 1 \sim r^0 , \qquad (3.9)$$

which expresses the conservation of mass. Therefore, it is not unreasonable to expect that in general

$$M(r, q) \sim r^{-\tau(q)} , \qquad (3.10)$$

where we may interpret $\tau(q)$ as a *generalized fractal dimension*. Knowledge of this quantity for $-\infty < q < +\infty$ allows a complete characterization of the fractal (or non-fractal) set. If all $\tau(q)$ are distinct, then we say that the system has multifractal properties. Note that $D_f = \tau(0)$ yields the fractal dimension of the system discussed above, while the limits $q >> 1$ and $q << -1$ yield information about the regions of the system having high or low probability of possessing a certain property, respectively. In general $\tau(0) > \tau(1) > \tau(2) > \cdots$, and thus D_f is the maximum fractal dimension of the system. Geometrically speaking, a multifractal object is a set which, if broken into many pieces, each piece may be a fractal system whose fractal dimension is not necessarily the same as that of any other piece of the object. Multifractal phenomena were first explicitly found in the analysis of non-linear dynamical systems (see, for example, Grassberger, 1983; Hentschel and Procaccia, 1983), but had been used implicitly in the study of turbulent flows (Mandelbrot, 1974). Their use for characterizing turbulent flows was made explicit by Frisch and Parisi (1985), Jensen *et al.* (1985), and Meneveau and Sreenivasan (1987). They have also been found in many other systems of practical interest (see, for example, Meakin *et al.*, 1986; Halsey *et al.*, 1986; Cates and Witten, 1988; Sahimi and Arbabi, 1989). The interested reader should consult Stanley and Meakin (1988) for a fuller exposition to this important conceptual advancement.

An alternative characterization of a multifractal system is possible in terms of the *multifractal spectrum*, or the so-called $f(\alpha)$ curve. This is obtained via the transformation

$$\alpha(q) = -\frac{d\tau(q)}{dq} , \qquad (3.11)$$

$$f[\alpha(q)] = q\alpha(q) - \tau(q) . \qquad (3.12)$$

The largest value of the function f is just the fractal dimension D_f. For simple self-similar fractals one has

$$\tau(q) = D_f(1 - q) , \qquad (3.13)$$

which is again indicative of the fact that only one parameter, the fractal dimension D_f, suffices for characterizing a simple fractal.

Let us now give an example of a multifractal system that has practical importance. A useful method for characterizing fluid flow and mixing through a porous medium is through the *first passage time distribution*. Suppose that we inject several thousand tracer particles into a fluid flowing through a porous medium, and measure the times that they spend to travel with the fluid, pass through the pore space, and arrive at the opposite face of the medium for the first time. Since the system is disordered, each tracer spends a different amount of time to pass through the medium, and thus one has a first passage time distribution, and its associated temporal moments. The time for passing through a pore is inversely proportional to the fluid velocity in the pore, which in turn is proportional to the pressure drop along the pore, or the fluid flux in the pore. It can be shown that the moments of the first passage time distribution (and thus the distributions of the pore flow velocities and fluxes) have multifractal properties. That is, characterization of this distribution requires calculation of *all* of its moments, since in general each moment is *not* related to any other moment. This is discussed in Chapter 9.

3.4 Fractional Brownian motion and long-range correlations

As discussed in Chapter 1, the geometrical and temporal properties of porous systems are characterized by long-range correlations in space and/or time. An example is the distribution of the permeabilities of heterogeneous rock that is discussed in Chapter 6. In this section we discuss one stochastic process that gives rise to fractal sets and long-range correlations, and has found many applications in various problems. We first consider a one-dimensional system, and define a stochastic process $B_H(x)$ with the following mean and variance

$$\langle B_H(x) - B_H(x_0) \rangle = 0 , \tag{3.14}$$

$$\langle [B_H(x) - B_H(x_0)]^2 \rangle \sim |x - x_0|^{2H} , \tag{3.15}$$

where x and x_0 are two arbitrary points, and H is called the Hurst exponent (Mandelbrot and Van Ness, 1968). This stochastic process is called *fractional Brownian motion* (fBm) and is self-affine.

A remarkable property of fBm is that it generates correlations whose extent is *infinite*. For example, if we define a correlation function $C(x)$ by

$$C(x) = \frac{\langle -B_H(-x)B_H(x) \rangle}{\langle B_H(x)^2 \rangle} , \tag{3.16}$$

then one finds that $C(x) = 2^{2H-1} - 1$, i.e., this correlation function is *independent* of x. Moreover, the type of correlations can be tuned by varying H. If $H > 1/2$, then fBm displays *persistence*, i.e., a trend (for example, a high or a low value) at x is likely to be followed by a similar trend at $x + \Delta x$. On the other hand, if $H < 1/2$, then fBm generates *antipersistence*, i.e., a trend at x is not likely to be followed by a similar trend at $x+\Delta x$. For $H = 1/2$ there are no correlations of the type defined above [$C(x) = 0$], and $B_H(x)$ is completely random. Thus, varying H allows us to generate infinitely long-range correlations or anticorrelations.

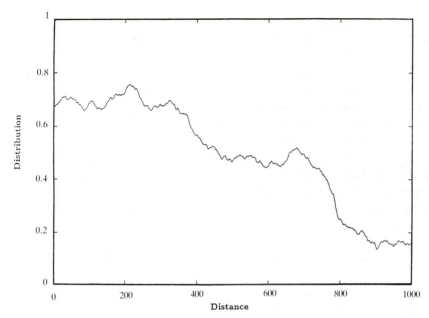

Figure 3.3: Typical pattern generated by a two-dimensional fBm with $H = 0.8$

Figures 3.3 and 3.4 show two typical patterns generated by fBm. One can also use fBm for generating spatial correlations in a two- or three-dimensional space. One now writes

$$\langle [B_H(\mathbf{r}) - B_H(\mathbf{r_0})]^2 \rangle \sim |\mathbf{r} - \mathbf{r_0}|^{2H} , \qquad (3.17)$$

where $\mathbf{r}=(x, y, z)$ and $\mathbf{r_0}=(x_0, y_0, z_0)$ are two arbitrary points in space. If in Eqs. (14) and (15) we replace x by time t, we obtain an fBm that generates long-range temporal correlations.

A convenient way of representing a distribution function is through its spectral density $S(\omega)$, the Fourier transform of its variance. For fBm in d-dimensions it can be shown that

$$S(\omega) \sim \frac{1}{(\sum_{i=1}^d \omega_i^2)^{H+d/2}} . \qquad (3.18)$$

where $\omega = (\omega_1, \cdots, \omega_d)$ is the Fourier-transform variable. This spectral representation also allows us to introduce a cutoff ξ_{co} such that

$$S(\omega) \sim \frac{1}{(\xi_{co} + \sum_{i=1}^d \omega_i^2)^{H+d/2}} . \qquad (3.19)$$

Introducing this cutoff allows us to control the length (or time) scale over which events or spatial properties of a system are correlated (or anticorrelated). Thus, for length (or time) scales $l < 1/\sqrt{\xi_{co}}$ they preserve their correlations (anti-correlations), but for $l > 1/\sqrt{\xi_{co}}$ they become random and uncorrelated. The spectral density representation also provides a convenient method for generating

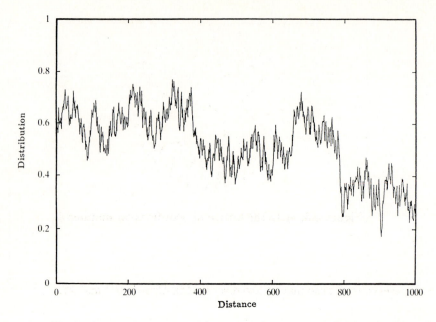

Figure 3.4: Typical pattern generated by a two-dimensional fBm with $H = 0.2$.

a sequence of numbers that obey a fBm using a fast Fourier transform technique. One generates first random numbers, uniformly distributed in (0,1), and assigns them to the sites or bonds of a d-dimensional network. The Fourier transform of the d-dimensional array of the numbers is then calculated. These Fourier-transformed numbers are then multiplied by the square root of $S(\omega)$. The results, when inverse Fourier-transformed back into the real space, obey the statistics of fBm. To avoid the problems associated with the periodicity of the numbers arising as a result of their Fourier transforming, one should generate the array for a much larger network than the actual size that is to be used, and use its central part.

Note that fBm is not differentiable, but it can be made so (in an approximate way) by smoothing it over an interval. Using such a smoothing, the "derivative" of a fBm, called *fractional Gaussian noise* (fGn), can be obtained whose spectral density in, e.g., one dimension, is given by

$$S(\omega) \sim \frac{1}{\omega^{2H-1}} \, . \tag{3.20}$$

A wide variety of systems have been shown to follow fBm or fGn statistics. For example, as we discuss in Chapter 6, the permeability and porosity of many heterogeneous rock masses follow fBm and fGn, respectively. Thus, such stochastic processes have been used for generating porosity and permeability fields of large

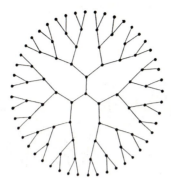

Figure 3.5: A Bethe lattice of coordination number 3.

scale reservoirs. These matter are discussed in Chapters 9 and 11.

3.5 Percolation processes

We now discuss the ideas and concepts of percolation theory that we shall use in the rest of this book. As we see in the subsequent Chapters, such ideas and concepts are invaluable tools for understanding rock properties and transport therein. Percolation processes were first developed by Flory (1941) and Stockmayer (1943) to describe polymerization in which small branching molecules react and form large macromolecules. However, Flory and Stockmayer did not call their theory a percolation process, and also developed their theory of polymerization for a special kind of lattice, namely, the Bethe lattice, an endlessly-branching structure without any closed loops, an example of which is shown in Fig. 3.5.

In the mathematical literature percolation processes were introduced by Broadbent and Hammersley (1957). They originally dealt with the concept of the spread of a hypothetical fluid through a random medium. Generally speaking, the spread of a fluid through a medium may involve some random elements. But one has to realize that the underlying mechanism of this randomness might be of two very different types. In one type, the randomness is dictated by the *fluid*, in that it is the fluid that decides what path to take in the system. This is the classical diffusion process. In the other type, the randomness in the path of the particle are imposed by the *medium*: this was the new situation that was considered by Broadbent and Hammersley (1957). This new phenomenon was called a *percolation process*, since they thought that the spread of the fluid through the random medium resembled the flow of coffee in a percolator!

3.5.1 Bond and site percolation

Since percolation deals with various properties of a disordered medium, and in particular a porous medium, we need to have a model of the medium in order to discuss percolation. Although models of porous media and fractured rock are discussed in detail in Chapters 6 and 7, for now we imagine that we can represent the medium by a network or lattice in which the bonds represent the pore throats (or the fractures, if the medium is fractured), and the sites or nodes, where the bonds meet, represent the pore bodies (or the intersections of two or more fractures). In their original paper, Broadbent and Hammersley focussed on two problems. One was the *bond percolation problem*, in which the bonds of the network are either occupied or intact randomly and independently of each other with probability p, or they are *vacant* or removed with probability $1 - p$. For a large network, this assignment is equivalent to removing a fraction $1 - p$ of all bonds at random. Figure 3.6 shows various stages of a bond percolation process on the square network. As discussed in Chapter 1, the intact bonds can represent the high-permeability regions of a pore space through which most of the fluid flow and transport take place, whereas the vacant bonds can represent the low-permeability (or impermeable) regions of the pore space that contribute very little to flow or transport. Of course, in natural rock the high- and low-permeability regions are not necessarily distributed randomly throughout the pore space, but for now we ignore such complications. In Chapter 9 we consider a more general percolation process in which the correlations between various regions of the pore space are not ignored.

Two sites are called *connected* if there exists at least one path between them consisting solely of occupied bonds. A set of connected sites bounded by vacant bonds is called a *cluster*. If the network is of very large extent and if p is sufficiently small, the size of any connected cluster is likely to be small. But if p is close to 1, the network should be entirely connected, apart from occasional small holes. At some well-defined value of p, there is a transition in the topological structure of the network; this value is called the *bond percolation threshold* p_{cb}. This is the largest fraction of occupied bonds below which there is no sample-spanning cluster of occupied bonds.

The second problem considered by Broadbent and Hammersley was a *site percolation problem*. In this case, sites of the network are occupied with probability p and vacant with probability $1 - p$. Two nearest-neighbor sites are called connected if they are both occupied, and connected clusters on the network are again defined in the obvious way. Similar to bond percolation, there is also a site percolation threshold p_{cs} above which an infinite (sample-spanning) cluster of occupied sites spans the network. Figure 3.7 shows site percolation clusters on a

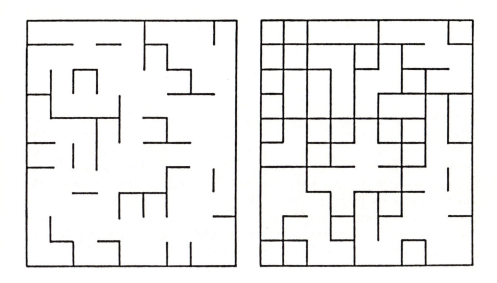

Figure 3.6: Bond percolation on the square network at $p = 1/3$ (left) and $p = 2/3$.

square network. Note that the percolation phenomenon as defined here is a static process; that is, once a percolation network is generated, its configuration does not change with time. Dynamic percolation processes have also been invented and are discussed in Chapter 12. Depending on a specific application of percolation theory to a problem of interest, many variants of the classical percolation processes have been developed. The interested reader is referred to Stauffer *et al.* (1982) and Sahimi (1994b) for a list of variants of the classical percolation problem and their applications.

The derivation of the exact values of p_{cb} and p_{cs} is an extremely difficult problem. In fact, this has been possible to date only for certain lattices related to the Bethe lattice and for a few two-dimensional networks. For the Bethe lattice Fisher and Essam (1961) showed that

$$p_{cb} = p_{cs} = \frac{1}{Z - 1} \qquad (3.21)$$

where Z is the coordination number of the lattice, i.e., the number of bonds connected to the same site. We compile the current estimates of p_{cb} and p_{cs} (and their exact values if they exist) for three common two-dimensional lattices in Table 3.1, while the most accurate numerical values of p_{cb} and p_{cs} for four common three-dimensional lattices are compiled in Table 3.2. Also shown in these tables is the product $B_c = Z p_{cb}$ and, as can be seen, this quantity is an almost invariant of percolation networks and $B_c \simeq d/(d-1)$. The significance of B_c is discussed below.

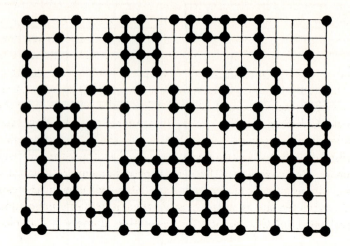

Figure 3.7: Site percolation clusters on the square network at $p = 0.5$.

Table 3.1

Values of bond percolation threshold p_{cb}, site percolation threshold p_{cs}, and $B_c = Zp_{cb}$ for three common two-dimensional networks.

Network	Z	p_{cb}	B_c	p_{cs}
Honeycomb	3	$1 - 2\sin(\pi/18) \simeq 0.6527^*$	1.96	0.6962
Square	4	$1/2^*$	2	0.5927
Triangular	6	$2\sin(\pi/18) \simeq 0.3473^*$	2.08	$1/2^*$

*Exact result

Table 3.2

Numerical estimates of bond percolation threshold p_{cb}, site percolation threshold p_{cs}, and $B_c = Zp_{cb}$ for four common three-dimensional networks.

Network	Z	p_{cb}	B_c	p_{cs}
Diamond	4	0.3886	1.55	0.4299
Simple-Cubic	6	0.2488	1.49	0.3116
BCC	8	0.1795	1.44	0.2464
FCC	12	0.119	1.43	0.199

3.5.2 Computer simulation of percolation on a network

Generating a percolating network by randomly removing sites or bonds is not actually suitable for engineering applications, because in addition to the sample-spanning cluster, this method also generates isolated finite clusters. In most applications one works only with the sample-spanning cluster (or at least the process of interest starts with a single cluster), and therefore we must first delete all isolated clusters from the system. Alternatively (and more efficiently), we can use a different method due to Leath (1976) and Alexandrowicz (1980) that generates only the sample-spanning (or the largest) cluster. In this method one starts with a single occupied site which is usually selected to be the center of the network. One then identifies the nearest-neighbor sites of the occupied site and considers them occupied and adds them to the cluster if random numbers R, uniformly distributed in (0,1) and attributed to the sites, are less than the fixed value p. The perimeter (the nearest-neighbor empty sites) of these sites are found and the process of occupying the sites continues in the same way. If a selected perimeter site is not occupied, then it remains unoccupied forever. The generalization of this method for generating clusters of occupied bonds is obvious.

An important task in computer simulations of percolation systems is to count the number of clusters of a given size. For example, during displacement of a fluid A by another immiscible fluid B we may need to know the number of islands or blobs of fluid A of a given size which are completely surrounded by B, which is equivalent to knowing the number of clusters of a given size within the context of a percolation model. The algorithm due to Hoshen and Kopelman (1976) can perform this task very efficiently. This algorithm is described in detail by Stauffer and Aharony (1992), where a computer program is also given for counting the clusters.

3.5.3 Characterization of a percolation system

In addition to p_{cb} and p_{cs}, the connectivity of percolation clusters, and hence their transport properties are characterized by several other quantities, the most important of which are as follows.

(i) The *percolation probability $P(p)$*, which is the probability that, when the fraction of occupied bonds is p, a given site belongs to the infinite cluster of occupied bonds.

(ii) The *accessible fraction $X^A(p)$*, which is that fraction of occupied bonds (or sites) belonging to the infinite cluster.

(iii) The *backbone fraction $X^B(p)$*, which is the fraction of occupied bonds in the infinite cluster which actually participate in flow or transport, since some of the bonds in the infinite cluster are dead-end and do not carry any current (flow) and, therefore, $X^A(p) \geq X^B(p)$. Figure 3.8 shows a sample-spanning percolation cluster and its backbone on the square network. An

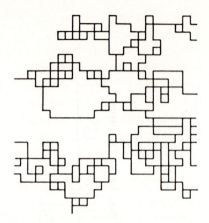

Figure 3.8: A sample-spanning bond percolation cluster (left) and its backbone on the square network at $p = 0.58$. Periodic boundary condition has been used in the vertical direction.

efficient algorithm for identifying the backbone of a percolation network is described by Rintoul and Nakanishi (1992).

(iv) The *correlation length* $\xi_p(p)$, which is the typical radius of percolation clusters for $p < p_c$, and the length scale over which a percolation network is macroscopically homogeneous for $p > p_c$. Thus, in any Monte Carlo simulation of percolation processes, the linear size L of the network must be larger than ξ_p for the results to be independent of L.

(v) The *average number of clusters of size s (per lattice site)* $n_s(p)$, which is an important quantity in some of the problems of interest here because it corresponds, for example, to the number of islands or blobs of a fluid of a given size that are formed during the displacement of one fluid by another, if the displaced fluid is incompressible (see Chapter 12).

(vi) The *effective conductivity* G, which is the conductivity of the network in which the occupied bonds are conducting and the vacant ones are insulating.

(vii) In a similar way, the effective diffusivity D and hydrodynamic permeability K of the system can be defined. Namely, if the occupied bonds represent pores through which flow or diffusion takes place, and the vacant bonds are closed to flow or diffusion, an effective permeability and diffusivity can be defined. In Chapter 8 various methods of estimating G, K and D are discussed. Figure 3.9 shows the typical behavior of some of these properties for a simple-cubic network in site percolation. Also shown is

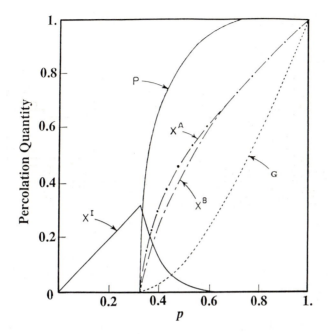

Figure 3.9: Typical variations of some of the percolation quantities in site percolation on the simple-cubic network.

$X^I(p) = p - X^A(p)$, the fraction of isolated occupied sites, which attains its maximum at p_{cs} and vanishes rapidly for $p > p_c$.

3.5.4 Scaling theory of percolation processes

One of the most important characteristics of percolation processes is the *universal* scaling laws that they obey. The behavior of many percolation quantities near the percolation threshold is insensitive to the network structure, and to whether the percolation process is a site or a bond percolation problem. The quantitative statement of this apparent universality is that many percolation properties obey scaling laws near the percolation threshold, and the *critical exponents* that characterize such scaling laws depend only on the Euclidean dimensionality d of the system. Even long, but finite, range correlations do not change this universality. In general, near the (site or bond) percolation threshold p_c, we have the following scaling laws

$$P(p) \sim (p - p_c)^{\beta_P} , \tag{3.22}$$
$$X^A(p) \sim (p - p_c)^{\beta_P} , \tag{3.23}$$
$$X^B(p) \sim (p - p_c)^{\beta_B} , \tag{3.24}$$
$$\xi_p(p) \sim |p - p_c|^{-\nu_p} , \tag{3.25}$$
$$G(p) \sim (p - p_c)^{\mu} , \tag{3.26}$$

Fractals and Percolation

$$K(p) \sim (p - p_c)^e \quad . \tag{3.27}$$

Although in most cases $\mu = e$, there are certain disordered media for which $e \neq \mu$. Such systems are discussed below. The scaling behavior of the effective diffusivity $D(p)$ is related to that of $G(p)$. According to Einstein's relation, $G \sim n_e D$, where n_e is the density of the electrons. Although a particle can diffuse on all clusters, only diffusion on the sample-spanning cluster contributes significantly to D, in which case, $n_e \sim X^A(p)$, i.e., $G(p) \sim X^A(p)D(p)$, and therefore

$$D(p) \sim (p - p_c)^{\mu - \beta_p} \quad . \tag{3.28}$$

For large clusters near p_c, $n_s(s)$ obeys the following scaling law

$$n_s \sim s^{-\tau_p} f[(p - p_c)s^{\sigma_p}] \quad , \tag{3.29}$$

where τ_p and σ_p are two more universal critical exponents, and $f(x)$ is a scaling function such that $f(0)$ is not singular. These exponents are *not* all independent. For example, one has, $\tau_p = 2 + \beta_p \sigma_p$ and $\nu_p d = \beta_p + 1/\sigma_p$, and in fact knowledge of ν_p and another exponent is sufficient for determining most of the percolation exponents. The implied prefactors in Eqs. (22)-(29) *do* depend on the type of lattice and are not universal. If two phenomena are described by two different sets of critical exponents, the physical laws governing the two phenomena must be fundamentally different. Thus critical exponents can help one to distinguish between different classes of problems and the physical laws that govern them. In Table 3.3 values of the critical exponents in two and three dimensions are compiled. For comparison, the values of the exponents obtained with Bethe lattices are also shown. It is clear that, as far as the critical exponents are concerned, a Bethe lattice is a rather poor representation of a three-dimensional system.

Table 3.3

Values of the critical exponents of percolation. The exponents at $d = 2$ and for Bethe lattices are exact. Values of μ for $d = 2$ and 3 are numerical estimates.

Exponent	$d = 2$	$d = 3$	Bethe Lattice
β_p	5/36	0.41	1
β_B	0.48	1.05	2
τ_p	187/91	2.18	5/2
σ_p	36/91	0.45	1/2
ν_p	4/3	0.88	1/2
μ	1.3	2.0	3

3.5.5 Formation of fractal structures in percolation systems

As mentioned above, the correlation length ξ_p has the physical significance that for length scales L *larger* than ξ_p the system is macroscopically homogeneous. However, for length scales *smaller* than ξ_p the system is *not* homogeneous, and the macroscopic properties of the system depend on L. In this regime, the sample-spanning cluster is statistically self-similar at all length scales less than ξ_p, and its mass M (its total number of occupied bonds or sites) scales with ξ_p as

$$M \sim \xi_p^{D_p} \ , \tag{3.30}$$

where D_p is the fractal dimension of the cluster [compare this equation with Eq. (3)]. However, D_p is not a totally new quantity and is given by

$$D_p = d - \frac{\beta_p}{\nu_p} \ , \tag{3.31}$$

so that $D_p(d = 2) = 91/148 \simeq 1.9$ and $D_p(d = 3) \simeq 2.52$. Similarly, for $L < \xi_p$, the backbone is a fractal object and its fractal dimension D_{BB} is given by

$$D_{BB} = d - \frac{\beta_B}{\nu_p} \ . \tag{3.32}$$

Note that at $p = p_c$ the correlation length is divergent, so that the sample-spanning cluster and its backbone are fractal objects at *any* length scale. Note also that if $L < \xi_p$ one should replace ξ_p in (30) by L, in which case one should also rewrite Eqs. (22)-(28) as

$$P(L) \ \sim \ L^{-\beta_p/\nu_p} \ , \tag{3.33}$$

$$X^A(L) \ \sim \ L^{-\beta_p/\nu_p} \ , \tag{3.34}$$

$$X^B(L) \ \sim \ L^{-\beta_B/\nu_p} \ , \tag{3.35}$$

$$G(L) \ \sim \ L^{-\mu/\nu_p} \ , \tag{3.36}$$

$$K(L) \ \sim \ L^{-e/\nu_p} \ , \tag{3.37}$$

$$D(L) \ \sim \ L^{-\theta_p} \ , \tag{3.38}$$

where $\theta_p = (\mu - \beta_p)/\nu_p$. Thus, *scale-dependent properties are a signature of a fractal structure*. Similarly, if a heterogeneous rock is characterized by a permeability distribution with a correlation length ξ_k, for any length scale $L \ll \xi_k$ we expect the rock to have a scale-dependent permeability. This point is crucial to our discussion of dispersion and displacement processes in heterogeneous rock in Chapters 9-12. Once it is established that a system is a fractal, many classical laws of physics have to be significantly modified. For example, Fick's law of diffusion with a constant diffusivity is no longer appropriate for describing diffusion processes in fractal systems. Instead, the diffusion coefficient is a time- and length-dependent quantity; this is called *anomalous* (Gefen *et al.*, 1983) or *fractal* diffusion (Sahimi *et al.*, 1983b). Therefore, when interpreting experimental data, one has to make sure that one is not in the regime of anomalous diffusion; otherwise, the interpretation of the data in terms of a constant diffusivity may be seriously in error. This is discussed in Chapter 8.

3.5.6 Percolation in finite systems and finite-size scaling

So far we have discussed percolation in disordered systems that are of infinite extent. However, percolation in finite systems deserves discussion since, both in practical applications and in computer simulations, one usually deals with a system of a finite extent. Fisher (1971) has developed a theory for the scaling properties of a finite system near a critical point, such as a percolation threshold, which is usually called finite-size scaling. According to this theory, in a finite system, as p_c is approached ξ_p eventually exceeds the linear size L of the network, in which case L becomes the dominant length scale of the system. Therefore, the variation of *any* property P_L of a system of size L is written as

$$P_L \sim L^{-\gamma_p} f(u) \qquad (3.39)$$

with $u = L^{1/\nu_p}(p - p_c) \sim (L/\xi_p)^{1/\nu_p}$, where $f(u)$ is a non-singular function. If in the limit $L \to \infty$, one has a scaling law such as $P_\infty \sim (p - p_c)^{\delta_p}$, then one must have $\gamma_p = \delta_p/\nu_p$. Therefore the variations with L of $P_L(p)$ in a *finite network* at p_c can be used to obtain information about the quantities of interest for an *infinite network* near p_c. This method has been used successfully by many authors to obtain accurate estimates of the critical exponents from simulation of finite systems. The finite size of the network also causes a shift in the percolation threshold (Levinshtein *et al.*, 1976; Reynolds *et al.*, 1980),

$$p_c - p_c(L) \sim L^{-1/\nu_p} \quad . \qquad (3.40)$$

In this equation p_c is the percolation threshold of the infinite system, and $p_c(L)$ is an *effective* percolation threshold for a finite system of linear dimension L. However, we should note that even Eqs. (39) and (40) are valid for large values of L. In practice, very large systems cannot easily be simulated, and therefore an equation such as (39) is modified to

$$P_L \sim L^{-\gamma_p}[a_1 + a_2 g_1(L) + a_2 g_2(L)] \quad , \qquad (3.41)$$

where g_1 and g_2 are two *correction-to-scaling* terms that are particularly important for small and moderate values of L, and where the as are constant. For *transport properties*, $g_1 = (\ln L)^{-1}$ and $g_2 = L^{-1}$ often provide an accurate estimate of γ_p (Sahimi and Arbabi, 1991). Equation (41) tells us how to estimate a critical exponent: Calculate P_L at $p = p_c$ for several network sizes L and fit the

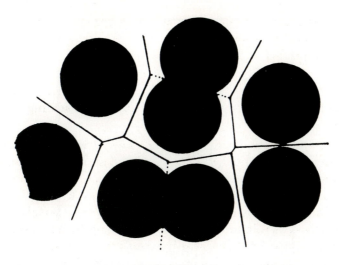

Figure 3.10: The two-dimensional Swiss-Cheese model.

results to Eq. (41) to estimate γ_p and thus δ_p.

3.5.7 Percolation in random networks and in continua

Although percolation in regular networks has been extensively used for investigating flow and transport in disordered systems such as porous media, percolation in continua and in topologically-random networks, those in which the coordination number changes from site to site, are also of great interest, since in most practical situations one has to deal with such systems. There are at least three ways of realizing a percolating continuum. In the first method, one has a random distribution of inclusions, such as circles, spheres or ellipses, in an otherwise uniform system (Pike and Seager, 1974; Haan and Zwanzig, 1977; Gawlinski and Stanley, 1981; Elam *et al.*, 1984; Thorpe and Sen, 1985; Sen and Torquato, 1988; Sevik *et al.*, 1988; Torquato *et al.*, 1988; Xia and Thorpe 1988). Figure 3.10 shows the so-called Swiss-Cheese model in which spherical (circular) inclusions are punched at random in an otherwise uniform system.

In such systems percolation is defined either as the formation of a sample-spanning cluster of the paths between untouching inclusions, or as the formation of a sample-spanning cluster of touching or overlapping inclusions. In the second method, one generates a percolating continuum by dividing the space into regular or random polyhedra (Winterfeld *et al.*, 1981), a fraction of which allow flow and transport, while the rest of them disallow them. Figure 3.11 presents an example of such a model. These models are discussed in detail in Chapter 6. Finally, in the third method, one distributes at random conducting sticks of a given aspect ratio, or plates of a given extent, and studies transport in such systems (Balberg

Figure 3.11: Two-dimensional Voronoi tessellation. The Voronoi network is obtained by connecting the centers of the neighboring polygons.

et al., 1983, 1984; Balberg and Binenbaum, 1983, 1985; Charlaix *et al.*, 1984, 1987a; Robinson, 1984a, b). This model has been proposed for representing fractured rock, and is discussed in Chapter 7.

One of the most important discoveries for continuum percolation (Scher and Zallen, 1970) is that a critical occupied volume fraction ϕ_c, which is defined as

$$\phi_c = p_{cs} f_l \quad , \tag{3.42}$$

where f_l is the filling factor of a lattice when each site of the lattice is occupied by a sphere in such a way that two nearest-neighbor impermeable spheres touch one another at one point, appears to be an almost invariant of the system, whose value is about 0.17 for three-dimensional systems. Shante and Kirkpatrick (1971) generalized this idea to permeable spheres, and showed that the average number of bonds per sites B_c at p_c is related to ϕ_c by

$$\phi_c = 1 - \exp\left(-\frac{B_c}{8}\right) \quad , \tag{3.43}$$

and that the continuum B_c is the limiting value of $p_{cs}Z$, when $Z \to \infty$. It is clear from Table 3.2 that in three dimensions, $B_c \simeq 1.5$. It has been established (Haan and Zwanzig, 1977; Gawlinski and Stanley, 1981; Elam *et al.*, 1984; Balberg and Binenbaum, 1985) that the geometrical exponents, defined by Eqs. (22)-(25), are the same for lattice and continuous systems.

However, flow and transport in percolating continua can be quite different from those in discrete networks. Consider, for example, the Swiss-Cheese model. If flow or transport takes place through the channels between the non-overlapping spheres, then the system can be mapped onto an equivalent problem on the *edges* of the Voronoi polygons or polyhedra (Kerstein, 1983; see Fig. 3.11). The Voronoi network, which is obtained from the Voronoi tessellation by connecting the centers of the polygons or polyhedra, was used by Jerauld *et al.* (1984b, d) to study transport in a random network. Its *average* coordination number is about 6 and 15.5 in two and three dimensions, respectively. Halperin *et al.* (1985) and Feng *et al.* (1987) used a scaling analysis and showed that the critical exponents μ_c and e_c, defined for the conductivity and permeability of the Swiss-Cheese model, are quite different from μ and e defined above. In particular, they showed that, in a three-dimensional Swiss-Cheese system, $\mu_c \simeq \mu + 1/2$, and $e_c \simeq \mu + 5/2$, whereas for the two-dimensional system, $\mu_c = \mu$ and $e_c \simeq \mu + 3/2$.

The model in which the matrix is insulating, but the spherical inclusions are very good conductors, was employed by Batchelor and O'Brien (1977) to study transport in granular porous media (see Chapter 13). This system can also be mapped onto an equivalent random network. Jerauld *et al.* (1984b,d) showed that the geometrical critical exponents for such random networks are the same as those for regular networks. Moreover, they established that as long as the average coordination number of a regular network and a topologically-random one (for example, the two-dimensional Voronoi and triangular networks) are about the same, many transport properties of the two systems are, for all practical purposes, identical.

3.6 A glance at history

Before closing this Chapter, it may be interesting to give a brief review of the history of application of percolation theory to modelling porous media problems. Despite the fact that Broadbent and Hammersley had expressed the hope that their theory would someday be used for solving some practical problems involving porous media, explicit use of percolation processes for describing flow phenomena in porous media gained popularity only in the 1980s. Since "who was the first to use percolation" has been a matter of some contention and controversy, it may be interesting to review the history to see "who said what and when," at least according to the published papers in the open literature.

To the best of our knowledge, Torelli and Scheidegger (1972) were the first who recognized the usefulness of percolation theory for modelling flow and dispersion phenomena in porous media. These authors were interested in hydrodynamic dispersion in porous media (see Chapter 9), and pointed out that percolation theory, if appropriately modified and applied, may provide some useful insight into the behavior of the phenomenon. However, they did not actually use percolation and, in fact, they did not even report any results in their paper!

Melrose and Brandner (1974) suggested that the entrapment of oil in reservoir rocks is similar to percolation processes, and proposed that an approach based on percolation may yield deeper insight into the problem. Again, these authors did not actually calculate anything using their idea! Davis *et al.* (1975), who studied transport processes in composite media, remarked at the end of their paper that, "Although, to our knowledge, no quantitative work has been done on the subject, we believe that 2-phase oil-water flow in oil fields is a percolation process in which the connectivity of each phase determines the relative permeability of that phase." However, these authors also did not report any result.

Larson (1977) and Larson, Scriven, and Davis (1977) also suggested that percolation theory may be useful for describing entrapment of one fluid phase by another in porous media. To demonstrate the usefulness of their idea, they calculated the percolation cluster size distribution n_s in a Bethe lattice for various coordination numbers and made a qualitative comparison between the results and relevant experimental data (see Chapter 12). Almost simultaneously, Chatzis and Dullien (1977) published a paper in which they calculated several percolation properties for various two- and three-dimensional networks and pointed out how they may be used for simulating two-phase flow in porous media. They compared their predictions with the measured capillary pressure curves. This is discussed in Chapter 5. Levine *et al.* (1977) discussed the application of percolation theory to wetting/dewetting phenomena in porous media, and pointed out how the effective-medium approximation (see Chapter 8) may be used for estimating the permeability of a porous medium.

Shortly after these three papers, de Gennes and Guyon (1978) also suggested that two-phase flow problems in porous media may belong to the class of percolation processes. They used visualization of mercury porosimetry (see Chapter 5) as an example, and proposed methods of using percolation concepts for modelling this phenomenon and other processes in porous media. They also pointed out how permeability and cluster size distribution in porous media may be calculated using percolation.

Finally, two papers in 1980 further established the applicability of percolation for modelling two-phase flow in porous media. Lenormand and Bories (1980) proposed a percolation model, now popularly known as invasion percolation (see Chapter 12), for modelling a drainage process, i.e., a process in which a non-wetting fluid displaces a wetting fluid from a porous medium. Golden (1980) discussed the application of percolation theory for studying two-phase flow problems and the associated hysteresis (history-dependent) phenomena that are routinely observed in porous media (see Chapters 5, 12, and 13).

After publication of these original papers, there was an explosion of new ideas and methods for modelling porous media problems using percolation theory. We shall review these concepts and methods in the appropriate Chapters of this book.

3.7 Conclusions

Fractal and percolation phenomena enable us to characterize many properties of disordered systems, which would otherwise seem very difficult to understand. Moreover, they point out the fundamental role of length scale for macroscopic homogeneity of the system. If the linear size of our system is less than the length scale at which it can be considered homogeneous, then the classical laws of physics have to be fundamentally modified. An example is diffusion which, as we discuss in Chapter 8, cannot be described by the classical diffusion equation with constant diffusivity, if the system is fractal.

Chapter 4

Diagenetic Processes and Formation of Rock

4.0 Introduction

Before we discuss characterization of porous media and fractured rock, it is necessary to understand the processes that give rise to their present structure. Thus, in this Chapter we consider *natural* porous media and fractured rock, such as oil reservoirs and aquifers, and study the processes that, over the past millions of years, have created them. These are called *diagenetic processes*, and what follows is a brief description of them. We do not, however, consider the formation of other types of natural or man-made porous media.

4.1 Diagenetic and metasomatic processes

In order to understand rock properties, one has to have an understanding of the diagenetic processes that lead to its formation. Rock formation starts with deposition of sediments and is followed by compaction and alteration processes that can cause drastic changes in the morphology of the reservoir. Consider, for example, sandstones which are assemblages of discrete grains with a wide variety of chemical compounds and mixtures. If the environment around the sandstone changes, the grains start to react and produce new compounds. The mechanical properties of the grains also change. The chemical and physical changes in the sand after its deposition constitute diagenetic processes. The main features of diagenetic processes are (i) mechanical deformation of grains; (ii) solution of grain minerals; (iii) alteration of grains; and (iv) precipitation of pore-filling minerals, cements, and other materials. The latter three features involve changes in the chemical composition of rock, which are usually induced by transport of some reactants in the pore space. These phenomena are called *metasomatic processes*, and have a key influence on the volume of the content of the reservoir because they control its porosity.

Immediately after deposition diagenesis starts; it continues during burial and uplift of the rock until outcrop weathering reduces it again to sediment. These changes produce an end product with specific diagenetic features, whose nature depends on the initial mineralogical composition of the system and also on the composition of the surrounding basin-fill sediments. Given a system with a particular mineralogical composition, its diagenetic history depends on several factors, including time-dependent exposures to varying temperatures and pressures and the chemistry of the pore fluid. All of these factors constitute the *historical* aspects of a reservoir and affect its quality. Therefore, the ability of reservoir rocks to produce, say, oil, is closely related to their diagenetic history. If appropriate relations between diagenesis and petrophysical properties of reservoir rocks can be found, one can use such relations in the analysis of reservoirs to predict their potential for producing oil or any other material that they may contain.

Porosity of natural rocks, i.e., the volume fraction of their open space, has either a *primary* or a *secondary* origin. Primary porosity is due to the original pore space of the sediment, whereas secondary porosity is due to the fact that unstable grains or cements have undergone chemical and physical changes through reaction with the formation water and have partially or entirely passed into the solution. Therefore, if the pore space is restored through dissolution of authigenic minerals, then the original porosity that had been protected from precipitation by deposition of minerals is converted into secondary porosity. According to Schmidt and McDonald (1979), solution pores provide more than half of all the pore space in many sedimentary rocks. The significance of secondary porosity in carbonate rocks has been recognized for a long time, but its importance to sandstones has only relatively recently been appreciated (Hayes, 1979).

As discussed by Schmidt and McDonald (1979), there are five classes of secondary porosity in sandstones, defined according to their origin: (i) fracturing; (ii) shrinkage; (iii) dissolution of sedimentary grains and matrix; (iv) dissolution of authigenic pore-filling cement; and (v) dissolution of authigenic replacive minerals. Five different kinds of pores can contain secondary porosity, namely, (i) intergranular pores; (ii) oversized pores; (iii) moldic pores; (iv) intraconstituent pores; and (v) open fractures. Of these, fractures are distinctly different from the other four types of pores, and therefore are discussed separately in this book. The existence of secondary porosity can sometimes be recognized even with the naked eye. Other indications of the occurrence of secondary porosity include oversized or elongated pores, corroded and fractured grains and several others.

The diagenetic processes discussed above lead to distinct morphologies for reservoir rocks. Pores can take on essentially any shape or size, and they can also be highly interconnected. Patsoules and Cripps (1983) used scanning electron microscopy to study rock and obtained information about the shapes, sizes and connectivity of the pores and the roughness of their surface. They reported that their rock, which was upper cretaceous chalk from East Yorkshire (in England) and the North Sea, contained highly interconnected pores. Some of the ring-shaped pores of the chalk were connected to at least 25-30 other pores. These pores remain connected even when the porosity of the system is very low, and therefore one important effect of the diagenetic process is to keep the pore space highly interconnected. As discussed above, rock formation involves compaction and alteration processes. During the alteration processes complex phenomena such as nucleation on the surface of the pores and mineral crystal growth take place. These are time-dependent phenomena, which reduce the porosity and permeability of the rock. If the permeability of the medium is reduced, the flow rate also decreases, which means that the rate of nucleation of mineral crystals increases. However, the crystals cannot grow indefinitely because they are limited by the growth rate at the time they are nucleated. Moreover, the growth of new mineral crystals inhibits that of the older ones. Thus there is competition between nucleation of new mineral crystals and the growth of the older crystals, which determines the distribution of the crystal sizes. Since diagenetic processes are similar for many different rocks, and since there appear to be many similarities between the geometries of various rocks, one may hope that many fundamental elements of pore formation processes are *universal*, independent of many microscopic properties of rocks. If so, one may be able to develop a general model of pore formation and growth processes that can explain, at a fundamental level, many features of various rocks. If such a model can be developed, its generality may be comparable to that of percolation processes discussed in Chapter 3, which explain the fundamental role of the interconnectivity of the microscopic elements of a disordered medium on its macroscopic properties. However, such a model has not yet been developed.

How can we model diagenetic processes? A study of the literature shows that there are essentially two approaches to this problem. The first approach, which we call *continuum modelling*, relies on the continuum equations of transport and reaction in order to model metasomatic processes. It largely ignores the morphology of the pore space and its time variations, and attempts to characterize the process by *average* macroscopic properties. The only morphological property of the rock that is included in this model is the porosity. The details of the kinetics of the surface reactions are usually included in the model. Such models contain several parameters, e.g., the diffusivity of each species, which have to be estimated independently or measured experimentally. This approach, which has been developed by several authors (Palciauskas and Domenico, 1976; Wood and Surdam, 1979; Cussler, 1982; Wood and Hewett, 1982; Cussler *et al.*, 1983; Walsh *et al.*, 1984; Lichtner *et al.*, 1986; Kim and Cussler, 1987; Kopinsky *et al.*, 1988; Novak *et al.*, 1989), is essentially the solution to the continuum equations of transport and reaction in a porous medium.

The second approach, in which the details of reaction kinetics and mass transfer are ignored, is what we call *geometrical modelling*. The diagenetic process is modeled by starting from a model of unconsolidated pore space and making several simplifying assumptions about the rate of change of grains and pores. This approach can take into account the effect of connectivity and percolation of pores and grains. In what follows we discuss both approaches.

4.2 Continuum models of diagenetic processes

The main goal in these models is determining the reaction rates of various reactants and the distribution of their concentrations in the porous medium.

Most authors have studied the case in which transport of the reactants is only by diffusion. The diffusion problem can be generally divided into two categories, which are *interdiffusion*, which is diffusion through both faces of a medium of finite length, and *unidirectional diffusion*, which is diffusion into a semi-infinite medium. We consider a porous system containing a dilute solution of N solutes in a solvent at constant temperature and pressure. The concentration C_i of the reactant i at time t is governed by a diffusion-reaction equation (see Chapter 2):

$$\frac{\partial}{\partial t}(\phi C_i) = \frac{\partial}{\partial x}\left(\phi D_i \frac{\partial C_i}{\partial x}\right) + \sum_{i=1}^{M} \alpha_{ij} r_i, \quad j = 1, \cdots, N \qquad (4.1)$$

where D_i is the diffusivity of the ith species, ϕ is the porosity of the medium, M is the number of chemical reactions occuring in the system, and α_{ij} are the stoichiometric coefficients for the reaction rate r_i. The medium is assumed to be one dimensional. The reaction rates are all non-zero.

We assume that the porous medium consists of an inert solid matrix in addition to a reactive solid phase with concentration C_s in equilibrium with the aqueous solution of the N solutes. If the volume fraction of the reacting solid is ϕ_s, then conservation of the solid phase requires that

$$\frac{\partial \phi_s}{\partial t} = -V_m r_s, \qquad (4.2)$$

where r_s is the reaction rate of the solid phase, and V_m is its molar volume. Moreover, because the volume of the porous medium is conserved, we must have

$$\phi(x,t) + \phi_s(x,t) = \phi_r \qquad (4.3)$$

where ϕ_r is the reactive volume fraction occupied by the solid C_s and the aqueous solution. In principle, both ϕ and ϕ_s vary in space and time, but it is usually assumed that, to a first approximation, the porosity remains constant during diffusion and reaction. This implies that $\phi_s << \phi$.

So far, the formulation of the problem is general. However, if N is large, then analytical solutions of the problem are difficult to derive. Thus, one has to make simplifying assumptions about the number of the reactants and the number of chemical reactions occuring in the system in order to derive an analytical solution for the quantities of interest, or else solve the problem by a numerical method. For example, Cussler (1982) considered the case in which the diffusivity D_i is the same for all solutes, and the solutes are subject to I chemical equilibria which means that $(N - I + 1)$ solutes can be considered as independent. Thus, the solution is treated as containing N solutes, $N - I$ of which are independent. The independent solutes are numbered from $n = N - I$ to N, in which case one can write the chemical equilibria in the form

$$C_i = C_i(C_n, \cdots, C_N), \quad i = 1, \cdots, n - 1. \qquad (4.4)$$

Then, since not all concentrations are independent, one can simplify the problem by, for example, writing

$$\frac{\partial C_i}{\partial t} = \sum_{j=n}^{N} \left(\frac{\partial C_i}{\partial C_j}\right)\left(\frac{\partial C_j}{\partial t}\right) , \qquad (4.5)$$

and similarly for $\partial^2 C_i/\partial x^2$. After some manipulations, one finds the reaction rates

$$r_i = -D \sum_{j=1}^{I} \sum_{k=n}^{N} \sum_{m=n}^{N} (A_{ij})^{-1} \left(\frac{\partial^2 C_j}{\partial C_k \partial C_m}\right)\left(\frac{\partial C_k}{\partial x}\right)\left(\frac{\partial C_m}{\partial x}\right) , \qquad (4.6)$$

where

$$A_{ij} = \alpha_{ij} - \sum_{k=n}^{N} \left(\frac{\partial C_i}{\partial C_k}\right)\alpha_{kj} . \qquad (4.7)$$

Although the reaction rates are nonzero, they are assumed not to depend on chemical kinetics. They arise because of the coupling between diffusion and reaction. Thus, the procedure to study this problem consists of finding the concentrations C_i by solving Eq. (1), and then calculating the reaction rates r_i. Cussler (1982) studied several limiting cases with non-linear chemical equilibria. As an example, consider the case in which one has two solutes and one reaction. Then, the chemical equilibrium has the form

$$C_1^{\alpha_{11}} = K_e C_2^{-\alpha_{21}} , \qquad (4.8)$$

where K_e is the equilibrium constant. Thus, Eq. (6) yields

$$r_1 = -\left[\frac{DC_1(\partial \ln C_2/\partial x)^2}{\alpha_{11} + (C_1/C_2)(\alpha_{21}^2/\alpha_{11})}\right]\left[\frac{\alpha_{21}}{\alpha_{11}}\left(\frac{\alpha_{21}}{\alpha_{11}} + 1\right)\right] . \qquad (4.9)$$

More complex expressions can be obtained for the case in which one has several solutes and reactions. For example, if one has three solutes and one reaction, then $C_1^{\alpha_{11}} = -K_e C_2^{-\alpha_{21}} C_2^{\alpha_{31}}$. Cussler *et al.* (1983) used this theory and developed a criterion that tells us whether a solid can dissolve or precipitate in a porous medium. For example, suppose that one has three solutes and one reaction. If

$$\frac{\alpha_{21}}{\alpha_{11}}\left(\frac{C_{2\infty}}{C_{20}} - 1\right) > 1 + \frac{\alpha_{31}}{\alpha_{11}} > 0 , \qquad (4.10)$$

then one has precipitation of the solid reactant in the solution near the surface of the system, and dissolution of it deep inside the pores. Here C_{20} and $C_{2\infty}$ are the concentrations at the surface of the system and deep within it, respectively.

As a simple example, consider the following reaction

$$Ca(OH)_2 + 2H^+ \rightleftharpoons Ca^{++} + 2H_2O \qquad (4.11)$$

If C_1 and C_2 are the concentration of Ca^{++} and H^+, respectively, then $C_1 = K_e C_2^2$ and

$$r_1 = \frac{-2DC_1(\partial \ln C_2/\partial x)^2}{1 + 4C_1/C_2} \qquad (4.12)$$

and $Ca(OH)_2$ always precipitates. As an example of a system with three reactants and one reaction, consider the following

$$CaCO_3 + H^+ \rightleftharpoons Ca^{++} + HCO_3^-\qquad(4.13)$$

If C_1, C_2 and C_3 are the concentration of Ca^{++}, HCO_3^- and H^+, respectively, then

$$r_1 = \frac{2DC_1[(\partial \ln C_2/\partial x)^2 - (\partial \ln C_2/\partial x)(\partial \ln C_3/\partial x)]}{1 + C_1/C_2 + C_1/C_3}\qquad(4.14)$$

and (10) predicts that the solid precipitates in some regions of the system, but dissolves in others. Thus, using these equations one can make qualitative as well as quantitative predictions about the behavior of the system.

Extending Cussler's theory, Kim and Cussler (1987) considered the case in which the chemical reaction is nonlinear and reversible. Kopinsky *et al.* (1988) studied a system in which there is a dissolution front. Internal dissolution is enhanced by unequal diffusivities of all solutes, and both precipitation and dissolution can occur simultaneously. Novak *et al.* (1989) considered the general problem of diffusion, reaction and dissolution in a one-dimensional porous medium. The medium they considered contained an initial distribution of minerals at a uniform concentration. The aqueous phase composition was fixed at one boundary. This simulates, for example, the case in which an aqueous phase flows parallel to one face of the medium, maintaining a constant composition at that face. Novak *et al.* developed analytical solutions for many different cases.

If transport of the reactants is also affected by a flow field (convection), then the same approach can be used except that, instead of a diffusion-reaction equation for each species, one should use a convective-diffusion and reaction equation discussed in Chapter 2 (see, for example, Walsh *et al.*, 1984). Thus, Eq. (1) is replaced by

$$\frac{\partial}{\partial t}(\phi C_i) + v\frac{\partial}{\partial x}(\phi C_i) = \frac{\partial}{\partial x}\left(\phi D_i \frac{\partial C_i}{\partial x}\right) + \sum_{i=1}^{M} \alpha_{ij} r_i , \quad i = 1, \cdots, n-1 , \quad(4.15)$$

where v is the average flow velocity through the medium. Therefore, the essence of the approach is the same as before, except that the flow field has to be determined before the concentration fields can be calculated. Except for some special cases (see below), this has to be done numerically. Note that, as mentioned above, the only property of the porous medium that enters such models is its porosity. Otherwise, the porous medium is represented by a continuum, and no information about its topology or geometry is incorporated in the model.

It should be pointed out that because of simultaneous transport and reaction of various species, one can have a sharp reaction front which can give rise to various forms of discontinuities in the solution. Such discontinuities make the numerical solutions of the governing equations difficult to calculate. Therefore, it is important to study the position of the reaction front and the speed with

which it moves, and how the front affects the solution of the problem. Lichtner
et al. (1986) have studied this extensively, and what follows is a summary of
their discussion.

For the sake of simplicity, we ignore the effect of the flow field. Suppose
that the instantaneous position of the reaction front is x_r, and thus its velocity is
$v_r = dx_r/dt$. At the reaction front we have a discontinuity in the volume fraction
of the reacting solid phase, since it is zero on one side of the front and non-zero
on the other side. On the other hand, the solute concentration is piecewise
continuous with a kink at the reaction front. This results in a jump in the solute
flux across the front. If A is the cross-sectional area of the porous medium, then
the number of moles of solid phase that have reacted can be expressed as

$$\Delta n_s = \frac{A\Delta x_r [\phi_s]}{V_s} \,, \qquad (4.16)$$

where $[\cdot]$ denotes the jump in the quantity across the front x_r, $[\phi_s] = \phi_s^\ell - \phi_s^r$.
Here ϕ_s^ℓ and ϕ_s^r denote, respectively, the solid volume fraction to the left and
right of the reaction front, and V_s is the molar volume of the solid reactant. If
we have precipitation at the front, then $\phi_s^\ell = 0$, but if we have dissolution at the
front, then $\phi_s^r = 0$. Since we have assumed that transport of the species is by
only diffusion, we can calculate the total number of moles Δn_i of the *i*th solute
that reacts in a time interval Δt. This quantity is given by

$$\Delta n_i = A\Delta t(Q_i^\ell - Q_i^r) = A\Delta t[Q_i] \,, \qquad (4.17)$$

where Q_i^ℓ and Q_i^r are the fluxes to the left and right of the reaction front, respectively. Because of mass conservation we must also have

$$\Delta n_i = \frac{A\Delta x_r \alpha_i [\phi_s]}{V_s} \,, \qquad (4.18)$$

where α_i is the number of moles of the *i*th species in one mole of solid. If we
divide Eqs. (17) and (18), then in the limit $\Delta t \to 0$ we obtain

$$\frac{dx_r}{dt} = \frac{V_s}{\alpha_i} \frac{[Q_i]}{[\phi_s]} \quad i = 1, \cdots, N \,. \qquad (4.19)$$

Equation (19), which relates the velocity of the reaction front to the jump discontinuities at the front, is called the *Rankine-Hugonoit* equation. Because of
such jump discontinuities at the reaction front, the integration of r_s, the reaction
rate of the solid-fluid reactions, over an interval $[x_r - \epsilon , x_r + \epsilon]$ containing the
reaction front, in the limit $\epsilon \to 0$, yields

$$\lim \int_{x_r-\epsilon}^{x_r+\epsilon} r_s \, dx = \lim \frac{1}{\alpha_i} \int_{x_r-\epsilon}^{x_r+\epsilon} \frac{\partial Q_i}{\partial x} \, dx = \frac{1}{\alpha_i}[Q_i] \,. \qquad (4.20)$$

To give the reader some idea about how exact analytical solutions are derived
for this type of problem, and how the reaction front affects such solutions, we

briefly discuss a system containing one or two solutes. The interested reader should consult Lichtner *et al.* (1986) or Novak *et al.* (1989) for more details and many analytical solutions for several limiting cases. First, consider a system that contains only one solute. The position of the reaction front x_r is assumed to follow the following equation

$$x_r(t) = 2c\sqrt{Dt} \ , \tag{4.21}$$

where c is a constant that is determined below. Thus the velocity of the reaction front $v_r = dx_r/dt$ is given by

$$v_r = c\sqrt{\frac{D}{t}} = \frac{v_d}{R} \ , \tag{4.22}$$

where $v_d = \sqrt{D/t}$ is the diffusion velocity, and $R = 1/c$, usually called the retardation factor, is the velocity of the reaction front relative to the diffusion velocity. The aqueous phase contains one component A which reacts with the solid phase. Because of the discontinuity at the reaction front, we need to determine the concentration C_A on both sides of the front separately. In the region to the right of the front, $x \geq x_r$, the concentration C_A^r is constant, and its value is given by the equilibrium constant K_e of the reaction between A and the solid phase, $C_A^r = K_e$. In the region to the left of the front, $x < x_r$, C_A^ℓ is governed by Eq. (1), coupled to a moving reaction front. It is well-known (see, for example, Bird *et al.*, 1960) that the solution of this type of problem is in terms of the error function. After applying the initial and boundary conditions one obtains

$$C_A^\ell(x,t) = C_0 + \frac{K_e - C_0}{\mathrm{erf}(c)}\mathrm{erf}\left(\frac{x}{2\sqrt{Dt}}\right) \ , \tag{4.23}$$

where $C_0 = C_A(x = 0, t)$, and erf denotes the error function. To determine the constant c we use the Rankine-Hugonoit equation and compute the flux jump across the reaction front using the solutions to the left and right of the front. We obtain $K_e - C_0 = c\sqrt{\pi}\phi_s^0\mathrm{erf}(c)e^{c^2}/(V_s\phi)$, where ϕ_s^0 is the initial volume fraction of the reactive solid. This equation has a physical root only if $K_e > C_0$, i.e., when the solution at the boundary is unsaturated with respect to the solid phase. If $c << 1$, then it is not difficult to show that

$$C_A^\ell(x,t) = C_0 + (K_e - C_0)\frac{x}{x_r(t)} \ . \tag{4.24}$$

Next, consider the case of a system with two solutes A and B with equal diffusivities. The initial conditions are $C_i(x,0) = C_i^0$ and $\phi_s(x,0) = \phi_s^0$, and the boundary condition is $C_i(0,t) = C_{i0}$ ($i =$ A or B). To obtain an analytical solution to the problem we define two functions $\chi(x,t) = C_A/\alpha_A - C_B/\alpha_B$ and $\eta(x,t) = (C_A/\alpha_A + C_B/\alpha_B)/2$. Then, it is not difficult to see that $\chi(x,t)$ satisfies the following diffusion equation

$$\frac{\partial \chi}{\partial t} = D\frac{\partial^2 \chi}{\partial x^2} \tag{4.25}$$

and $\eta(x,t)$ satisfies the diffusion-reaction equation

$$\frac{\partial(\phi\eta)}{\partial t} - \phi D \frac{\partial^2 \eta}{\partial x^2} = r_s \ . \tag{4.26}$$

The position of the reaction front x_r is given by

$$\frac{d\dot{x}_r}{dt} = \frac{V_s[Q_\eta]}{[\phi_s]} \ , \tag{4.27}$$

where $Q_\eta = (Q_A/\alpha_A + Q_B/\alpha_B)/2$. Note that, because $\chi(x,t)$ satisfies a diffusion equation with no reaction, the flux Q_χ is continuous across the reaction front. The solution to $\chi(x,t)$ is in terms of an error function

$$\chi(x,t) = \chi_1 + \chi_2 \text{erf}\left(\frac{x}{2\sqrt{Dt}}\right) \ , \tag{4.28}$$

where χ_1 and χ_2 are two constants given by $\chi_1 = C_{A0}/\alpha_A - C_{B0}/\alpha_B$ and $\chi_2 = (C_A^0 - C_{A0})/\alpha_A - (C_B^0 - C_{B0})/\alpha_B$. On the other hand, since $\eta(x,t)$ satisfies a diffusion-reaction equation, its solution has to be found by considering it separately in the regions to the left and right of the reaction front. In the region to the left of the front, the solid phase has completely dissolved, and thus $\eta(x,t)$ satisfies a diffusion equation with *no* reaction. Thus, $\eta^\ell(x,t)$ is given by an equation similar to (28) but with two constants η_1 and $\eta_2(t)$. η_1 is determined from the boundary condition at $x = 0$, $\eta_1 = (C_{A0}/\alpha_A + C_{B0}/\alpha_B)/2$. $\eta_2(t)$ is determined from the fact that $\eta(x,t)$ is continuous across the reaction front, yielding $\eta_2(t) = [\eta_+(t) - \eta_1]/\text{erfc}$, where $\eta_+(t)$ is given below. The solution $\eta^r(x,t)$ to the right of the front is determined by using the fact that the aqueous solution is in local equilibrium with the solid phase. Then, for $x \geq x_r$ one finds that η_+ is given as a function of χ

$$K_e = \alpha_A^{\alpha_A} \alpha_B^{\alpha_B} [\frac{1}{2}\chi + \eta_+(t)]^{\alpha_A} [-\frac{1}{2}\chi + \eta_+(t)]^{\alpha_B} \ . \tag{4.29}$$

Note that we have two identities

$$\alpha_A \alpha_B (\eta_+^2 - \frac{1}{2}\chi^2) = C_A C_B \ , \tag{4.30}$$

$$(\alpha_A + \alpha_B)\eta_+ - \frac{1}{2}(\alpha_A - \alpha_B)\chi = \frac{\alpha_A}{\alpha_B} C_A + \frac{\alpha_B}{\alpha_A} C_B \ . \tag{4.31}$$

These equations predict that for $\chi_2 \neq 0$ one has precipitation to the right of the reaction front. As can be seen, even with two solutes the problem is already quite complex. It becomes increasingly difficult to obtain an analytical solution for a system with more than two solutes, and thus one has to resort to numerical methods for obtaining the solution of the problem.

Such numerical solutions are usually based on a finite-difference approximation (see Chapter 14) to the governing equations. However, the existence of a sharp reaction front, the relatively large number of solutes, and the nonlinear chemical kinetics conspire to make the numerical solutions difficult to obtain. Lichtner *et al.* (1986) have discussed such finite-difference approximations, to whom the interested reader is referred.

4.3 Geometrical models of diagenetic processes in granular rock

A review of the literature indicates that there have been a few attempts for modelling of diagenetic processes using the concepts of percolation discussed in Chapter 3 and their appropriate generalizations, and the statistical physics of disordered media. Notable among these are the works of Wong *et al.* (1984) and Roberts and Schwartz (1985). In the model of Wong *et al.* (1984) one represents the initial porous medium (before the diagenetic process) by a three-dimensional network of sites and bonds. The bonds represent the pore throats or channels through which a fluid flows, and the sites represent the pore bodies where the pore throats meet. Such models of porous media are discussed in detail in Chapter 6. Each bond i is assumed to be a cylindrical fluid-filled pore to which a randomly-selected radius R_i is assigned. Note that the pore permeability is proportional to R_i^2, and its hydraulic conductance is proportional to R_i^4. To mimic the consolidation process and the reduction in the porosity and permeability of the system during the diagenetic process, a tube is selected at random and its radius is reduced by a fixed factor x

$$R_i \rightarrow x R_i \quad , \tag{4.32}$$

where $0 < x < 1$. Of course, this simple model cannot really simulate the effect of deposition of irregularly shaped particles in an irregularly shaped pore, or that of thin lubricating films of fluid which, if present, inhibit grain contact. However, the model has two attractive features, namely, (i) that it preserves for any $x > 0$ the network connectivity, even when the porosity has almost vanished; and (ii) that the amount of change in the pore radius R_i at any step of the simulation (time) depends on the value of R_i at that time. Wong *et al.* (1984) used this model qualitatively to explain empirical laws such as Archie's law and the Kozeny equation for permeability of a porous medium. These are discussed in Chapter 8. Note that the limit $x = 0$ represents a percolation process.

The second geometrical model for diagenesis of granular porous media is due to Roberts and Schwartz (1985); it was studied further by Schwartz and Kimminau (1987). In this model, the initial porous medium is a dense pack of randomly distributed spherical grains of random radii R. Such a model was originally proposed by Bernal (1959, 1960) for studying liquids. Thus, the coordinates of the centers of the spheres follow the Bernal distribution (Bernal, 1959, 1960; see also Alben *et al.*, 1976); see Fig. 4.1. The initial volume fraction of the spheres in this model is 0.636, and thus the initial porosity of the system–the volume fraction of the empty space between the spheres–is 0.364.

 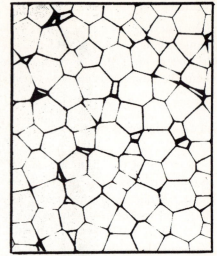

Figure 4.1: Two stages of the grain consolidation model at porosities $\phi = 0.36$ (left) and $\phi = 0.03$. Dark area is the pore space (after Schwartz and Kimminau, 1987)

The radii of the particles are then allowed to increase simultaneously, as a result of which the system's porosity and permeability decrease. In the region where the spheres overlap, the grains are truncated. This process can be continued to yield a series of porous media with the desired porosity. The percolation threshold of the system, i.e., the critical porosity below which no fluid flow through the system can take place, is $\phi_c = 0.03 \pm 0.004$. The system at this porosity is also shown in Fig. 4.1. Given that the initial porosity of the system with the Bernal distribution is 0.364, this model generates models of porous media whose porosities span more than one order of magnitude. They also do resemble natural sandstones, an example of which is shown in Fig. 4.2. Because the porosity of sandstones and similar rocks is usually less than 0.4, this algorithm provides a reasonable model of the diagenetic process. Schwartz *et al.* (1989a) also considered a model in which the initial grains were not spherical, in order to simulate anisotropic or stratified media.

Two points are worth mentioning here. First, if instead of the Bernal distribution of the spheres one starts with a simple-cubic lattice in which spherical grains of unit radius are placed at its nodes (thus each sphere is a neighbor to six other spheres) and follow the same algorithm, then the percolation threshold or the critical porosity of the system is 0.349, which is close to that of the random sphere packing. If one starts with a body-centered cubic lattice of spheres (in which each sphere is a neighbor to eight other spheres), then one obtains $\phi_c \simeq 0.0055$, one order of magnitude smaller than what can be achieved with the Bernal distribution. This indicates the relative flexibility of the model for obtaining the desired porosities. Secondly, the sedimentation and diagenetic processes that give rise to many sedimentary rocks such as sandstones tend to favor

Figure 4.2: Cemented Devonian sandstone from Illinois. Compare this with the model shown in Fig. 1 at porosity $\phi = 0.03$ (after Roberts and Schwartz, 1985).

a distribution of particles that are roughly equal in size (Pittman, 1984). In this region, the model of Roberts and Schwartz (1985) is much more efficient than one in which the porosity is reduced by adding additional spheres with smaller and smaller radii to progressively fill the pore space of the original packing. To obtain a comparable porosity range by this method, one has to use spheres whose radii vary over many orders of magnitude. But, the final configuration would bear little resemblance to most naturally occurring granular porous materials.

4.4 A geometrical model of carbonate rock

The two models described above have been useful in developing a unified framework for a description of many properties of granular media such as sandstones. Such porous media possess pore and solid phases that have many simple characteristics. However, other porous media, e.g., carbonate rocks (such as those of Iran), are more complex and their pore and solid phase geometries are not as simple as those of granular porous media such as sandstones. The major differences between carbonate and sandstone reservoirs, as discussed by Pittman

(1984), are (i) mineralogy; (ii) origin of grains; (iii) size and shape of grains; and (iv) influence of early diagenesis on carbonate rocks. For example, most minerals in carbonate rocks are relatively soluble carbonate materials, whereas sandstone's grains originate through erosion of existing rocks with transportation of the minerals by fluid flow to site of deposition, as discussed above. The grains in carbonate rocks pack more loosely than those in sandstones, and they are usually large with shapes like twigs, rods and flakes. The pores of carbonate rocks tend to be sheet-like, rather than cylindrical or tube-like. Early diagenesis had a much stronger effect on carbonates than on sandstones. It is, therefore, clear that carbonate rocks have a more complex heterogeneous pore system than sandstones.

The diagenesis of carbonate rocks such as crystalline dolomite usually starts with a high porosity packing of $CaCO_3$ grains (Wardlaw, 1976; Blatt *et al.*, 1980). The initial grain sizes are in the range 1-10 μm. Nucleation at random sites of $CaMg(CaCO_3)_2$ rhombohedral crystals starts the dolomitization process. These centers are usually at the surface of the $CaCO_3$ grains and, after several million years, grow into grains whose sizes are of order of tens of microns. Gradually, $CaCO_3$ is replaced by $CaMg(CaO_3)_2$, where Mg ions come from the brine that saturates the pore space. This replacement introduces intercrystalline porosity and, at the same time, the original porosity is decreased by compaction and cementation processes which are similar to those during diagenesis of sandstones. At the end of diagenesis, the solid matrix is made of random $CaMg(CaO_3)_2$ "rhombs," with a broad pore size distribution.

To model this process, Crossley, Schwartz, and Banavar (1991) proposed the following model. One starts with a three-dimensional random number array $\Phi_0(x, y, z)$, distributed uniformly over (0, 1). This initial "configuration" is then smoothed by convolving it with a kernel $W(x, y, z|w)$ with a correlation length w:

$$\Phi(x, y, z) = \int_{-\infty}^{\infty} \int_{-\infty}^{\infty} \int_{-\infty}^{\infty} W(x - x', y - y', z - z'|w)\Phi_0(x', y', z')dx'dy'dz' \quad .$$

(4.33)

To generate a pore structure similar to those in carbonate rocks, Crossley *et al.* (1991) showed that the Gaussian distribution

$$W(x, y, z|w) \sim \exp\left(-\frac{x^2 + y^2 + z^2}{w^2}\right) \quad ,$$

(4.34)

leads to pore structures similar to those in crystalline dolomites. The Gaussian convolution described by (33) and (34) mimics the disordered nucleation and growth of $CaMg(CaCO_3)_2$ rhombs. After $\Phi(x, y, z)$ is obtained, it is binarized by setting a threshold to distinguish between pore and solid phase voxels. The size of the pores is controlled by w. A pore can be rendered with higher resolution if w increases. This process is much more time consuming than that used for generating a model of sandstones described above. Figure 4.3 shows a comparison

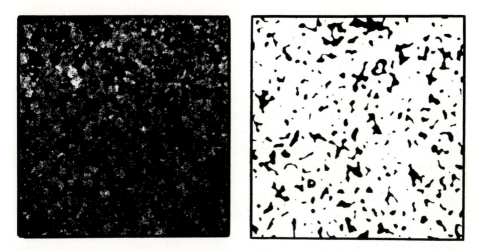

Figure 4.3: Comparison of a thin section optical photograph of crystalline dolomite (left) with the Gaussian model. Dark area is the pore space (after Crossley *et al.*, 1991).

between a thin section optical photograph of crystalline dolomite, and an image of it generated by this model. The critical porosity for this system is about 0.1.

4.5 Diagenetic processes of fractured rock

So far our discussion has been limited to unfractured porous media. However, reservoir rock in which a network of interconnected fractures may exist is also of considerable importance, since in practical applications, such as field-scale displacement of oil by a displacing fluid, or groundwater flow, one has to deal with such reservoirs. The presence of fractures, natural or man-made, is crucial to the economics of oil recovery from underground reservoirs, and to the development of groundwater resources. In both cases, fractures provide high permeability patterns for fluid flow in reservoirs that are otherwise of very low permeabilities and porosities, and would not be able to produce at high rates. Fractures and faults represent what we called megascopic heterogeneities in Chapter 1. The effect of such heterogeneities is so severe that many of the smaller scale heterogeneities, such as those at the pore or laboratory scales, may seem "simple" by comparison. Despite the obvious significance of fractures, the field of characterization of fractured rocks is not as well-developed as that of unfractured porous media. The existence of a heterogeneous framework of reservoir rock interpenetrated by a network of fractures poses a difficult setting for the estimation of recoverable hydrocarbons and the implementation of improved recovery methods, or for modelling groundwater flow. Here we give a brief discussion of diagenetic processes of fractured rock to highlight the differences between these processes and those for unfractured porous media discussed above. Characterization of fractured rock is discussed in Chapter 5.

Many fractured reservoirs, such as Monterey sediments in California, were originally complex mixtures of primary biogeneous components, finely disseminated organic matter, fine terrigeneous sediment and authigenic mineral formed during early diagenesis (Issacs, 1984). As discussed above, the accepted mechanism for diagenesis relies on a dissolution-precipitation sequence. An additional and important feature of diagenetic origin is the extensive lamination of the rock. Rich spatial patterns of alternating layers of high and low porosity (delineating low and high extents of diagenesis, respectively) are prevalent in many fractured reservoirs over many length scales ranging from millimeter to tens of millimeters. This spatial microlamination and layering is an integral aspect of the formation in fractured rocks, and represents a significant obstacle to reservoir characterization. Various theories have been advanced to explain such phenomena, based on cyclic sedimentation triggered by climatic events. This traditional approach should be contrasted to more modern theories (see, for example, Feeney *et al.*, 1983) that attribute the laminations to self-organization, driven by the type of the competition between non-linear reaction and diffusion that was discussed above.

The sedimentologic, tectonic, and diagenetic histories of fractured formations are complex, making results from conventional measurements difficult to interpret. It is well-known that many of such formations have had complex biodegradation, maturation, and migration histories. Depositional environments could also influence the primary composition of the sediment and also affect diagenesis and the development of reservoir-related properties, such as brittleness and dolomitization. Sedimentary cycles on scales from millimeters to meters are common throughout many of such fractured formations. These cycles appear to record important fluctuations in environmental conditions, such as oxygen levels, and may also serve as indicators of depositional environments. Diverse rock compositions have also been observed in many fractured reservoirs. Silica phase transformation and dolomitization can lead to the production of fractured reservoirs characterized by a wide range of physical properties. Production data from many fractured reservoirs indicate that high productivities must be associated with fractures. Fracture porosities are generally considered low, usually in the range 1-6%, whereas the pore porosity is usually larger than 10%.

4.6 Conclusions

Now that we have gained some understanding of diagenetic processes, it should be clear to us that they play an all important role in the formation of rock heterogeneities, from the smallest scales (grains, pores, and pore surfaces), to the largest (fractures and faults) and their subsequent modifications. It is not clear yet how the present structure of natural rock has survived over the past millions of years. However, we should be happy that the present structure of natural rock that allows fluid flow to take place has somehow survived, because if it had not, we could not have porous media that allow fluid flow at extremely low porosities, we would not have had large oil and gas reservoirs (and thus we would not have had oil and gas industries as we know them today), and we would not have had aquifers that provide us the water we need in our daily lives. What is important to remember, however, is that any realistic modelling of fluid transport and displacement processes in porous media and fractured rock has to take into account the effect of their heterogeneities that are the result of diagenetic processes.

Chapter 5

Morphology of Porous Media and Fractured Rock

5.0 Introduction

In Chapter 4 we learned how natural porous media and fractured rocks are formed. In this Chapter we discuss their geometrical and topological properties. The geometry of a porous system describes the shapes and sizes of its pores and/or fractures and their surface roughness, while its topology describes the way the pores and/or fractures are connected to one another. Various experimental methods are used for measuring such properties. However, the interpretation of the data is not straightforward and requires proper modelling. Thus we also review and discuss various models that have been developed for interpreting such data.

5.1 Porosity, specific surface area and tortuosity

Perhaps the simplest property of a porous system is its porosity ϕ, which is the volume fraction of its pores and/or fractures. In Chapter 3, where we discussed percolation properties of a system, we distinguished p, the fraction of open bonds or sites of the system, regardless of whether they can be reached from the external surface of the system, from the *accessible* fraction $X^A(p)$ of the open bonds or sites. Clearly, we always have $X^A(p) \leq p$. Likewise, we must distinguish the *total* porosity ϕ of a system, regardless of whether all parts of the void space that contribute to ϕ can be reached, from the *accessible* porosity ϕ_A which is the volume fraction of that void part of the system that can be reached from its external surface. In this sense p and X^A are the analogs of ϕ and ϕ_A.

The porosity of a system can be measured by several methods. The simplest of them is a direct method in which the total volume of the system is measured. Then the system is crushed to remove all the void space, and the volume of the solid is measured. Clearly, this method measures the total porosity of the system. One of the most widely used methods for measuring the accessible porosity of a system is the so-called gas expansion method. In this method the porous system is enclosed in a container filled with a gas such as air. The container is then connected with an evacuated container, which causes a change in the pressure of the system. The accessible porosity of the system is then estimated from

$$\phi_A = 1 - \frac{V_1}{V_s} - \frac{V_2}{V_s}\frac{P_f}{P_f - P_i} , \tag{5.1}$$

where V_1 is the volume of the container in which the porous system is enclosed, V_2 is the volume of the evacuated container, V_s is the volume of the porous sample, P_i is the initial pressure of the system, and P_f is the final pressure. There are several other methods of measuring ϕ and ϕ_A that are discussed by Collins (1961) and Scheidegger (1974), and therefore they are not discussed here.

Another morphological property of interest is the specific surface area Ξ, which is the ratio of the surface area of the pores and the bulk volume of the porous medium, and thus it is expressed as a reciprocal length. There are several methods of measuring Ξ which are discussed by Scheidegger (1974). For example, one can use a photomicrograph of polished sections of a sample porous medium with sufficient contrast between the pores and the matrix. From the relation between two-dimensional (surface) measurements and the properties of the three-dimensional system an estimate of Ξ can be obtained [see Eq. (74) below]. Adsorption experiments (see below) can also be used for measuring Ξ, although they depend on the size of the probing molecules, and therefore they may underestimate Ξ.

A third characteristic of a porous medium is its tortuosity T_p, which is usually defined as the ratio of the true length of the flow path of a fluid particle and the straight-line distance between the starting and finishing points of the particle's motion. Clearly, T_p depends on the porosity. If ϕ is very low (near the percolation threshold), then T_p is very large. In fact, we expect T_p to *diverge* at the percolation threshold. In classical models of flow and transport in porous media, T_p is often treated as an adjustable parameter.

5.2 Fluid saturation, capillary pressure, and contact angle

The next quantity to be considered is the saturation of a fluid that partially or totally fills the void volume of a porous system. The saturation of a fluid is the volume fraction of the void space that is filled with the fluid. Similar to porosity, the saturation of a fluid can be measured by several methods. For example, one may weigh a porous sample before and after it is filled by a fluid, from which the fluid saturation can be easily estimated.

Consider now a situation in which a drop of water is placed on a surface immersed in oil. Then the interface between the two fluids intersects the solid surface at an angle θ, which is called the contact angle; see Fig. 5.1. Associated with this system are three surface tensions, corresponding to the two fluid-solid interfaces, and the water-oil interface. These surface tensions are related to each other through the Young-Dupre equation

$$\sigma_{ow} \cos \theta = \sigma_{os} - \sigma_{ws} , \tag{5.2}$$

where σ_{os} and σ_{ws} are the surface tension between oil and the solid surface, and water and the solid surface, respectively. If $\theta < 65°$, then we say that the system

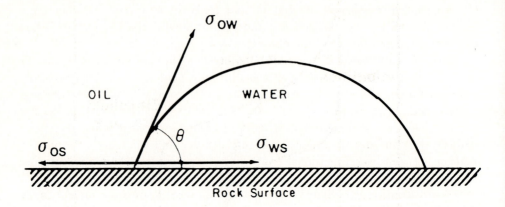

Figure 5.1: The contact angle formed between a pair of liquids and the rock surface.

is preferentially *water-wet*, while for $105° < \theta < 180°$ the surface is *oil-wet*. In between, for $65° < \theta < 105°$, the surface is said to be *intermediately-wet*, and has no strong preference for any of the two fluids. The quantity $SC = \sigma_{os} - \sigma_{ws} - \sigma_{ow}$ is sometimes called the *spreading coefficient*.

An important quantity for characterization of a porous system is the capillary pressure P_c. Suppose that two immiscible fluids, e.g., water and oil, are in contact in a bounded system such as a cylindrical tube, and are separated by an interface. Then there is a discontinuity in the pressure field as one moves across the interface from one fluid phase to the other. That is, if we consider two points, one in the water phase at pressure P_w and one in the oil phase at pressure P_o, both in the immediate vicinity of the interface, then $P_w \neq P_o$. The pressure difference $P_c = P_o - P_w$ is called the capillary pressure (the pressure is usually higher in the non-wetting fluid, oil in this case), and is given by the Laplace equation

$$P_c = \sigma_{ow} \cos \theta \left(\frac{1}{r_1} + \frac{1}{r_2} \right) . \tag{5.3}$$

Here r_1 and r_2 are the principal radii of curvature of the interface. For an interface in a cylindrical tube $r_1 = r_2 = r$, where r is the radius of the tube. If we consider the capillary pressure in a simple system such as a cylindrical or a conic tube, then it is clear that P_c should depend uniquely on the amount of the fluid in the system, and hence on its saturation.

However, the situation is different in a porous system with many irregularly shaped pores and/or fracture, because different saturations may yield the same P_c, and vice versa, since there are several interfaces in the system at different locations. This is indeed what one finds in an experiment measuring P_c in a porous system. For example, suppose that we initially fill our porous system

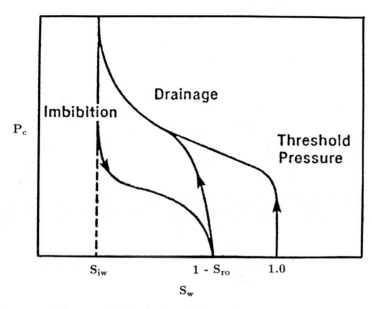

Figure 5.2: Typical capillary pressure curves.

with water, and then expel it gradually by injecting oil into the pore space. At each stage of the experiment, and for each value of the water saturation S_w, we have a corresponding capillary pressure P_c. If we continue this experiment for a long enough time, we reach a point where no more water is produced, although there is still some water left in the pore space. The saturation of water at this stage, i.e., at the *highest* value of P_c, is called the *irreducible* saturation S_{iw}, and the process of expelling water by oil is called *drainage*. Now, if we start with the porous system at the end of drainage (at the highest value of P_c) and expel the oil gradually by injecting water into the system, a process called *imbibition*, we obtain another capillary pressure-saturation curve, which is very different from what we obtained during drainage, and thus there is *hysteresis* in the capillary pressure-saturation curves. Figure 5.2 shows a typical capillary pressure-saturation diagram during both drainage and imbibition. As can be seen, even when $P_c = 0$, there is still some oil left in the system. The saturation of oil at this point is called the *residual* saturation S_{ro}. Obviously, drainage or imbibition does not have to start at $S_w = 1$ or at $S_w = S_{iw}$, but can start at any saturation in between the two, i.e., we can have secondary drainage, secondary imbibition, and so on. If we do such experiments, we again find different capillary pressure curves, also shown in Fig. 5.2. We discuss contact angles and capillary pressure curves in greater details in Chapter 12, where we consider multiphase flow in porous media. Our goal in introducing them in this Chapter is to use them for characterizing porous systems (see below).

Capillary pressure curves can also be measured by several methods. One of the most popular methods is based on mercury porosimetry, which is discussed separately, as it is a popular method of measuring the pore size distribution of a porous system. Another technique is the centrifuge method in which a small sample is saturated by a wetting fluid and is placed in a container filled with a non-wetting fluid. The system is then rotated at an angular velocity ω. The density of the wetting fluid ρ_w is usually larger than that of the non-wetting fluid ρ_{nw}, and thus the wetting fluid leaves the system at the outer radius. At the same time, the wetting fluid is replaced by the non-wetting fluid at the inner radius. Then the capillary pressure at the inner radius of rotation R_1 of the sample is given by

$$P_c = \frac{\omega^2(\rho_w - \rho_{nw})}{2}(R_2^2 - R_1^2) , \tag{5.4}$$

where R_2 is the outer radius of rotation of the sample where the pressure has the same value in both phases, i.e., where $P_c = 0$. Thus, if we change ω, different amounts of water will be expelled from the system which can be measured, and therefore the capillary pressure-saturation diagram can be obtained. The saturation that is used in such a diagram is an *average* saturation, defined as

$$\langle S_w\rangle(R_2 - R_1) = \int_{R_1}^{R_2} S_w dr . \tag{5.5}$$

This equation is often replaced by an approximate equation due to Hassler and Brunner (1945) and Slobod *et al.* (1951)

$$\langle S_w\rangle P_{c1} = \int_0^{P_{c1}} S_w(P_c)dP_c , \tag{5.6}$$

so that if we differentiate Eq. (6), we obtain

$$\langle S_w\rangle + P_{c1}\frac{d\langle S_w\rangle}{dP_{c1}} = S_w(P_{c1}) , \tag{5.7}$$

which can be used for constructing a saturation-capillary pressure curve.

5.3 Pore size distribution

Practically every book or article about porous media mentions the "pore size distribution" of porous systems, but often it is not clear what is meant by this distribution. In a porous medium consisting of particles, the spaces between the particles are called voids, whereas if the particles themselves are porous, then the void space in the particles is called pore. However, careful examination of natural porous media reveals that what are usually referred to as pores can in fact be divided into two groups. In the first group are *pore bodies* where most of the porosity resides, while in the second group are *pore throats*, which are the channels that connect the pore bodies. One usually assigns *effective* radii to pore bodies and throats which, in reality, are nothing but the radii of spheres that have the same volume. Thus pore bodies and pore throats are defined in terms of approximate maxima and minima of the largest-inscribed-sphere radius. In a network representation of the pore space (see Chapter 6), the pore bodies are represented by the sites or nodes of the network, and the pore throats are represented by its bonds. All of the volume of a pore body can be assigned to the corresponding node; alternatively it can be apportioned among the network bonds, which is what is done in most network modelling of transport processes in porous media. Obviously, if the pore size distribution is known, then an average pore size, which is just the first moment of the distribution, can also be defined. However, a popular but empirical method of characterizing a porous medium is by its *mean hydraulic diameter d_H*, defined as

$$d_H = 4\frac{V_p}{S_p} \tag{5.8}$$

where V_p and S_p are the volume and surface of the pores, respectively. In view of the discovery that the pore volume and surface of many porous media have fractal properties (see below), d_H may not be such a useful quantity as it was thought of before, because d_H may now depend on the length scale since, as discussed in Chapter 3, fractal properties are scale-dependent.

The pore size distribution is defined as follows: It is the *probability density function* that gives the distribution of pore volume by an effective or characteristic pore size. Even this definition is somewhat vague, because if the pores could be separated, then each pore could be assigned an effective size, in which case the pore size distribution would become analogous to the particle size distribution. But, because the pores are interconnected, the volume that one assigns to a pore can be dependent upon both the experimental method and the model of pore space that one employs to interpret the data. Four methods of measuring pore size distribution are mercury porosimetry, adsorption-desorption experiments, small-angle scattering, and nuclear magnetic relaxation methods. The first two methods have been used extensively, while the latter two are newer and seem to be more accurate.

5.3.1 Mercury porosimetry, network simulation and percolation

In mercury porosimetry, the porous medium is first evacuated and then immersed in mercury. The pressure is then increased and the volume of mercury injected into the porous medium is measured as a function of the applied pressure. This corresponds to the drainage process discussed above, as mercury acts as a non-wetting fluid. The pressure is usually increased either incrementally or continuously. Larger and larger pressures are needed to penetrate progressively smaller pores. Very high pressures can even damage the internal structure of the medium, but we ignore them here. The pressure is then lowered back to atmospheric pressure, as a result of which the mercury is retracted from the pores. This corresponds to imbibition. As discussed above, during imbibition or retraction there is a characteristic shift, or hysteresis, between the injection and retraction curves. Similar to the drainage and imbibition experiments described above, some mercury stays in the medium even after the pressure is lowered back to atmospheric pressure. This technique was first developed by Ritter and Drake (1945) and has remained popular ever since. It is usually used for pores between 3 nm and 100 μm.

A precise apparatus for measuring mercury-injection curves is described by Thompson *et al.* (1987a,b). It consists of four components: (i) a mercury reservoir positioned on an elevator raised by a stepper-motor-driven screw; (ii) a sample holder on a pan balance connected to the reservoir by stainless-steel tubing; (iii) stainless-steel electrodes located on the top and bottom of the cylindrical sample; and (iv) electronics for measurement of the AC resistance, the temperature, and the atmospheric pressure. The apparatus is shown in Fig. 5.3. The experiment is automated by computer control. Before injection is started, the pore space is evacuated to a pressure of 10^{-3} Pa. During the measurements, the elevator height is changed by typically 0.1-10 mm and the sample weight is monitored until equilibrium is reached. Typical experimental sensitivities are 10^{-5} cm^3 for volume, 0.5 Pa for pressure, and 0.1 $\mu\Omega$ for resistance, which result in resolutions of better than 1 part in 10^4 for all the parameters of interest. A typical experiment consists of 30,000 observations taken at 3 sec. intervals.

While mercury porosimetry is a relatively straightforward experiment, the interpretation of the data is not simple. The data are usually interpreted using a variation of Eq. (3) due to Washburn (1921)

$$P_c = \frac{2\sigma_{mv}}{r}\cos(\theta + \varphi) \quad , \tag{5.9}$$

where P_c, the applied pressure, is just the capillary pressure between mercury and the vaccum, σ_{mv} the interfacial tension between mercury and the vacuum, θ the contact angle between mercury and the surface of the pore, and φ is the wall inclination angle at which the pore radius is r, with $r_t \leq r \leq r_b$, where r_t and r_b are, respectively, the pore throat and the pore body radii. Equation (9) is of course the same as (3) in the limit $r_1 = r_2 = r$ and $\varphi = 0$, and results from a capillary force balance on a cylindrical tube. Up until 1977, the interpretation of porosimetry was based on modelling the pore space as a bundle of nonintersecting capillary tubes. However, it is now clear that such a model of a pore space is hopelessly wrong, and cannot be used for interpreting mercury porosimetry data.

Leverett (1941) defined a reduced capillary pressure function which is usually used for correlating data and is defined by

$$J = \frac{P_c}{\sigma_{mv}\cos\theta}\sqrt{\frac{K}{\phi}} \quad , \tag{5.10}$$

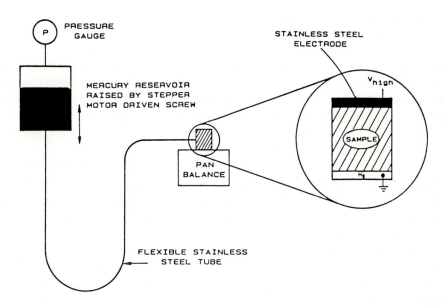

Figure 5.3: Schematic diagram of the apparatus for mercury-porosimetry experiments (after Thompson *et al.*, 1987a).

where K is the permeability of the pore space. This function was named the Leverett J-function by Rose and Bruce (1949). It has been found that the J-function is successful in correlating capillary pressure data originating from a specific lithologic type *within the same formation*, but it is not of general applicability. Perhaps the reason for this lack of generality is that $\sqrt{K/\phi}$ is not an adequate scale factor for taking into account the individual differences between pore structures of various porous media. Capillary pressure curves have been reported by a large number of authors, a long list of whom is given by Dullien (1979).

As Fig. 5.2 indicates, there is hysteresis between the injection (drainage) and retraction (imbibition) curves. At the end of retraction, mercury can be reinjected into the medium and a second injection curve can be obtained. In some cases the hysteresis depends on the *history* of the system, or the way the experiment has been carried out. Thus in some cases hysteresis can be eliminated by performing the experiment very slowly, while in other cases it *cannot* be eliminated. The latter type of hysteresis was called *permanent hysteresis* by Everett (1967).

Although we are discussing capillary pressure curves during mercury porosimetry, the phenomenon belongs to the general class of two-phase flow problems in porous systems which are discussed in Chapter 12. For now, it suffices to say that, in general if a non-wetting fluid (in this case mercury) is to displace a perfectly wetting fluid, it must overcome a capillary pressure at the pore throat [see

Eq. (3) or (9)]

$$P_c = 2\frac{\sigma_{nw}}{r_t} \quad , \tag{5.11}$$

where $P_c = P_{nw} - P_w$, P_{nw} and P_w are the pressures in the non-wetting and wetting phases, respectively, and σ_{nw} is the interfacial tension between the two phases. Similarly, for the wetting phase to displace the non-wetting phase in the pore body, the capillary pressure must be

$$P_c = 2\frac{\sigma_{nw}}{r_b} \quad . \tag{5.12}$$

In general, as Eqs. (3) and (9) indicate, capillary pressure curves depend on the contact angles, and therefore their shapes can be characteristic of the wettability of the pore space. This is discussed in Chapter 12 where we consider the effect of wettability on multiphase flow in porous media.

By now it should be clear to the reader that the sequence of the pores that are penetrated by mercury during mercury porosimetry depends not only on the pore size distribution of the pore space, but also on its topology—the way its pores are connected to each other. Therefore, mercury porosimetry belongs to the class of percolation phenomena discussed in Chapter 3. Although the effect of pore space connectivity on mercury porosimetry, or more generally, the capillary pressure curves for any two-phase flow problem in porous media, had been appreciated for some time, it was only in 1977 that the connection between this phenomenon and percolation was recognized and appreciated. Chatzis and Dullien (1977), Larson (1977), de Gennes and Guyon (1978), Larson and Morrow (1981) and Wall and Brown (1981) were among the first to recognize the possibility of developing a percolation model for interpreting mercury porosimetry data and capillary pressure phenomena in porous media. Androutsopoulos and Mann (1979) used two-dimensional networks of interconnected pores to model such phenomena, although they did not mention percolation explicitly. All of these authors recognized that a bundle of non-intersecting capillary tubes is completely inadequate for interpreting mercury porosimetry data. Over 40 years ago, Meyer (1953) had already recognized the effect of the interconnectivity of the pores when he stated that, "There may, for example, be large pores which one would expect to fill at a low pressure, which have no connection with the mercury except through smaller pores. The effect of these ink-bottle pores is to assign too small a portion of the pore space to the large pores and too large a part to the small pores, if the mercury injection data is taken at its face value."

As discussed by Larson and Morrow (1981) and Wall and Brown (1981), pores that are close to the external surface of a porous medium can be reached more easily than those in the middle of the medium, since if a pore in the interior of the medium is to be penetrated by the mercury, a connection with the external surface via the already penetrated pore bodies and pore throats has to be established. If this effect is not taken into account, one obtains a wrong pore size distribution. This was nicely demonstrated by Dullien and Dhawan (1975),

who compared pore size distributions obtained by photomicrographic techniques
with those inferred from mercury porosimetry data interpreted with the above
assumptions. Of course, one way of decreasing the effect of interior pores is to
use thin or small samples, which also reduces the measurement time. However,
before this is done, one has to establish that the pore size distribution obtained
with small or thin samples is in fact representative of the actual (much larger)
porous medium. Larson and Morrow (1981) developed a model that could take
into account the effect of sample size.

Once one recognizes the importance of interconnectivity and pore accessibility, then the application of percolation concepts to mercury porosimetry seems
natural. Many authors have used such concepts to calculate the capillary pressure curves of porous media (Larson and Morrow, 1981; Wall and Brown, 1981;
Conner *et al.*, 1984, 1988; Conner and Lane, 1984; Neimark, 1984a; Chatzis
and Dullien, 1985; Heiba, 1985; Lane *et al.*, 1986; Ramakrishnan and Wasan,
1986). Some recognized that, although mercury porosimetry is a percolation
process, there are certain differences between this process and the random percolation described in Chapter 3 (see, for example, Lane *et al.*, 1986). Some used
two-dimensional networks which are actually not suitable for simulating mercury
porosimetry, since this is a two-phase flow problem, and no two-dimensional bicontinuous structure exists (that is, one cannot have two fluid phases flowing
simultaneously in a two-dimensional space), while others (for example, Larson
and Morrow, 1981) represented the pore space by a Bethe lattice to take advantage of the analytical expressions for the percolation properties of Bethe lattices
derived by Fisher and Essam (1961). Tsakiroglou and Payatakes (1990) developed a more general three-dimensional network simulator in which percolation
was not used explicitly, although pore interconnectivity played an essential role
in their model.

Before describing a percolation model of mercury porosimetry, let us first
mention a few earlier works which, although did not use the terminology of percolation theory explicitly, were more or less percolation models, since they took
into account the effect of pore interconnectivity. The earliest work appears to
be that of Ksenzhek (1963), who used a cubic lattice of pores to investigate the
penetration of a porous medium by a non-wetting liquid. The pores were assumed to be capillary tubes of a given radius. The pore radii were distributed
according to a probability density function. By making a few simplifying assumptions, Ksenzhek derived several formulae for various quantities of interest.
In particular, a quantity essentially equivalent to the percolation probability was
calculated and its dependence on the network size was investigated. Topp (1971)
noticed that, in the early theories of hysteresis phenomena developed by Everett (1954) and others, only the shapes of the pores determined the shape of
the curves and the sequence by which pores are filled by the penetrating fluid,
whereas in reality both pore geometry and the *state of its neighboring pores*
should be important. This is also what Meyer (1953) had stated 18 years earlier,

but the significance of Topp's work is that he developed integral convolutions of the pore size distribution, and a quantity that is essentially equivalent to the accessibility function of percolation discussed in Chapter 3. Similar to the work of Ksenzhek (1963), Topp made several simplifying assumptions, but it seems that he clearly recognized the significance of *both* the pore size distribution *and* the pore space accessibility. Pis'men (1972) developed elegant integral expressions describing capillary equilibrium, which included the effects of branching of bifurcating pores and pore distribution. Finally, Chatzis and Dullien (1977) used percolation concepts to calculate capillary pressure curves, and Androutsopoulos and Mann (1979) used network simulations, without mentioning percolation, to achieve the same end. Complete details of the work of Chatzis and Dullien (1977) are given in their 1985 paper.

Let us now describe two approaches to modelling mercury porosimetry in a porous medium. In one approach one represents the pore space by a network and models the phenomenon in great details. The most sophisticated of such approaches was developed by Tsakiroglou and Payatakes (1990), whose work we discuss. In the second approach one uses percolation concepts to model the phenomenon. Of course, a percolation model cannot take into account the effect of all the pore-level phenomena, but as we see below, this approach provides valuable information about the behavior of the phenomenon.

Tsakiroglou and Payatakes (1990) represented the porous medium by a simple-cubic network of cylindrical bonds (representing the pore throats) and spherical sites (representing the pore bodies). The effective diameters of the bonds and sites were selected from Gaussian distributions. The length of the bonds was adjusted so that the porosity of the network matched that of the porous medium for which the model was intended. Most authors have assumed that the injection process is controlled solely by the diameter of the bonds (throats), which implies that once a bond is filled by mercury, so also is the downstream site. However, some experiments have shown that this may not be the case, and that mercury menisci can reach equilibrium not only at the entrance to a throat (bond), but also at the entrance to a pore body (site). Tsakiroglou and Payatakes allowed such a possibility in their model. We now consider a meniscus that is entering a spherical site (see Fig. 5.4). The capillary pressure P_c can be determined as a function of the position z of the contact line in the sphere, and is given by $P_c(z) = 2\sigma_{mv}\cos\alpha / r(z)$, where $r(z)$ is the radius of the pore at z and is given by

$$r(z) = \left[\frac{d_t^2}{4} + z\sqrt{d_s^2 - d_t^2} - z^2 \right]^{1/2} \tag{5.13}$$

The angle α is given by $\alpha = \theta_1 - \theta$, where

$$\tan\theta_1 = \frac{dr(z)}{dz} = \frac{\sqrt{d_s^2 - d_t^2} - 2z}{2r(z)} \tag{5.14}$$

Here d_s and d_t are the diameters of the sphere and the throat, respectively. If

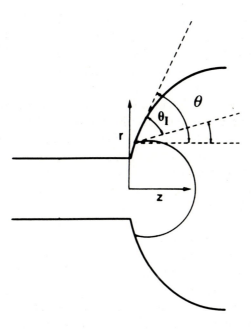

Figure 5.4: A meniscus leaving a pore throat and entering a spherical pore body (after Tsakiroglou and Payatakes, 1990).

$P_c(z) \geq 4\sigma_{mv} \cos\theta/d_t$, then the external pressure for moving the meniscus into the sphere is larger than that for moving it into the throat.

In general, during mercury retraction collars of the wetting fluid (mercury vapor and/or air) are formed in the throats which are saddle-shaped (see Fig. 5.5). As the pressure is lowered during retraction, the curvature of these collars decreases, until some interfaces become unstable and rupture. This phenomenon is called *snap-off* (Mohanty *et al.*, 1980; see Chapter 12 for a detailed discussion). In a cylindrical throat the shape of a collar can be determined by solving the Young-Laplace equation in cylindrical coordinates

$$r\frac{d^2r}{dz^2} + \frac{rP_c}{\sigma_{mv}}\left[1 + \left(\frac{dr}{dz}\right)^2\right]^{3/2} - \left(\frac{dr}{dz}\right)^2 = 1 \qquad (5.15)$$

with the boundary condition that $dr/dz = -\tan\theta_R$ at $z = 0$. Thus, varying P_c allows us to determine $r(z)$ by solving Eq. (15) numerically. If the volume of the collar is V_c, then the collar is stable if $dV_c/dP_c \leq 0$ (Everett and Haynes, 1972). Thus, if we set $dV_c/dP_c = 0$, we can determine the critical capillary pressure for snap-off and the corresponding shape of the interface. Note that the length of a throat must be long enough for a collar to form, otherwise no collar would form which would may not be realistic.

Now we can start simulating a mercury porosimetry process. We first consider the injection process and place mercury menisci at the entrance to all the throats

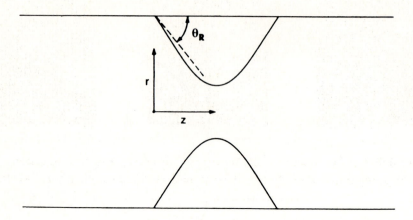

Figure 5.5: Saddle-shaped collars of the wetting phase around the non-wetting phase during slow imbibition (after Tsakiroglou and Payatakes, 1990).

on the external surface of the network. The applied pressure is set to be small enough that no meniscus can enter a throat yet. Then the applied pressure is increased by a small amount, and all menisci at the throat entrances are examined. If the applied pressure exceeds the capillary pressure of a throat, that throat is filled by mercury, and the meniscus is placed at the entrance to the downstream spherical site. After all menisci at the pore entrances are examined and moved if necessary, the menisci at the entrances to the downstream spheres are examined. If the applied pressure exceeds the capillary pressure for entering a sphere, that sphere is filled with mercury, and new menisci are placed at the entrance to the throats that are connected to that sphere and do not contain mercury. If a sphere has several menisci placed at its entrances, the smallest capillary pressure determines whether the sphere is filled with mercury. The new menisci are examined to determine if any new throat can be filled with mercury. If the applied pressure is not large enough, it is increased by a small amount, all the menisci are examined again, and so on. The simulation continues until mercury-filled bonds and sites form a sample-spanning cluster. This completes the simulation of the injection process. Figure 5.6 shows typical configurations of a two-layer network of coordination number 5 at various stages of the process.

At the end of this simulation, we can start simulating the retraction process. The applied pressure is lowered by a small amount, and a search is carried out to find (i) those pores that contain menisci that are connected to the external mercury sink through continuous mercury paths, and can move under the present conditions, and (ii) throats that contain mercury that must snap-off under the present conditions. For such pores Eq. (15) is solved numerically to determine the shape of the interface. Snap-off can leave pockets of isolated mercury in some throats or spheres, which are then ignored for the rest of the simulation.

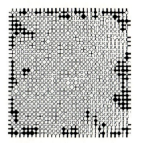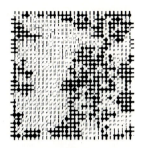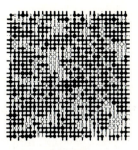

Figure 5.6: Configurations of the pore network during mercury injection. The applied pressure increases from left to right (after Tsakiroglou and Payatakes, 1990).

The applied pressure is lowered again, and the search continues. As can be seen, simulation of the retraction process is much more difficult than the injection process, because one has to deal with the snap-off phenomenon. Figure 5.7 shows typical configurations of the same network during the retraction process. Figure 5.8 shows typical capillary pressure-saturation curves that are obtained with a $20 \times 20 \times 20$ network, which resemble those obtained experimentally. This figure also shows that the network size effect is not very strong. The advantage of this model is that, most of the pore-level phenomena are taken into account, and therefore one obtains a clear understanding of the phenomena involved during mercury porosimetry. Moreover, one can vary the pore throat and pore body size distributions, and the coordination number of the network to study their effect on mercury porosimetry. To obtain the pore throat and pore body size distributions, one can assume the two distributions to have certain functional forms with one or two adjustable parameters, and vary the parameters until the simulation results match the data. However, this procedure is very tedious and time consuming.

Let us now use percolation concepts and describe a simple model for mercury porosimetry. We should first note that a percolation model for describing any two-phase flow phenomenon in porous media is appropriate if the capillary pressure across a meniscus separating the two fluids (e.g., mercury and the vacuum) is greater than any other pressure difference in the problem, e.g., that due to buoyancy. The second condition is that frictional losses due to viscosity must be small compared with the capillary work. If we define a capillary number Ca by

$$Ca = \frac{\eta v}{\sigma_{mv}} \quad , \tag{5.16}$$

where v is the average fluid velocity and η the average viscosity, then one must have $Ca \ll 1$ in order to fulfill this criterion.

Figure 5.7: Configurations of the pore network during mercury retraction. The pressure decreases from left to right (after Tsakiroglou and Payatakes, 1990).

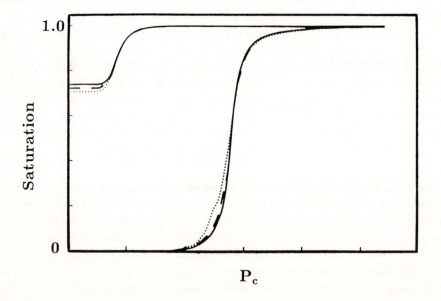

Figure 5.8: Typical capillary pressure curves obtained with the network model of Tsakiroglou and Payatakes (1990). Dotted curves are for a $10 \times 10 \times 10$ network, dashed curves for size $15 \times 15 \times 15$, and solid curves for size $20 \times 20 \times 20$.

The porous medium is again represented by a three-dimensional network of cylindrical bonds and spherical sites. We ignore the size of the pore bodies and consider a pore size distribution $f(r)$ for the pore throats. We describe two percolation models that are, however, closely related. In the first model, developed by Larson and Morrow (1981), one writes the capillary pressure as $P_c = \sigma_{mv}C$, where C can be interpreted as the curvature of a meniscus, $C \sim 1/r$ [see Eq. (9)]. Now consider the injection process. The network is exposed to an applied pressure, which is equivalent to some curvature C. As C increases it exceeds the entry curvature C_i for a larger and larger fraction of the pores. In fact, C exceeds C_i for a fraction $Y_i(C)$ of the pore volume given by

$$Y_i(C) = \int_0^C y_i(C_i)dC_i \qquad (5.17)$$

Here $y_i(C_i)$ is the pore entry-curvature distribution given by

$$y_i(C_i) = \int_0^\infty y(C_i, C_w)dC_w \qquad (5.18)$$

where $y(C_i, C_w)$ is the joint probability distribution for pore entry and pore withdrawal curvatures. It has the properties that $y(C_i, C_w) \geq 0$ for $C_i \geq 0$ and $C_w \geq 0$, and $y(C_i, C_w) = 0$ for $C_i < C_w$. Moreover, it has to be renormalized which means that

$$\int_0^\infty \int_0^{C_i} y(C_i, C_w)dC_idC_w = 1 \qquad (5.19)$$

With this interpretation, $Y_i(C)$ is the fraction of the pores with entry curvature less than C. However, not all pores with an entry curvature C are accessible from the external surface of the network, where accessibility is defined in the percolation sense discussed in Chapter 3. Thus, Larson and Morrow (1981) assumed that the saturation of the non-wetting fluid (mercury) is given by

$$S_{nw} = X^A[Y_i(C)] \qquad (5.20)$$

As discussed below, Eq. (20) is only a rough approximation, since one has to take into account the effect of the pore size distribution $f(r)$. In any event, for any given capillary pressure $P_c = \sigma_{mv}C$ the fraction $Y_i(C)$ is calculated from which S_{nw} is determined. This completes the calculation of the injection curve.

Now consider the retraction process. As the applied pressure at the end of the injection process is decreased, the curvature C also decreases and the non-wetting fluid is ejected from the pore space from the pores for which the withdrawal curvature is between, say, C_w and $C_w - dC_w$, and the entry curvature C_i or less. The fraction of such pores is

$$y_r(C_i, C) = \int_0^{C_i} y(C_i', C)dC_i' \qquad (5.21)$$

However, not all such pores actually expel the non-wetting fluid, because some of them were not invaded during the injection process in the first place, since they were not accessible, and some of them, although containing mercury, cannot expel it because they are not connected to the external surface by a path of mercury-filled pores. Thus, during retraction the fraction of the pores with injection curvature less than or equal to C_i and withdrawal curvature less than or equal to C_w is given by

$$Y_r(C_i, C) = \int_0^C y_r(C_i, C_w) dC_w \qquad (5.22)$$

But only a portion $X^A(Y_r)$ of such pores still contain mercury that is accessible, which implies that the fraction of such pores is $X^A(Y_r)/Y_r$. Thus, as the curvature C_w is reduced by an amount dC_w (as the applied pressure is reduced), the saturation of the non-wetting fluid is also decreased by dS_{nw}, given by

$$-dS_{nw} = \frac{X^A[Y_r(C_i, C_w)]}{Y_r(C_i, C_w)} y_r(C_i, C_w) dC_w \qquad (5.23)$$

which after integration yields

$$S_{nw} = S_i - \int_C^{C_i} \frac{X^A[Y_r(C_i, C_w)]}{Y_r(C_i, C_w)} y_r(C_i, C_w) dC_w \qquad (5.24)$$

where S_i is the initial saturation at which the retraction process started. Thus, for any given P_c one can determine the corresponding S_{nw}. Obviously, when the applied pressure P_c is zero, the corresponding curvature is also zero. At this point the saturation of the non-wetting fluid is at its residual value S_{rnw}, which is determined from (24) by setting $C = 0$. This completes the calculation of the retraction curve.

At the end of the retraction process, we can consider a second injection process. It is not difficult to see that for the second injection curve we have

$$y_i(C_i, C) = \int_0^C \frac{X^A[Y_r(C_i, C_w)]}{Y_r(C_i, C_w)} y(C, C_w) dC_w \qquad (5.25)$$

using the fact that for $C < C_w$ we have $y(C, C_w) = 0$. The corresponding saturation is given by

$$S_{nw} = S_{rnw} + \int_0^C y_i(C_i, C') dC' \qquad (5.26)$$

Obviously, as Eq. (9) indicates, for a given curvature distribution one can calculate a corresponding pore size distribution $f(r)$, since $C \sim 1/r$. Thus, one assumes a functional form for $f(r)$, with perhaps one or two adjustable parameters and varies them until the predicted curves agree with the data. The advantage of this model is that, if the accessibility function X^A is available (which depends on the coordination number of the network), all of the computations can be done analytically. However, for three-dimensional networks X^A is not available in closed form, and only its numerical values have been calculated.

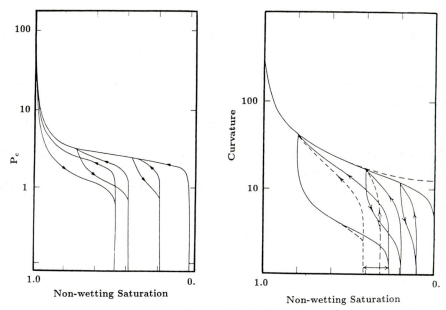

Figure 5.9: Comparison of experimental capillary pressure curves (left) with the predictions of the percolation model. Dashed curves are for a pore network that is 10 pores thick, while the predicted solid curves are for an infinitely large network (after Larson and Morrow, 1981).

Thus, the integrals in Eqs. (23)-(25) have to be evaluated numerically, which is still much simpler than the computations associated with the network simulation described above. Larson and Morrow (1981) assumed that the pore space can be represented by a Bethe lattice of coordination number 4 (see Chapter 3) which is not very realistic, but has the advantage that its X^A is available in closed form (Fisher and Essam, 1961). They also assumed that

$$y(C_i, C_w) = 6g(C_i)g(C_w) \int_{C_w}^{C_i} g(C)dC , \quad C_i \geq C_w \tag{5.27}$$

which has the properties that were mentioned above, and used $g(x) = 2x\exp(-x^2)$ with $x = \log C$ (this results in a broad distribution). Figure 5.9 compares their predictions with the data for a Becher dolomite with a porosity $\phi = 0.174$. The qualitative agreement between the predictions and the data is striking. Also shown is the effect of sample size on the capillary pressure curves. The most important reason for this agreement is the fact that the concept of accessibility which determines which pores can be invaded by mercury, and from which pores mercury can be withdrawn, has been used. This concept is fundamental to modelling multiphase flow phenomena in porous systems, and is emphasized in this book.

This model was refined and extended by Heiba *et al.* (1982, 1992). As already discussed, during both injection and withdrawal or retraction, the subdistributions of the pore space, accessible to and occupied by mercury, are different.

Consequently, the pore size distribution of the subset of pore space occupied by mercury differs from the overall pore size distribution. Heiba *et al.* derived analytical formulae for such subdistributions. Thus, during injection of mercury into the pore space, the fraction of pores that are *allowed* to it is (Heiba *et al.*, 1982, 1992)

$$Y_i(r_{min}) = \int_{r_{min}}^{\infty} f(r)dr \quad , \qquad (5.28)$$

where r_{min} is the minimum pore radius into which the mercury can penetrate. The fraction of those pores that are accessible to and thus occupied by the mercury is $X^A(Y_i)$. Therefore, during injection the distribution $f_i(r)$ of the pore radii that are occupied by the mercury is (Heiba *et al.*, 1982, 1992)

$$f_i(r) = f(r)/Y_i(r_{min}) , \quad r \geq r_{min} \qquad (5.29)$$

and obviously $f_i(r) = 0$ for $r < r_{min}$. The idea behind Eqs. (28) and (29) is that during the injection process, the *largest* pores (i.e., those with the *smallest* entry curvature) will be occupied [which can be understood by examining Eq. (9)]. During injection the non-wetting fluid (mercury) saturation is given by

$$S_{nw} = \frac{X^A[Y_i(r)] \int_{r_{min}}^{\infty} f(r)V_p(r)dr}{Y_i(r_{min}) \int_0^{\infty} f(r)V_p(r)dr} \quad , \qquad (5.30)$$

where $V_p(r)$ is the volume of the pore throat of radius r. Equation (30) should be compared with Eq. (20).

During retraction, as the pressure is lowered the mercury is first expelled from the *smallest* pores [see Eq. (9)]. The *allowed* fraction of such pores is (Heiba *et al.*, 1982, 1992)

$$Y_r = \int_0^{r_0} f(r)dr + \left[1 - \frac{X^A(Y_{i,t})}{Y_{i,t}}\right] \int_{r_0}^{\infty} f(r)dr \quad , \qquad (5.31)$$

where r_0 is the radius of the pore at a given capillary pressure P_c such that the mercury is expelled from all pores for which $r \leq r_0$, $Y_{i,t} = Y_i(r_{min,t})$, and $r_{min,t}$ is the pore radius at the end of injection process. The first term of the right side of Eq. (31) is clearly the fraction of pores from which the mercury is expelled, if at the end of injection there were no pores that were not inaccessible to it. However, at the end of injection a fraction $1 - X^A(Y_{i,t})/Y_{i,t}$ of the pores could not be reached by the mercury and, consequently, the second term of the right side of Eq. (31) is the fraction of pores that were not invaded by the mercury at the end of injection. Hence, the size distribution of the pores from which the mercury is expelled is given by (Heiba *et al.*, 1982, 1992)

$$f_r(r) = \begin{cases} \dfrac{f(r)}{Y_r}\left[1 - \dfrac{X^A(Y_{i,t})}{Y_{i,t}}\right] & , \; r > r_0 \\[3mm] \dfrac{f(r)}{Y_r}\left\{1 - \dfrac{X^A(Y_{i,t})}{Y_{i,t}}\left[1 - \dfrac{X^A(Y_r)}{Y_r}\right]\right\} & , \; r_{min,t} < r < r_0 \end{cases} \qquad (5.32)$$

Clearly $f_r(r) = f(r)/Y_r$ for $r < r_{min,t}$. Thus, the model of Heiba *et al.* classifies the pores more carefully than that of Larson and Morrow (1981), and moreover calculates the saturation correctly, as it takes into account the effect of a pore size distribtion and the dependence of the volume of a pore on its effective radius.

The procedure for using this model to extract the pore size distribution of the pore space is as follows. As the first step a functional form for $V_p(r)$ and hence a pore shape have to be assumed. Next, one has to calculate the accessibility function of the pore space, which means that either the average coordination number $\langle Z \rangle$ of the pore space has to be known from measurements, or it must be treated as an adjustable parameter of the system in order to fit the percolation model to the data. Later in this Chapter we discuss how a combination of a percolation model and adsorption-desorption isotherms in porous media can be used for estimating $\langle Z \rangle$. Alternatively, $\langle Z \rangle$ may be estimated from serial sectioning of the porous sample (see below). Observe that S_{nw} can be measured and and P_c is set during the experiment, and thus r_{min} can be calculated. Thus, assuming an $f(r)$ as an initial guess, Eq. (30) is iterated many times until a satisfactory $f(r)$ can be found. As before, a particular form of $f(r)$ with one or two adjustable parameters is assumed and the parameters are varied until a satisfactory fit is found. However, note that since before mercury forms a sample-spanning cluster, $X^A = 0$, one *cannot* obtain the complete pore size distribution: no information about the largest pores that are penetrated by the mercury at the very initial stages of the process can be obtained. On the other hand, if we use the measurements during the retraction process, we obtain information about the size distribution of the pore bodies, but this information is again not complete. One way of resolving this difficulty is to do the measurements in small samples, so that the effective percolation threshold would be small and, as a result, more information could become available. Care should, however, be taken to ensure that the small sample is representative of the true porous medium.

It is important to understand all the assumptions that are made in order to arrive at these formulae. (i) The pore space is infinitely large. (ii) The entire process can be described by a random bond percolation. (iii) Entrapment of the mercury in isolated clusters is ignored. The first assumption is essential if we are to use the results for percolation on infinitely large lattices. However, for a given network, one can calculate X^A as a function of its linear size L. Figure 5.10 shows the dependence of X^A on L for site percolation on a simple-cubic network. For the Bethe lattice, the accessible fraction X^A can be calculated analytically for a given lattice size, which is why this lattice was used by some authors. The main effect of the size or thickness of a porous medium is increased accessibility of the pore space for thinner samples, which causes a reduction in the sharpness of the injection-curve knee. This effect is also evident in Fig. 5.10 where we show the dependence of X^A on L. Injection curves for unconsolidated packings indicate rather strong dependence on sample thickness for systems up to about 10 particle diameters or about 30 pore-throat diameters. For thicker media, the

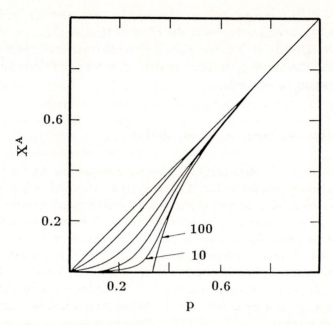

Figure 5.10: Effect of network linear size on the accessibility function X^A. Numbers refer to the linear size of the network.

dependence is relatively weak, and if the thickness exceeds 20 particle diameters, no appreciable sample size can be detected.

The second assumption is not, strictly speaking, correct. As already discussed, there is some evidence that once mercury fills a pore throat, the corresponding meniscus does not necessarily enter the downstream pore body, and moreover pore bodies largely control the retraction process, since this process is controlled by the sizes of the pore bodies [see Eq. (12)], and therefore retraction is a sort of site percolation process. Therefore a correct percolation model of mercury porosimetry should involve a mixture of bond and site percolation, and size distributions for *both* pore bodies and pore throats, whereas the above formulae were derived assuming a size distribution for the pore throats and ignoring that of the pore bodies. The assumption that the entire process is a classical *random* percolation phenomenon is also, strictly speaking, not correct, since in practice the pore space is invaded by the mercury from its external surface, and therefore the phenomenon is an *invasion percolation* process which is discussed in Chapter 12. However, as discussed there, the error caused by this assumption is often small and can be neglected. Finally, although the third assumption is not completely correct, the resulting error is not large. Although one has to consider a percolation problem in which trapping of clusters of one kind is allowed, if they are completely surrounded by clusters of another kind, a problem that was first studied by Sahimi (1985) and Sahimi and Tsotsis (1985) in the context of catalytic pore plugging, computer simulations (Dias and Wilkinson, 1986) showed that for three-dimensional networks the effect of trapping is so small that can be neglected. Despite such assumptions, shortcomings, and criticism, as Fig. 5.10 indicates, the above percolation models have been relatively successful in describing mercury porosimetry.

5.3.2 Sorption isotherms and percolation

Another method of determining the pore size distribution of a porous medium is based on adsorption-desorption isotherms (Barrett *et al.*, 1951). Normally, liquid nitrogen is used in such an experiment although, in principle, one can also use gases. Let us consider first nitrogen adsorption in a single pore. During the adsorption experiment the pressure is increased as a result of which an adsorbed film of nitrogen forms on the pore walls whose thickness increases with increasing pressure. At condensation pressure P_{co} the pore is filled with a (liquid-like) condensed phase which results in a step increase in the adsorption isotherm. The condensation pressure is given by the Kelvin equation which, for a pore of radius r, is given by

$$\frac{P_{co}}{P_0} = \exp\left[-\frac{2\sigma_{lv}V_L}{RT(r-\ell)}\right] \tag{5.33}$$

in which P_0 is the saturation pressure, σ_{lv} the liquid-vapor surface tension, R the gas constant, T the temperature, V_L the molar volume of the liquid, and ℓ is the thickness of the adsorbed layer present on the surface of the pore when condensation occurs. Thus, Eq. (33) tells us that for any value of P_{co}/P_0 the adsorption process can be uniquely parameterized by an effective radius which from here on we denote by r_a. Hence, adsorption or desorption processes correspond to an increase or decrease in r_a, respectively. During adsorption, all pores are equally accessible, vapor condenses in all pores of size $r > r_a$, and liquid nitrogen fills the pores. For $r < r_a$ the pores fill rapidly and continuously with nitrogen. Thus, during this process, often called *primary* adsorption, connectivity of the pores plays no role. All that matters is the effective size of the pores.

Consider now the primary desorption process. At the beginning, as the pressure is reduced, the desorption isotherm does not retrace that of adsorption but, similar to mercury porosimetry, forms a hysteresis loop before rejoining the adsorption isotherm. It is generally believed that two mechanisms contribute to this hysteresis. The first mechanism is thermodynamic in nature. A metastable phase may exist beyond the coexistence pressure during adsorption and/or desorption. This means that during adsorption a vapor phase may exist at pressures above P_{co}, or during desorption a liquid phase may exist below P_{co}. Thus, this is a pore-level mechanism and has nothing to do with the connectivity of the pore space. In the second mechanism, the geometry and interconnectivity of a pore do matter. A pore with an effective radius r is allowed to desorb (to contain vapor)

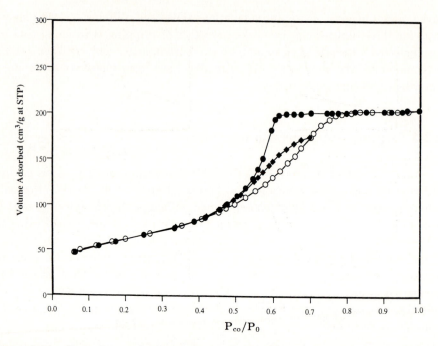

Figure 5.11: Experimental nitrogen sorption isotherms for a porous alumina sample: Adsorption (open circles), desorption (solid circles), and secondary desorption (solid diamonds) (after Liu *et al.*, 1993).

if $r > r_a$ (so that at a given pressure the liquid phase is metastable with respect to the vapor phase), *and* if it has access to either the bulk vapor in primary desorption, or the isolated vapor pockets in secondary desorption which occurs after the secondary adsorption, and thus this is a network-level (or pore-space level) mechanism. It has been argued by Ball and Evans (1989) that, unless the pore size distribution of the pore space is very narrow, the network mechanism is more important in the formation of the hysteresis loops. Typical adsorption-desorption isotherms are shown in Fig. 5.11. The International Union of Pure and Applied Chemistry has classified the various hysteresis loops that can be observed experimentally as types H1, H2, H3 and H4, which are shown in Fig. 5.12. According to the report of Sing *et al.* (1985) to IUPAC, at least for types H1, H2 and H3 the connectivity plays an important role.

Computer simulation of adsorption and desorption processes in porous media is very similar to, but simpler than, mercury porosimetry described above. As usual, we represent the pore space by a network of interconnected pore bodies (sites) and pore throats (bonds). At the end of primary adsorption and at the maximum pressure, all pores of the network are filled by liquid nitrogen. This is, for example, the case for porous media that have hysteresis loops of types H1 and H2. As the pressure decreases nitrogen vaporizes from some of the pores

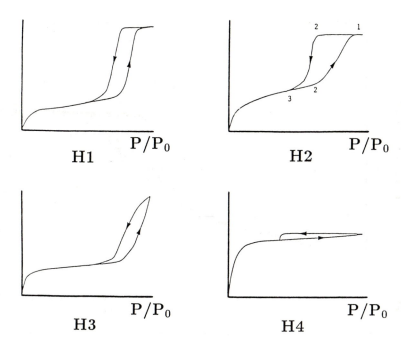

Figure 5.12: Typical hysteresis loops observed in adsorption-desorption experiments.

adjacent to the external surface of the network. This is the linear and nearly horizontal part of the desorption curve shown in Fig. 5.11. After nitrogen in the pores at the surface has vaporized, some of the internal pores gain access to the vapor phase, so that with decreasing pressure the vapor phase becomes sample-spanning. Clearly, desorption is a percolation process in which the analog of p, the probability that a bond is open in percolation, is the fraction Y of the bonds in which nitrogen is below its condensation pressure, i.e., it is the fraction of the pores from which nitrogen would vaporize if all such pores had access to the vapor phase. Therefore, $1 - Y$ is the fraction of the pores that would contain liquid if all the pores had access to the vapor phase. The analog of the accessible fraction X^A is the fraction of the pores from which nitrogen has actually vaporized. That is, $1 - X^A$ is the fraction of the pores that contain liquid nitrogen.

Thus, a simple model for simulating primary desorption is as follows. We ignore the pore bodies and concentrate only on the pore throats, to which we assign effective radii selected from a pore size distribution. As the pressure is lowered, we examine the bonds to see whether nitrogen can vaporize in them using the two criteria discussed above. The simulation continues until a sample-spanning cluster of the vapor-filled pores is formed. Typical results obtained with a simple-cubic network are shown in Fig. 5.13, which have a striking similarity to those shown in Fig. 5.11. The secondary desorption process can also be simulated

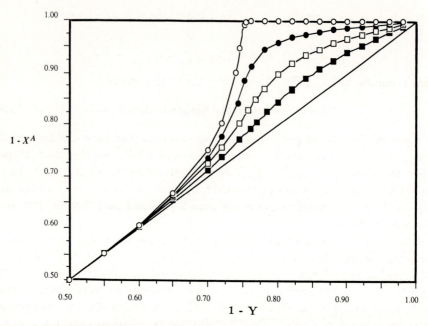

Figure 5.13: Typical isotherms obtained with the percolation model of adsorption and desorption (after Liu *et al.*, 1993).

by a similar model, except that the simulation starts with a network in which a fraction Y_v of the bonds are already occupied by the vapor phase, which have remained in the network at the end of secondary adsorption. Various versions of this model have been used by several authors to study adsorption and desorption phenomena in porous media (Wall and Brown, 1981; Mason, 1982, 1983, 1988; Neimark, 1984b; Zhdanov *et al.*, 1987; Parlar and Yortsos, 1988, 1989; Ball and Evans, 1989; Mayagoitia *et al.*, 1989; Zgrablich *et al.*, 1991; Liu *et al.*, 1993). For example, Mason and Parlar and Yortsos used Bethe lattices to take advantage of the analytical expressions for the accessibility function. Zhdanov *et al.* (1987) assumed that the radius of any pore body is greater than all pore throats in order to simplify the problem.

Similar to mercury porosimetry, one can derive analytical formulae for the size distributions of the pore bodies and pore throats occupied by the vapor or liquid phase during adsorption and desorption. Consider, for example, primary desorption during which a pore filled with liquid vaporizes if $r > r_a$, and if it is accessible to the vapor phase. The fraction of the pore (bodies or throats) that are actually occupied by the vapor is given by

$$Y_i = X_i^A \tag{5.34}$$

where X_i^A is the usual percolation accessibility function, and i denotes a site or a bond. The size distribution of the liquid-filled pores is simply given by

$$f_{Li}(r) = \begin{cases} f_i(r)/(1 - Y_i) & , \quad r < r_a \\ f_i(r)(1 - Y_i/p_i)/(1 - Y_i) & , \quad r > r_a \end{cases} \tag{5.35}$$

which is similar to Eq. (32), where the quantity p_i given by

$$p_i = \int_{r_a}^{\infty} f_i(r)dr \tag{5.36}$$

is simply the fraction of pore bodies or pore throats that have a radius greater than r_a. Similar to mercury porosimetry, if we assign effective sizes to both pore bodies and pore throats, then since the effective radii of all the pore throats that are connected to the same pore body must be smaller than that of the pore body itself, the size distributions of the pore bodies and pore throats must obey certain restrictions.

Adsorption and desorption isotherms in a porous medium can be used for extracting size distributions for the pore bodies and pore throats. If we use a Monte Carlo simulation method, then the two distributions with one or two adjustable parameters are used. The parameters are varied until the simulation results agree with the experimental data. This is a systematic but very tedious and time consuming process. One can also use Eq. (33) to obtain a pore size distribution (Barrett *et al.*, 1951). However, this equation becomes inaccurate if the pore sizes are small, and breaks down completely if the pores are so small that the microstructure of the fluid inside the pores plays a role. A refined version of this idea is to use the adsorption isotherm, since the primary adsorption isotherm does not depend on the connectivity of the pore space, and therefore the isotherm can be thought of as the aggregate of the isotherms for the individual pores that make up the pore space. Thus, if r is the effective radius of a pore, one can write

$$N(P) = \int_{r_{min}}^{r_{max}} f(r)\rho(P, r)dr \tag{5.37}$$

where $N(P)$ is the number of moles adsorbed at a pressure P, r_{min} and r_{max} are the effective radii of the smallest and largest pores present in the pore space, and $\rho(P, r)$ is the molar density of the adsorbed species at pressure P in a pore of radius r. $N(P)$ can be directly measured, while one has to use a molecular model to predict $\rho(P, r)$. Then, a numerical method is used to find $f(r)$ from Eq. (37). This method was used successfully by Seaton *et al.* (1989), and later by Lastoskie *et al.* (1993). Finally, similar to mercury porosimetry, one can use the analytical formulae derived for the saturation of the liquid or gas phase during desorption and obtain a pore size distribution for the pores.

5.3.3 Small-angle scattering

Mercury porosimetry is applicable in the range 3 nm to 100 μm, and is of limited accuracy in the small pore size range. Adsorption-desorption methods can provide pore size distributions in the range 1-60 nm. However, as we discussed above, the isotherms show hysteresis and it is not guaranteed that the pore size distribution calculated from primary desorption will agree with that obtained from the secondary adsorption. Moreover, in both mercury porosimetry and adsorption-desorption methods one needs to have some information on the connectivity or the average coordination number of the pore space. Now we discuss a method that appears to be free of such limitations. The method is based on small-angle scattering methods, either small-angle x-ray scattering (SAXS), or small-angle neutron scattering (SANS).

The basic idea is as follows. One measures the scattering intensity $I(q)$, where q is the magnitude of the scattering vector given by

$$q = \frac{4\pi}{\lambda} \sin \left(\frac{\theta_s}{2}\right) \quad . \tag{5.38}$$

Here λ is the wavelength of the radiation scattered by the sample through an angle θ_s. One then assumes a pore shape, e.g., a sphere, a cylinder or a sheetlike structure. Suppose that the effective size of a pore (for example, its radius) is r with a number density n_p. Then, according to Vonk (1976), one has

$$I(q) = \bar{\rho}^2 \sum_{i=1}^{n} n_p V_p^2 |S_F(qr)|^2, \tag{5.39}$$

where V_p is the volume of a pore of effective radius r. Here $\bar{\rho}$ is the difference in scattering amplitude densities of the solid matrix and the pore space, and $S_F(qr)$ is a *form factor* which depends on the shape of the pores. For pores of any shape, one must have $S_F \sim 1$ as $q \to 0$, and $S_F \to 0$ as q becomes sufficiently large. Thus one measures $I(q)$, assumes a pore shape, fits the measurements to Eq. (39), and calculates n_p by a constrained least-squares fitting procedure.

Using this idea and SAXS and SANS, Hall *et al.* (1986) measured the pore size distributions of eight different rocks. Three of them were fractured rocks, while two of them were sandstone. Figure 5.14 shows their typical results, obtained with SANS and compared with the results obtained with mercury porosimetry and adsorption-desorption isotherms. The rock studied was a shale outcrop from Eastern Kentucky with a porosity $\phi = 4\%$. In general, the pore size distributions obtained by scattering methods tend to agree with those that secondary adsorption isotherms yield. Note that adsorption and desorption isotherms show significant hysteresis, resulting in significantly different cumulative pore volumes, and that mercury porosimetry results are in between adsorption and desorption results. This figure also reveals a basic dilemma for anyone who wishes to measure the pore size distribution of a porous medium: What method should one use and when? How can one know *a priori* which method of measuring the pore size distribution yields the most accurate and realistic results? These are questions that, despite their significance, have not yet found definitive answers. While both mercury porosimetry and adsorption-desorption methods suffer from

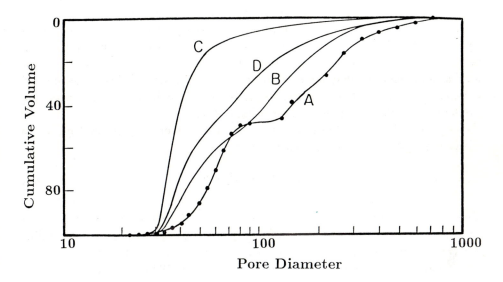

Figure 5.14: Comparison of cumulative pore-size distribution of an oil shale, obtained by SANS (A), with those obtained by nitrogen adsorption (B), desorption (C), and mercury porosimetry (D) (after Hall *et al.*, 1986).

the fact that *a priori* knowledge of the connectivity of the pore space and pore shapes is essential to their success, the scattering method also has its own shortcomings. First of all, it contains the unknown shape factor S_F, the specification of which requires specifying the pore shape, and secondly, even if the pore shape is specified, the resulting pore size distribution appears to be sensitive to the pore shape. The conclusion is that all of the above methods of determining the pore size distribution have their own strengths and weaknesses. At present, scattering and adsorption methods seem to be more reliable, although the range of pore sizes that can be detected by the former method seems to be broader than that of the latter method.

5.3.4 Nuclear magnetic resonance

Application of nuclear magnetic resonance (NMR) for determining the pore size distribution of a porous medium such as rock seems to have been pioneered by Cohen and Mendelson (1982). We should, however, mention the work of Brownstein and Tarr (1979) who used the method to study proton-spin relaxation in water in biological cells, and delineated the separate influence of diffusion and surface relaxitivity (see below). In this method, the porous medium is first saturated with a suitable fluid such as water. An appropriate pulse is then applied and the magnetization relaxation with time is measured. Magnetization relaxation is caused by the interaction of the solid surface of the pores with the fluid near the surface, as well as with that in the bulk. Therefore the relaxation rate can provide direct information about the surface-to-volume ratio and, hence, an effective pore size. If the porous medium is characterized by a pore size distribution, and if there are regions of the pore space that are separated by more than one diffusion length (by which the molecules move in the pore space), then such regions can be distinguished in the relaxation data. If the pore space of the medium is too complex, the NMR relaxation may not be able to reveal all of its complexities. Moreover, if the ratio of signal to noise is finite, then extracting a pore size distribution may be too difficult. Despite these difficulties, NMR relaxation has been used for probing the pore space of various rocks and other porous media, and obtaining their pore size distributions. Let us now describe how the NMR data are analyzed for determining the pore size distribution. To do this, we follow Cohen and Mendelson (1982) and Schmidt *et al.* (1986).

One assumes that each pore contains two kinds of fluid. One is a layer of thickness d, adsorbed on the pore surface with relaxation time t_a, and the other is the fluid in the bulk away from the surface with relaxation time t_b. In the presence of a fluid applied from the surface, t_a is shorter than t_b because the applied field hinders the diffusion of the fluid. The ratio t_a/t_b depends on the nature of the adsorbent and the surface geometry. The NMR relaxation, together with diffusion, acts to smooth any spatial gradient in the magnetization, which exists between the adsorbed and bulk fluids, as well as between fluids in adjacent pores. The governing equation for the magnetization M_z is given by

$$\frac{\partial M_z}{\partial t} = \gamma_p (\mathbf{M} \times \mathbf{H})_z - \frac{M_z - M_\infty}{t_r} + D\frac{\partial^2 M_z}{\partial z^2} \qquad (5.40)$$

where $\mathbf{H}$ is the magnetic field, γ_p the proton gyrometric ratio, D the diffusivity, t_r some relaxation time, and M_∞ the equilibrium magnetization.

In a pore of effective radius r, the magnetic field gradients between the surface and the bulk are smoothed by diffusion in a time

$$t_d = \left(\frac{r^2}{6D}\right)\left(\frac{S_p l}{V_p}\right) \ , \qquad (5.41)$$

where S_p and V_p are the surface and pore volume, respectively. Each pore is characterized by a relaxation time t_p. If $t_d < t_p$, then one will observe an averaged signal for that pore. However, if $t_d > t_p$, then one will observe a complex signal because of the spatial inhomogeneities. Thus there is a critical pore radius r_c such that if $r < r_c$ one will observe an averaged signal, whereas for $r > r_c$ one will obtain a complex or multicomponent signal. Therefore, for pores with $r < r_c$, the average relaxation time $\langle t \rangle$ is given by

$$\frac{1}{\langle t \rangle} = \frac{1 - S_p l/V_p}{t_b} + \frac{S_p l/V_p}{t_a} \ , \qquad (5.42)$$

and by measuring $\langle t \rangle$ for a given pore one can obtain S_p/V_p.

So far we have discussed only diffusion within a pore. One should also consider diffusion between the pores, which depends on the distance L_p between them. For a porous medium of spherical pores of radius r one has $L_p \sim r\phi^{-1/3}$, where ϕ is the porosity. If diffusion between pores totally dominates the rock response, only one relaxation time is observed for the entire medium. Normally, however, diffusion between pores is not significant, and the porous medium behaves as a collection of isolated pores. In this situation, each pore has its own relaxation time, which depends on its surface-to-volume ratio. Thus, if one groups pores of the same effective radius together, one can write

$$M_z(t) = M_\infty + (M_0 - M_\infty) \int_{\omega_{min}}^{\omega_{max}} P(\omega)e^{-\omega t}d\omega \quad , \tag{5.43}$$

where $\omega = t_r^{-1}$ is the frequency of relaxation and $P(\omega)$ the fraction of fluid that resides in pores with relaxation frequency ω. Equation (43) can be rewritten as

$$M_z(t) = M_\infty - \int_{\omega_{min}}^{\omega_{max}} P_1(\omega)e^{-\omega t}d\omega \quad , \tag{5.44}$$

where $P_1(\omega) = (M_\infty - M_0)P_\omega$. Because one measures $M_z(t)$ at various discrete times τ_j $(j = 1, 2, \ldots, N)$, Eq. (44) is written in a discretized form

$$M_z(\tau_j) = M_\infty - \sum_{\omega_i=\omega_{min}}^{\omega_{max}} e^{-\omega_i \tau_j} P_1(\omega_i) \quad . \tag{5.45}$$

Note that $P(\omega)$ is normalized, and therefore $\sum_{\omega_i=\omega_{min}}^{\omega_{max}} P(\omega_i) = 1$. Equation (45) is then solved for $m + 2$ unknowns and N data points, where the interval $(\omega_{min}, \omega_{max})$ has been divided into m subintervals of length $\Delta\omega = (\omega_{max} - \omega_{min})/m$. If $N \geq m + 2$, then Eq. (45) is used to calculate S_p/V_p for pores with frequency ω_i. If a pore shape is assumed, the effective size of the pore can then be calculated.

The NMR method is based on the assumption that diffusion between pores is not important and hence the pores can be treated independently. Cohen and Mendelson (1982) and Mendelson (1982) discussed the conditions under which this assumption is valid for an NMR experiment. One geometrical requirement for the validity of this assumption is that the pore throats be relatively narrow, because then diffusion between pores will be severely restricted. For some porous media this assumption is valid, while for some others it is not. In the latter case, one can still obtain a pore size distribution, but the effective sizes that are obtained are only rough estimates of the true values. Since a pore shape has to be assumed anyway, which is an approximation by itself, the calculated pore size distribution will be based on two approximations. Latour *et al.* (1992) presented some data on the temperature dependence of decay of the spectra as evidence for the concept of isolated or uncoupled pores. McCall *et al.* (1991) actually implemented the method of Cohen and Mendelson (1982), and showed how the spectrum of decay narrows as the diffusivity increases. Mendelson (1986) extended the above analysis to a fractal porous media discussed above.

The NMR method was used by Schmidt *et al.* (1986), Lipsicas *et al.* (1986) and Bilardo *et al.* (1991) for measuring the pore size distributions of various sandstones. Schmidt *et al.* (1986) compared their results with those obtained by mercury porosimetry, and showed that the NMR technique is more sensitive to the pore structure and can also reveal a bimodal pore size distribution, if there is one. Note that, while mercury porosimetry and desorption isotherms depend critically on the interconnectivity of the pore space (and hence percolation concepts), NMR and small-angle scattering methods are independent of this effect, and thus are more flexible and presumably more accurate.

5.4 Topological properties of porous media

Any porous medium consists of a pore space and a solid matrix. Parts of the pore space may be isolated and inaccessible from the external surface of the medium, whereas the solid matrix is mostly connected and accessible. The solid matrix and the pore space are separated by a pore wall, which is essentially an oriented surface. One of the simplest concepts for characterizing the topology of a porous medium is the coordination number Z, which is loosely defined as the number of pore throats that meet at a given point or pore body of the medium. For regular pore structures, such as cubic arrays of spheres, it is easy to determine Z. For an irregular pore space one has to define an average coordination number $\langle Z \rangle$, and this average has to be taken over a large enough sample. For microscopically-disordered, macroscopically-homogeneous media, $\langle Z \rangle$ is independent of sample size. Moreover, topological properties of porous media are invariant under any deformation of the pore space and solid matrix.

How can we estimate $\langle Z \rangle$ and other topological properties of a porous medium? Stereology (Underwood, 1970) and serial sectioning (Pathak *et al.*, 1982; Lin and Hamasaki, 1983; Koplik *et al.*, 1984; Yanuka *et al.*, 1984; Lin *et al.*, 1986; Kwiecien *et al.*, 1989) have been used in the past to deduce the three-dimensional structure of porous media. In particular, Kwiecien *et al.* (1989) developed computer programs that take data, analyze them, and generate the computer image of a porous medium and its various properties such as pore body and pore throat size distribution and the average coordination number. However, neither of these methods is used routinely at present. More popular are indirect methods by which only *statistical* information about the structure of the system is obtained. Some of the indirect methods are NMR, porosimetry, and sorption experiments, already discussed, which may yield parts or all of the pore size distribution, and if $\langle Z \rangle$ is treated as an adjustable parameter, it can also simultaneously be estimated with the pore size distribution. Mason (1988) and Seaton (1991) (see also Liu *et al.*, 1992 for further elaboration) developed a direct method for estimating $\langle Z \rangle$ which is based on the percolation model of adsorption-desorption phenomena discussed above. In what follows, we describe Seaton's method and comment briefly on Mason's technique which is closely related to Seaton's.

Seaton's method is based on finite-size scaling analysis discussed in Chapter 3, according to which we can write

$$X^A(p) = L^{-\beta_p/\nu_p} f[(p - p_c)L^{1/\nu_p}] \quad , \qquad (5.46)$$

which can be rewritten as

$$\langle Z \rangle X^A(p) = L^{-\beta_p/\nu_p} f[(\langle Z \rangle p - B_c)L^{1/\nu_p}] \quad , \qquad (5.47)$$

using $B_c = \langle Z \rangle p_{cb}$ (see Chapter 3). Accurate values of $X^A(p)$ were obtained by Kirkpatrick (1979) for a simple-cubic network for various sizes L, and can be shown to follow Eq. (46).

Consider now, as an example, the H2 loop in Fig. 5.12. The desorption curve has three segments indicated by 1, 2 and 3 in the figure. In the 1-2 interval, the isotherm is almost linear, and occurs because of decompression of the liquid nitrogen in the pores. In the corresponding percolation network, $Y(= p)$, the fraction of open pores (i.e., those in which the nitrogen pressure is below the condensation pressure) increases, but X^A is still zero because a sample-spanning cluster of open pores has not been formed yet. At point 2, the network reaches its percolation threshold , a sample-spanning cluster of open pores is formed, and the metastable liquid nitrogen in the pores of the cluster vaporizes. If one decreases pressure further, the number of pores containing metastable nitrogen, and the number of pores whose nitrogen has vaporized, both increase. At point 3, almost all the pores in which the nitrogen pressure is below their condensation pressure can also vaporize, and therefore $X^A \simeq Y$. Note that in a finite percolation network one has a *smeared out* percolation (see Fig. 5.11) in which the discontinuity in the desorption isotherm causes a rapid increase in the slope, and can even be detected in practice. A similar analysis can be used for interpreting the H1 loop.

Thus Seaton's method consists of two steps: (i) $X^A(Y)$ is determined from the adsorption-desorption data; and (ii) $\langle Z \rangle$ and L are determined by fitting Eq. (47) to $X^A(Y)$. At this point, and similar to the other methods, it is necessary to assume a relation between the pore radius and length. For example, one may assume that the length and the radius of a pore are uncorrelated. Note that $X^A(Y)/Y$, which is the ratio of the number of pores in the percolation cluster and the number of pores below their condensation pressures, can also be written as, N_p/N_b, where N_b is the number of moles of nitrogen which would desorb if all the pores containing nitrogen below its condensation pressure had access to the vapor phase, and N_p is the number of moles of nitrogen which actually have desorbed at that pressure. Now, if N_A is the number of moles of nitrogen which are present in the pores at a given pressure during the adsorption experiment, N_D the number of moles of nitrogen which are present in the pores at that pressure during the desorption experiment, and N_F the number of moles of nitrogen which would have been present in the pores at that pressure during the desorption experiment if no nitrogen had vaporized from the pores which contain nitrogen below its condensation pressure, then it is clear that $N_p = N_F - N_D$, and $N_b = N_F - N_A$, and therefore

$$\frac{X^A(Y)}{Y} = \frac{N_F - N_D}{N_F - N_A} \quad , \tag{5.48}$$

so that $X^A(Y)/Y$ is written in terms of measurable quantities. The final step is to determine Y, so that $X^A(Y)$ can be calculated form Eq. (48). But this is straightforward, because if $f(r)$ is the normalized distribution of pore numbers of pore radius r, then for a given pressure one has

$$Y = \int_r^\infty f(x)dx \quad , \tag{5.49}$$

where r is the pore radius in which nitrogen condenses at that pressure. Therefore, given the pore size distribution $f(r)$ determined from mercury porosimetry, desorption isotherms, or any other method, Y and hence $X^A(Y)$ can be determined.

Conceptually, Mason's method (1988) has many similarities with Seaton's. However, Mason (1988) adopted the Bethe lattice as the network model of the pore space, and although this enabled him to write down several analytical formulae for his theoretical adsorption-desorption isotherms, it is not clear how his estimate of $\langle Z \rangle$ can be related to the average connectivity of the pore space, since a Bethe lattice is not expected to be a reasonable model of any real pore space. However, one can establish an approximate relation between his $\langle Z \rangle$ and that of an equivalent three-dimensional network. Since for three-dimensional networks the bond percolation threshold p_{cb} is given by (see Table 3.2)

$$p_{cb} \simeq \frac{1.5}{\langle Z \rangle} \tag{5.50}$$

and because for a Bethe lattice of coordination number Z_b, $p_{cb} = (Z_b - 1)^{-1}$, then if we fix p_{cb} on a Bethe lattice and its "equivalent" three-dimensional network, we find a relation between $\langle Z \rangle$ and Z_b

$$\langle Z \rangle = \frac{3}{2}(Z_b - 1) \quad , \tag{5.51}$$

so that the average coordination number of the actual porous medium will be larger than what Mason's method predicts. For large values of Z_b, the difference between $\langle Z \rangle$ and Z_b can become significant.

A more precise method of characterizing the connectivity of a pore space relies on Betti numbers. These numbers were discussed by Barrett and Yust (1970) for metallurgical systems, and by Lin and Cohen (1982) and Pathak *et al.*

(1982) for porous rocks. A fundamental theorem of topology (see, for example, Alexandroff, 1961) states that two structures are topologically equivalent if and only if their Betti numbers are *all* equal. For a given structure one can define many Betti numbers, and their precise definition requires considerable knowledge of topology. According to Barrett and Yust (1970), "The *n*th Betti number β_n of a complex··· (is)··· the maximum number of *homologly-independent n*-cycles," which is a quite complex statement! For our present purpose though, we only need the first three Betti numbers. The zeroth Betti number β_0 is the number of isolated clusters in a structure. In other words, β_0 is the number of *separate* components that make a structure. For example, the grain space of a single, finite sandstone has $\beta_0 = 1$. Thus $\beta_0 > 1$ may indicate that the structure contains isolated porosity. The first Betti number β_1 is the number of holes through a structure, or the maximum number of non-intersecting closed curves that can be drawn on the surface of the structure *without separating it*. It is given by $\beta_1 = E - N_V + 1$, where E is the number of edges, and N_V the number of vertices (sites) of the network equivalent of the pore space. For example, if a torus is cut along a closed curve, the resulting solid can be deformed into a cylinder, whereas if the cylinder is cut along a closed curve, it separates into two disconnected clusters. Thus, the first Betti number of the torus is one, whereas that of the cylinder is zero.

The notion of the *genus* of a surface is also used for characterizing the topology of a complex system. Also called *holeyness*, the genus G and the first Betti number are equal for graphs lying on surfaces in complexes. One can use a genus per unit volume G_V by normalizing it over the volume over which it is measured. For large systems $\beta_1 \simeq E - N_V$, and therefore $G_V = \beta_1/N_V = (E/N_V) - 1$. Note that for graphs or a network equivalent of a porous medium, G_V is half of the coordination number, but the notion of genus and genus per unit volume are more general than the coordination number. It is clear that the first Betti number or genus is also a measure of multiplicity of independent paths in a structure. Finally, the second Betti number β_2 is a measure of the sidedness of a structure. For example, a solid structure containing n isolated pores is n-sided.

The Betti numbers can be defined for both the solid matrix, β_0^s, β_1^s and β_2^s, and for the pore space β_0^p, β_1^p and β_2^p, but they are related through the following relationships

$$\beta_0^p = 1 + \beta_2^s \ , \tag{5.52}$$
$$\beta_1^p = \beta_1^s = G \ , \tag{5.53}$$
$$\beta_0^s = 1 + \beta_2^p \ , \tag{5.54}$$

and therefore

$$\beta_0^p + \beta_2^p = \beta_0^s + \beta_2^s \ , \tag{5.55}$$

which means that the topologies of pore space and solid matrix are conjugate, and one need measure only one of them. For a microscopically disordered rock,

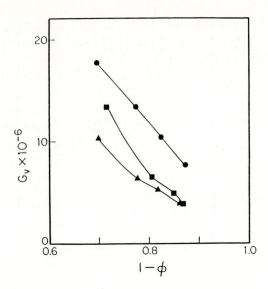

Figure 5.15: Variations of genus per unit volume G_V (in cm^{-3}) with the porosity of a sintered copper powder. The nominal diameter of the particles was 60μ, and the results are for spherical (triangles), polyhedral (squares), and irregular (circles) particles (after Pathak *et al.*, 1982).

the Betti numbers have to be averaged over a large enough sample. Although, as mentioned above, one may also use topological measures per unit volume, these measurements suffer from the disadvantage that they depend on the unit chosen for volume. For example, a heavily-consolidated rock with many large, irregular grains that have many contacts with one another may have the same genus per unit volume as a lightly consolidated rock that consists of small, well-rounded grains with few grain-to-grain contacts.

Topology and geometric shapes are related through the Gauss-Bonnett theorem (see, for example, Kreyszig, 1959). The local Gaussian curvature of a surface $\mathcal{H}_G$ is given by $\mathcal{H}_G = \mathcal{H}_1\mathcal{H}_2$, where $\mathcal{H}_1$ and $\mathcal{H}_2$ are the local principal curvatures of the surface. $\mathcal{H}_G$ is negative if the surface is saddle shaped and positive if the surface is convex or concave. One defines the integral Gaussian curvature $\langle\mathcal{H}_G\rangle$ by

$$\langle\mathcal{H}_G\rangle = \int_s \mathcal{H}ds \quad . \tag{5.56}$$

According to the Gauss-Bonnett theorem, one has

$$\langle\mathcal{H}_G\rangle = 4\pi(1 - G_V) \quad . \tag{5.57}$$

Reservoir rocks are highly porous and have high genus. They also have large negative $\langle\mathcal{H}_G\rangle$. Therefore they must be riddled with pore wall areas that are saddle shaped.

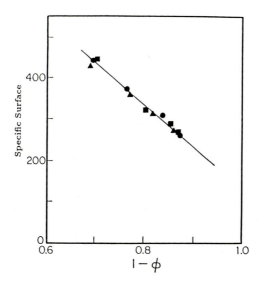

Figure 5.16: The same as in Fig. 15, but for the specific surface Ξ (in cm^2/cm^3) (after Pathak *et al.*, 1982).

These topological properties were measured by Pathak *et al.* (1982) for artificial porous media. They sintered three different copper powders: (i) spherically shaped grains in the range 30-90 microns; (ii) electrolytically prepared grains of less regular shape in the range 30-90 microns; and (iii) electrolytically prepared grains in the size range 250-300 microns. The sintering process parallels, in many important aspects, the diagenesis of sedimentary rocks described in Chapter 4. By using cold compression of spherical grains, they also prepared polyhedral-shaped particles. Using serial sectioning, they measured the genus per unit volume G_V and surface area per unit volume Ξ of their porous media. Figure 5.15 shows how G_V varies with the porosity ϕ for the three different porous media, while Fig. 5.16 presents the variations of Ξ with ϕ. As Fig. 5.15 indicates, with increasing sintering the initial rough surface and edges of the original powders are smoothed out; the surface areas per unit volume no longer depend on the original shape and show a universal dependence on ϕ.

Lin and Cohen (1982) studied six different Berea sandstones and measured by serial sectioning and image analysis several of their topological properties. Only for the main pore subsystem β_1 was measured: its minimum was 91, whereas its maximum was found to be 280, while β_0 was measured to be about 23, indicating large amounts of isolated porosity. Also measured were the number of contacts per pore section which had a broad distribution. The connectivity of the pore or grain system of Berea was found to be lower than the connectivities of regular monosized sphere packs with similar porosities and the same mean grain diameter.

To summarize, all of these studies indicate that for sandstones an average coordination number between 4 to 8 is a reasonable estimate (Lin and Cohen, 1982; Koplik *et al.*, 1984; Yanuka *et al.*, 1984).

5.5 Fractal properties of porous media

In the last section we discussed the pore size distribution and connectivity of natural porous media such as rock. The average coordination number of a sedimentary rock can vary anywhere from 4 or 5 to 15. Many other types of porous media, e.g., catalyst particles, coals, and membranes can also have an average coordination number roughly in the same range. Therefore what distinguishes natural porous media such as rock from other types of porous media is their geometry, i.e., the shapes and sizes of its pores and its possible fractal properties. In the last several years, fractal properties of porous media and fractured rock have attracted considerable attention, and many theoretical, computer simulation, and experimental studies have been undertaken in an attempt to understand them. In this section, we discuss and review fractal properties of porous media, while those of fractured rock are discussed in the next section. There are six basic methods of measuring fractal properties: the box-counting method, adsorption studies, chord-length measurements, correlation function measurements, small-angle scattering, and spectral methods. Of these, the box-counting method was already discussed in Chapter 3 as a fundamental method of measuring fractal properties and dimensions.

5.5.1 Adsorption methods

In a pioneering work, Avnir *et al.* (1983) measured, using gas adsorption methods, pore surface properties at the nanometer scale. The monomolecular coverage n, e.g., moles/adsorbent weight, for various species of different molecular weight, and hence different surface coverage per site σ_s, was found to satisfy the following relationship

$$n \sim \delta^{-D_s} = \sigma_s^{-D_s/2} \tag{5.58}$$

where δ is the radius of the non-overlapping spheres with which the surface is covered (see Chapter 3). Under the assumption that surface coverage per site σ_s is uniquely determined by the adsorbed gas species, we recognize D_s in Eq. (58) as the fractal dimension obtained with the box-counting method discussed in Chapter 3. It is called the *surface* fractal dimension of the pores, since it is a measure of the surface roughness. For smooth surfaces $D_s = 2$, but surface fractal dimensions as high as nearly 3 have been obtained for various rough surfaces. Avnir *et al.* (1983, 1985) extended the range well beyond molecular sizes by studying adsorption properties of fractal surfaces in larger particles, and by considering their scaling with the particle Euclidean size R_p. Contrary to Eq. (58), a single species was used. The following equation is then expected to hold

$$n \sim R_p^{D_s-3} \qquad (5.59)$$

if we assume that the surface area is proportional to $R_p^{D_s}$, and that the particle weight varies with the volume as R_p^3. In order to measure D_s, the system under study is sieved into several fractions. For each fraction the apparent monolayer value n is determined by any convenient method, e.g., adsorption from solution. If D_s is very close to 3, which is indicative of very wiggly porous material, then n becomes independent of R_p. One can also express Eq. (59) in terms of an *apparent* surface area S_a,

$$S_a = Nn\sigma_s \sim R_p^{D_s-3} \quad , \qquad (5.60)$$

where N is the Avogadro's number. The range of self-similar and fractal behavior can also be found with this method. If a fractal dimension D_s is found from the measurements of monolayer values of sieved fractions of particle diameter from R_{min} to R_{max}, with a probe molecule of cross-sectional area σ_0, then $\sigma_{max} = \sigma_0(R_{max}/R_{min})^2$, and the range of self-similarity is

$$\sigma_0 \leq \sigma_s \leq \sigma_{max} \quad . \qquad (5.61)$$

It is clear that in order to get maximum information on the molecular size geometry of the surface, one should select σ_0 to be as small as possible. This is the case in practice since nitrogen or argon is usually used. Note that σ_0 dictates the finest resolution in probing a surface, and therefore if a large σ_0 is used, then the question of self-similarity below σ_0 can only be speculated.

Measurements of Avnir *et al.* (1983, 1985) revealed interesting results. Six carbonate rocks were found to have fractal pore surface with $2.16 \leq D_s \leq 2.97$, seven types of soils with $2.19 \leq D_s \leq 2.99$, and a number of crushed rocks from nuclear test sites with fractal dimensions in the range $2.7 \leq D_s \leq 3$. These results will be compared with those obtained by other methods discussed below.

Adsorption methods are *not* free of limitations or potential problems. If, as discussed by de Gennes (1985), chemical disorder on the pore surface is important, or if molecular conformation and orientation are functions of the structure of the pore surface, then adsorption methods can yield estimates of D_s that are biased. Moreover, if D_s is close to 3, which is indicative of a highly rough surface, some parts of the surface can shadow neighboring surfaces. This leads to incomplete adsorption and a lower bound to D_s, rather than its true value. However, while such problems may be important when one studies porous media such as catalysts and coal particles, they do not seem important as far as reservoir rocks are concerned, since for them adsorption methods have yielded estimates of D_s that are in general agreement with those obtained by other methods. Aside from these issues, one major shortcoming of adsorption methods is that the size range of the adsorbates is very narrow, usually from 0.2 to 1 nm. One could use probes with high molecular weights, but the problems of conformation and orientation of the molecules mentioned above can emerge.

5.5.2 Chord-length measurements

There are two basic methods of measuring chord lengths, namely, on fracture surfaces and on thin sections. A description of each method follows.

(i) *Chord-length measurements on fracture surfaces.* This method was described in the papers of Krohn and Thompson (1986) and Krohn (1988a), which we summarize. At the outset, however, we should mention that these authors do not distinguish between a fractal pore surface (fractal dimensionality D_s) and a fractal pore space (fractal dimensionality D_f). In fact, Katz and Thompson (1985) argued that for sandstones $D_s = D_f$. This will be discussed below. One counts features in a large number of horizontal lines (e.g., a hundred or more) across a digitized image of a fracture surface (see Fig. 5.17). The counting is then repeated for a number of magnifications and locations. One starts by selecting a highly structured location on the surface and digitizing the images at several different magnifications. A constant resolution for feature detection is then set using a digital low-pass filter. Feature sizes, defined as the distance between local maxima, are then measured. This generates a histogram, which is linearized and placed on a log-log plot. That this can be done is because the probability of detecting a feature at each magnification is known. The effect of various factors on the construction of the histograms and the resulting plots were thoroughly investigated by Krohn and Thompson (1986) and Krohn (1988a). For example, they showed that the same fractal measures are obtained whether one uses an analog or a digital filter, that the results are independent of the filter frequencies, and that signal-to-noise ratios do not have any important effect on the result. One sets an amplitude threshold to make sure that features that correspond to noise are not counted. This threshold sets a cutoff for the resolution, which sets a limit to the high frequencies that are counted. However, to ensure a constant frequency cutoff, the threshold must be set as a fraction of the signal size at each magnification. The signal size is measured by counting the features using a constant threshold, and measuring the average amplitude difference between neighboring minima and maxima for features of a size less than the cutoff of the filter. The amplitude is usually measured for features with sizes between 15/512 and 20/512 of the field of view.

This technique does not depend on the delineation of the pore or grain space. It is an automatic method, which statistically measures structural features using scanning electron microscopy (SEM) images of the surface. A change in contrast in the secondary electron intensity of the SEM that results in a local maximum in intensity, is defined as the edge of a feature. The technique makes

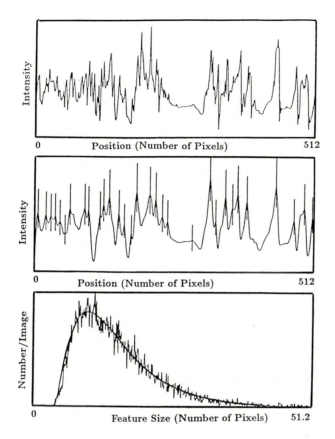

Figure 5.17: The size distribution of surface features (roughness) of a sandstone (after Krohn and Thompson, 1986).

it possible to decide whether features of a given size dominate the geometry of the pore space. Ehrlich *et al.* (1980) and Orford and Whalley (1983) also used SEM measurements of grain roughness to analyze the results in terms of fractal concepts. However, they measured the roughness of individual grains by analyzing the outline of the grains in a grain mount, whereas fracture surface technique measures the pore-grain interface without isolating individual grains. As a result, while the fracture surface technique yields a single fractal dimension for *all* lengths, the fractal analysis of Orford and Whalley (1983) does not.

The next step is to analyze the feature distribution. For fractal behavior, the number of features counted per centimeter $N_{cm}(l)$ for features of size l can be expressed as

$$N_{cm}(l) \sim l^{2-D_f}, \qquad (5.62)$$

where $l_1 \leq l \leq l_2$, and l_1 and l_2 are the limits of fractal behavior. For $l > l_2$ the samples are homogeneous and $D_f = 3$, which is the case if the geometrical features appear only as statistically random noise. Because all measurements are made from images, one expresses the feature sizes in terms of pixels, where a pixel is $1/512$ of the image. One obtains a sequence of intensities $I(J)$ for representing the digitized data, where J is a pixel ranging from 1 to 512. If one edge of a feature is at J_1 and the other is at J_2, then the feature size l is $J_2 - J_1$. For each image the width in centimeters of the field of view is $12/Ma$, where Ma is the magnification. Therefore

$$N_{cm}(l) = a \left(\frac{12l}{512Ma} \right)^{2-D_f} \quad , \tag{5.63}$$

However, the true number of features counted, $N(l)$, is written as

$$N(l) = N_{cm}(l) P_f(l) L(l) \quad , \tag{5.64}$$

where P_f is the probability of finding a feature, and $L(l)$ is the distance in centimeters over which the features are counted. The digital filter sets $P_f(l)$, which can be determined by performing the Fourier transform of the impulse response and expressing the amplitude as a function of L. The probability of resolving a feature is directly dependent on the amplitude of the filter, and equals 1 at the largest feature sizes. This probability is set to zero for $l < l_0$, where the amplitude of the filter becomes less than the signal-to-noise threshold, in order to simulate the amplitude threshold for the removal of the noise. The final expression for $N(l)$ is

$$N(l) = a P_f(l) \left(\frac{12l}{512Ma} \right)^{2-D_f} \left(\frac{12}{Ma} \right) [1 - F(l-1)] \quad , \tag{5.65}$$

where $F(l-1)$ is the fraction of the field of view occupied by features of size less than l,

$$F(l-1) = \frac{1}{512} \sum_{i=1}^{l-1} i N(i) \quad , \tag{5.66}$$

and $N(i)$ is the number of features of size i. Thus, the model contains two adjustable parameters, namely, the prefactor a and the fractal dimension D_f.

The chord lengths that are measured by this technique could represent either pore-surface structure or fracture-surface structure. The method does have the drawback that the fracturing process may introduce unwanted structures. Thus one has to make sure that a section of the surface is measured and not a projection. Figure 5.18 shows typical results for Berea sandstone with a porosity of about 20%. For this sample, it was found that $D_f \simeq 2.85$, and $l_2 > 32$ μm. In general, after examining a dozen sandstones, it was found that D_f is *not* universal and is in the range $2.55 < D_f < 2.85$.

(ii) *Chord-length measurements on thin sections.* This method is not as accurate as the fracture-surface technique and was essentially developed to provide

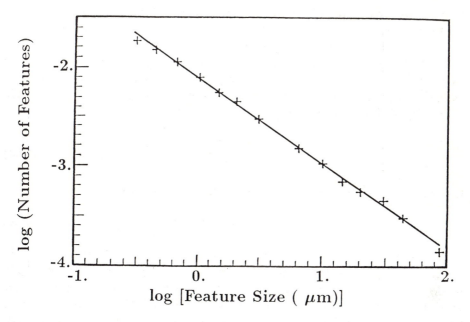

Figure 5.18: Typical fractal plot for Berea sandstone (after Krohn and Thompson, 1986).

data that are complementary to those obtained with the fracture-surface method. However, it also has its own advantages. For example, it can be used for measuring the amount of porosity and its distribution, which may or may not be fractal. But let us first discuss the method itself by following Krohn (1988a) and summarizing her work.

In this method one digitizes SEM images and delineates the pore space whenever the intensity is less than a set gray level. Usually, the edge of a feature appears bright on SEM images. If one examines the gray level histograms of images, one finds that the distribution of grains always appears to be brighter than the pore distribution. The gray level for pore fill is between those of grains and pores, and therefore it is important to measure the pores within the pore fill. Once the SEM images are digitized, chord lengths are measured from the interception of horizontal lines with the surface of pores. Using a logarithmic bin size, one constructs a histogram of the number of chords whose lengths are in a given range. The results are not dependent on the specific choice of gray level, so long as the method is consistent from magnification to magnification.

Typical results are shown in Fig. 5.19 for Coconino sandstone, which has a porosity of about 10%. The measured fractal dimension is about 2.75, which is close to that of Berea sandstone. The results with thin sections generally agree with those obtained with the fracture surface technique. Note that from both methods one can obtain estimates for l_2, the upper limit of fractal behavior.

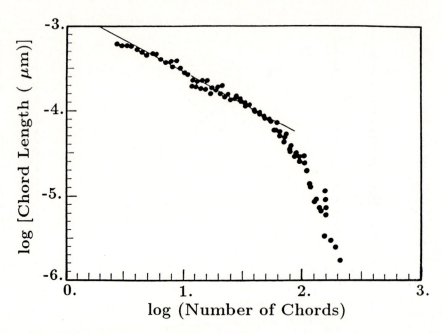

Figure 5.19: Fractal plot for Coconino sandstone, showing the lower and upper limits of fractal behavior (after Krohn, 1988a).

The chord-length methods do not contain any information on the correlation functions, and therefore the results obtained with these two methods are *not* unambiguous evidence for fractal behavior (see below).

The linear intersection of the pore space that one uses in chord-length measurements on thin sections can be used for characterizing the pore space. For example, one can measure a pore volume distribution, which is defined as the porosity associated with each chord-length and can be expressed as

$$\phi(L) = N_C(l)l(\Delta l)^2 \quad , \tag{5.67}$$

where $N_C(l)$ is the number of chords per unit volume of length l, and $(\Delta l)^2$ is the cross-sectional area associated with each chord which is equal to one pixel. To obtain $N_C(l)$ one counts the chord lengths on the thin section and assumes that the thin section is representative of the core. Figure 5.20 shows the pore volume distribution for Coconino sandstone. It is clear that most pores are in the fractal regime, but there are also some which are not. Thus there are generally two types of behavior for sandstones which are Euclidean and fractal, and the pore volume of the rock may include any amount of the two types of porosity. There is almost no sedimentary rock which does not have any fractal component. The fractal component is the result of diagenetic processes discussed above which results in the deposition of clays on the surface of the grains, making it rough. In a second paper, Krohn (1988b) measured the fractal properties of carbonate rocks

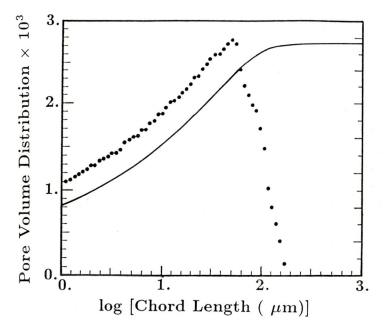

Figure 5.20: Pore volume distribution for Coconino sandstone, showing the upper limit of fractal behavior (compare with Fig. 19) (after Krohn, 1988a).

and shales, and found qualitatively the same behavior as that of sandstones. The only case for which fractal behavior was not observed was Arkansas stone which is a recrystallized quartz sandstone that is almost pure quartz with single crystal grains which have well-defined crystal faces. Note that Krohn's results are consistent with those of Avnir *et al.* (1983, 1985) discussed above.

5.5.3 Correlation function method

Measuring fractal properties of a given system in terms of autocorrelation functions, already mentioned in Chapter 3, is the most unambiguous method of establishing whether a system is fractal or not. In this method one measures the density-density autocorrelation function $C(r)$ at a distance $r = |\mathbf{r}|$

$$C(r) = \frac{1}{V} \sum_{r'} s(r')s(r + r') \quad , \tag{5.68}$$

where V is the volume of the system. The origin of the coordination system is in the pore space, $s(r) = 1$ if a given point at a distance r from the origin belongs to the pore space, and $s(r) = 0$ otherwise. Thus, the geometrical meaning of $C(r)$ is the probability of finding a given point at a distance r in the pore space. For a d-dimensional system and large values of r we must have $C(r) \sim r^{D_f - d}$, and therefore D_f can be estimated from a logarithmic plot of $C(r)$ versus r. Fara and Scheidegger (1961) were the first to use such statistical properties for characterizing porous media. Their method consisted of the following elements. One draws an arbitrary line through a porous medium. Points on this line are defined by giving them an arc length κ from an arbitrarily-selected origin. Certain values of κ correspond to the pore space, while other values represent the solid matrix. A function $f(\kappa)$ is then defined such that $f = 1$, if the line at κ passes through the pore space, and $f = -1$, if the line at κ passes through the matrix. It is easy to see that, $\langle f \rangle = 2\phi - 1$, where ϕ is the porosity of the medium, $\langle f^n \rangle = \langle f \rangle$ if n is odd, and $\langle f^n \rangle = 1$ if n is even. One then carries out a spectral analysis of f by calculating its Fourier transform, from which some information about the structure of the pore space and the solid matrix can be obtained. These basic ideas were later used by others for obtaining the fractal properties of porous media (see below).

Berryman and Blair (1986) investigated the statistical properties of the function $s(\mathbf{r})$ used in Eq. (68). If we define the following quantities

$$S_1 = \langle s(\mathbf{r}) \rangle , \tag{5.69}$$

$$S_2(\mathbf{r}_1, \mathbf{r}_2) = \langle s(\mathbf{r} + \mathbf{r}_1)s(\mathbf{r} + \mathbf{r}_2) \rangle , \tag{5.70}$$

$$S_3(\mathbf{r}_1, \mathbf{r}_2, \mathbf{r}_3) = \langle s(\mathbf{r} + \mathbf{r}_1)s(\mathbf{r} + \mathbf{r}_2)s(\mathbf{r} + \mathbf{r}_3) \rangle , \tag{5.71}$$

then because two points lie along one line and three points lie in a plane, these quantities can be measured by using images of cross sections of the porous medium. If one assumes that the porous medium is macroscopically homogeneous and isotropic, then it is easy to show that $S_2(\mathbf{r}_1, \mathbf{r}_2) = S_2(\mathbf{r}_2 - \mathbf{r}_1) = S_2(|\mathbf{r}_2 - \mathbf{r}_1|)$. Moreover,

$$S_1 = S_2(0) = \phi , \tag{5.72}$$

$$\lim_{r \to \infty} S_2(r) = \phi^2 , \tag{5.73}$$

$$S_2'(0) = -\frac{\Xi}{4} , \tag{5.74}$$

where Ξ is the specific surface area, internal surface area per unit volume, discussed above. Equation (74) was first derived by Debye *et al.* (1957).

The next step is to obtain images of the porous medium in order to analyze them. The standard method consists of the following steps (Berryman and Blair 1986). Samples of the porous medium are saturated with a low viscosity epoxy, from which petrographic thin sections are prepared, which are then polished. A SEM is used in backscatter mode for viewing the thin sections and producing high contrast images of the pore space and the solid matrix. Various magnifications of the images are produced and are digitized with a raster scanning digitizer. The resulting digital images are then stored on arrays of given sizes. These arrays are then processed using digital image techniques. From this an image of zeros and ones that closely approximate the matrix and pore space of the working image is created, which is then used to calculate various correlation functions.

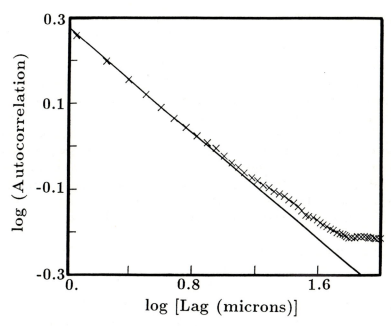

Figure 5.21: Autocorrelation function for Pico River sandstone from Utah, showing the upper limit of fractal behavior (where the straight line ends) (after Thompson *et al.*, 1987a).

Katz and Thompson (1985) used an optical technique to measure the correlation functions. In their method, backscattered micrographs of polished thin sections are photographically enhanced to produce a binary image. Two identical negatives are made on 35-mm film format, and are placed in an optical microscope to measure the transmitted light through both films. The transmitted intensity is measured. The correlation function $C(r)$ is calculated as the transmitted intensity as a function of the distance one film is translated relative to the other. Because of polishing, resolution is limited. The polishing is usually done by a $1/4\ \mu$m abrasive that leaves scratches of $1\ \mu$m dimension on the surface of the thin section.

Figure 5.21 shows the results for a Price River sandstone from Utah (Thompson *et al.*, 1987a). The plot has been made on a log-log scale in order to reveal the possible fractal behavior. On such a plot, the deviations from a straight line reveal the limits l_1 and l_2 of fractal behavior. The porosity of this sandstone is very low, and it has been highly altered by the diagenetic processes discussed in Chapter 4, so much so that the original sedimentary sandstone grains are difficult to recognize. If the alterations by the diagenetic process is not very severe, then the pore space may not be a fractal, and only the pore surface may have fractal properties. In such a case, the correlation function has a complex structure, even on a log-log plot. In some cases, the pore space is fractal, but a variety of complicating factors make the shape of the correlation function look complex. For example, the Coconino sandstone has a fractal pore space, but it is *anisotropic*. As a result, even the log-log plot of its correlation function is not a straight line, and chord-length measurements are a better way of revealing its fractal properties (see Figs. 5.19 and 5.20). Another complicating factor is the presence of pores that are not connected, or are separated by more than a distance l_2. Such pores are uncorrelated and their presence can complicate correlation function measurements. In such cases, even the appearance of straight lines on a log-log plot of correlation functions is *not* unambiguous evidence for fractal behavior of the pore space. Thus although methods that use thin sections of the porous medium can yield important information about the structure of a porous medium, they also have their limitations.

5.5.4 Small-angle scattering methods

As already mentioned, small-angle scattering methods have been used to obtain pore size distributions of shaly rocks (Hall *et al.*, 1986). In this section we outline how these methods are used to study fractal properties of a pore space and its pore surface. Small-angle scattering provides a measure of fractal behavior at length scales between 0.5 and 50 mm. In a scattering experiment the observed scattering density $I(\mathbf{q})$ is given by the Fourier transform of $C(\mathbf{r})$

$$I(\mathbf{q}) = \int_0^\infty C(\mathbf{r}) \exp(i\mathbf{q} \cdot \mathbf{r}) d^3\mathbf{r} \quad , \tag{5.75}$$

where $\mathbf{q}$ is the scattering vector whose magnitude is given by Eq. (38). For a scattering experiment, $C(\mathbf{r})$ refers to spatial variations in scattering amplitude per unit volume, rather than its physical density. For a porous medium with sufficiently low porosity, it is not unreasonable to assume that, to a good approximation, there will be no interference scattering, and therefore the total scattering intensity is the sum of the scattering from all pores. For an isotropic medium, $C(\mathbf{r}) = C(r)$, where $r = |\mathbf{r}|$, and Eq. (75) becomes

$$I(q) = \int_0^\infty 4\pi r^2 \frac{\sin(qr)}{qr} C(r) dr \quad . \tag{5.76}$$

As we already discussed for fractal objects, the correlation function for a three-dimensional system is given by, $C(r) \sim r^{D_f - 3}$. Substitution of this into Eq. (76) yields

$$I(q) \sim q^{-D_f} \Gamma(D_f - 1) \sin[(D_f - 1)\pi/2] \quad , \tag{5.77}$$

where Γ is the Gamma function. Both light scattering and small-angle x-ray scattering from silica aggregation clusters confirmed Eq. (77) (Schaefer *et al.*, 1984). As already discussed, in real systems the range of scale invariance and fractal behavior may be limited by lower and upper cutoffs l_1 and l_2. Finite size of a system can also limit this behavior. Under such conditions, the assumption of scattering by individual pores may break down and lead to interference scattering. To remedy this situation, Sinha *et al.* (1984) introduced into $C(r)$ an exponentially decaying term, incorporating a scattering correlation length ξ_s which reflects this upper limit, namely

$$C(r) \sim r^{D_f-3}\exp(-r/\xi_s) \quad, \tag{5.78}$$

which, when used in Eq. (76), yields

$$I(q) \sim q^{-1}\Gamma(D_f - 1)\xi_s^{D_f-1}\sin\left[(D_f - 1)\tan^{-1}(q\xi_s)\right][1 + (q\xi_s)^2]^{(1-D_f)/2} \quad. \tag{5.79}$$

This equation was also confirmed by Sinha *et al.* (1984) for silica particle aggregates. Note that in the limit $\xi_s \to \infty$, we recover Eq. (77), and for small values of $q\xi_s$ and $D_f = 3$ (homogeneous systems) we recover

$$I(q) \sim 8\pi\xi_s^2[1 + (q\xi_s)^2]^{-1} \quad, \tag{5.80}$$

the classical result of Debye *et al.* (1957).

If r is small, i.e., scattering at larger values of q but still within the small-angle approximation, then the scattering reflects the nature of the boundaries between the pores and their surfaces. If the pore surface has a fractal structure with a fractal dimension D_s, then the scattering technique may be used to estimate D_s. The surface fractal dimension D_s may or may not be the same as the fractal dimension D_f of the pore space itself. One may also have a non-fractal object with a fractal surface and vice-versa. Bale and Schmidt (1984) showed that for rough surfaces described by a fractal dimension $D_s > 2$, the correlation function takes on the following form

$$C(r) \sim 1 - ar^{3-D_s} \quad, \tag{5.81}$$

in which $a = n_0[4\phi(1 - \phi)V]^{-1}$, where V is the sample volume, ϕ the porosity, and n_0 a nonuniversal constant having the dimensions of area, which becomes the pore surface area if it is smooth and nonfractal. Substitution Eq. (81) into (76) gives

$$I(q) \sim q^{D_s-6}\Gamma(5 - D_s)\sin[(D_s - 1)\pi/2] \quad, \tag{5.82}$$

which reduces to $I(q) \sim q^{-4}$, the classical result of Porod (1951) for smooth surfaces for which $D_s = 2$, which is valid for the *shortest* length scales. Bale and Schmidt (1984) were able to confirm this for pores in lignites and sub-bituminous coals using SAXS; see Fig. 5.22. If both the pore space and pore surface are fractal and $D_f \neq D_s$, it is not difficult to show that

$$I(q) \sim q^{D_s-2D_f} \quad. \tag{5.83}$$

Therefore one has a crossover from the q^{-D_f} dependence to the $q^{D_s-2D_f}$ dependence to the q^{D_s-6} dependence. The crossover between q^{-D_f} and q^{D_s-6} occurs at

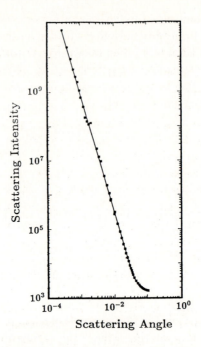

Figure 5.22: The scattering intensity for lignite coal. The scattering angle is in radians (after Bale and Schmidt, 1984).

a value of ξ_s such that $q \sim \xi_s^{-1}$. If D_s is close to 3, as found for some shaly rocks (Mildner *et al.*, 1986), the crossover between the q^{-D_f} and q^{D_s-6} regimes may be difficult to discern. Finally, if the correlation function is not isotropic, but only possesses rotational symmetry around a unique axis, one would obtain a scattering law which has an elliptically symmetric dependence on the azimuthal orientation of $\mathbf{q}$. This elliptical dependence can be removed by averaging the scattering law in terms of a reduced scattering vector. Small-angle scattering methods encounter difficulty at the smallest length scales. For example, it is difficult to find in small-angle scattering data fractal rock data which span one order of magnitude in length scale.

Wong *et al.* (1986) used SANS and studied 26 different porous media, 12 of which were sandstones, 4 were shales, 4 were limestones, and 6 were dolomites. Of the 16 sandstones and shales, 15 were found to have a *fractal pore surface* but *not* a fractal pore volume with $2.25 \le D_s \le 2.9$. The largest value was found for a Coconino sandstone, consistent with Krohn's (1988a) result discussed above (Fig. 5.20). The lowest value was found for a Fountainebleau sandstone. It was found in SEM images of Coconino sandstone that the quartz grains are covered by clay which results in a convoluted surface and a high value of D_s. On the other hand, the SEM images of Fountainebleau sandstone showed that the quartz grains were very clean. Thus, as discussed in Chapter 4, diagenetic processes give rise to highly convoluted surfaces and large values of D_s. The fact that the data of Wong *et al.* (1986) indicated that $D_s \neq D_f$ for many samples is significant in view of the proposal of Katz and Thompson (1985) that $D_s = D_f$.

Wong *et al.* (1986) also found that the carbonate rocks they studied had quite different behavior than that of their sandstones and shaly rocks. The carbonate rocks they studied were quite "clean," showing almost no trace of clays, and therefore no diagenetic alteration. The scattering intensity indicated a q^{-4} behavior for $q \leq 0.02 \text{ Å}^{-1}$, indicating a smooth surface [see Eq. (82) with $D_s = 2$]. What is the reason for this? As discussed in Chapter 4, the formation process of carbonate rocks is similar to conventional crystal growth, in which carbonates can dissolve in water and reprecipitate later. The roughness of the surface would be determined by the competition between thermal fluctuations and the surface tension if the water is clean. Wong *et al.* argued that under these conditions the width of the interface w grows with the length scale r as $w \sim (\ln r)^2$, a growth so slow that a fractal structure may be hard to detect. They also argued that, even if the water does contain impurities, the same phenomenon should exist. Note that the results of Wong *et al.* (1986) for carbonate rocks are not inconsistent with those of Krohn (1988b), who found a *fractal pore volume* for such rocks, although in her formulation of data analysis, she does not seem to distinguish between a fractal pore surface and fractal pore volume. The reason for this apparent fractality of the pore volume of carbonates is that, their grain size distribution is broad and this, together with their packing, can lead to a fractal pore volume.

We should mention two other studies of reservoir rocks and their fractal properties. Lucido *et al.* (1988) used SANS on 18 different volcanic rocks and concluded that: (i) the pore volumes of these rocks are *not* fractal; and (ii) it is not possible to determine from their data whether the pore surfaces were fractal. Hansen and Skjeltorp (1988) used the box-counting method and studied sandstones from 0.5-200 μm. They found that $D_f \simeq 2.7 \pm 0.05$, and $D_s \simeq 2.56 \pm 0.07$, which, to within the estimated errors, are almost consistent with the equality of D_f and D_s.

5.5.5 Spectral method

This method was proposed by Voss (1985) and was further discussed by Hough (1989). It is applicable to self-affine fractals discussed in Chapter 3, rather than self-similar fractals, such as pore surfaces and pore volumes that we have discussed so far. As discussed in Chapter 3, self-affine fractals are used for describing systems that have different scale-invariant properties parallel and perpendicular to a surface. A set is self-affine if it is invariant under the so-called affinity transformation, $x_i \rightarrow a_i x_i$, where the scaling parameter a_i is distinct for each direction. Thus, such surfaces are anisotropic where the anisotropy may be caused by, e.g., an external force such as gravity. Such anisotropies are usually seen in large-scale geological systems.

A function $z(t)$ has a Fourier transform $\hat{z}(\omega, T)$ in the interval $0 < t < T$ given by

$$\hat{z}(\omega, T) = \frac{1}{T} \int_0^T z(t) \exp(2\pi i \omega t) dt \quad , \tag{5.84}$$

and a spectral density $S(\omega)$

$$S(\omega) = T|\hat{z}(\omega, T)|^2 \quad \text{as } T \to \infty. \tag{5.85}$$

For a self-affine fractal one has

$$S(\omega) \sim \omega^{-\beta_f} \quad . \tag{5.86}$$

One can define a two-point autocorrelation function by

$$C(\tau) = \langle z(t) z(t+\tau) \rangle^2 - \langle z(t) \rangle^2 \quad , \tag{5.87}$$

which is somewhat similar to $S_2(\mathbf{r}_1, \mathbf{r}_2)$ defined by Eq. (70). In the case of a random and stationary process, $C(\tau)$ is related to $S(\omega)$ through the Wiener-Khintchine relation

$$C(\tau) = \int_0^\infty S(\omega) \cos(2\pi \omega \tau) d\omega \quad . \tag{5.88}$$

Equation (88) can also be extended to non-stationary processes. Then for a self-affine fractal Eqs. (86) and (88) yield

$$C(\tau) \sim \tau^{\beta_f - 1} \quad . \tag{5.89}$$

On the other hand, $C(\tau)$ is related to the mean square increments $M_i(\tau)$ of the function $z(t)$ by

$$M_i(\tau) = \langle |z(t+\tau) - z(t)|^2 \rangle = 2(\langle z^2 \rangle - \langle z \rangle^2) - 2C(\tau) \quad . \tag{5.90}$$

If the fractal dimension of a two-dimensional profile is D_f, then Eq. (90) is used to relate β_f to D_f, with the result being

$$D_f = \frac{1}{2}(5 - \beta_f) \quad , \quad 1 < \beta_f < 3 \quad . \tag{5.91}$$

Hough (1989) discussed the mathematically rigorous conditions under which Eq. (91) can be derived. If $\beta_f > 3$, D_f sticks to $D_f = 1$ and does not change. Instead of Eq. (84), which is a one-dimensional Fourier transform, one can also perform a two-dimensional Fourier transform and obtain the fractal dimension of two-dimensional topography. In this case $2 < D_f < 3$, as opposed to Eq. (91) which yields $1 < D_f < 2$.

Brown and Scholz (1985) used this technique to study the topography of various rock surfaces up to wavelengths of nearly 1 m. The estimated fractal dimensions varied between 1 and 1.7. The fact that the fractal dimension was found to vary with the wavelength means that these surfaces are not self-similar or self-affine on all length scales.

5.6 Porosity and pore size distribution of fractal porous media

Katz and Thompson (1985) proposed that the porosity of fractal porous media can be estimated from

$$\phi = c \left(\frac{l_1}{l_2}\right)^{3-D_f} , \tag{5.92}$$

where c is a constant of order unity, and l_1 and l_2 are the lower and upper limits of fractal behavior. The predictions of this equation seem to agree well with the measured values, indicating the usefulness of D_f for estimating porosity of porous media.

On the other hand, Pfeifer *et al.* (1984) proposed that for fractal porous media, the total volume V of pores of diameter $\geq 2r$ obeys

$$-\frac{dV}{dr} \sim r^{2-D_s} \tag{5.93}$$

from which a pore size distribution can be determined.

5.7 Morphology of fractured rocks

So far our discussion has been limited to unfractured porous media. However, fractured rocks are also of considerable importance, since in practical applications, such as field-scale displacement of oil by a displacing fluid, or groundwater flow, one has to deal with such rocks. Hence, in this section we discuss morphological and fractal properties of fractured rocks.

5.7.1 Characterization of fractured rock

Characterization of fracture properties of rock has to be done at two different scales. (i) A single fracture, because a fracture is not just a pair of parallel flat plates, as has been assumed in many modelling efforts. The internal surface of a fracture is usually rough, and this roughness can have a strong influence on its flow properties. (ii) A network of interconnected fractures, which can have a quite complex topology. In particular, the orientation of fractures is not completely random, and usually there is some correlation between the orientations of some of the fractures that are close to each other. For example, it is often observed that many of the fractures observed in a single sample or outcrop are more or less parallel to one or a few planes. Those fractures that have essentially the same orientation are usually grouped in one fracture set, and an important task in characterization of fractured rock is identification of such fracture sets. The orientation of the fractures in any set depends on the tectonic history of the rock. Let us discuss first some properties of a single fracture.

(i) *Fracture surface roughness* is important in reservoir modelling because it controls the aperture variation, and therefore channeling of flow between the fracture walls. It also controls the closure of fractures under lithostatic stress (Brown and Scholz, 1985), and can be important in working out the temporal development of the fracture network by identifying fractures with a common mechanical and temporal origin. It has been shown by several authors (Brown and Scholz, 1985; Brown, 1987a,b; Power *et al.*, 1987; Schmittbuhl *et al.*, 1993) that the surface of fractures of rock has fractal properties. However, it is clear that the directions along the mean fracture plane and perpendicular to it should be treated differently. Therefore, it is not unreasonable to expect that, if the fracture surface does have fractal properties, they have to be self-affine (anisotropic) rather than self-similar. This expectation is confirmed by the field measurements of Schmittbuhl *et al.* (1993) who measured the height of a granitic fault surface as a function of position along one-dimensional profiles, and showed that the profiles exhibit self-affinity or anisotropic scale invariance. We discuss this in more details in the next section where we consider fractal properties of fractures and fracture networks.

(ii) *Fracture aperture b* is a crucial parameter which determines its permeability; the volumetric flow rate Q through a fracture is given by $Q \sim b^\zeta$, where $\zeta = 3$ for a fracture with a smooth surface. For a fracture with rough surface, ζ can be as large as 6. One possible source for this variation in ζ may be the roughness of the fracture surface, which may make it essentially meaningless to define an effective aperture for the fracture. In general fracture aperture is not constant, because in addition to its surface roughness there are also voids and contact areas. Moreover, apertures measured at the surface may be wider than those at depth due to stress release. For some rocks no correlation has been observed between fracture-trace length and its aperture. Fractures in a network thus appear to have different characteristics than isolated fractures, where the aperture is expected to correlate with fracture length. For some reservoirs it has been found that the frequency of inverse aperture y, when plotted against the inverse aperture, follows a power-law, a strong indication of self-similar and fractal distribution of the aperture. Laboratory measurements by Gale (1987) and others for some other rocks have indicated that the distribution of apertures may be log-normal

$$f(b) = \frac{1}{\sqrt{2\pi}(\ln 10)\sigma_b b} \exp\left[-\frac{(\log b - \log b_0)^2}{2\sigma_b^2}\right] \qquad (5.94)$$

where b_0 and σ_b are the parameters of the distribution. If the aperture distribution is fractal, it implies long-range correlations between the aperture values, and the correlation length ξ_b is as large as the linear size of the system. Even if this distribution is not fractal, the aperture values are not completely random, and the aperture correlation length ξ_b may be non-zero.

(iii) *Fracture-trace length* is another important parameter, since its distribution is one of the most important factors in the connectivity of a fracture network, the frequency by which the fractures intersect one another, and even the flow properties of the network. For example, a few very long fractures may provide better flow paths for the network than numerous shorter fractures. However, measurements of fracture-trace lengths suffer from various biases. As pointed out by Baecher and Lanney (1978), some of these biases are as follows. (1) If one or both ends of a fracture cannot be observed, the measured length is less than the true length of the fracture. (2) If some sort of cut-off length is used in the determination of fracture-trace length distribution, then fractures that are shorter than the cut-off may be ignored. Moreover, sometimes it is difficult to discriminate between a natural fracture and one caused by, e.g., mining operations. Because of these biases, various authors have suggested methods for constructing the true distribution of fracture-trace lengths (see, for example, Cruden, 1977; Baecher and Lanney, 1978; Pahl, 1981). For some rock, e.g., the Yucca Mountain in Nevada, it has been found that the fracture-trace lengths can best-fit with a power-law of the form $y = ax^c$, where y is the frequency, x is the trace length, and a and c are constants. This is also indicative of a fractal distribution of fracture-trace lengths.

(iv) *Fracture orientation* is usually quantified by its *strike* and *dip*. The strike is the trace of the intersection of a fracture with a horizontal plane. The direction of the strike is specified by its azimuth, usually counted in degree clockwise from the north. The dip is the magnitude of the angle between the fracture and a horizontal plane, also expressed in degree. If the strike is oriented so that the dip plunges on the right-hand side, the azimuth of the strike, which takes on a unique value between 0 and 360°, and the dip angle, which is between 0 and 90°, specify the orientation of the fracture. As can be seen, characterization of even a single fracture is not an easy task.

Now that we know various characteristics of individual fractures, we should also gain some idea about the quantities that characterize a fracture network. But before discussing these, let us point out how experimental data for such characteristics are obtained. There are three types of surveys for characterizing rock. The first one is a *borehole survey*, in which measurements are usually made on oriented cores. Alternatively, they can be deduced from the resistivity of various samples (resistivity logs). The location, orientation, aperture, surface geometry, and other properties are recorded for each fracture. This type of survey provides information about the fracture density and its regionalization, the orientation distribution, and the presence of fracture sets. In the second type of surveys, called *scanline survey*, which is an extension of borehole surveys to two dimension, one paints and collects the fractures that intersect a horizontal line on an outcrop. Each fracture is represented by its trace. This survey is usually used for pit batters and high bench walls where it is possible to observe long traces. Finally, in an *areal survey* one collects all fracture traces that intersect a rectangle. If the traces continue beyond the rectangle, only that portion of them that is within the rectangle is recorded. This survey is usually used for drift or tunnel walls, but can also be used for outcrops. Thus, although sampling of the spatial properties of fracture networks should, in principle, be three dimensional, the surveys discussed above are essentially one or two dimensional, since three-dimensional surveys and samplings are difficult. They require geophysical imaging and mapping of the fracture surfaces in a volume of the rock. One can interpolate between a sequence of closely spaced two-dimensional fracture-trace maps that are parallel to one another. Now we discuss a few characteristics of fracture network of rocks.

(i) *Fracture density* is an important parameter in modelling fracture networks of rock. The two-dimensional fracture density f_2 is defined as the sum of fracture-trace lengths per unit area. The three-dimensional fracture density f_3 is defined as the average fractured surface per unit volume of rock. If the fracture network is isotropic (which is often not the case), then $f_2 = f_3$. One can even define a one-dimensional fracture density f_1 along a straight line as the number of fracture intersections per unit length. The three-dimensional density can be obtained from a borehole or scanline survey, with the help of a weighting of the observed fractures. However, the correct weight is important. Suppose that N fractures intersect the line, and that the ith fracture intersects it with an angle ϑ_i. Then it is not difficult to show that the fracture density f_3 is given by

$$f_3 = \frac{1}{L} \sum_{i=1}^{N} \frac{1}{\sin \vartheta_i} \qquad (5.95)$$

If the angles ϑ_i are completely random, then the average value of $\sin \vartheta$ between a fracture plane and a survey line is $1/2$, and between a fracture plane and a survey plane is $\pi/4$. Thus, for this particular case one has $f_1 = f_3/2$ and $f_2 = f_3/\pi$.

(ii) *Connectivity* of a fracture network, similar to the average coordination number of a porous medium, has an important effect on its permeability. Similar to the roughness of fracture surfaces, it has been shown that the connectivity of a network of fractures has fractal properties. Thus we now discuss fractal properties of fractures and fracture networks.

5.7.2 Fractal properties of fractures and faults

Prior to the first application of fractal geometry to the characterization of fracture networks, collection of data was limited mostly to one-dimensional sampling of spacing between fractures intersected along a traverse. This was done without any regard for or knowledge of orientation or size of the fractures (see, for example, Priest and Hudson, 1976, 1981). The mathematical analysis of the data was limited to calculation of arithmetic averages of the properties of interest. Priest and Hudson (1976, 1981), Baecher *et al.* (1977), Baecher and Einstein (1978), and Hudson and Priest (1983) used log-normal or exponential distributions to analyze frequency diagrams of fracture spacing. However, these analyses did not provide deep insight into the properties of fracture networks. LaPointe (1980) took a step in the right direction by exploring the use of semivariograms as a method of analyzing fracture spacing along a scanline. Semivariograms plot the second moment of the distribution of the number of fractures per unit length of the scanline as a function of the length of sampling increment over some range of increment size. This is similar to the box-counting method (see Chapter 3) for measuring fractal dimension of a set. In fact, such semivariograms can be recast into fractal plots if we plot logarithm of semivariance versus logarithm of sample increment size. The slope of the resulting diagram provides an estimate of the fractal dimension of the set.

Application of fractal concepts to characterization of fracture network of rocks was first explored in 1985 in a study carried out by the U.S. Geological Survey as part of the program to characterize the geologic and hydrologic framework at Yucca Mountain in Nevada (Barton and Larsen, 1985; Barton *et al.*, 1987; summarized in Barton and Hsieh, 1989). The site was evaluated by the U.S. Department of Energy as a potential underground repository for high-level radioactive waste. Barton and Larsen (1985) developed the *pavement method* of clearing a subplanar surface and mapping the fracture surface in order to measure its connectivity, trace length, density, and fractal scaling in addition to orientation, surface roughness, and aperture. An example of one of their mapped pavements is given in Fig. 5.23.

The most significant observation of the Yucca Mountain study was that the fractured pavements have a fractal geometry and are scale-invariant. The importance of this result was that it was possible to represent the distribution of fractures ranging from 20 cm to 20 m by a single parameter, the fractal dimension D_f, calculated by a box-counting method (see Chapter 3)

$$D_f = \frac{\log N(l)}{\log(1/l)}, \tag{5.96}$$

where $N(l)$ is the number of fractures of length l. For the fractured surfaces

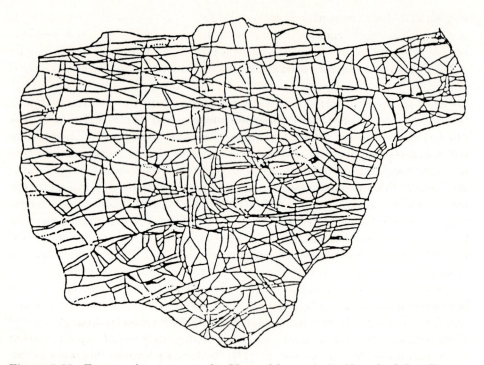

Figure 5.23: Fractured pavements for Yucca Mountain in Nevada (after Barton and Larsen, 1985).

analyzed, fractal dimensions were found to be $D_f \simeq 1.6 - 1.7$. This is the same range of fractal dimensions found over a wider range of scales in fault-gouge by Sammis *et al.* (1985), who also proposed a simple physical reason why materials fractured in shear zones evolve toward self-similarity with a fractal dimension of 1.6. It is possible that the mechanisms which produce fractal gouges are also responsible for fractal fracture networks (which may be viewed as poorly-developed gouges). Note that a fractal dimension $D_f \simeq 1.6 - 1.7$ for a fractured surface implies $D_f \simeq 2.6 - 2.7$ for the corresponding three-dimensional fracture network. LaPointe (1988) carried out a careful analysis of three fracture-trace maps of Barton and Larsen (1985). His analysis indicated that for the corresponding three-dimensional fracture networks $D_f \simeq 2.37$, 2.52, and 2.68, with an average of about $D_f \simeq 2.52$. Two-dimensional maps of fracture traces spanning nearly *ten* orders of magnitude, ranging from micrifractures in Archean Albites to large fractures in South Atlantic seafloors were analyzed by Barton (1992). The resulting fractal dimensions ranged from 1.3 to 1.7, with an average of about 1.5, implying a fractal dimension $D_f \simeq 2.5$ for the corresponding three-dimensional fracture networks.

A similar result was obtained for the Geysers geothermal field in northeast California (Sahimi *et al.*, 1993). This field, from which heat is extracted for generating electrical power, covers an area of more than 35,000 acres and represents one of the most significant geothermal fields in the world. The heterogeneous nature of the reservoir, its fracture network, and *non-sedimentary* rock distinguish it from ordinary sandstone reservoirs in terms of reservoir evaluation (Stockton *et al.*, 1984). While the fractures are the main conduits for fluid transport through the reservoirs, tight rocks containing very small pores between the major fractures contain more than 90% of the fluid reserves. The fractures of the reservoir can be detected during drilling since they produce a sudden and measurable increase in steam pressure. The average spacing between steam producing fractures is of the order 100-500 ft. Sahimi *et al.* (1993) argued that at *large* length scales, the fracture network must have the structure of percolation clusters, and developed a model for nucleation and formation of fracture networks which seems to support their view. This may explain the value $D_f \simeq 2.5$ for three-dimensional fracture networks, the same as that of percolation clusters at the percolation threshold (see Chapter 3). We come back to this point in Chapter 7, where we discuss modelling of fracture networks of rocks.

Nolen-Hoeksema and Gordon (1987) studied the fracture patterns in Stockbridge dolomite marble. This marble is from Unit "A" of the Stockbridge formation (near Canamn, Connecticut), and is a white, high quality, dolomite marble whose average grain size is about 0.3 mm. Its properties are isotropic and there is no discernable texture or fabric. The fracture pattern in this rock is very branched and appears to be a highly interconnected network. Chelidze and Gueguen (1990) analyzed the fracture pattern in this rock, and showed that the three-dimensional fracture network is a fractal object with a fractal dimensionality of about 2.5, essentially the same as that of percolation clusters. Finally, the distribution of contact areas in single, natural fractures in quartz monzonite (Stripa granite) was measured by Nolte *et al.* (1989) and was found to be fractal with a fractal dimensionality of about 2.

In many cases, fracture networks are products of fragmentation processes. Rocks are fragmented by joints and weathering. Explosives are often used to fragment rocks. Another mechanism for fragmenting a porous medium is dissolving it in a reactant (e.g., an acid). Large fractures can form in all cases. If the fragmentation process can give rise to a fractal fragment size distribution, then the fractures that are formed by the fragmentation process may also be expected to be fractal objects, and this has been found to be so in many cases. Turcotte (1986) analyzed the size distribution of rocks that had been impacted by an explosion, and also basalt rocks that had been impacted by polycarbonate projectiles and fragmented, and many other systems, and showed that the number n_m of fragments of mass m scales with m as

$$n_m \sim m^{-\tau} \ , \tag{5.97}$$

where τ was found to be about 0.85. Since $D_f = 3\tau$ can be interpreted as a fractal dimension, his results imply a fractal dimension of about 2.5. Note that Eq. (97) is similar to Eq. (3.29) for cluster size distribution of percolation clusters, discussed in Chapter 3. Likewise, Poulton *et al.* (1990) found that the length and spacing of discontinuities in rock masses also follow power-laws, typical of fractal systems. Finally, Sahimi (1991) showed how fractal fragment size distribution and fracture patterns can arise as a result of the consumption of a porous medium by a reactant.

As mentioned above, it has been shown by several authors that the surface of rock fractures has fractal and self-affine properties. For example, if we consider a two-dimensional roughness profile, then the average height h of the profile is related to its length L by

$$h \sim L^H \tag{5.98}$$

where H is the roughness or Hurst exponent introduced in Chapter 3, where we discussed fractional Brownian motion. A value $H \simeq 0.85$ was found by Schmittbuhl *et al.* (1993) for granitic faults, implying strong positive and long-range correlations. Of course such roughnesses make simulation of fluid flow in fractures by conventional methods very difficult. However, a new method of simulating fluid flow, based on lattice-gas automata, is ideally suited for this problem. This method is discussed in Chapter 14.

We now discuss briefly fractal properties of faults. In simulation of fluid transport and displacement processes in a reservoir, it is usually assumed that the entire system is stratified and continuous. However, the presence of faults severely undermines this assumption because faults are usually created when two strata or layers move with respect to each other, as a result of some mechanical process, and the interface between the two displaced layers is what constitutes a fault. So, in some sense, faults are similar to fractures and one often finds large faults in almost any kind of reservoir. However, unlike fractures, which can be created by a variety of processes, ranging from diagenetic to mechanical processes, faults are usually manifestations of tectonic processes that reservoirs experienced in the past. Moreover, unlike fractures which usually provide large permeability zones and facilitate transport of fluids in reservoir rocks, faults may or may not do so. Sometimes they hinder fluid transport in the reservoir, because nonintersecting faults can compartmentalize reservoirs and isolate large portions of them. They can also interfere with fluid flow in the reservoir. On the other hand, faults are generally recognized as the largest-scale heterogeneities of any reservoir, and therefore they can be easily detected.

Tchalenko (1970), who studied the structure of shear deformation zones over many length scales, observed that over many orders of magnitude in length scale, ranging from millimeters to hundreds of meters, shear deformation zones are similar. This strongly suggests that the fault patterns are fractal objects. Others (Andrews, 1980; Aki, 1981; King, 1984) also suggested that fault patterns are fractal systems. Okubo and Aki (1987) and Aviles *et al.* (1987) analyzed maps of the San Andreas fault system in California, and obtained fractal dimensions for fault surfaces varying from 1.1 to 1.4.

5.8 Conclusions

By now it should be clear to the reader that, (i) percolation plays an important role in the interpretation of experimental data on geometrical and topological properties of porous systems, and that (ii) heterogeneities, from the smallest scales (grains, pores, and pore surfaces), to the largest (fractures and faults), possess fractal properties which are important to *both* the geometry and topology of the porous systems. Why should natural porous media and fractured rocks have fractal properties? This is not completely understood yet, but there is no doubt that diagenetic processes discussed in Chapter 4 play an important role in the formation of fractal rock. As we emphasized in Chapter 4, what is important to remember is that any realistic modelling of fluid transport and displacement processes in porous media and fractured rocks has to take into account the effect of such fractal properties.

Chapter 6
Models of Porous Media

6.0 Introduction

Now that we know how various properties of porous media are measured, and how the experimental data are interpreted, the natural question that comes to mind is: How can we model porous media? Any realistic modelling of flow and transport phenomena in a porous medium has to include, as the first ingredient, a realistic model of the medium itself. But any model of porous media that we decide to use should depend on the type of porous media that we wish to deal with, *and* also on our computational limitations. For example, in Chapter 4 we discussed two fundamental models for porous media such as sandstones and carbonate rocks. However, our present computational capabilities do not allow us to make routine use of these models for simulating complex phenomena such as displacement processes. Thus, we still have to use models that are simple enough that allow us simulation of various flow and transport phenomena with reasonable computation time, while they also contain the essential features of the porous medium of interest. In this Chapter we mostly review and discuss various models of *consolidated* porous media. We consider models in which one can incorporate three fundamental degrees of heterogeneities discussed in Chapter 1, namely, microscopic (pore level), macroscopic (core plug level) and megascopic (field-scale level) heterogeneities. Porous media with the first two types of heterogeneities constitute what we refer to as *macroscopic* (laboratory-scale) porous media, while those with the third type of heterogeneities represent megascopic porous media. Models of unconsolidated porous media, such as packed beds of particles, are discussed in Chapter 13, while those of fractured rock are discussed in Chapter 7.

6.1 Models of macroscopic porous media

Pore space models are needed for evaluating the transport coefficients and other important dynamical properties of porous media. The simplest of such properties are perhaps the permeability K and electrical conductivity σ of a fluid-saturated porous medium. Thus one major goal of modelling a pore space has always been to predict such properties, given some morphological properties of the pore space. The simplest property of a pore space is its porosity ϕ, and therefore an obvious goal for many years was to find a relationship between ϕ and K, whose existence had seemed so obvious that in the early literature on flow of oil through reservoirs no distinction had been made between K and ϕ; it had been assumed that they are proportional. Later on, many empirical correlations between K and ϕ were suggested, the best-known of these was perhaps that of Rose (1945) who proposed that $K \sim \phi^{m'}$, where m' is some undetermined constant. This relation is similar to the famous Archie's law (Archie, 1942) for the electrical conductivity of a fluid-saturated porous medium, which states that

$$\sigma = \sigma_f \phi^m \quad , \tag{6.1}$$

where σ_f is the fluid conductivity. However, it must be clear to the reader that there *cannot* be any general relationship between K and ϕ, because one can obviously have two porous media having the same ϕ but very different K. This obvious example prompted Cloud (1941) to conclude that, "there is no sensible relation between porosity and permeability." Thus it became obvious that one has to develop a realistic model of the pore space before trying to estimate its permeability and transport coefficients. Over the years, many models of porous media have been developed, most of which have been motivated by a certain phenomenon, and often the model could be used to study that particular phenomenon and predict some of its properties. However, most of these models are not general enough to be useful for studying other types of problems, and they often contain parameters that either are defined very vaguely, or have no physical meaning whatsoever and their sole purpose is to make the models' predictions agree with experimental data for a particular phenomenon. In what follows we describe various models of porous media, first for macroscopic media, and then for megascopic porous media.

6.1.1 One-dimensional models

In this class of models the pore space is envisioned to be made of a bundle of parallel capillary tubes, or a collection of tubes in series. The radii of the tubes can be the same for all, or they can be selected from a pore size distribution. The tubes can be all cylindrical, or have converging-diverging segments. Sometimes the tubes are deformed in order to give them a tortuosity, or they may be given periodic constrictions. None of such models can take into account the effect of the interconnectivity of the pores, the existence of pore loops of various extents, etc. As a results, many predictions of such models are grossly in error, and they are so hopelessly simple that they cannot be modified or improved. Scheidegger (1974) and Van Brakel (1975) give lucid discussions of such models.

6.1.2 Spatially-periodic models

These models have been described and discussed by Nitsche and Brenner (1989) and Adler (1992). In this class of models the pore space is represented by a periodic structure, the unit cell of which can be a capillary periodic network or some other geometrical element. An example is shown in Fig. 6.1. A spatially-periodic model is also characterized by an associated lattice that contains the translational symmetries of the porous medium for which the model is intended. Because of its periodic structure, the lattice is of infinite extent and is generated from any one lattice point by discrete displacements of the form, $\mathbf{R} = i_1\mathbf{e}_1 + i_2\mathbf{e}_2 + i_3\mathbf{e}_3$, where $\mathbf{I} = (i_1, i_2, i_3)$ is a triplet of integers, and $\{\mathbf{e}_1, \mathbf{e}_2, \mathbf{e}_3\}$ is a triad of basic lattice vectors. This triad is *not* unique because, by applying any unimodular 3×3 matrix whose entries are integer to the basis $\{\mathbf{e}_1, \mathbf{e}_2, \mathbf{e}_3\}$, one can obtain another equally valid basis. It is often convenient to use cells that are parallelepipeds built on a given choice of basic lattice vectors. Given the flexibility that one has with the choice of the unit cells, their shapes are often ambiguously defined.

Spatially-periodic systems are also characterized by two length scales. One is the microscopic length scale of the lattice l_m defined as, $l_m = max[d_{min}(\mathbf{r})]$, where $d_{min}(\mathbf{r})$ is the distance between $\mathbf{r}$ and the nearest lattice points. For example, for a cubic lattice of size a, $l_m = \sqrt{3}a/2$. The second scale is the length scale L over which the averages of some physical fields, e.g., the pressure or concentration field, vary in a reasonable manner. This length scale is typically of order of the linear size of the porous medium for which the model in intended. For a porous medium to be macroscopically homogeneous, one must have $L \gg l_m$.

The simplest spatially-periodic lattice model consists of a two-dimensional array of circular cylinders. Despite its simplicity, no rigorous results for transport in this model were obtained until Sangani and Acrivos (1982) used square and hexagonal arrays of circular cylinders, calculated the permeability of the system, and discussed their results in the context of heat transfer in porous media. Later, Larson and Higdon (1987) considered flow in the same lattices in both the axial and transverse directions. Hasimoto (1959) obtained the first results for three-dimensional lattices of spheres in the limit of small sphere concentrations. The first results for the full range of sphere concentrations were obtained by Zick and Homsy (1982) and by Sangani and Acrivos (1982). Their results are discussed in Chapter 13 where we consider flow in unconsolidated porous media such as packed beds.

The analysis of transport processes in such models is a relatively simple problem, when numerical or analytical calculations are confined to a unit cell. In principle, the unit cell can have an arbitrary shape, but if one were to analyze a disordered unit cell of arbitrary shape, the analysis would be no easier than that of other models of porous media discussed below. In a sense, spatially-periodic models represent a sort of mean-field approximation to the true disordered system because they do not contain any real heterogeneities, and attempt to mimic the properties of the system in some average way. In some cases, the predicted effective properties come close to those of some real disordered media. For example, Ryan *et al.* (1980) showed that the predicted effective reaction rate of

Figure 6.1: A disordered but spatially-periodic porous medium.

a spatially-periodic model provides a useful estimate for some highly unconsolidated porous media such as packed beds. Many years ago, Philip (1957)
stated that, "The particular case of flow through a cubical lattice of uniform
spheres ... appears capable of providing information on permeability-geometry
relations." This statement turns out to be true in the case of the systems studied
by Hasimoto (1959), Zick and Homsy (1982), and Sangani and Acrivos (1982).
Brenner (1980), Carbonell and Whitaker (1983) and Eidsath *et al.* (1983) studied hydrodynamic dispersion in spatially-periodic models and found qualitative
agreement between some of their results and the experimental data of Gunn and
Pryce (1969). Their results are discussed in Chapter 13. However, we should
point out that the main reason for the agreement between the predictions and
the experimental data in all of these studies is that the geometry of the models
used closely resembles that of the experimental systems. For example, Gunn
and Pryce (1969) performed their dispersion experiments in a spatially-periodic
porous medium. Nitsche and Brenner (1989, p. 244) argue that, "while any given
model of sample porous rock cannot generally be expected to possess perfect geometrical order, this does not mean that a spatially-periodic model is not useful
for understanding the fundamentals of a penetrant fluid flow through its inter-
stices." Nonetheless, the usefulness of such models for predicting the effective
transport properties of and simulation of various phenomena in disordered and
consolidated porous media that are the main focus of this book is very limited.
Nitsche and Brenner (1989) and Adler (1992) provide an extensive list of ref-

erences for spatially-periodic models; the interested reader should consult these references.

What are the main shortcomings of spatially-periodic models of porous media? Three major shortcomings limit the usefulness of such models. The first is the fact that regular arrays of spheres (or other regularly-shaped particles or inclusions) are limited to relatively low maximum concentrations of spheres, which are significantly *below* the solid volume fraction of many real porous media that are of interest here. The second is the fact that flow in regular lattices of isolated spheres occurs *around* the spheres instead of flow *through* narrow pores found in real porous media of interest here. Finally, such models may be useful for *unconsolidated* porous media in which the solid phase is *not* a sample-spanning percolation cluster, whereas in consolidated porous media, such as sandstones, both the solid and the fluid phases are macroscopically connected (at least in single-phase flow). This effect may not be very important for estimating the absolute permeability of the porous medium if the heterogeneities are not broadly distributed, but it is important for other transport phenomena in porous media, such as two-phase flow, hydrodynamic dispersion, etc., and even for single-phase flow when the medium is highly disordered. Moreover, for heat transfer in geothermal reservoirs the effect of heat conduction through the solid matrix is important, and obviously heat conduction through a sample-spanning solid matrix is completely different from that in isolated solid inclusions.

To extend such spatially-periodic models to consolidated porous media, Larson and Higdon (1989) made a simple extension. They started from a regular lattice of spheres, but then allowed the sphere radii to increase beyond the point of touching in order to form overlapping spheres. Obviously, the solid fraction of this model can be anywhere between the original fraction, before the growth of the spheres is started, and unity. Different lattices result in different pore shapes and sizes. This model is similar to the grain consolidation model of Roberts and Schwartz (1985), discussed in Chapter 4, except that Roberts and Schwartz (1985) mostly used a random distribution of spheres, whereas Larson and Higdon (1989) only used a regular lattice as the starting point. The advantage of this model is that it is amenable to certain analytical and semianalytical calculations and, at the same time, it mimics certain features of consolidated porous media discussed in Chapter 5.

6.1.3 Bethe lattice models

Next to spatially-periodic models are branching network models. These are nothing but Bethe lattices of a given coordination number (see Chapter 3) that have been used routinely in the statistical mechanics literature for investigating critical phenomena in the mean-field approximation. As far as their applicability to modelling porous media is concerned, branching networks suffer from two major shortcomings. First, although they contain interconnected bonds that can mimic the interconnectivity of a pore space, they lack closed loops of bonds, which are a major element of the topology of any real pore space. Second, for a Bethe lattice of coordination number Z, the ratio of the number of sites on the external surface of the network and the total number of sites is $(Z - 2)/(Z - 1)$ (Ziman, 1979), which takes on finite values for any $Z \neq 2$, whereas for large three-dimensional networks this ratio is essentially zero. Thus surface effects may strongly affect any property of a Bethe lattice, which sometimes lead to anomalous phenomena such as those discussed by Hughes and Sahimi (1982), who investigated diffusion processes on Bethe lattices. The advantage of Bethe lattice models is that it is often possible to derive analytical formulae for the properties of interest, and sometimes, surprisingly, the predictions of such formulae agree well with those of three-dimensional systems. Examples include diffusion and conduction in disordered Bethe lattices that are discussed in Chapter 8.

Liao and Scheidegger (1969) and Torelli and Scheidegger (1972) were the first to use Bethe lattices for modelling transport in porous media. These authors studied hydrodynamic dispersion in a porous medium, modeled by a Bethe lattice of a given coordination number. In particular, Torelli and Scheidegger showed that such a model is fairly successful in predicting the flow velocity-dependence of longitudinal dispersion coefficient (see Chapter 9). Others have also used Bethe lattices to model transport and reactions in porous catalysts (for a review see Sahimi *et al.*, 1990).

6.1.4 Network models

Without discussing their theoretical foundations, we used network models of macroscopic porous media in Chapter 5 for describing mercury porosimetry and adsorption-desorption phenomena. The reason for this is that a network model of porous media is intuitively appealing, since it is clear to everyone that fluid paths in a porous medium may branch and, later on, join one another. This prompted many people to think of a network model of pore space in which the bonds represent in some sense the pore throats or the narrow channels that connect the sites which represent the pore bodies. To each bond is assigned an effective radius which can be selected from a probability density function, or an experimentally-measured pore size distribution (see Chapter 5). In principle, the bonds do not have to be cylindrical. They can represent sheet-like pores (as in carbonate rocks) or have converging-diverging segments. Normally, however, one chooses a pore configuration for which a given transport process can be solved analytically, and this is the reason why cylindrical pores have been used in most of such network modellings. Most authors also ignore sites and assign no volumes to them, although there have been several papers in which this assumption has been relaxed.

Although the idea of using a network to represent the pore space is intuitively clear and has been used for a long time, it was only in the beginning of 1980s that Mohanty (1981) and Lin and Cohen (1982) provided a firm mathematical foundation to such modelling approaches. In particular, Mohanty (1981) developed the procedure for deriving a network model for a given pore space. We follow his work and describe the mapping of a porous medium onto a network, but delete all the mathematical details. One has to go through three intermediate steps to get to the final network. In the first step, the pore space is reduced to a *stripwork*, which is a connected assemblage of strips or surfaces which run through the pore space. To do this, one grows a tangent sphere at each point of the pore wall until the sphere first makes contact with another point or points on the pore wall. The loci of the centers of such spheres form a continuous–insofar as the pore space is connected– multiply creased surface. This surface is what we call the stripwork.

The next step is the reduction of a stripwork to its *primitive* network. Consider the curves that bound the stripwork, except for those on the faces of the porous sample, and imagine shrinking this stripwork by moving its boundary points along the geodesics orthogonal to a strip-boundary. As the stripwork is shrunk, two possibilities may arise. First, one boundary point may meet another boundary point. In such a case, the two point are no longer moved. The loci of such points form the primitive network. The second possibility arises because more than two strips in a stripwork can join at a higher contact line. In the process of shrinking, a boundary point may merge onto such a line and become an interior point. It ceases to be a boundary point at such a point and it is not kept track of any further. The cusps and corners on the stripwork can be looked upon as limits of geodesic circular arcs of the surface for which this shrinking process is well defined. An important property of the primitive network is that only three lines emanate from each branch point in the network ordinarily and, as a rule, in completely random media. This implies that the coordination number of the primitive network is *three*. The final step is the reduction of a primitive network to its *working* network, which is what is actually used in modelling porous media. First of all, the dead-ends on the primitive network with cross-sectional radius monotonically increasing into the network are eliminated from the network. These eliminated pieces represent some details of end pores and do not have to be included in the final working network.

A physical entity, e.g., a fluid-fluid meniscus in two-phase flow, does not distinguish between two neighboring primitive nodes if they are close by and the primitive throat in between is not sufficiently pronounced. Depending on the physical process a criterion can be set up as to how pronounced a primitive throat need be. Based on such a criterion, all primitive nodes with sufficiently shallow minima or no minima in between are clumped together to form a node or pore body, which takes the position and the cross-sectional radius of the primitive node with the maximum cross-sectional radius in the clump. The branches coming out of the clumps are assigned as emanating from the node. The cross-sectional area associated with a node is called the pore body radius. Between every two nodes is a pore segment or a branch called the pore throat. The minimum of the cross-sectional radius along a throat is called the pore throat radius. The arc length of a branch on the network is defined to be the length of the corresponding pore throat. The number of branches emanating from a node is called the coordination number of the node.

In this manner network models of pore space are given precise meaning. The concept of a pore size distribution that we took for granted in Chapter 5 can now be precisely defined. The network that results from this mapping usually has a random topology, and its coordination number varies from node to node. Thus random networks such as the Voronoi network (see Chapter 3) have also been used in the literature as models of porous media. Note that if we map the model of Roberts and Schwartz (1985) for sandstones, which was discussed in Chapter 4, onto an equivalent network, the resulting network has a random topology. Over sixty years ago, Bjerrum and Manegold (1927) used a random network, made of randomly distributed points in space, connected to one another by cylindrical tubes, to study transport in porous media, although the computational limits of their time severely limited their ability for doing any extensive computations. Extensive and analytical calculations with such models were first carried out by de Josselin de Jong (1958) and Saffman (1959) in the context of hydrodynamic dispersion in porous media which are discussed in Chapter 9. As already mentioned in Chapter 3, computer simulations of Jerauld *et al.* (1984b,d) showed that, as long as the average coordination number of a random network is very close or equal to the coordination number of a regular network, the effective transport properties of the two networks are essentially identical.

The network models that were just described are *mathematical models* which are used in computer simulations of flow phenomena in porous media. Another class of such models are *physical network models* which are man-made and transparent networks of pore bodies and pore throats. These models have been developed for flow visualization studies and have been particularly useful for gaining a deeper understanding of displacement of one fluid by another. The first of such models was constructed by Chatenever and Calhoun (1952), who made bead packs from single layers of glass and Lucite beads. They used this model to study displacement of oil with brine. Mattax and Kyte (1961) made the first etched glass network to study displacement processes in porous media, and Davis and Jones (1968) improved significantly their technique for constructing etched glass networks by introducing photoetching techniques. Finally, Bonnet and Lenormand (1977) developed a resin technique for controlling the geometry of the network. Currently, etched glass and molded resin are used for constructing most of the physical networks. Lenormand (1990) and Buckley (1991) reviewed various techniques of constructing such physical networks, and

Figure 6.2: A two-dimensional continuum model of randomly-distributed disks.

discussed the results of flow studies using such micromodels.

6.1.5 Continuum models

These models were already mentioned in Chapter 3, where we discussed per-colation in continuum systems. Now we consider them in more details and sum-marize some of their structural properties. Consider first a statistical distribution of N identical d-dimensional spheres of radius R, which we call it phase 2, in volume V distributed in an otherwise uniform system, which represents phase 16. Without loss of generality, we assume that phase 1 represents the pore space of a porous medium through which a fluid flows, while phase 2 represents the solid matrix of the medium. If the particles are not allowed to overlap, one ob-tains the classical model for an unconsolidated porous medium, to be discussed in Chapter 13. If the particles do overlap and cluster, one can generate a wide variety of models with complex microstructures. The one-dimensional ($d = 1$) model (i.e., a system of rods) can be used to model certain laminates. The two-dimensional model (i.e., a system of distributed disks) may be employed to model fiber-reinforced materials and thin films, widely-used classes of porous media mentioned in Chapter 1. An example is shown in Fig. 6.2, where the sizes of the disks are distributed according to a uniform probability density function.

This model is quite general if the spheres are allowed to overlap. The inter-section of the spheres does not have to represent a certain physical phenomenon, but can be only a way of generating a disordered porous medium with a cer-tain microstructure. An example is the *penetrable concentric shell* model or the *cherry-pit* model (Torquato, 1984), in which each d-dimensional sphere of diam-eter $2R$ is composed of a hard impenetrable core of diameter $2\lambda R$, encompassed

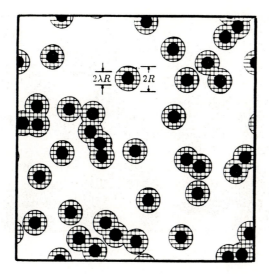

Figure 6.3: A two-dimensional example of the cherry-pit model in which a shell of thickness $(1 - \lambda)R$ of the disks is fully penetrable (after Torquato, 1991).

by a perfectly penetrable shell of thickness $(1 - \lambda)R$, an example of which is shown in Fig. 6.3. The limits $\lambda = 0$ and 1 correspond, respectively, to the cases of fully penetrable sphere model, or the Swiss-Cheese model discussed in Chapter 3, and the totally impenetrable sphere model. Varying λ allows one to tune the connectivity of the particles (i.e., the number of particles with which a given particle is in contact), and obtain a wide variety of models with desired properties.

In the fully penetrable sphere model (the limit $\lambda = 0$) there is no correlation between the particles. If ϕ_p is the volume fraction of the particle phase, then this phase becomes sample spanning for $\phi_p \simeq 0.68$ and 0.3, for $d = 2$ and 3, repetively (i.e., these values represent the percolation thresholds of the particle phase), and thus the porosity of the system is $\phi = 1 - \phi_p$. Note that no bicontinuous structure can exist in two dimensions, and therefore as soon as the particle phase becomes sample spanning, the pore space becomes discontinuous. On the other hand, for $d = 3$ the system is bicontinuous (i.e., both the particle phase and the pore space are sample spanning) for $0.3 \leq \phi_p \leq 0.97$. As mentioned in Chapter 4, where we discussed geometrical models of diagenetic processes, for $d = 3$ the pore space loses its macroscopic connectivity for $\phi \leq 0.03$, which explains why for $\phi_p \geq 0.97$ the system is no longer bicontinuous. Note that the sphere model can be generalized to a system composed of oriented cylinders (Bergman, 1980) or ellipsoids (Lado and Torquato, 1990).

How is a system of particles simulated? The classical method of computer generation of a packing of spherical particles is that of Visscher and Bolsterli (1972) which is discussed in Chapter 13. As a simple method consider, for example, a system of d-dimensional spheres with hard cores. The number of the particles N and the volume of the system V are fixed. Particles are initially placed in a cubical cell, whose volume is $V = L^d$, on the sites of a regular lattice, e.g., the face-centered or body-centered lattice. No hard core overlap is assumed initially. The particles are then moved randomly by a small distance to new positions. These new positions are either accepted or rejected according to whether or not the hard cores overlap. One usually uses *periodic boundary conditions* which means that if a particle exits from an external face of the system, an idential particle enters the system from the opposite face of the system. This essentially simulates an infinitely-large system. After the particles have been moved a sufficiently large number of times, the system reaches equilibrium, and its configuration no longer changes.

Let us now summarize some microstructural properties of these models. Their transport properties are discussed in Chapters 8 and 13. Torquato (1991) has reviewed many other properties of these models. Consider first a system of identical particles of *arbitrary* shapes with number density ρ_p. We define a dimensionless density $\eta_p = \rho_p v_p$, where v_p is the volume of a particle which, e.g., for a sphere of radius R is given by $v_p = 2R$ for $d = 1$, $v_p = \pi R^2$ for $d = 2$, and $v_p = 4\pi R^3/3$ for $d = 3$. Lee and Torquato (1988) calculated by computer simulations the porosity of the penetrable concentric shell model as a function of the parameter λ. The following approximate, but very accurate, formulae for the porosity of these systems with spherical particles have been derived

$$\phi(\eta_p, \lambda) = (1 - \lambda^d \eta_p) \exp\left[-\frac{(1 - \lambda^d)\eta_p}{1 - \lambda^d \eta_p}\right] \Psi_d(\eta_p, \lambda) , \qquad (6.2)$$

where $\Psi_1(\eta_p, \lambda) = 1$, and

$$\Psi_2(\eta_p, \lambda) = \exp\left[-\frac{\lambda^2 \eta_p^2 (1 - \lambda)^2}{(1 - \lambda^2 \eta_p)^2}\right] , \qquad (6.3)$$

$$\Psi_3(\eta_p, \lambda) = \exp\left[-\frac{3\lambda^3 \eta_p^2}{2(1 - \lambda^3 \eta_p)^3}(2 - 3\lambda + \lambda^3 - 3\lambda^4 \eta_p + 6\lambda^5 \eta_p - 3\lambda^6 \eta_p)\right] . \quad (6.4)$$

These results were derived by Rikvold and Stell (1985). If the particles are completely impenetrable ($\lambda = 1$), then $\eta_p = \phi_p = 1 - \phi$. However, when a degree of penetrability is allowed ($\lambda \neq 1$), we have $\eta_p(\lambda) \geq \phi_p(\lambda)$. On the other hand, for a model of fully penetrable particles of arbitrary shapes ($\lambda = 0$) we have

$$\phi = \exp(-\eta_p) \qquad (6.5)$$

These results are also predicted by Eqs. (2)-(4) in the appropriate limits. For the case of fully penetrable spheres, the specific surface area Ξ of the system is given by

$$\Xi = \rho_p \phi \frac{\partial v_p(R)}{\partial R} , \qquad (6.6)$$

while for the case of totally impenetrable spheres one has

$$\Xi = \rho_p \frac{\partial v_p(R)}{\partial R} . \tag{6.7}$$

If the spheres are polydispersed with a radius distribution $f(R)$, then v_p must be replaced by its average value $\langle v_p \rangle$. Thus, for polydispersed impenetrable spheres we have

$$\phi = 1 - \rho_p \langle v_p(R) \rangle , \tag{6.8}$$

and its specific surface area is given by [compare this with Eq. (7)]

$$\Xi = \rho_p \frac{\partial \langle v_p(R) \rangle}{\partial R} . \tag{6.9}$$

For fully penetrable spheres Eq. (5) is rewritten as

$$\phi = \exp[-\rho_p \langle v_p(R) \rangle] , \tag{6.10}$$

and Eq. (6) is modified to

$$\Xi = \rho_p \frac{\partial \langle v_p(R) \rangle}{\partial R} \exp[-\rho_p \langle v_p(R) \rangle] , \tag{6.11}$$

In all cases the averages are taken with respect to $f(R)$. Equations (10) and (11) were derived by Chiew and Glandt (1984).

The attractive feature of continuum models is that many real porous media do look like a continuum. The main disadvantage of these models is the complexities that are involved if one is to use such continuum models for simulating flow and displacement processes in porous media. Thus, although the effective transport properties of such continuum models can be and have been calculated using a variety of techniques (see Chapters 8 and 13), these models are still too complex for routine use in simulation of many important phenomena in porous media.

6.2 Models of pore surface roughness

As discussed in Chapter 5, in most natural rocks, and even in man-made porous media such as catalyst particles, the interface between the matrix and the pores is very rough. It is covered by features or overhangs which often give rise to a fractal surface. For some phenomena occurring in porous media the presence of such overhangs and the fractal nature of the surface have very little effect and can be ignored, while some other phenomena are affected strongly by them. Examples include flow of fines (small, solid and electrically-charged colloidal particles) and deep-bed filtration in porous media, and the distribution of a wetting phase on the pore surface of a porous medium. How can we modify a network model to include the effect of a rough or fractal pore surface? Three

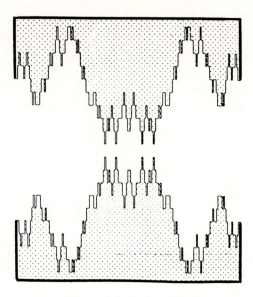

Figure 6.4: Rough fractal pore surface generated by a random walk (after Schwartz *et al.*, 1989b).

approaches have been proposed that are in essence similar, but their details are different. These models are as follows.

The first model that we discuss is due to Sahimi and Imdakm (1991) and Imdakm and Sahimi (1991). In their model, the pore space is represented by a three-dimensional network of cylindrical tubes. However, the surface of the cylindrical pores is not smooth, but covered by protrusions, overhangs, or features of distributed heights h. Sahimi and Imdakm showed that such features have a crucial effect on flow of fines in porous media and filtration processes. The inclusion of such overhangs in the model resulted in good agreement between the predicted and measured quantities of interest, whereas their exclusion resulted in unphysical results.

The second model is due to Schwartz *et al.* (1989b), who were interested in the effect of pore surface roughness on electrolytic conduction. A two-dimensional model was used in which the pore surface roughness was created by one of the following two methods. In the first method, the roughness was generated by a random walk. An example of the generated roughness is shown in Fig. 6.4. Here the interface between the pore and the matrix is a self-affine fractal curve (see Chapters 3 and 5). In the second method, the interface is a triadic-Koch curve discussed in Chapter 3. As discussed there, the Koch curve can be iterated repeatedly to increase the roughness of the interface. An example is shown in Fig. 6.5 where the interface after the third iteration is presented. Although unlike the first method the Koch curve is a self-similar fractal, the qualitative aspects of the two interfaces are similar. Of course, both are idealized models

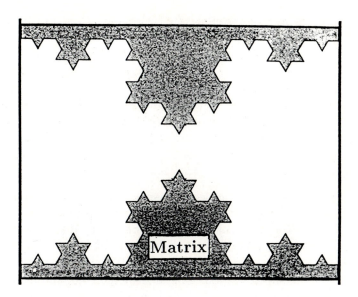

Figure 6.5: Rough fractal pore surface generated by a Koch curve (after Schwartz *et al.*, 1989b).

of a rough interface between a pore and the matrix, but they can capture some features of such interfaces.

The third model is due to Katz and Trugman (1988) and is intended for proper modelling of distribution of a wetting fluid over the rough surface of a pore. In their model, the rough surface is represented by a triangular network in which an independent random number h_j, representing the height of the vertex, is assigned to the jth vertex of the network, which is distributed according to a continuous and bounded distribution. Bonds between the vertices lie along creases or folds in the surface. Now consider the angle α formed by the two triangular faces defining the crease, where α is defined in terms of the plane perpendicular to the crease. If $\alpha < 180°$, then the crease is defined to be a $(+)$ crease; otherwise it is a $(-)$ one. The $(+)$ and $(-)$ creases reverse roles when one considers the backside of the surface, and therefore there are equal numbers of $(+)$ and $(-)$ creases. The relative heights of four vertices that define the parallelogram enclosing a bond dictate whether the bond is a $(+)$ or a $(-)$ crease.

Consider now the distribution of a strongly wetting fluid (contact angle $\simeq$ 0) on a rough surface. The roughness of the surface provides channels capable of retaining the fluid up to very high capillary pressures. Thus, in the context of Katz and Trugman's model, the wetting fluid occupies only the $(+)$ creases because they are the energetically-favorable states. If one assumes that the $(+)$ creases are strongly conducting, then the hydraulic conductivity of the network of the wetting fluid can be calculated. Because the state of a given crease depends on its neighbors, the question of whether the (+) creases that make the conducting phase form a percolating network is one of a correlated percolation. The correlation is of course short range, but it usually decreases the percolation threshold of the networks. If the correlation range increases, then the percolation threshold can decrease to small values, so that the network of (+) creases can be conducting down to very small values of fraction of (+) creases, consistent with experimental observations (Dullien *et al.*, 1986) that a strongly-wetting fluid can retain its macroscopic network structure down to very small values of its saturation. If the contact angle is finite, the problem becomes more complex and the model of rough surfaces described above has to be modified.

Before closing this section, let us mention a few of the pioneering works that used network models to study transport phenomena in porous media. We already mentioned Bjerrum and Manegold's work. Benner *et al.* (1943) introduced a pore doublet model which was in fact nothing but a hexagonal network of channels through which fluid transport could take place. The same model was also used by Rose and Witherspoon (1956). But, due to the computational limitations of their times, no extensive calculations were carried out and the two papers received very little attention. Owen (1952) used a cubic network of bonds and sites to investigate the origin of Archie's law, Eq. (1). In this work, which was very sophisticated for its time, the nodes of the network represented pore bodies to which a volume was assigned. The bonds were narrow channels, representing pore throats and connecting pore bodies, to which no significant volume was assigned. Owen (1952) calculated the tortuosity factor for such a network and showed that it is affected strongly by the structure of the network.

The first application of network models to modelling two-phase flow in porous media was pioneered by Fatt (1956). He used various two-dimensional networks of bonds representing the pore throats. The radii of the bonds were selected from a probability density function, representing the pore size distribution of the medium. No volume was assigned to the nodes. The length of each bond was assumed to be proportional to the inverse of its radius. Using this model and an analogy between laminar flow in tubes and Ohm's law of electrical currents, Fatt investigated flow of two immiscible fluids in porous media, and calculated the relative permeability to each fluid phase, i.e., the permeability of the sample-spanning cluster of the bonds filled with a fluid, divided by the overall permeability of the network (see Chapter 12). He showed that, consistent with experimental data, the relative permeability to each phase effectively vanishes at a *finite* value of the phase saturation which, in the language of percolation theory, means that the relative permeabilities vanish at a non-zero percolation threshold. Later, Rose (1957) and Dodd and Kiel (1959) used such network models to study immiscible displacement processes in porous media. We already mentioned in Chapter 5 the work of Ksenzhek (1963) who used a network model to predict capillary pressure curves for porous media. Thus although in the condensed matter literature two seminal papers of Kirkpatrick (1971, 1973) are generally credited for popularizing the use of resistor networks for investigating transport and percolation in disordered systems, the above pioneering works had already used such models to study transport processes in disordered porous media.

6.3 Models of megascopic porous media

At the next level of complexity are field-scale, or megascopic porous media. Our discussion in Chapter 5 indicated that laboratory-scale, or macroscopic porous media can be analyzed in great detail, and reasonable understanding of their structure and properties can be and have been obtained. This is not, however, the case with megascopic porous media. Experimental data for the morphology of such porous media usually are grossly incomplete, their effective properties such as permeability and elastic constants vary greatly at different length scales, often by at least a few orders of magnitude, and the relations between these properties and the volume or linear size of the porous media which we wish to study are often unknown. These complexities make the transition from laboratory-scale to field-scale porous media very difficult. Even if we have detailed information about the variability of all quantities of interest at the macroscopic level, we still may not be able to obtain complete information and meaningful average properties at the megascopic scale. For these reasons, modelling field-scale porous media usually involves stochastic techniques. Thus, although field-scale porous media are, in principle, intrinsically deterministic, because of the reasons just discussed one often has to speak of the stochastic nature of such systems, and describe their properties in terms of statistical quantities. This also makes it clear why a continuum approach to flow phenomena in field-scale porous media, based on the classical equations of transport, is often unsuccessful, because such equations can provide information only about quantities that vary in a deterministic manner. Of course, one can modify the continuum approach by developing *stochastic equations of transport* in which one or more variables, such as the permeability, porosity, and fluid velocity vary stochastically. We discuss this approach in Chapter 9 where we study dispersion in porous media. Here it suffices to point out that two major problems restrict the usefulness of stochastic equations of transport. One is that often the appropriate form of such equations is not obvious. The second problem is that, even if one uses appropriate stochastic equations, they cannot take into account the effect of long-range correlations that often exist in field-scale porous media.

For these reasons, stochastic discrete models have been developed for describing field-scale porous media. These models are in some sense an extension of network models for macroscopic porous media, except that now each block of the network represents a laboratory-scale portion of the porous medium. Moreover, it is relatively straightforward to include the effect of long-range correlations in such discrete models. A complete discussion of such models would be too long to be given here. The interested reader should consult Haldorsen *et al.* (1988). Here we discuss briefly three basic discrete models of megascopic porous media. We restrict our attention to two-dimensional models, as the thickness of field-scale porous media is usually very small compared with their length or width, and thus such porous media are essentially two-dimensional. However, it is not difficult to extend these models to three dimensions.

6.3.1 Random hydraulic conductivity models

In this approach the porous medium is represented by a rectangle which is divided into many smaller rectangular blocks that are supposed to represent a portion of the reservoir which is homogeneous on the scale of the block's size. To each block a randomly-selected hydraulic conductivity is assigned. This type of model was pioneered by Warren and Skiba (1964) and Heller (1972). In both studies it was assumed that there are no correlations between the conductivities of various blocks. Schwartz (1977) modified this model by inserting blocks of lower conductivities in an otherwise homogeneous two-dimensional region. One can also accommodate a non-random spatial structure by controlling the density and mode of aggregation of the inserted blocks. In principle, the blocks do not have to be rectangular, but non-rectangular blocks cause a lot of difficulties for numerical computations of quantities of interest.

Smith and Freeze (1979) and Smith and Schwartz (1980, 1981a,b) modified this basic model by including correlations between the blocks' hydraulic conductivities which usually exist in field-scale porous media. In their model, it is assumed that the spatial variations of hydraulic conductivities are described by a statistically homogeneous stochastic process. The spatial structure of the conductivity field is represented by a first-order nearest-neighbor stochastic process. It is assumed that the hydraulic conductivity G_b of the blocks is log-normally distributed, but of course any other distribution can be used. If $Y = \log G_b$, then the first-order nearest-neighbor stochastic process implies that Y_{ij}, the random variable for the block whose center's coordinates are (i, j), is given by

$$Y_{ij} = \alpha_x(Y_{i-1,j} + Y_{i+1,j}) \;+\; \alpha_z(Y_{i,j-1} + Y_{i,j+1}) + \epsilon_{ij} \;, \qquad (6.12)$$

where α_x and α_z are, respectively, autoregressive parameters expressing the degree of spatial dependence of Y_{ij} on its two neighboring values in the x and z directions, and ϵ_{ij} is a normal random variable uncorrelated with other ϵ_{ij}s. If $\alpha_x = \alpha_z$, then the medium has a statistically isotropic covariance structure. Otherwise, the medium has an anisotropic structure and the covariance between conductivity values is dependent upon orientation. The random variables ϵ_{ij} are distributed according to a normal distribution with a zero mean and a given variance.

6.3.2 Fractal models

The models of Smith and Freeze (1979) and Smith and Schwartz (1980, 1981a,b) with short-range correlations were significantly generalized by Hewett (1986), who argued that permeability distribution and porosity logs of megascopic porous media obey fractal statistics, and therefore there are infinitely-long-range correlations between the permeabilities, porosities and other properties of various regions of a reservoir, rather than the short-range correlations that were considered by Smith, Freeze, and Schwartz. More precisely, Hewett (1986) proposed that the porosity logs and the permeability distributions are fractional Gaussian noise (fGn) and fractional Brownian motion (fBm), respectively. These fractal distributions were already introduced and discussed in Chapter 3.

How do we analyze our data in order to uncover long-range correlations that may exist in the data? Analysis of fGn data is best done by using the rescaled range $R(l)/S(l)$, where $R(l)$ is the range of the accumulated departure from the mean of the variable, and $S(l)$ is the standard deviation (see Feder, 1988). This method was first proposed by Hurst *et al.* (1965). In mathematical terms, if a variable e takes the value $e(l)$ at position l, $R(l)$ is given by

$$R(l) = X_{max}(l, L) - X_{min}(l, L) , \quad 1 \le l \le L , \tag{6.13}$$

where

$$X(l, L) = \sum_{u=1}^{l} [e(u) - \langle e \rangle_L] , \tag{6.14}$$

$$\langle e \rangle_L = \frac{1}{L} \sum_{l=1}^{L} e(l) , \tag{6.15}$$

and $S(l)$ is given by

$$S = \sqrt{\frac{1}{L} \sum_{l=1}^{L} [e(l) - \langle e \rangle_L]^2} . \tag{6.16}$$

It can be shown that for fGn one has

$$\frac{R}{S} = \left(\frac{L}{2} \right)^H , \tag{6.17}$$

so that a logarithmic plot of R/S versus L should yield a straight line with the slope H. A similar analysis can be carried out for fBm data.

Based on an analysis of extensive data for the permeability and porosity distributions of typical reservoirs, Hewett (1986) showed that vertical porosity logs are samples of fGn, while the lateral distribution of these properties follows fBm, but with the same H. His analysis also indicated that $H > 1/2$, indicating

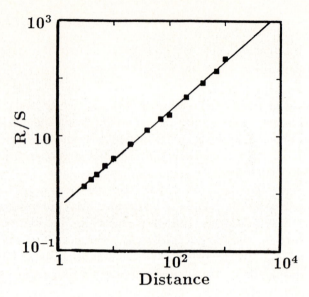

Figure 6.6: R/S analysis of the porosity as a function of distance (depth, in feet).

long-range positive correlations in the data (see Chapter 2). Figure 6.6 shows such a plot for a porosity log of a Middle East reservoir, which yields $H \simeq 0.89$. Figure 6.7 shows a two-dimensional permeability field generated by a fBm with $H = 0.89$. Using a fBm and a fGn, the permeability and porosity distributions can be generated and used in the simulation of fluid transport in heterogeneous media. This is discussed in Chapters 8, 9, and 11.

6.3.3 Multifractal models

Multifractals were introduced in Chapter 3 as a tool of generating a disordered structure in which various regions of the system have distinct properties, which are not necessarily the same as those of the system as a whole. For this reason, multifractals were suggested by Meakin (1987) and Lenormand *et al.* (1990) as a tool for generating highly heterogeneous porous media with long-range correlations. Consider a two-dimensional system, such as a square grid, and a probability p which can be related at the end of the construction of a model to a measure such as permeability or porosity, and is distributed uniformly in the interval $(1 - a, 1 + a)$, with $0 \le a \le 1$. In the first step of constructing the model, a value p_{11} is selected at random and is given to all the pixels of the initial square. The first dichotomy is then carried out to make four squares of size $2^{n-1} \times 2^{n-1}$, and four values p_{21}, p_{22}, p_{23}, and p_{24} are selected at random and attributed to each of the four squares. The same process is repeated n times. At the end of the process, each pixel is characterized by n values of the probability

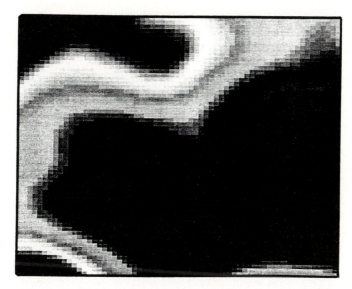

Figure 6.7: A two-dimensional permeability field generated by a fBm with $H = 0.89$. Darkest and lightest areas represent the least and the most permeable regions, respectively. Observe that, because of the positive correlations ($H > 0.5$), such regions are clustered together.

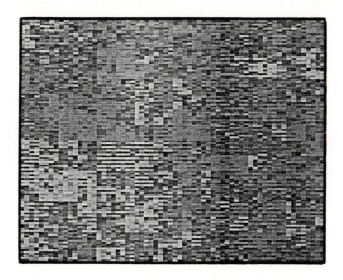

Figure 6.8: A two-dimensional isotropic permeability field generated by a multi-fractal distribution with $a = 0.8$. Darkest and lightest areas represent the least and the most permeable regions, respectively.

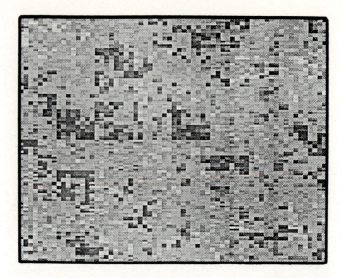

Figure 6.9: A two-dimensional anisotropic permeability field generated by a multifractal distribution.

p. A new measure P is now defined which is the product of the n random values of p. The main property of this process is that, similar to any fractal distribution, it introduces correlations between pixels at all scales. A lower cut off can also be introduced into this model by considering m steps of dichotomy, where $m < n$, and n is the dimension of the pattern. One first makes $2^{n-m} \times 2^{n-m}$ independent multifractal patterns of size m_0, and then computes $2^{n-m} \times 2^{n-m}$ independent products $p_{m+1} \times p_{m+2} \times ... \times p_n$. The pixel values of each multifractal are then multiplied by these products in order to obtain products of order n. An example is shown in Fig. 6.8 with $m = 4$ and $n = 7$. This procedure can be further generalized to anisotropic systems by considering two probabilities p_x and p_y for x- and y-axes, distributing p_x and p_y uniformly in $(1 - a_x, 1 + a_x)$ and $(1 - a_y, 1 + a_y)$, and identifying p by the product $p_x \times p_y$. An example is shown in Fig. 6.9. For simulating flow in a field-scale porous medium, the measure p can be thought of as the effective permeability of a portion of a porous medium. Given a pattern and a permeability distribution, flow and displacement processes in such a medium can be investigated.

6.4 Interpolation schemes and conditional simulation

No matter which model we use, we have to confront the fact that most of our experimental data for various properties of field-scale porous media are along a vertical line (well) in a three-dimensional reservoir. If information along many lines can be obtained, then we can compare the data along lines that are close to each other with the hope of gaining an understanding of how they might be correlated with each other. If we have some insight about the diagenetic process of our reservoir (see Chapter 4), and if we also have some idea about its structure, our interpretation of the data will be facilitated. Many field-scale porous media are stratified (layered), and the vertical lines or wells along which the data are collected run through them. Several statistical techniques have been developed for splitting the wells into layers. These techniques maximize the property variance between strata and minimize the variance within them. One then has to find the average properties within a layer, and such averages can be defined in various ways, e.g., arithmetic, geometric, etc.

In any event, lack of extensive experimental data is handled in two different ways. In one method one uses interpolation, by which from a limited amount of experimental data one draws smooth maps of the property of interest. There are two types of interpolation schemes. The first one is a smooth interpolation which provides the most accurate estimate of a property at a given location. It is called smooth because this interpolation leads to a smoother variance than the original distribution. This process is called *kriging*. This is a local estimation technique which provides the most accurate linear unbiased estimator of a property. Suppose that we have n measured values of a property $P(x)$. The linear estimator $P^*(x)$ is given by

$$P^*(x) = \sum_{i=1}^{n} \alpha_i P(x_i) . \tag{6.18}$$

Here the weights α_i are calculated to ensure that the estimator is unbiased and the estimation variance is minimum. This nonbias condition implies that $\langle P^*(x) \rangle = \langle P(x_i) \rangle$, which means that

$$\sum_{i=1}^{n} \alpha_i = 1 . \tag{6.19}$$

The variance of estimation can be written in terms of the α_is, which is then minimized subject to the constraint (19), using the method of Lagrange multipliers. Once the weights α_i are known, interpolation is carried out using Eq. (18). In this manner, kriging not only provides the most accurate (with minimum variance) estimate of the interpolated property, it also gives us some idea about the reliability of the estimate.

The second interpolation scheme is called *stochastic interpolation*, since this method interpolates between measured values with a realization of a random function that has a variance structure similar to that of the original data. However, this method may or may not provide an accurate estimate of a property at any given location. There are several methods of stochastic interpolation. For example, in one method that has been used for reservoirs that obey fractal and fBm statistics, values are smoothly interpolated to the midpoint of the intervals between the data, and a random number obeying fBm statistics (see Chapter

3 for the method of generating such numbers), or any other statistics deemed appropriate, is added to all the values. We call this number *noise*. This is repeated on successively finer intervals to produce a stochastic field realization of the property of interest. The smooth interpolation can be done by, e.g., kriging.

However, when we add the noise to the interpolated values, we have to keep in mind that at certain locations in the system we do have the real and measured data. Therefore, one does not add the noise to the values at the locations at which measured data are available. The resulting map usually looks very realistic. Simulation of flow phenomena using such stochastically-generated maps of the property values is called *conditional simulation*, since the simulation honors the observed values. As in any stochastic method, one has to generate many realizations of maps of the properties of interest, carry out the simulation, average the results over all realizations, and analyze the results for their statistical properties and uncertainties. These are relatively routine computations and analyses, and thus are not discussed here.

6.5 Conclusions

Realistic models of porous systems, for both macroscopic and megascopic media, are now available. Depending on the available computational power, one can generate a model of a pore space with a great amount of detail, and use techniques such as lattice-gas automata that are discussed in Chapter 14, for simulating flow phenomena in such media.

Chapter 7

Models of Fractured Rock

7.0 Introduction

In Chapter 5 we discussed several characteristics of porous media and fractured rock, and in Chapter 6 we learned how to model porous media. In this Chapter we review and discuss models of fractured rock. A study of the literature indicates that there are at least five classes of models of fractured rock. The first class consists of classical multi-porosity models which were proposed in the early 1960s. The second class includes network models of fractured rocks which are in essence extentions of network models of porous media discussed in Chapter 6, and are to some extent similar to percolation networks. In the third class are models that are based on simulated annealing concepts. The fourth class includes fractal models that are motivated by some of the experimental data for the morphology of fracture networks (see Chapter 5) that indicate that such networks have fractal characteristics. Finally, in the last class are mechanical models of fractured rock that attempt to model a fracture network by modelling deformation of the matrix of rock, and the nucleation and propagation of the resulting fractures. Note that aside from the multi-porosity models, all other models that are discussed here are based on discrete networks of fractures. The main difference between them is in the method that is used for generating the networks, which is why we discuss them separately.

7.1 Continuum approach: The multi-porosity models

Aside from the early work of Pollard (1959) and its extension by Pirson and Pirson (1961), some of the earliest papers on continuum modelling of fractured rock are those of Barenblatt *et al.* (1960), Warren and Root (1963), and Odeh (1965), whose approach was very different from that of Pollard. Barenblatt *et al.* introduced what is popularly known as the *double-porosity model*. In their model a fractured rock is envisioned to consist of two porous systems, the matrix with high porosity and low permeability, and the fractures with low porosity but high permeability. Fluid flow takes place through the two systems separately, and there can be exchange of fluids between the systems at their interface. As in every continuum model, the behavior of a given flow phenomenon is modeled by the classical continuum equations of transport. As discussed in Chapter 1, these equations represent the average behavior of the system, where the averaging is taken with respect to a representative elementary volume (REV). For fractured rock one has to have two such REVs, one for the matrix and one for the fractures.

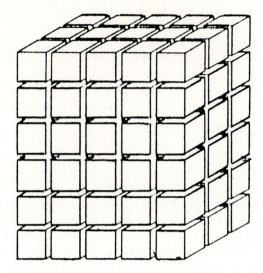

Figure 7.1: The double-porosity model of Warren and Root. The blocks, which are not sample-spanning, represent the matrix (after Warren and Root, 1963).

If one wishes to average the behavior of the system over only one REV, then it has to be much larger than the REV used in modelling flow through unfractured porous media, and has to include many matrix blocks and fractures. Such an approach is usually called *large scale averaging*, and is fully discussed in Chapter 9 where we consider hydrodynamic dispersion.

Warren and Root (1963) proposed an idealization of the oringinal model of Barenblatt *et al.* In their model, fractured rock is represented by a regular, fully-connected fracture network, embedded in a porous matrix represented by parallelepiped blocks, an example of which is shown in Fig. 7.1. The matrix is assumed not to be sample spanning, so that fluid flow occurs only through the fracture network. It has been claimed that this model is not limited to the idealized structure envisioned by Warren and Root, but in practice only the idealized model has been used. A notable feature of the Warren-Root model is that some of its parameters can be estimated from the matrix properties, and size and shape of the blocks. More sophisticated variations of this model were developed by Kazemi (1969) and Kazemi *et al.* (1976). In particular, Kazemi (1969) developed a double-porosity model in which the fractures are distributed uniformly throughout the system. Kazemi *et al.* (1969) also proposed methods for estimating various parameters of the model.

Closmann (1975) and Abdassah and Ershaghi (1986) extended the Warren-Root model by including in it *three* degrees of porosity. They considered fluid flow in a fractured system where there are two kinds of matrix having good (high) and poor (low) structural and flow properties. It was assumed that flow between the two matrices was negligible. This model was motivated by investigation of the structure of fracture network of carbonate rock (see Chapter 4), and the observations of actual well tests in such reservoirs that indicate anomalous behavior that cannot be explained by the double-porosity model of Warren and Root (see Chapter 10).

Although multi-porosity models have helped us gain a better understanding of some of the flow phenomena in fractured rock, and although they may be appropriate for studying fluid flow in a *uniform* set of fractures, they are not suitable for modelling flow phenomena in fracture networks of natural rocks because, as discussed in Chapter 5, such fracture networks have been shown to have complex morphological properties, such as incomplete fracture connectivity, fracture surface roughness, and fractal characteristics over certain length scales. There is no continuum multi-porosity model that can take into account the effect of such complex factors. Moreover, these models contain too many adjustable parameters, often with no clear physical meaning and no method of estimating them.

7.2 Network models

The idea that a fractured rock can be represented by a network of interconnected and finite fractures is intuitively clear and appealing, and is also supported by the experimental data discussed in Chapter 5. Thus representing fractured rock by a discrete network has a relatively long history. Romm (1966), Parsons (1966), Snow (1969), Caldwell (1972), and Wilson and Witherspoon (1974) appear to be among the first to have used such an approach. They used electrical analog models to study flow through a network of fractures. In particular, Parsons (1966) used square and triangular networks of resistors in which each bond (or resistor) represented a finite fracture. However, his model had the drawback that the current in each resistor was assumed to be proportional to the width of the conductor, whereas flow rate in a fracture is proportional to some power of the fracture's aperture that, as discussed in Chapter 5, can be as high as six (see also Tsang and Witherspoon, 1981). Snow (1969) used a three-dimensional model in which fractures were idealized as infinitely long and parallel ducts. As such, his model was in the same spirit as the bundle of parallel capillary tubes discussed in Chapter 6. Similar models were also developed by Castillo *et al.* (1972) and Krizek *et al.* (1972). We consider now two- and three-dimensional fracture networks separately, and study their properties.

7.2.1 Two-dimensional fracture networks

Over the past decade many authors have developed two-dimensional models of fractured rock in the form of networks of fractures of finite extent. In these

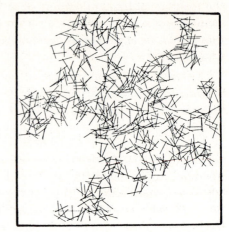

Figure 7.2: A two-dimensional random fracture network (left) and its sample-spanning portion (right) (after Robinson, 1984b).

models (Dienes, 1980; Long *et al.*, 1982; Englman, *et al.*, 1983; Schwartz *et al.*, 1983; Smith and Schwartz, 1984; Charlaix *et al.*, 1984, 1987a; Endo *et al.*, 1984; Robinson, 1984a,b; Long and Witherspoon, 1985; Ross, 1986; Long and Billaux, 1987; Shimo and Long, 1987; Gueguen and Dienes, 1989; Hestir and Long, 1990; Robinson and Gale, 1990; Mukhopadhyay and Sahimi, 1992) fractures are represented by one-dimensional finite line segments. As such, these fracture networks are similar to two-dimensional networks of pores already discussed. These models can even be a reasonable representation of a three-dimensional system if most of the hydraulic conductivity is in the intersections between fractures, or if fluid flow is channelized in the fractures. Flow channeling in the fractures does happen sometimes, in which case a two-dimensional model of fracture networks can be useful.

In the two-dimensional models, one distributes the fractures at random in a plane. One of the simplest models is the Poisson model which was first used by Long *et al.* (1982). In this model, in a square block of size $L \times L$ one chooses x- and y-coordinates for a specified number of lines or fractures centers from a uniform distribution in $(0, L)$. Once the coordinates of the centers of the lines are selected, the orientation of the lines are also selected from a given distribution. Then the lines are assigned randomly-selected lengths and hydraulic conductances. If the fractures cross the boundary of the system, they are truncated, but no truncation is done inside the $L \times L$ box. Figure 7.2 shows a typical network obtained by this method. As such, these models are practically identical with percolation networks of sticks studied by Balberg *et al.* (1983, 1984), and Balberg and Binenbaum (1983, 1985) that were mentioned in the discussion of continuum percolation given in Chapter 3. In particular, Balberg and Binenbaum (1983) studied two-dimensional anisotropic systems of conducting sticks which may be more realistic models for two-dimensional fractured reservoirs, since such system are usually anisotropic. Moreover, Balberg *et al.* (1984) considered a three-dimensional fracture network in which the fractures were finite cylinders of length L and radius r, and studied the dependence of the percolation threshold of the system on the aspect ratio L/r and on macroscopic anisotropy of the system.

One major difference between fracture networks and percolation networks is as follows. In a percolation network, an upper bound to all properties is the case when p, the fraction of conducting bonds, is unity. All quantities of interest are normalized against the $p = 1$ case, because otherwise any property of the network would depend on its linear size. This is true about *any* network simulation. On the other hand, for random fracture networks, there is theoretically no end to the degree of fracturing. If we add one more fracture to the network, its permeability or hydraulic conductivity increases infinitum. Thus, one cannot determine how "filled" a fracture network is, and there is no analog of the $p = 1$ case in percolation networks. As Hestir and Long (1990) pointed out, one can study systems in which λ_ℓ, the average frequency of fractures intersecting a sample line, is held constant. Since, any fracture network can be rescaled to a given constant λ_ℓ if it is held fixed, then the permeability of the network has its maximum value when all fractures are infinitely long, which is the system that Snow (1969) studied. Thus, if λ_ℓ is held fixed, Snow's analytical results for the permeability of a fracture network is equivalent to $p = 1$ case for percolation networks.

The next issue to resolve is how to relate the parameters of fracture networks to those of percolation networks such as p and $\langle Z \rangle$, the average coordination number of the network. Robinson (1984a,b) and Charlaix *et al.* (1987a) used the average number of intersections per fracture ζ as the measure of the connectivity (see below). Suppose now that the average fracture length is $\langle \ell \rangle$, the orientation distribution is $g(\theta)$, and the density of fracture centers is λ_A. It is easy to show that $\lambda_\ell = \langle \ell \rangle \lambda_A$, and

$$\zeta = \lambda_\ell \langle \ell \rangle \Theta(\theta) \, , \tag{7.1}$$

where $\Theta(\theta)$ is defined by

$$\Theta(\theta) = \int_0^\pi \int_0^\pi \sin|\theta - \theta_0| g(\theta) g(\theta_0) d\theta d\theta_0 \, . \tag{7.2}$$

For example, if the orientations of the fractures are uniformly distributed, $g(\theta) = 1/\pi$, and $\Theta(\theta) = 2/\pi$. Now, for every ζ, there is a $p_f(\zeta)$, which is the analog of p for percolation networks. For example, in the fracture network of Englman *et al.* (1983) one has $p_f(\zeta) = 1 - \exp(-\zeta)$. Therefore, a critical value of ζ, ζ_c, can also be defined, and thus all results of percolation theory discussed in Chapter 3 can be written in term of ζ, if we replace p everywhere with $p_f(\zeta) =$

$1 - \exp(-\zeta)$. As the second example, consider Robinson's (1984a,b) model that was analyzed by Hestir and Long (1990). In this model, $p_f(\zeta)$ is the average fraction of a fracture which is available for flow. Now, consider a fracture of length ℓ with $n(\ell)$ intersections. If the fracture length is constant, then $n(\ell)$ will be a Poisson process. The average fraction of a fracture available for flow, i.e., the fraction which is between the two end sites separated by ℓ, is $[n(\ell) - 1]/[n(\ell) + 1]$. Therefore, if P_n is the probability that $n(\ell) = n$, then one simply has, $P_n = \zeta^n e^{-\zeta}/n!$ and

$$p_f(\zeta) = \sum_{n=2}^{\infty} \frac{n-1}{n+1} P_n = \left(1 + \frac{2}{\zeta}\right)(1 + e^{-\zeta}) - \frac{4}{\zeta} \ . \qquad (7.3)$$

Again, all standard results of percolation can now be written down for Robinson's model by replacing p everywhere with p_f. Hestir and Long (1990) worked out several other examples relating p_f (or p) to ζ. They also considered the case in which the fracture lengths were not constant, but were distributed according to a given distribution. The goal of Hestir and Long (1990) was to use the effective-medium approximation (see Chapter 8) and the scaling theory of permeability near p_c (see Chapter 3) to predict hydraulic conductivity of fracture networks, and in order to do that one has to relate p (or p_f) to ζ.

Long and Billaux (1987) developed a two-dimensional network model of fractures by incorporating field data into the model. The network was generated subregion by subregion where the properties of each subregion were predicted through geostatistics. The region in which the data were collected was divided into statistically-homogeneous subregions. The fractures were divided into five sets based on their tectonic history. It was observed that in each set fractures were spaced close together and had similar orientations. This information was incorporated into the network model. Another piece of information that was used in the simulation was the aperture distribution. The criterion for accepting a model was its ability for reproducing the measured permeability of the subregion. After all subregions were created, they were joined together to create the entire region. When the model was applied to Fanay-Augeres, a uranium mine in France, it was found that macroscopically the region was barely connected. Only 0.1% of the fractures contributed to the permeability of the system. This implied that, (i) the system had the structure of a percolation network, and was extremely close to its p_c; and (ii) the fracture network was a fractal object over the scale of the observation. Both of these support the ideas of Sahimi *et al.* (1993) (Chapter 5), discussed below, that at large length scales the fracture network of rock is similar to a percolation network. These results also appear to be typical of many fractured rocks, and indicate the significance of percolation concepts to the modelling of such systems.

The applicability of two-dimensional networks of fractures to modelling natural fractured rocks is limited. One main reason for this is that a two-dimensional model cannot realistically describe the fracture network connectivity, because fractures which do not connect in a planar cut may connect in the third dimension. Moreover, whenever two-dimensional data have been used with a three-dimensional model, it has been found that one has a non-uniqueness problem in the sense that, many three-dimensional models can account for the same two-dimensional data. However, despite its significance, three-dimensional models have received less attention than their two-dimensional counterparts, perhaps because of the complex computations that are involved.

7.2.2 Three-dimensional fracture networks

In the three-dimensional models, fractures are represented by either discs of finite radius (Long *et al.*, 1985; Shapiro and Andersson, 1983; Andersson and Dverstorp, 1987; Charlaix *et al.*, 1987a; Liggett and Medina, 1988; Billaux *et al.*, 1989; Tsang *et al.*, 1988; Piggott and Elsworth, 1989; Cacas *et al.*, 1990a,b; Nordqvist *et al.*, 1992), or by flat planes of finite dimensions (Wilke *et al.* 1985). There is experimental evidence that three-dimensional fractures are either roughly elliptical or discs-shaped (Pollard, 1976). In some cases, compared with the actual system we wish to study, the fractures are very long, so that they can be considered as essentially infinite. In this case, a simple stochastic model based on the Poisson distribution, which is an extension of the two-dimensional networks discussed above, may suffice. The Poisson process can also be used for generating clusters of fractures that are concentrated in a given subregion. The center of each subregion is generated by the Poisson process, and then around each center a cluster of fractures are generated. Similar to conditional simulation of flow in megascopic porous media discussed in Chapter 6, any realistic fracture model should also honor the available experimental data. Thus, if, for example, the conditioning ought to be done with respect to the observed fracture density, one first generates the fracture network, rejects all the fractures that intersect the surveyed lines or surfaces, and adds the actual fractures that have been observed. This is particularly simple with a Poisson model, since in this model each fracture is generated independently of all others.

There have been some attempts to use a three-dimensional fracture network for simulating the hydraulic behavior of a given field and match the measured data. Billaux *et al.* (1989) extended the model of Long and Billaux (1987), discussed above, to three dimensions. The fractures were represented as disks placed randomly in space. The diameter of each disc was selected independently from a probability distribution which was assumed to be log-normal. To locate the fractures in the space, a point process, called the *parent-daughter process*, was used. In this method one starts from a Poisson process and places a cloud of points (or daughters) around each Poisson point (called a parent or seed). The number of points in each cloud is a Poisson random variable, and each point is placed in a given cloud independently of all other points. The motivation for

Figure 7.3: A three-dimensional network of disk-like fractures (after Billaux *et al.*, 1989).

doing this is the fact that experimental data indicate that fractures of real rocks often occur in swarms. Figure 7.3 shows a typical swarm of disk-like fractures. As in the case of the two-dimensional model (Long and Billaux, 1987), the fractures were divided into five different sets. The orientation of the discs in each set was characterized as a fluctuation about the mean orientation for the set. This fluctuation had a spatial structure that could be simulated with geostatistics. After each set was generated, a model of a given fractured field was created by putting together all of the sets. Using this model, the hydraulic and transport properties of the fractured rock were simulated. In a work somewhat similar to that of Billaux *et al.* (1989), Dverstorp and Andersson (1989) used a three-dimensional network of disc-like fractures and showed that the model can be calibrated by one set of data, and then be used to predict and match another set of data. These are discussed in Chapter 10.

7.2.3 Continuum representation of fracture networks

We already discussed the multi-porosity models of fractured rock. However, as discussed above, such models are not appropriate for representing fracture networks of rock when the networks have incomplete (disordered) connectivity, fractal properties, etc. On the other hand, fracture networks of the type discussed above are very useful for gaining a deeper understanding of flow phenomena in

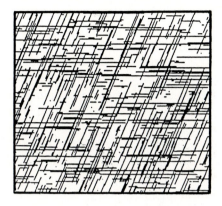 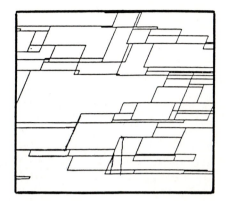

Figure 7.4: A two-dimensional fracture network (left) and its equivalent dual permeability model (after Clemo and Smith, 1989).

fractured rock. The main disadvantage of such models is that, if they are to be used for modelling fractured rock at large scales (of order of at least several hundreds of meters), the necessary computations become prohibitive. For this reason, a continuum representation of fracture networks would be desirable. However, in order not to be hampered by the same type of shortcomings that the multi-porosity models suffer, one has to first use network simulations to gain a better understanding of a given phenomenon at smaller scales, and then find an appropriate REV for averaging the simulation results. This allows one to transfer the smaller-scale results to larger-scale, continuum model. In doing so, one has to make the assumption that the morphological properties of the smaller-scale network used in the simulations are representative of the much larger fracture network whose simulation is not feasible. As discussed below in Section 7.5, this assumption is not always justified, but has been used in the past. For example, although the fracture network is sample-spanning and above its percolation threshold, it may be quite close to the percolation threshold, a situation that is encountered quite often in fractured rock (see below). Thus, the correlation length of the network associated with percolation (see Chapter 3) is very large, and therefore a REV, even if it can be defined, would be so large that averaging the results of the network simulations, and transferring the results to larger scales would be meaningless. Moreover, if the fracture network is a fractal object, the linear size of the REV must be larger than the length scale at which the network has fractal characteristics, and often such length scales are very large.

An alternative method was suggested by Clemo and Smith (1989). They proposed the use of a hierarchy of fractures, from the smallest ones to the largest. In this scheme, only those fractures that are deemed more important to flow are modeled individually. Clemo and Smith (1989) called such fractures *primary*. It should be clear from our discussion of percolation networks in Chapter 3 that the primary fractures have to be part of the backbone of the fracture network, i.e., the multiply connected part of the network through which flow and transport take place. Moreover, such fractures have the largest apertures in the network. The vast majority of the fractures are not as important to flow and transport through the network as the primary fractures. These fractures are thus classified in groups and are represented by some effective properties. These groups are called network blocks. Together, the combination of the primary fractures and network blocks constitute what Clemo and Smith called the *dual permeability model*. However, this should not be confused with the double-porosity model discussed above, because the dual permeability model is just a way of classifying the fractures according to their importance to flow and transport in the entire fracture network. Note that although the individual fractures that are grouped in the blocks are not as important as the primary fractures, the overall contribution of the blocks is about the same as that of the primary fractures. Figure 7.4 shows a typical fracture network and its equivalent dual permeability model. The effective properties of the blocks, e.g., their hydraulic conductance and porosity, are calculated from those of the fractures that they contains. Simulation of flow and transport in such a network is much less time consuming than the original fracture network, and is discussed in Chapter 10.

7.2.4 Percolation properties of fracture networks

Charlaix *et al.* (1984) argued that *at the percolation threshold* of a three-dimensional fracture network made of equal flat disks of radius r with a density of λ_A of disks per unit volume, one must have

$$\lambda_A r^3 \sim 0.15 - 0.3 \quad . \tag{7.4}$$

Note that r^3 essentially represents the volume of a disk. For polydisperse disks with a distribution of the radii one must replace r^3 with $\pi^2 \langle r^2 \rangle \langle r \rangle$ in Eq. (4), which is nothing but $\langle \text{surface} \rangle \langle \text{perimeter} \rangle / 2$, where $\langle \cdot \cdot \rangle$ represents the average of the quantity with respect to the distribution of the radii. These predictions were confirmed by Charlaix (1986) using Monte Carlo simulations. Robinson (1984b) found that for three-dimensional networks of planar fractures, the percolation threshold, i.e., the number of intersections per plane is about 2, whereas for two-dimensional networks of fractures with constant length the percolation threshold, i.e., the average number of fractures intersecting a given fracture, was found to be about 3.1. Wilke *et al.* (1985) considered percolation in networks of planar fractures and showed that, the geometrical critical exponents of percolation in their model are the same as those of random percolation (see Chapter 3). Madden (1983) represented a three-dimensional fracture network by a tessellation of cubes and studied the connectivity properties of the network using renormalization group theory (for principles of renormalization group theory see Chapter 8).

7.3 Simulated annealing model

This approach was developed by Long *et al.* (1991), and appears to be a promising way for constructing models of fractured rocks. We already discussed the models that were developed by Long and Billaux (1987), Billaux *et al.* (1989), and Dverstorp and Andersson (1989). There is a basic problem that may limit, at least in some cases, the usefulness of these models. This is the fact that, although there may be a large number of fractures in a given rock or field, usually only an extremely small fraction of the fractures contribute to the fluid flow. That is, most of such fractured rocks are at or near their percolation threshold, have fractal properties, and therefore cannot be treated with the classical methods of analysis. To overcome this difficulty, Long *et al.* (1991) developed a simulated annealing method. Such methods were originally developed by Kirkpatrick *et al.* (1983) and others for optimization processes.

In the work of Long *et al.* (1991) rock is represented by a three-dimensional network of finite fractures. Then, simulated annealing is used to find an appropriate pattern of connected fractures given some experimental information about rock. The most important issue to be resolved is to find how the fractures are connected, so that the transport behavior of the fracture network can mimic that of the actual system and honors the data. Given a problem, simulated annealing can be used if one has, (i) a set of possible configurations of the system; (ii) a method for systematically changing the configuration; (iii) an "energy" function, in analogy with the work of Kirkpatrick *et al.* (1983), to minimize; and (iv) an annealing schedule of changing a temperature-like variable, so that the system can reach its minimum. At this point the fracture network takes on its optimal configuration that can honor the experimental data.

For fluid flow in fractured rock the requirement (i) is obvious: it is a network of fractures in which some of the fractures allow fluid flow to take place (they are "on"), and some are closed to the transport process (they are "off"). Suppose $\{C\}$ denotes the set of all possible configurations of on and off fractures (note that such networks are similar to percolation networks of open and closed bonds discussed in Chapter 3). Requirement (iii) is a method according to which a given configuration of the fracture network is changed. Suppose that one assigns a probability function for randomly selecting a fracture. Then, if the selected fracture is on, it is turned off and vice-versa. This generates a new configuration of the fracture network. One now defines the "neighborhood" N_C of C as the set of all configurations that are very close to C (in the language of Long *et al.*, they are one step away from C). Annealing the system means picking a configuration from N_C and comparing it with C. In order to make the comparison precise, one needs to define an "energy" E, the third ingredient of simulated annealing.

Long *et al.* (1991) defined E as

$$E = \sum_j |f(\mathbf{O}_j) - f(\mathbf{S}_j)|^\gamma \quad , \tag{7.5}$$

where $\mathbf{O}_j$ and $\mathbf{S}_j$ are vectors of *observed* and *simulated* responses, respectively, γ a constant and $f(\mathbf{x})$ a real monotonic function. The measurements could be hydrological, geophysical, or geological. A probability distribution of the configurations was then assumed to be expressable as a Gibbs' distribution

$$P(C) = a \exp \left[\frac{E(C)}{T} \right] \quad , \tag{7.6}$$

where a is the renormalization constant which is very difficult to estimate, because one must know the energy of *all* configurations, which is impossible, and T is a temperature-like variable to be defined below.

Because $P(C)$ is a Gibbs' distribution, C, the current configuration, can be modeled as a Markov random field, which means that the transition probability for moving from C to C' depends only on C and C' and not on the previous configurations from the set $\{C\}$. Thus, the transition probability can change from configuration to configuration, but it does not depend on the previous configurations that were examined. Therefore, given C and N_C, the transition probability for moving from C to C' (given our current configuration C) is equal to the probability that we select C', times the probability that the system would make the transition to a given configuration C'. Therefore,

$$P\{C \to C'|C\} = \begin{cases} 0 & C' \notin N_C \\ P(C'|C) \times 1 & \begin{array}{l} C' \in N_C, \\ E(C') - E(C) \le 0 \end{array} \\ P(C'|C) \exp\{[E(C) - E(C')]/T\} & \begin{array}{l} C' \in N_C, \\ E(C') - E(C) > 0 \end{array} \end{cases} , \tag{7.7}$$

where for the last equation we must have $C \ne C'$. The final ingredient of the model is a schedule for lowering the temperature as annealing progresses. This means that as annealing continues, one is less likely to keep those configurations that increase E. The temperature schedule that Long *et al.* (1991) used was *ad hoc* but effective, and was actually suggested by Press *et al.* (1986), who proposed that the temperature should be changed after a number of configuration iterations which is sufficient to produce a fixed number of acceptable changes. At the end of each iteration i, the temperature T_i is decreased using a geometric series

$$T_{i+1} = T_i R_i^i \quad , \tag{7.8}$$

where $0 < R_i < 1$. Thus, one only needs to select the initial temperature. This is selected such that it is of the same order of magnitude as the energy difference between the first two configurations, so that the energy difference between successive configurations remains (most of the time) between zero and 1.

What does one do if, for example, quantitative information, such as the range of possible responses, is available, but there are no actual measurements? For example, suppose one wants to predict the flow rate Q at a point in the system far from a point at which a measurement was done, and has the information that the flow rates were observed to be between a and b, $Q \in [a, b]$. In this case, each time a configuration is changed, the following steps are also taken: (i) the new point is added to the configuration and Q is calculated at that point; (ii) a new energy function E' is calculated such that $E' = 0$, if the calculated Q is in $[a, b]$, $E' = (Q - a)^\gamma$, if $Q < a$, and $E' = (Q - b)^\gamma$ if $Q > b$; (iii) if $E(C') + E'(C') < E(C) + E'(C)$, then C' is kept. Otherwise, the usual annealing probability is used to keep or reject C'; and (iv) finally, the new point is removed and the process continues. This completes the simulated annealing method for selecting a configuration of a network of fractures that can mimic important features of a given rock, and reproduces its measured properties. Long *et al.* (1991) showed that even if the initial configuration is a completely connected network of open ("on") fractures, the final configuration is usually a network which is *barely connected*, although it contains a very large number of fractures. This implies that the fracture network is a sample-spanning percolation cluster just above its percolation threshold, in agreement with the proposal of Sahimi *et al.* (1993), which was mentioned in Chapter 5 and is discussed below. However, Long *et al.* did not note the similarity between their results and percolation.

7.4 Synthetic fractal models

As discussed in Chapter 5, in many situations fracturing is a product of a fragmentation process. The distribution of fragment sizes is often a power law [see Eq. (5.97)], a signature of a fractal system. A power-law distribution of fragment sizes has been known since 1940s, when it was first discovered in grinding operations. Gilvarry (1964) used an exponential distribution for describing repetitive fracturing. He assumed that imperfections, which are randomly distributed in the rock, initiate fracturing. An important parameter of his analysis was $p_f(\ell)\delta\ell$, the probability that a fragment whose size is between ℓ and $\ell + \delta\ell$ can be fractured. Turcotte (1986) was probably the first to document many rock fracturing applications of fragmentation, in which the size distribution is described by a power law. He also showed that such power laws can arise in Gilvarry's model if $p_f(\ell)$ is constant. For an idealized fragmentation process in which the fragments of a given generation are all of the same relative size, the size distribution is a power law with an exponent τ that depends on p_f [see Eq. (5.97)].

On the other hand, as discussed in Chapter 5 fracture networks of natural

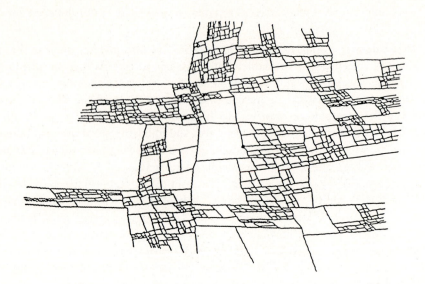

Figure 7.5: A two-dimensional synthetic fractal model of fracture networks with $p_f = 0.9$ (after Acuna and Yortsos, 1994).

rock have fractal properties, at least over certain length scales. Thus, a fragmentation process may be used for generating fractal fracture patterns. To do this, Barnsley (1988) proposed the method of iterative function system, by which a fractal-like pattern is obtained from an initial simple shape, called the *initiator*, by applying iteratively a set of numerical transformations, called the *propagators*. Each iteration creates multiple sets of smaller, transformed images that occupy the place of the previous image. After several generations, the set converges to a pattern that may be a fractal. In essence, this method creates fragments of various sizes and shapes. For creation of two-dimensional patterns, two transformations are used that are given by

$$x_n = a_1 x_{n-1} + a_2 y_{n-1} + a_3 x_{n-1} y_{n-1} + a_4 , \qquad (7.9)$$

$$y_n = a_5 x_{n-1} + a_6 y_{n-1} + a_7 x_{n-1} y_{n-1} + a_8 . \qquad (7.10)$$

Here x_n and y_n are the coordinates of a given point of the nth generation, and the a_is are constants of the transformations. If $a_3 = a_7 = 0$, we have linear transformations that can generate self-affine fractals (see Chapter 3). If, in addition $a_1 = a_6$, and $a_2 = -a_5$, the transformations can generate self-similar fractals. Similar transformations can be used for generating three-dimensional patterns.

   To make the resulting fracture patterns look more realistic, Acuna and Yortsos (1994) used the fragmentation probability p_f with the above transformations. Specifically, as fragmentation of the blocks continues, only a fraction p_f of the newly generated blocks is allowed to further fracture. Figure 7.5 shows a two-

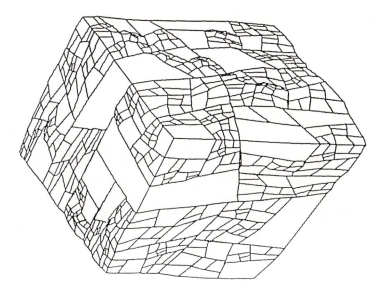

Figure 7.6: A three-dimensional synthetic fractal model of fracture networks (after Acuna and Yortsos, 1994).

dimensional example with $p_f = 0.9$, while Fig. 7.6 presents a three-dimensional example which has a fractal dimension $D_f = 2.78$.

Although this method does generate fracture patterns that look realistic, and although as discussed by Acuna and Yortsos (1994), some of their hydraulic properties can mimic those of fractured rock, the method has several shortcomings. There is no clear physical interpretation of the probability p_f, the key parameter for generating fractal fracture patterns, and thus it is not clear how it can be measured. The model also contains many adjustable parameters, the a_is, and it is not clear how they can be estimated if some experimental data for the fracture pattern, or any other property of the rock are available. This model ignores completely the origins of fracturing process, which are the deformation of rock and accumulation of large stresses that result in the nucleation and propagation of the fractures (see below). Finally, the fact that the fractal dimensions of the patterns can be made to be similar to those of fracture networks of rock does not imply that flow and transport properties of the patterns are also similar to those of fractured rock.

A more realistic synthetic model was proposed by Barton *et al.* (1987). In their model values are selected randomly from frequency distributions of fracture-trace length, spacing, orientation, crossing, and dead-end fracture intersection distributions obtained from the analysis of fracture trace maps discussed in Chapter 5. This is very different from the model of Acuna and Yortsos (1994) and from the models that are based on a Poisson distribution and were discussed above. It is a more realistic approach because it incorporates experimental data in the model. Thus, not only the distributions of various characteristics of the fracture network are in agreement with the data, its fractal dimension for a two-dimensional model is found to be in the range 1.3-1.7, in agreement with the experimental data discussed in Chapter 5.

7.5 Mechanical fracture models

In this model one attempts to generate the fracture network by modelling deformation of rock which results in nucleation and propagation of fractures. Traditional approaches of fracture mechanics are based on the important criterion developed by Griffiths (1921) who proposed that a *single* crack becomes unstable to extension when the elastic energy released in the crack extension by a small length $\Delta\ell$ becomes equal to the surface energy required to create a length $\Delta\ell$ of crack surface. However, this criterion is most useful for solids that are more or less homogeneous, and thus it cannot be expected to yield reasonable results for heterogeneous rock. In natural rock the presence of distributed porosity in the form of a large number of flaws with various sizes, shapes, and orientations makes the fracture problem far more complex. Disorder comes into play in many ways during a fracture process, and even small, initially present disorder can be enormously amplified during fracture. This makes fracture a collective phenomenon in which heterogeneities play a fundamental role. In fact, due to disorder, brittle solids generally exhibit large statistical fluctuations in fracture strengths, when nominally identical laboratory samples are tested under identical loading conditions. Because of these statistical fluctuations, it is insufficient, and indeed inappropriate, to represent the fracture behavior of heterogeneous rock by only its *average* properties: fluctuations are important and must not be neglected.

A relatively simple network model was introduced by Sahimi and Goddard (1986) for fracture of disordered solids. In this model the solid is represented by a network in which each bond represents a small portion of the system (for example, the solid matrix of rock). In classical mechanics a solid is represented by a spring or a beam, and thus each bond of the network also represents a spring (or a beam). These springs follow the laws of linear elasticity (i.e., force is proportional to the displacement) up to a critical threshold (defined below), beyond which they break irreversibly and create a microfracture in the system. Heterogeneities may be introduced in the elasticity and ultimate strength of the bonds. Thus, the sequence of breaking bonds and the spatial patterns they form model the regional fracture process. Various properties of such fracture phenomena have been investigated (see, for example, Sahimi and Goddard, 1986; de Arcangelis *et al.*, 1989; Arbabi and Sahimi, 1990b; Sahimi and Arbabi, 1992, 1993; for reviews see Herrmann and Roux, 1990; Sahimi, 1992c). One of the main advantages of this model is that any kind of disorder may be incorporated and its effect on the fracture process evaluated.

To see how the model generates a fracture network, consider a two- or three-dimensional network whose linear size is L, where L is the number of sites in a given direction. Every site of the network is characterized by a displacement vector $\mathbf{u}_i = (u_{ix}, u_{iy})$, or by $\mathbf{u}_i = (u_{ix}, u_{iy}, u_{iz})$. We consider here the case of brittle fracture for which a linear approximation is valid up to a threshold (defined below). The displacement $\mathbf{u}_i$ is computed by minimizing the elastic energy of the system with respect to $\mathbf{u}_i$, which is given by

$$E = \frac{\alpha}{2} \sum_{\langle ij \rangle} [(\mathbf{u}_i - \mathbf{u}_j) \cdot \mathbf{R}_{ij}]^2 e_{ij} + \frac{\beta}{2} \sum_{\langle jik \rangle} (\delta\theta_{jik})^2 e_{ij} e_{ik} . \qquad (7.11)$$

The first term of the right-hand side of Eq. (11) represents the contribution of stretching or central forces (Hooke's law), while the second term is due to the forces that change the angle between two bonds that have a site in common. α and β are the central and angle-changing force constants, respectively, $\mathbf{R}_{ij}$ is a unit vector from site i to site j, and e_{ij} is the elastic constant of the bond (spring) between i and j. Here $\langle jik \rangle$ indicates that the sum is over all triplets in which the bonds j-i and i-k form an angle whose vertex is at i. Equation (11) can be obtained from a discretization of a continuum equation, and has rigorous theoretical foundation (Arbabi and Sahimi, 1990a). $\delta\theta_{jik}$ is given by

$$\delta\theta_{jik} = \begin{cases} (\mathbf{u}_{ij} \times \mathbf{R}_{ij} - \mathbf{u}_{ik} \times \mathbf{R}_{ik}) \cdot (\mathbf{R}_{ij} \times \mathbf{R}_{ik})/|\mathbf{R}_{ij} \times \mathbf{R}_{ik}| & \mathbf{R}_{ij} \text{ not } || \text{ to } \mathbf{R}_{ik}, \\ |(\mathbf{u}_{ij} + \mathbf{u}_{ik}) \times \mathbf{R}_{ij}| & \mathbf{R}_{ij} \,||\, \text{ to } \mathbf{R}_{ik}, \end{cases}$$
$$(7.12)$$

where $\mathbf{u}_{ij} = \mathbf{u}_i - \mathbf{u}_j$. For *all* two-dimensional networks Eq. (12) is simplified to

$$\delta\theta_{jik} = (\mathbf{u}_i - \mathbf{u}_j) \times \mathbf{R}_{ij} - (\mathbf{u}_i - \mathbf{u}_k) \times \mathbf{R}_{ik} . \qquad (7.13)$$

One can think of αe_{ij} as the elastic constant of the bond $i - j$, and $\beta e_{ij} e_{ik}$ as the angle-changing force constant of the pair of bonds $i - j$ and $i - k$ (since angle changing involves two bonds). However, the elastic energy is expressed by the particular form given by Eq. (11) since it allows one to vary the parameters α and β independently. More generally, we may consider a network in which the bonds themselves can bend, i.e., a bond represents a beam, in which case one has to add the contribution of the torsional or bending forces to Eq. (11). Such a model was considered by de Arcangelis *et al.* (1989). However, it has been shown (Arbabi and Sahimi, 1990b) that many properties of fracture networks, such as those discussed below, are independent of whether or not such torsional forces are included in the elastic energy of the system.

In general, there are at least four different ways of incorporating rock heterogeneities in the model. (i) Deletion or supression of a fraction of the bonds at random or in a correlated manner. The removed bonds may represent the porosity of the system *before* the fracture process begins. (ii) Random or correlated distribution of elastic constants e of the bonds. The idea is that in natural rock the shapes, sizes, and strengths of the regions through which stress transmission takes place are broadly distributed, resulting in a different e for each microscopic region (bond). (iii) Random or correlated distribution of the critical thresholds. For example, each bond can be characterized by a critical displacement or length ℓ_c, such that if it is stretched beyond ℓ_c, it breaks irreversibly. The idea is that, even if rock is made up intrinsically of the same material (same e everywhere), it contains regions that offer different resistances to breakage under an imposed external stress or strain, because of, for example, defects or pre-cracked porosity in its structure. (iv) These methods generate microscopic and macroscopic heterogeneities. To generate *megascopic* heterogeneities, we divide the network into blocks of bonds within which the bonds may have the same properties, but between which ℓ_c, e, or the porosity vary spatially. The properties of the blocks are correlated with each other, where the correlations can be induced by, e.g., a fractal distribution.

A threshold value ℓ_c is now introduced for the length of a bond, which is selected from a probability density function. While any distribution function can be used, in our discussions we use the following

$$f(\ell_c) = (1 - \eta)\ell_c^{-c} , \qquad (7.14)$$

where $c = 0$ corresponds to a uniform distribution in $(0, 1)$. The advantage of this distribution is that by varying c one can obtain very narrow $(c \simeq 0)$ and very broad $(c \simeq 1)$ distributions. The same type of distribution can be used for the elastic constants e of the bonds. We then initiate the fracture process by applying a fixed external strain (or stress) to the network in a given direction. As mentioned above, to calculate the nodal displacements $\mathbf{u}_i$ we minimize the elastic energy E with respect to $\mathbf{u}_i$, $\partial E/\partial \mathbf{u}_i = 0$. Writing down this equation for every interior site of the network results in a set of dN simultaneous linear equations for nodal displacements $\mathbf{u}_i$ of a d-dimensional network of N internal nodes. This set is solved by a numerical method, usually an iterative technique such as the adaptive accelerated Jacobi-conjugate gradient that uses an acceleration parameter which is optimized for each iteration.

After determining the nodal displacements, the fracture process is initiated by breaking one or more bonds according to a given criterion. If the elastic constants of the bonds are distributed, we can simply break the most stretched bond, since such a bond is the weakest part of the system, and is therefore the most probable bond to break first. If the critical thresholds ℓ_c are distributed, we may break the bond for which $\ell_c - \ell_s$ is maximum, where ℓ_s is the current length of the bond in the deformed network. This would be equivalent to assuming that the rate at which the elastic forces relax throughout the network is much faster than the breaking of one bond. One can also break *all* bonds whose current lengths exceed their critical lengths. Each broken bond represents a microcrack in the system. An advantage of this model is that *any* criterion for the nucleation and propagation of the cracks can be used. Moreover, unlike the traditional models of fracture, we do *not* have to start the simulation with a system in

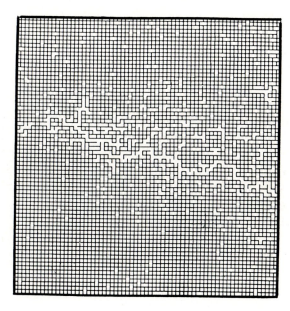

Figure 7.7: Two-dimensional fracture network obtained with $c = 0.1$. Broken bonds represent the fractures.

which a microcrack has been inserted to ensure the nucleation and propagation of the cracks. However, while such breaking criteria may be appropriate for a system under shear or tensile forces, they cannot be used for a system under compression. For such a case, we have to use a network of beams since they can also bend. Then, if the bending forces exerted on a beam exceed a critical value, that beam is broken. Alternatively, we can break the beam that suffers the largest bending force. However, as mentioned above, those characteristics of fracture networks that are of interest to us in this Chapter are independent of such details. After a bond is broken, we recalculate the nodal displacements u_i for the new configuration of the network, select the next bond to break, and so on. If the external stress or strain is not large enough to break any new spring, we gradually increase it. The simulation continues until a sample-spanning fracture pattern is formed and the network finally becomes macroscopically disconnected.

We are interested mainly in the structure of the fracture network that results from this fracturing process. Moreover, we would like to know whether the fracture patterns of macroscopic (laboratory-scale) porous media differ from those of megascopic (field-scale) porous media. It has been found that if c is small, a single crack is formed that spans the network and breaks it into two parts, with few smaller cracks that are not connected to the main sample-spanning fracture. This is shown in Fig. 7.7 for which $c = 0.1$. However, as c increases towards unity, many more cracks, in addition to the main sample-spanning fracture, are formed. This is shown in Fig. 7.8 for which $c = 0.9$. The reason for this difference is as

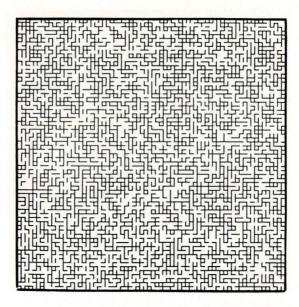

Figure 7.8: Two-dimensional fracture network obtained with $c = 0.9$.

follows. If c is small, the distribution of the thresholds is narrow, and the system is more or less homogeneous. As soon as the first microfracture is nucleated, the stress enhancement at its tip is larger than at any other part of the network. As a result, it is much more likely that the next microfracture is formed at the tip of the existing fracture. Therefore, a sample-spanning fracture is formed, with the possible exception of a few other microcracks distributed throughout the network. However, when c is large, the distribution of the bond stength is very broad, implying that there are many weak as well as strong regions in the network. Thus, although stress enhancement at the tip of the growing fracture is large, fracture growth may also take place at other regions of the system which are very weak. Thus, formation of the fracture network is more gradual.

The fracture patterns that are formed in this model are fractal objects, but their fractal dimension depends on the way the patterns are analyzed, and on the spatial distribution of the heterogeneities. One generally finds three distinct fractal dimensions. If one analyzes *only* the fractal behavior of the *single* fracture that spans the system, one obtains a fractal dimension $D_f \simeq 1.2 \pm 0.1$, consistent with the data for the surface of a *single major* fault or fracture discussed in Chapter 5. However, if one analyzes the statistics of *all* fractures of the system (whether or not they are connected to the sample-spanning crack), one finds $D_f \simeq 1.7 \pm 0.05$; see Fig. 7.9. This is also consistent with the experimental data discussed in Chapter 5. These results are obtained without any megascopic (block) heterogeneities. But what happens if megascopic heterogeneities are included in the model?

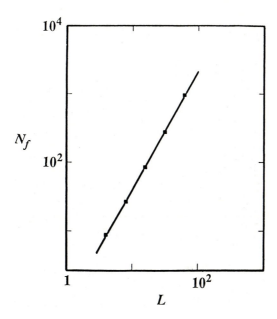

Figure 7.9: Logarithmic plot of the number of fractures N_f in an $L \times L$ network. The slope of the line, the fractal dimension of the fracture set, is 1.7.

To answer this question, we have to re-examine the distribution of rock heterogeneities. At small scales nucleation and propagation of the fractures do not take place at random, but depend on the stress field in the rock, because there are not regions with large differences in their mechanical strengths. Once a fracture nucleates in the rock, stress enhancement at its tip is larger than at any other point of the medium, and therefore the next microcrack almost surely develops at its tip. As a result, the fractal dimension of the sample-spanning crack is very low, $D_f \simeq 1.2$, since this fracture is almost like a line. However, at much larger length scales rock is *megascopically* heterogeneous. In such a rock fracture growth continues until a much stronger region is encountered. When this happens, fracture growth stops and another fracture nucleates in a weaker region of the rock. The growth of the new crack also stops when it encounters another strong region, and so on. This effect is accentuated if there are correlations between various regions which, as discussed in Chapters 5 and 6, is in fact the case. Thus, it is proposed (Sahimi *et al.*, 1993) that, viewed from such large scales, nucleation and propagation of fractures take place at essentially randomly selected regions of the rock, since they are more strongly dependent on the spatial distribution of the heterogeneities than on fluctuations in the stress field. Therefore, the fractal dimension of the fracture network at the largest scales should be close to that of percolation clusters since, as discussed in Chapter 3, in percolation the bonds of the networks are cut at random, i.e., fractures are created randomly.

To simulate fracture in rock with megascopic heterogeneities, one divides the

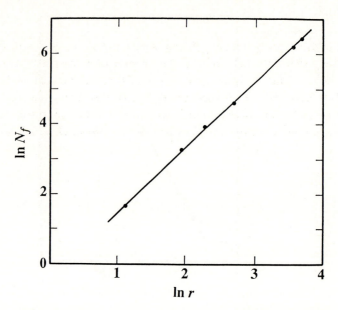

Figure 7.10: Number of fractures N_f versus the radius of the fracture network r in a two-dimensional system with block heterogeneities. The slope of the line, the fractal dimension of the fracture set, is 1.85.

network into $b \times b$ blocks, where $b << L$. Within each block e_{ij} is set to be the same for all bonds, but for each block e_{ij} is selected from a fractal distribution which reflects the long-range correlations that exist in rock (see Chapters 5 and 6). The threshold l_c of the bonds is also selected by the same method. Nucleation and propagation of the fractures are modeled as before. The statistics of the fractures are shown in Fig. 7.10, where we show the number of fractures versus the linear size of the fracture network. From this figure one obtains $D_f \simeq 1.85$, in good agreement with the theoretical expectation $D_f \simeq 1.9$ for two-dimensional percolation networks. This indicates that with the proper distribution of heterogeneities, the mechanical model of fracture described here can reproduce experimental data for fracture patterns observed in rock. If the percolation interpretation of fracture patterns is correct, then it should predict that for three-dimensional fracture networks of heterogeneous rock at large scales, $D_f \simeq 2.5$ (the same as that of percolation networks), consistent with the experimental measurements of the fractal dimensions of three-dimensional fracture networks discussed in Chapter 5. This again supports the validity of the argument. Finally, note that simulated annealing model of Long *et al.* (1991) discussed above also indicated that the optimal configuration of the fracture network found by the model is similar to a percolation network, although Long *et al.* themselves did not note this similarity.

7.6 Conclusions

Several methods of generating fracture network of heterogeneous rocks are now available. Although they differ in their degree of complexity, they all rely on discrete models of fractured rocks, and some of them do provide realistic models of the fracture networks. Therefore, simulation of flow and transport in fracture network of rocks is now possible, although one still has to spend considerable amount of computer time in order to carry out a reasonable simulation.

Chapter 8
Flow and Transport in Porous Media

8.0 Introduction

We already defined in Chapter 2 the concept of the permeability of a porous medium. In this Chapter we study and discuss single-phase flow and transport in porous media. We focus mainly on low-Reynolds-number flow, where Darcy's law is applicable, but we also discuss the deviations from it. We first give a theoretical derivation of Darcy's law, and then review and discuss various methods of estimating the permeability, electrical conductivity and diffusivity of fluid-saturated porous media. Of course, as a result of Einstein's relation (see Chapter 3) the electrical conductivity is proportional to the effective diffusivity D of the system, and therefore we do not need to discuss them separately.

8.1 The volume-averaging method and derivation of Darcy's law

As already mentioned in Chapter 2, Darcy's law is expressed as

$$\langle \mathbf{v} \rangle = -\frac{K}{\eta}(\boldsymbol{\nabla} P - \rho \mathbf{g}) \quad , \tag{8.1}$$

where $\langle \mathbf{v} \rangle$, η, and ρ are, respectively, the average velocity, viscosity, and density of the fluid, K is the permeability, P the pressure and $\mathbf{g}$ the gravity vector. Many authors have given various derivations of Eq. (1) (Bear, 1972; Gray and O'Neil, 1976; Neumann, 1977; Keller, 1980; Tartar, 1980; Larson, 1981; Whitaker, 1986a; Rubinstein and Torquato, 1989). The one-dimensional version of Eq. (1) was discovered empirically by Darcy in 1856. However, it is the extension of Darcy's law to more than one dimension which is of practical importance. In this section, we present a summary of Whitaker's (1986a) derivation of Eq. (1), which is based on the method of volume-averaging developed by him and his co-workers. This will also show us how continuum models of transport in porous media, of which Whitaker's method is perhaps the best known, are formulated.

The porous medium under consideration is shown in Fig. 8.1. The macroscopic length scale of the system is L, while the averaging volume is V. Other characteristic length scales of the system are l_β and l_σ, the characteristic length scales of the fluid and solid phases, respectively. We consider the most general case in which the medium can be anisotropic, i.e., have direction-dependent permeabilities. The boundary value problem that one has to solve is expressed by the continuity and momentum equations (see Chapter 2),

$$\boldsymbol{\nabla} \cdot \mathbf{v}_\beta = 0 \quad , \tag{8.2}$$

$$-\boldsymbol{\nabla} P_\beta + \rho_\beta \mathbf{g} + \eta_\beta \nabla^2 \mathbf{v}_\beta = 0 \quad . \tag{8.3}$$

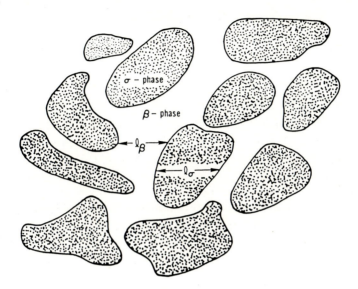

Figure 8.1: Schematic view of the porous medium used in the volume averaging method (after Whitaker, 1986a).

The interfacial area between the fluid and solid phases is $S_{\beta\sigma}$, and thus one boundary condition is that $\mathbf{v}_\beta = 0$ on $S_{\beta\sigma}$, the so-called *no-slip* boundary condition. As the second boundary condition one may specify $\mathbf{v}_\beta$ on the entrance and exit surfaces of the porous medium. One can now use the spatial averaging theorem (Anderson and Jackson, 1967; Marle, 1967; Slattery, 1967; Whitaker, 1967) which states that, for any quantity ψ_β associated with the β-(fluid) phase, one has

$$\langle \boldsymbol{\nabla} \psi_\beta \rangle = \boldsymbol{\nabla} \langle \psi_\beta \rangle + \frac{1}{V} \int_{A_{\beta\sigma}} \mathbf{n}_{\beta\sigma} \psi_\beta d\mathbf{A} \quad , \tag{8.4}$$

where $A_{\beta\sigma}$ is the interfacial area contained between the averaging volume, and $\mathbf{n}_{\beta\sigma}$ the unit outwardly-directed normal vector for the β-phase. Note that the averaging in Eq. (4) is over V. One can also define an average over phase volume, called the *intrinsic phase average*,

$$\langle \psi_\beta \rangle^\beta = \frac{1}{V_\beta} \int_{V_\beta} \psi_\beta dV \quad , \tag{8.5}$$

where V_β is the volume of the β-phase within V. Equation (4) and the no-slip boundary condition imply that $\boldsymbol{\nabla} \cdot \langle \mathbf{v}_\beta \rangle = 0$, whereas if we use the averaging defined by Eq. (5), we obtain $\boldsymbol{\nabla} \cdot \langle \mathbf{v}_\beta \rangle^\beta = -\phi_\beta^{-1} \boldsymbol{\nabla} \epsilon_\beta \cdot \langle \mathbf{v}_\beta \rangle^\beta$, where ϕ_β is the volume fraction of the β-phase (or the porosity of the porous medium). On the other hand, using Eq. (4) in the Stokes' equation, Eq. (3), yields

$$-\boldsymbol{\nabla} \langle P_\beta \rangle - \frac{1}{V} \int_{A_{\beta\sigma}} \mathbf{n}_{\beta\sigma} P_\beta d\mathbf{A} + \epsilon_\beta \rho_\beta \mathbf{g} + \eta_\beta \langle \boldsymbol{\nabla} \cdot \boldsymbol{\nabla} \mathbf{v}_\beta \rangle = 0 \quad . \tag{8.6}$$

If we use the obvious relation $\langle P_\beta \rangle = \phi_\beta \langle P_\beta \rangle^\beta$, the decompositions $P_\beta = \langle P_\beta \rangle^\beta + \tilde{P}_\beta$, and $\mathbf{v}_\beta = \langle \mathbf{v}_\beta \rangle^\beta + \tilde{\mathbf{v}}_\beta$, and the relation

$$\frac{1}{V} \int_{A_{\beta\sigma}} \mathbf{n}_{\beta\sigma} dA = -\nabla \phi_\beta \quad ,$$

which can be obtained from Eq. (4) using $\psi_\beta = 1$, then after some tedious manipulations and repeated applications of Eqs. (4) and (5), Eq. (6) finally becomes

$$-\nabla \langle P_\beta \rangle^\beta + \rho_\beta \mathbf{g} - \frac{1}{V_\beta} \int_{A_{\beta\sigma}} \mathbf{n}_{\beta\sigma} \tilde{P}_\beta dA + \frac{\eta_\beta}{V_\beta} \int_{A_{\beta\sigma}} \mathbf{n}_{\beta\sigma} \cdot \nabla \tilde{\mathbf{v}}_\beta dA = 0 \quad , \qquad (8.7)$$

where several terms have been deleted because their contributions are negligible; see Whitaker (1986a). We now need to develop governing equations for $\tilde{P}_\beta$ and $\tilde{\mathbf{v}}_\beta$. This is done by substituting the above decompositions into Eqs. (2) and (3) to obtain

$$-\nabla \tilde{P}_\beta + \eta_\beta \nabla^2 \tilde{\mathbf{v}}_\beta = -(-\nabla \langle P_\beta \rangle^\beta + \rho_\beta \mathbf{g} + \eta_\beta \nabla^2 \langle \mathbf{v}_\beta \rangle^\beta) \quad , \qquad (8.8)$$
$$\nabla \cdot \tilde{\mathbf{v}}_\beta = -\nabla \cdot \langle \mathbf{v}_\beta \rangle^\beta \quad , \qquad (8.9)$$

with the boundary conditions that $\tilde{\mathbf{v}}_\beta = -\langle \mathbf{v}_\beta \rangle^\beta$ on $S_{\beta\sigma}$ (since $\mathbf{v}_\beta = \mathbf{0}$ there), and specified $\tilde{\mathbf{v}}_\beta$ on the entrance and exit surfaces of the system. Since all terms on the left-hand side of Eq. (8) are of the order of $\langle \mathbf{v}_\beta \rangle^\beta / l_\beta$, whereas those on the right-hand side are of the order of $\langle \mathbf{v}_\beta \rangle^\beta / L$, the right-hand side can be safely neglected (since $l_\beta \ll L$) and Eq. (9) becomes $\nabla \cdot \tilde{\mathbf{v}}_\beta = 0$. Then, use of the averaging expressed by Eq. (5) in Eq. (8) yields

$$\frac{1}{V_\beta} \int_{V_\beta} (-\nabla \tilde{P}_\beta + \eta_\beta \nabla^2 \tilde{\mathbf{v}}_\beta) dV = -(-\nabla \langle P_\beta \rangle^\beta + \rho_\beta \mathbf{g} + \eta_\beta \nabla^2 \langle \mathbf{v}_\beta \rangle^\beta) \quad , \qquad (8.10)$$

which, when compared with Eq. (8), yields

$$-\nabla \tilde{P}_\beta + \eta_\beta \nabla^2 \tilde{\mathbf{v}}_\beta = \frac{1}{V_\beta} \int (-\nabla \tilde{P}_\beta + \eta_\beta \nabla^2 \tilde{\mathbf{v}}_\beta) dV \quad . \qquad (8.11)$$

We now search for a solution of the form

$$\tilde{\mathbf{v}}_\beta = \mathbf{B} \cdot \langle \mathbf{v}_\beta \rangle^\beta + \psi \quad , \qquad (8.12)$$
$$\tilde{P}_\beta = \eta_\beta \mathbf{b} \cdot \langle \mathbf{v}_\beta \rangle^\beta + \eta_\beta \zeta \quad , \qquad (8.13)$$

where ψ and ζ are arbitrary functions, which means that $\mathbf{B}$ (a second-order tensor) and $\mathbf{b}$ can be specified in any way one wishes. Whitaker (1986a) goes on to show that ψ and ζ can be ignored altogether. Thus, if we use Eqs. (12) and (13) in Eq. (7), define a tensor $\mathbf{T}$ by

$$\mathbf{T} = -\frac{1}{V_\beta} \int_{A_{\beta\sigma}} \mathbf{n}_{\beta\sigma} \cdot (\nabla \mathbf{B} - \mathbf{Ib}) dA \quad , \qquad (8.14)$$

where $\mathbf{I}$ is the identity tensor, and let $\mathbf{K} = \phi_\beta \mathbf{T}^{-1}$, Eq. (7) becomes

$$\langle \mathbf{v}_\beta \rangle = \frac{\mathbf{K}}{\eta_\beta}(\nabla \langle P_\beta \rangle^\beta - \rho_\beta \mathbf{g}) \quad , \tag{8.15}$$

which is Darcy's law, where $\mathbf{K}$ is the permeability tensor. For isotropic porous media, Eq. (15) reduces to Eq. (1).

8.2 The Brinkman and Forchheimer equations

As in any fluid flow problem, the range of validity of Darcy's law is expressed in terms of a Reynolds number Re. The Reynolds number is normally defined in terms of a characteristic length of the system in which fluid flow occurs. However, for flow through unconsolidated porous media it is customary to express Re in terms of a characteristic grain size, and for flow in consolidated porous media in terms of a mean pore size. Some authors have suggested $\sqrt{K/\phi}$ as the characteristic length, as K has the units $(\text{length})^2$, while others have used $\sqrt{K}$. In any event, it is generally believed that Darcy's law is applicable as long as $Re \leq 10$. This is the flow regime in which viscous forces are predominant. Beyond $Re = 10$ there is a transition zone in which inertial forces start to be predominant. Roughly speaking, for $Re > 100$ the transition zone ends and we enter the turbulent flow regime.

Darcy's law may also break down for flow of gases at low pressures through porous media, and since many gases such as air are used for determining permeability of porous media, the deviations from Darcy's law may be important. For example, if a pore diameter is comparable with the mean free path of a molecule, i.e., the mean distance that it travels before it collides with another molecule, then deviations from Darcy's law become significant. This breakdown may be due to the fact that at low pressures the no-slip boundary condition on the pore surface is no longer valid. Gas molecules move (diffuse) on the pore surfaces and contribute to their overall flux. This phenomenon is usually called the *Klinkenberg effect* or *Knudsen flow*.

If the fluid velocity is high enough that Darcy's law is not valid, then one has to add some correction terms to the basic Darcy's law in order to take into account the effect of the deviations. However, in the past this has been done mostly in an empirical manner. For example, Forchheimer (1901) suggested that for a pressure gradient $\Delta P/L$ one must write

$$\frac{\Delta P}{L} = a_1 \text{v} + a_2 \text{v}^2 \quad , \tag{8.16}$$

where v is the magnitude of the fluid velocity vector, and the *a*s are constant. To obtain a better representation of experimental data, a third-order term was added later

$$\frac{\Delta P}{L} = a_1 \text{v} + a_2 \text{v}^2 + a_3 \text{v}^3 \quad . \tag{8.17}$$

Other empirical or semi-empirical modifications of Darcy's law are discussed by Scheidegger (1974).

Brinkman (1947) proposed a more plausible modification of Darcy's law given by

$$\boldsymbol{\nabla} P = -\frac{\eta'}{K}\mathbf{v} + \eta'\nabla^2\mathbf{v} \,, \tag{8.18}$$

where η' is a *renormalized viscosity* which, in principle, can be different from η, the pure-fluid viscosity used in Eq. (1). Actually, in his original paper Brinkman (1947) used η, and in most applications of Eq. (18) this assumption has been made, and the permeability K has been estimated by some technique, with the results being in good agreement with the experimental data. But Koplik *et al.* (1983) suggested that η and η' are not necessarily the same. If the length scale over which the fluctuations in the velocity are important is much larger than $O(\sqrt{K})$, then the second term of Eq. (18) is negligible, and one recovers Darcy's law. Moreover, as the porosity of the system approaches unity, we must have $K \to \infty$ and $\eta'/\eta \to 1$, and we recover the *microscopic* Stokes' equation (3) (in which gravity has been neglected). Thus, Brinkman's equation is essentially an interpolation scheme between the microscopic Stokes' equation and the *macroscopic* Darcy's law. Note that, similar to K, the ratio η'/η is a property of the porous medium. The ratio η'/η can be related to a phenomenological *slip velocity* parameter that was found experimentally by Beavers and Joseph (1967), and was justified mathematically by Saffman (1971).

Koplik *et al.* (1983) considered an unconsolidated porous medium modeled by a random collection of spherical and non-overlapping particles, whose volume fraction is ϕ_p, and suggested that for small values of ϕ_p one has

$$\frac{\eta'}{\eta} = 1 - \frac{\phi_p}{2} \,. \tag{8.19}$$

However, the problem of calculating η' at high values of ϕ_p has not been solved yet. Wilkinson (1985) assumed that $\eta' = \eta$, and obtained a non-linear differential equation for the permeability K that should be used in Eq. (18). This equation can be integrated numerically if the porosity and grain size distribution are specified. For the latest developments regarding η' see Martys *et al.* (1994).

8.3 Predicting the permeability, conductivity and diffusivity

The above derivation of Darcy's law allows one, in principle, to calculate the permeability tensor $\mathbf{K}$, if the unknown functions $\mathbf{B}$ and $\mathbf{b}$ can be determined. In practice, however, one can *never* find exact expressions for $\mathbf{B}$ and $\mathbf{b}$, unless the structure of the porous medium is simple enough to allow exact calculation of $\mathbf{B}$ and $\mathbf{b}$ to be carried out. Most of the classical empirical or semi-empirical methods of estimating permeability of porous media are discussed by Bear (1972) and

Scheidegger (1974), and are not discussed here. Here we review and discuss more recent methods of estimating permeability, electrical conductivity, and effective diffusivity of fluid-saturated porous media. Most of these methods are applicable to determinination of all of these properties.

8.3.1 Continuum models: Exact results and rigorous bounds

Before discussing exact results, we should define precisely what is meant by exact. There are currently no exact expressions for the permeability and the conductivity of porous media with an *arbitrary microstructure*. Therefore, when we refer to exact results, we mean the results that have been obtained *for a given morphology of the pore space*. Some of the rigorous bounds that are discussed here can be used with porous media with arbitrary morphology, although in most cases they become useful if they are applied to a specific model porous medium with relatively simple morphology.

Most of the exact results for the permeability of porous media are for periodic arrays of spherical particles of radius r which are placed at the nodes of a regular lattice such as a simple-cubic lattice. These models are idealizations of unconsolidated porous media and are discussed in Chapter 13. A notable exception is the work of Larson and Higdon (1989), which was mentioned in Chapter 6. These authors considered the case in which the spheres in the cubic packing were allowed to grow. Figure 8.2 shows the model they considered. For this model, Larson and Higdon expressed the velocity and pressure fields in terms of harmonic expansions in spherical coordinates, as described by Happel and Brenner (1983). In this method, one writes

$$P = \sum_{n=-\infty}^{\infty} p_n \quad , \tag{8.20}$$

$$\mathbf{v} = \sum_{n=-\infty}^{\infty} [\nabla \times (\mathbf{r}\chi_n) + \nabla \psi_n + \alpha_n r^2 \nabla p_n + \beta_n \mathbf{r} p_n] \quad , \tag{8.21}$$

where $\alpha_n = (n+3)/2(n+1)(2n+3)\eta$, and $\beta_n = n/(n+1)(2n+3)\eta$, and p_n, χ_n, and ψ_n are solid spherical harmonics, which are written in terms of the associated Legendre functions in the form $r^n P_n^m(\cos\theta) \exp(im\psi)$, where P_n^m is the Legendre function. These expressions are used in the continuity and Stokes' equations, the geometry of the packing is specified and the numerical solution of the problem is obtained. For high concentrations of the spheres (low porosities) an analytical asymptotic expression was also derived by Larson and Higdon.

Childress (1972), Howells (1974), and Hinch (1977) considered flow through a *random* array of spheres and obtained an asymptotic expression for the permeability of the system for low values of ϕ_p, the volume fraction of the spheres.

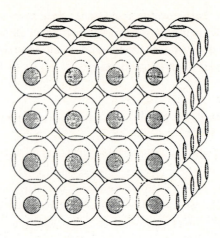

Figure 8.2: Schematic view of the consolidated model of Larson and Higdon.

Hinch's (1977) results can be summarized by the following equation

$$\frac{K_s}{K} = 1 + \frac{3}{\sqrt{2}}\phi_p^{1/2} + \frac{135}{64}\phi_p \log \phi_p + 16.456\phi_p \ldots \quad , \qquad (8.22)$$

where $K_s = 2r^2/(9\phi_p)$ is called the Stokes permeability. Actually, Hinch's result contained a few numerical errors which were corrected by Kim and Russel (1985), who also derived a few higher order terms of the above expansion. Adler and Brenner (1984a,b,c) studied a variety of transport processes in spatially-periodic capillary networks, and derived some rigorous results. However, due to the periodic nature of their networks, the usefulness of their results is limited.

There are also a variety of rigorous results for the conductivity of a model random medium, which is usually a two-component mixture of spheres (or another relatively simple geometrical object such as an ellipse or a cylinder) of conductivity σ_s, and the matrix (the channels between the spheres) in which the spheres are embedded, whose conductivity is σ_m. Usually the system has a periodic structure, but it can also have a more complex configuration. Various methods have been used to determine rigorously the conductivity σ of the system as a function of σ_s/σ_m (see, Bonnecaze and Brady, 1990, and Torquato, 1991, for extensive lists of references). For our problem, if we assume that the matrix is where transport takes place and the spheres represent the solid matrix of the medium, we would be interested in the limit $\sigma_s/\sigma_m = 0$. However, almost none of the results obtained with these models are applicable to this limit. They cannot provide an accurate estimate of the conductivity of the fluid-saturated porous medium in this limit.

Another set of rigorous results is obtained when, instead of trying to solve the problem completely and exactly, one obtains upper and lower bounds to the properties of interest. Prager (1961) and Weissberg and Prager (1962, 1970) pioneered this approach for calculating upper and lower bounds to K. [Note that Berryman and Milton (1985) corrected an error in the original results of Weissberg and Prager.] They proposed that certain bounds on K, which depend upon various distribution functions that statistically characterize the medium, may be used for estimating K for a wide range of sphere volume fraction ϕ_p. Weissberg and Prager (1962, 1970) evaluated these bounds for a model in which the centers of the spheres are distributed at random, the so-called *fully-penetrable sphere model*, or the Swiss-Cheese model discussed in Chapters 3 and 6. Torquato and Beasley (1987) considered the general case in which the spheres have an arbitrary degree of impenetrability, characterized by a parameter λ, where $\lambda = 0$ corresponds to the fully-penetrable model, whereas $\lambda = 1$ represents the case of completely-impenetrable spheres (see Chapter 6). Torquato and Beasley (1987) derived useful upper and lower bounds for the permeability of the system as a function of λ. In a later paper, Torquato and Lu (1990) derived upper and lower bounds for a polydispersed system of spherical particles.

We now summarize the most recent and accurate bounds for K, and refer the interested reader to Torquato (1991) for more details and bounds for other transport properties. For fully-penetrable sphere model, or the Swiss-Cheese model, discussed in Chapters 3 and 6, with identical spheres the lower bound of Weissberg and Prager is given by

$$\frac{K_s}{K} \geq -\frac{\ln\phi}{\phi\phi_p} \; . \tag{8.23}$$

For *dilute* random arrays of identical spheres in the penetrable-concentric-shell model, or the cherry-pit model discussed in Chapter 6, Torquato and Beasley (1987) derived the following lower bound

$$\frac{K_s}{K} \geq 1 + \left[\frac{3}{2} - \lambda^6 + \frac{11}{4}\lambda^4 + \frac{5}{6}\lambda^2 - \frac{9}{8}\lambda + \frac{3}{4}(1 + 3\lambda^2)\ln(1 + 2\lambda) - \frac{\lambda(1 + 7\lambda)}{16(1 + 2\lambda)^2} \right] \phi_p \tag{8.24}$$

which is a relatively accurate bound for $0.52 \leq \lambda \leq 1$, while for $\lambda \leq 0.52$ they derived the following lower bound

$$\frac{K_s}{K} \geq 1 + \frac{5}{8}(3 + 5\lambda) + O(\phi_p^2) \; , \tag{8.25}$$

which is more accurate than (24). Note that these bounds can be written in terms of ϕ, the porosity of the system, using the relations between ϕ and ϕ_p discussed in Chapter 6. For a system of polydispersed and overlapping spheres whose radii follow the Schulz distribution

$$f(r) = \frac{1}{\Gamma(\alpha + 1)} \left(\frac{\alpha + 1}{\langle r \rangle} \right) r^2 \exp\left[-\frac{(\alpha + 1)r}{\langle r \rangle} \right] \; , \quad \alpha > -1 \tag{8.26}$$

where $\Gamma(x)$ is the gamma function, $\langle r \rangle$ is the mean sphere radius, and α is a measure of polydispersivity ($\alpha = \infty$ corresponds to monodispersed spheres),

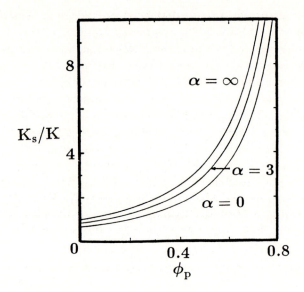

Figure 8.3: Lower bounds on the permeability K of a model of polydispersed and overlapping spheres whose radii follow the Schulz distribution, Eq. (26), versus the spheres volume fraction ϕ_p. $\alpha = \infty$ corresponds to a monodispersed system (after Torquato and Lu, 1990).

Torquato and Lu (1990) calculated a lower bound to K. Their results are shown in Fig. 8.3. An interesting result of Torquato and Beasley (1987) was that the bounds that incorporate a certain level of statistical information on the medium are *not* necessarily more accurate than those which incorporate *less* information. It should be mentioned that the Kozeny-Carman empirical formula

$$\frac{K_s}{K} = \frac{10\phi_p}{(1 - \phi_p)^3} \ , \tag{8.27}$$

is relatively close to the lower bound for the permeability of the random sphere model. For a comprehensive discussion of Eq. (27) see Scheidegger (1974).

Consider now a problem in which M_0 fluid particles, initially distributed uniformly in the pore space, are allowed to diffuse randomly (with diffusivity D) towards the pore surface, but react and are removed as soon as they reach the surface. If we denote the time dependent population of the particles as $M(t)$, the decay of the particles can be represented by the series

$$M(t) = \sum_{n=1}^{\infty} I_n \exp\left(-\frac{t}{t_n}\right) \ , \quad t_1 \geq t_2 \geq t_3 \cdots \tag{8.28}$$

where I_n and t_n are the amplitude and lifetime associated with the nth mode. The average lifetime $\langle t \rangle$ can be written as

$$\langle t \rangle = \sum_{n=1}^{\infty} I_n t_n \leq t_1 \ . \tag{8.29}$$

Torquato (1990) and Avellaneda and Torquato (1991) proposed the following upper bounds

$$K \leq \phi D \langle t \rangle \,, \tag{8.30}$$

$$K \leq \frac{\phi D t_1}{T_p} \,, \tag{8.31}$$

where T_p is the tortuosity factor of the porous medium discussed in Chapter 5. Since $\langle t \rangle$ is not directly related to the connectivity of the pore space, we do not expect (30) to provide a sharp upper bound to K. To see this, note that if larger pores of the medium are gradually disconnected from one another, K would decrease rapidly, whereas $\langle t \rangle$ would not change much. For this reason, bound (31) is more attractive because the tortuosity factor does depend on the connectivity of the pore space and, as discussed in Chapter 5, diverges as the percolation threshold is approached. This observation led Torquato and Kim (1992) to propose that

$$K \leq \frac{\phi D \langle \dot{t} \rangle}{T_p} \tag{8.32}$$

which in some sense is a hybrid of (30) and (31). Although these bounds establish interesting connections between K and reaction properties of a porous medium, they provide weak estimates of K, so that their practical usefulness is limited; see Kostek *et al.* (1992) and Schwartz *et al.* (1993) for a discussion of this.

The main problem with much of such rigorous bounds is that they are not very useful for the highly disordered porous media (some with fractal properties) that are of interest here. For example, the above upper and lower bounds are not useful if the porous medium is close to its percolation threshold. In fact, no bound, no matter how accurate, can predict the existence of a non-trivial (not zero or unity) percolation threshold, i.e., it cannot predict that K or σ would vanish at a porosity other than zero.

8.3.2 Continuum models: Field-theoretic and perturbation methods

It has been recognized for a long time that disorder is equivalent to a field, i.e., for flow in a porous medium heterogeneities of the medium can be thought of as random forces that act on fluid particles flowing through the medium. This fact has been exploited in the condensed matter literature for a long time for estimating transport properties of disordered materials. However, its use for flow phenomena in porous media is relatively new and has been attempted only recently. King (1987a) was the first to develop an approach based on this idea, usually called a *field-theoretic approach*, for estimating permeability of heterogeneous porous media.

For steady flow, Darcy's law together with the equation of continuity yield

$$\nabla \cdot (K \nabla P) = 0 \,. \tag{8.33}$$

Without loss of generality, the permeability can be taken to be isotropic, since the permeability tensor $\mathbf{K}$ [see Eq. (15)] is real and symmetric, and therefore can be diagonalized by using normal coordinates, which may be rescaled to ensure that it is isotropic. We may define a Green function G for Eq. (33) by

$$\nabla_r \cdot [K(\mathbf{r})\nabla G(\mathbf{r},\mathbf{r}')] = \delta(\mathbf{r}-\mathbf{r}') \quad . \tag{8.34}$$

If we use the Neumann condition of constant flux and the Green theorem, we obtain

$$P(\mathbf{r}) = \mathbf{v} \cdot \int G(\mathbf{r},\mathbf{r}')d\mathbf{S}' \quad . \tag{8.35}$$

Suppose that we have a homogeneous porous medium with permeability K_0. For this system the Green function is given by

$$K_0 \nabla_r^2 G_0(\mathbf{r},\mathbf{r}') = \delta(\mathbf{r}-\mathbf{r}') \quad . \tag{8.36}$$

Thus if we write $K(\mathbf{r}) = K_0 + \epsilon(\mathbf{r})$, where $\epsilon(\mathbf{r})$ is the perturbation, we obtain

$$K_0 \nabla_r^2 G = \delta(\mathbf{r}-\mathbf{r}') - \nabla \cdot (\epsilon \nabla G) \quad . \tag{8.37}$$

We now recognize that, according to the definition of a Green function, G_0 is the inverse of the operator $K_0 \nabla^2$, and exploit this to rewrite Eq. (37) as an integral equation

$$G(\mathbf{r},\mathbf{r}') = G_0(\mathbf{r},\mathbf{r}') - \int G_0(\mathbf{r},\mathbf{r}'')K_0\nabla_{r''} \cdot \epsilon(\mathbf{r}'')G(\mathbf{r}',\mathbf{r}'')d\mathbf{r}'' \quad . \tag{8.38}$$

If we take the Fourier transform of Eq. (38), we obtain

$$G(\mathbf{i},\mathbf{j}) = G_0(\mathbf{i})\delta(\mathbf{i}+\mathbf{j}) + G_0(\mathbf{i})\int M(\mathbf{i};\mathbf{l},\mathbf{m})\epsilon(\mathbf{l})G(\mathbf{m},\mathbf{i})d\mathbf{l}d\mathbf{m} \quad , \tag{8.39}$$

where

$$M(\mathbf{i};\mathbf{l},\mathbf{m}) = K_0[(\mathbf{l}+\mathbf{m})\cdot\mathbf{m}]\delta(\mathbf{l}+\mathbf{m}-\mathbf{i}) \quad . \tag{8.40}$$

This provides an iterative scheme by which a perturbation expansion for the Green function may be developed. Such expansions can be truncated at any order and have been tried by several authors (see, for example, Gutjahr *et al.*, 1978, and references therein). However, what one is interested in is not the permeability itself but its average, where the averaging is taken with respect to the distribution of the permeabilities. Hence, one has to average the perturbation series, as was done by King (1987a). If one is to average the nth term of Eq. (38), one needs to know the nth moment of $\epsilon(\mathbf{r})$. Having calculated the average Green function, the average permeability can be estimated, which can provide valuable information about the mean pressure field in the medium, since from Eq. (35) we can obtain

$$\langle P(\mathbf{r}) \rangle = \frac{\mathbf{v}}{\langle K \rangle} \cdot \int |\mathbf{r}-\mathbf{r}'|^{-1}d\mathbf{S}' \quad , \tag{8.41}$$

where $\langle K \rangle$ is the average permeability, and $|\mathbf{r}-\mathbf{r}'|^{-1}$ is the Green function for this problem in three dimensions. We should keep in mind that $G(\mathbf{r}, \mathbf{r}') = \ln|\mathbf{r} - \mathbf{r}'|$ in two dimensions.

In principle, this method is applicable to *both* macroscopic and megascopic porous media. King (1987a) showed that this method can reproduce several known exact results. For example, it is known (Matheron, 1967) that if the permeability distribution of a megascopic porous medium is log-normal, then the average permeability in two dimensions is exactly the geometric mean of the distribution, which this model also reproduces. Several other results obtained by various variants of perturbation expansions (Bakr *et al.*, 1978; Dagan, 1981, 1982a,b; Gutjahr and Gelhar, 1981; Mizell *et al.*, 1982) can also be reproduced by this formulation.

The main condition for the validity of this method is that the Green function $G(\mathbf{r}, \mathbf{r}')$ must be analytic around $G_0(\mathbf{r}, \mathbf{r}')$, i.e., it must have a Taylor-series expansion. This condition is satisfied if the perturbed system is qualitatively similar to the unperturbed one. The condition for this is that the porous medium should not be close to its percolation threshold (i.e., it should not have a lot of nearly impermeable regions), where the unperturbed system is not similar to the true, perturbed medium, or there should be no long-range correlations in the medium. Thus, for example, the permeability distribution cannot be fractal (see Chapter 6), if this method is to be useful.

8.3.3 Exact formulation and effective-medium approximations

The idea that in order to calculate effective transport properties of a disordered medium, one may replace the medium with a hypothetical homogeneous one which somehow mimics the behavior of the disordered medium, and then calculate the homogeneous medium's properties, has a long history. Of course, if such a substitution is possible, the problem of calculating effective transport properties of the homogeneous medium is much simpler than the original problem, as this replacement is only an approximation to the original problem. There are two approaches for implementing this idea. Maxwell-Garnett (1904) developed an approach which is applicable to the case when isolated inclusions are embedded in a continuous matrix consisting of a single phase; the effective properties of the disordered system are derived by placing a sphere (or an ellipse) of the homogeneous or the effective medium in this matrix. This is usually called the average *t*-matrix, or *non-self-consistent approximation*. In the second approach developed by Bruggeman (1935) each inhomogeneity is embedded in the effective medium itself, the unknown properties of which are determined in such a way that the volume average over all inhomogeneities yields no extra fields in the medium. This is usually called the *effective-medium approximation* (EMA), and it is this kind of approximation that is the focus of this section. In effect

EMA is an ingenious way of transforming a many-body problem into a one-body problem. Bruggeman's EMA, which was derived for a continuum model of disordered media, was rederived independently by Landauer (1952) [see Landauer (1978) for a history of EMAs].

We now consider network models of porous media, and derive an exact but implicit solution for the problem of unsteady state conduction and diffusion in a such networks. We then show that EMA is simply the simplest (the lowest-order) approximation to this implicit solution in a heirarchy of approximate solutions. Consider diffusion in a disordered porous medium. Because the medium is disordered, the diffusivity D varies in space, but we assume that it is independent of the time t. Then the governing equation for concentration C of the diffusants is given by (see Chapter 2)

$$\frac{\partial C}{\partial t} = \nabla \cdot (D \nabla C) . \tag{8.42}$$

If we discretize this equation by finite difference or finite element, we obtain

$$\frac{\partial C_i}{\partial t} = \sum_j W_{ij}[C_j(t) - C_i(t)] , \tag{8.43}$$

where $C_i(t)$ is the concentration at site or grid point i of the system, W_{ij} is proportional to the *local* diffusivity D_{ij} between i and j, as we have assumed that the diffusivity varies in space but is independent of t, and the sum is over all sites j that nearest neighbors of i. For example, for a one-dimensional system, $W_{ij} = (D_i + D_j)/(2\ell^2)$, where D_i is the diffusivity at i and ℓ is the distance between i and j. Observe that Eq. (43) is exactly equivalent to a pore network in which $W_{ij}[C_j(t) - C_i(t)]$ is just the diffusive flux in the pore connecting sites i and j, and $\partial C_i/\partial t$ represents the accumulation of concentration at i. Suppose that the initial concentration is $C_i(t = 0) = C_0 \delta_{i0}$. If we normalize C_i by C_0 and define $P_i = C_i/C_0$, then Eq. (43) can be rewritten as

$$\frac{\partial P_i}{\partial t} = \sum_j W_{ij}[P_j(t) - P_i(t)] . \tag{8.44}$$

Equation (44) can be given a probabilistic interpretation. Since P_i is positive and less than unity, it can be interpreted as the probability that a diffusing particle will be found at i at time t. With this interpretation W_{ij} is the *transition rate*, i.e., the probability that the diffusing particle diffuses from i to j, which we take it to be non-zero only when sites i and j are nearest neighbors. Equation (44) is usually referred to as the *master equation*. This correspondence between a discretized diffusion or conduction equation and the master equation indicates that diffusion and conduction in disordered porous media can be simulated by a random walk process. We now explain how this simulation is done.

In order to solve a diffusion or conduction problem in a disordered network by a random walk, one places a particle or diffusant at a site i of the network at position $\mathbf{r_0}$ at time $t = 0$. The particle is then allowed to move throughout the system and explore it. At every time step the particle selects one of the bonds $i-j$ (pore throats in a porous medium) that are connected to its present site i (pore body in a porous medium), with a probability proportional to the conductance of the bond (which is proportional to W_{ij}), and makes a jump (transition) to site j. Once at the new site, the particle selects its next step by the same method, makes another transition, and so on. This constitutes the random walk process. Each time the particle makes a transition from one site to another, its current position $\mathbf{r}$ is determined and $R^2(t) = (\mathbf{r} - \mathbf{r_0})^2$ is calculated. One repeats this procedure for a large number of particles and a large number of time steps, and calculates $\langle R^2(t) \rangle$, the mean square displacement of the particles at time t, where the averaging is taken with respect to all particles. In general, it can be shown (see, for example, Hughes, 1994) that for a d-dimensional system

$$\langle R^2(t) \rangle = 2dDt . \tag{8.45}$$

Thus, if we calculate $\langle R^2(t) \rangle$ we can also determine D, and thus the effective conductivity σ of the system. If as $t \to \infty$, $R = \sqrt{\langle R^2(t) \rangle}$ is much larger than any other characteristic length scale of the system, then the mean square displacement will vary linearly with t, we obtain a constant diffusivity (and conductivity), diffusion is Fickian, and the long-time behavior of the system is described by the classical diffusion equation, $\partial C/\partial t = D\nabla^2 C$. For example, if the conductivities of a megascopic porous medium are distributed according to a distribution with a correlation length ξ_σ, then for $R >> \xi_\sigma$ diffusion will be Fickian. However, if R cannot become larger than the dominant length scale of the system as t becomes large, then $\langle R^2(t) \rangle$ will not vary linearly with t, D will be time dependent, and the long-time behavior of the system cannot be described by the classical diffusion equation. In this section we use an EMA to develop approximate analytical expressions for $\langle R^2(t) \rangle$. Later in this Chapter we describe how an efficient computer simulation of random walk solution of diffusion and conduction in heterogeneous media is carried out.

Taking the Laplace transform of Eq. (44) yields

$$\omega \tilde{P}_i(\omega) - \delta_{i0} = \sum_j W_{ij}[\tilde{P}_j(\omega) - \tilde{P}_i(\omega)] , \tag{8.46}$$

where ω is the Laplace transform variable conjugate to t. We now introduce an *effective-medium network* with all transition rates equal to $\tilde{W}_e(\omega)$, and site occupation probabilities $\tilde{P}_i^e(\omega)$, so that

$$\omega \tilde{P}_i^e(\omega) - \delta_{i0} = \sum_j \tilde{W}_e(\omega)[\tilde{P}_j^e(\omega) - \tilde{P}_i^e(\omega)] , \tag{8.47}$$

where the effective transition rate $\tilde{W}_e$ is yet to be determined. Observe that $\tilde{W}_e(\omega)$ is assumed to be time dependent. The reason for this becomes clear shortly. If we subtract Eqs. (46) and (47), after some rearrangements we obtain

$$(Z_i + \epsilon)[\tilde{P}_i(\omega) - \tilde{P}_i^e(\omega)] - \sum_j [\tilde{P}_j(\omega) - \tilde{P}_j^e(\omega)] = -\sum_j \Delta_{ij}[\tilde{P}_i(\omega) - \tilde{P}_j(\omega)] , \quad (8.48)$$

where Z_i is the coordination number of site i, $\epsilon = \omega/\tilde{W}_e$, and $\Delta_{ij} = (W_{ij} - \tilde{W}_e)/\tilde{W}_e$. A *site-site* Green function G_{ij} is now introduced by the equation

$$(Z_i + \epsilon)G_{ik} - \sum_j G_{jk} = -\delta_{ik} , \quad (8.49)$$

in terms of which Eq. (48) is rewritten as

$$\tilde{P}_i(\omega) = \tilde{P}_i^e(\omega) + \sum_j \sum_k G_{ij}\Delta_{jk}[\tilde{P}_j(\omega) - \tilde{P}_k(\omega)] . \quad (8.50)$$

Equation (50) is an exact but implicit solution of (44) which applies to *any* network, irrespective of its dimensionality or topological structure (since we have not specified d or Z_i). It tells us that the solution for a disordered system is the sum of two terms, one of which corresponds to the solution of the same problem in an effective or homogeneous network (which somehow mimics the behavior of the disordered network), while the other term represents the fluctuations in the solution that arise as the result of having inhomogeneities in the system. For topologically-random networks, such as the Voronoi network (see Chapter 3), further progress requires a statistical treatment of the Green function appropriately coupled to the disorder of the network. Thus, we limit our attention to regular networks for which the coordination number is constant. For a simple-cubic network in d dimensions ($Z = 2d$), the construction of the Green function is particularly straightforward. If we label by d integers $(m_1, m_2, \cdots, m_d)$ the relative positions of two sites i and j on the network, separated by m_d bond lengths along the principal axis d, then the Green function G_{ij} is given by

$$G_{ij} = G(m_1, m_2, \cdots, m_d) = -\frac{1}{2}\int_0^\infty \exp[-x(Z + \epsilon)/2] \prod_{i=1}^d I_{m_i}(x) dx . \quad (8.51)$$

Here $I_{m_i}(x)$ denotes the modified Bessel function of order m_i. The Green function for many other networks is given by Sahimi *et al.* (1983b). Equation (50) can also be rewritten in terms of the "flux" $\tilde{Q}_{ij}(\omega) = \tilde{P}_i(\omega) - \tilde{P}_j(\omega)$. If we denote bonds with Greek letters and assign directions to them and let $\gamma_{\alpha\beta} = (G_{il} + G_{jk}) - (G_{jl} + G_{ik})$ be a *bond-bond* Green function, where i and l (j and k) are the network sites with tails (heads) of arrows on bonds α and β, respectively, then Eq. (50) can be written as (Sahimi *et al.*, 1983b)

$$\tilde{Q}_\alpha(\omega) = \tilde{Q}_\alpha^e(\omega) + \sum_\alpha \Delta_\beta \gamma_{\alpha\beta}\tilde{Q}_\beta(\omega) . \quad (8.52)$$

The effective transition rate $\tilde{W}_e$ is calculated by the following *self-consistency condition*. Solve Eq. (52) for an arbitrary set of individual transition rates, i.e., for arbitrary Δ_β. Select any bond α and require that

$$\langle \tilde{Q}_\alpha \rangle = \tilde{Q}_\alpha^e , \quad (8.53)$$

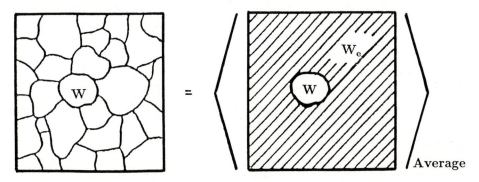

Figure 8.4: Schematic implementation of the EMA, transforming a disordered medium (left) to a homogeneous one.

where $\langle \cdot \rangle$ denotes an averaging over all possible transition rates of all bonds. Equation (53) requires that the fluctuation in the flux in bond α from its value in the effective network vanish on the average. If we substitute the effective transition rate so obtained in Eq. (46) and invert the Laplace transform, we obtain

$$\frac{\partial P_i}{\partial t} = \sum_j \int_0^t W_e(t - \tau)[P_j(\tau) - P_i(\tau)]d\tau . \qquad (8.54)$$

Thus, if we can determine $\tilde{W}_e$ exactly or approximately, we can replace the master equation (44) by a *generalized master equation*, Eq. (54). If, instead of a network, we use a continuum model and carry out an analysis similar to the above, then the analog of Eq. (54) for a continuum model would be

$$\frac{\partial C}{\partial t} = \int_0^\infty D_e(t - \tau)\nabla^2 C(\mathbf{x}, \tau)d\tau , \qquad (8.55)$$

where D_e is the analog of W_e and $\mathbf{x}$ is a point in space. Equations (54) and (55) indicate that matching a heterogeneous medium to a uniform medium *induces memory*. However, this does not necessarily imply that the long-time diffusivity of the system is time dependent. As we show below, whether the long-time diffusivity of a porous medium is constant or depends on t is dependent upon the distribution of the heterogeneities of the medium. The Laplace transform of mean square displacement $\langle \tilde{R}^2(\omega) \rangle$ of a diffusing particle is given by (Sahimi *et al.*, 1983b)

$$\langle \tilde{R}^2(\omega) \rangle = \frac{Z}{\omega^2}\tilde{W}_e(\omega) , \qquad (8.56)$$

and therefore Eqs. (45) and (56) enable us to calculate the effective diffusivity of the medium.

In practice, however, we cannot solve Eq. (53) for an *arbitrarily large* set of bonds, as the computations become too complex, and thus we have to develop approximate schemes. In an EMA approach, one assigns to all but a finite cluster of bonds in the network the effective transition rate $\tilde{W}_e(\omega)$ (so that $\Delta_\beta \neq 0$ only for a finite set of bonds), and proceeds as above, now averaging over the transition rates of the bonds in the cluster to determine $\tilde{W}_e(\omega)$. Figure 8.4 shows the schematic representation of this idea. In the simplest case the cluster contains only one bond, and obviously one can also consider a cluster with two bonds, three bonds, etc., each of which corresponds to a distinct level of approximation. These approximations constitute the heirarchy that was mentioned above. To obtain rapid convergence of the results with increasing cluster size to the infinite system results, it is important to choose a suitably symmetrical cluster of bonds whose transition rates are to be allowed to fluctuate. These are discussed in detail by Sahimi (1984). In the simplest approximation, only a single bond α has a transition rate W differing from $\tilde{W}_e(\omega)$ (this is called *single-bond* EMA), in which case Eq. (53) simplifies to

$$\langle \frac{1}{1 - \gamma_{\alpha\alpha}\Delta_\alpha} \rangle = 1 , \tag{8.57}$$

where $\gamma_{\alpha\alpha} = -2/Z + (2\epsilon/Z)G(\epsilon)$ and $G(\epsilon) = -G_{ii}(\epsilon)$. If $f(W)$ is the transition rate distribution, Eq. (57) becomes

$$\int_0^\infty \frac{f(W)}{1 - \gamma_{\alpha\alpha}\Delta_\alpha} dW = 1 . \tag{8.58}$$

How accurate is this EMA? We consider percolation disorder, i.e., a system in which a randomly-selected fraction p of the bonds are open (that is, they allow diffusion and flow to occur), and the rest are closed (they do not allow diffusion and flow to take place). Thus, we use

$$f(W) = (1 - p)\delta_+(W) + ph(W) , \tag{8.59}$$

where $h(W)$, the distribution of the transition rates of the open bonds, can be any normalized probability density function. In this limit Eq. (58) becomes (Odagaki and Lax, 1981; Webman, 1981; Sahimi *et al.*, 1983b)

$$p\int_0^\infty \frac{h(W)}{1 - 2/Z + 2G\epsilon/Z + 2(1 - \epsilon G)(W/\tilde{W}_e)/Z} dW = \frac{p - 2/Z + 2G\epsilon/Z}{1 - 2/Z + 2G\epsilon/Z} . \tag{8.60}$$

Equation (60) predicts that as long as $p > p_c$, $W_e = \tilde{W}_e(\omega \to 0)$ exists and is a constant, so that Eq. (56) predicts that

$$\langle R^2(t) \rangle = ZW_e t , \tag{8.61}$$

which, together with Eq. (45) imply that $D = ZW_e/2d$. That is, we have classical diffusion with a constant diffusivity which is described by the usual diffusion equation. Figure 8.5 shows typical results for the time dependence of $\tilde{W}_e$ obtained with Eq. (60).

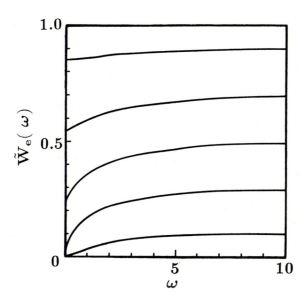

Figure 8.5: The behavior of $\tilde{W}_e$ as a function of the Laplace transform variable ω. The curves are for, from top to bottom, $p = 0.9$, 0.7, 0.5, 0.3, and 0.1.

What happens at $p = p_c$? At this point the sample-spanning cluster of the network is a fractal object, and the correlation length ξ_p of the network is divergent (see Chapter 3). Thus, $R = \sqrt{\langle R^2(t) \rangle}$ is always less than ξ_p, and since ξ_p is the dominant length scale of the system, our discussion given above indicates that diffusion cannot be Fickian. Indeed, it is not difficult to show that in this limit Eq. (60) predicts, for example, that for a simple-cubic network (Sahimi *et al.*, 1983b)

$$\tilde{W}_e(\omega) \sim \left[\frac{p_c G(0)}{(1 - p_c)h_{-1}} \right]^{1/2} \omega^{1/2} , \qquad (8.62)$$

where $h_{-1} = \int_0^\infty dW \, h(W)/W$. Therefore, if we use Eq. (56), we immediately obtain the following equation for large times

$$\langle R^2(t) \rangle \sim \frac{12}{\pi^{1/2}} \left[\frac{p_c G(0)}{(1 - p_c)h_{-1}} \right] t^{1/2} , \qquad (8.63)$$

which indicates that the mean square displacement of the diffusants grows *slower than linearly* with t. Therefore, our EMA predicts correctly that diffusion in a percolation network at $p = p_c$ is not Fickian. If we attempt to obtain an effective diffusivity by writing $D = \langle R^2(t) \rangle/(6t)$, Eq. (63) would predict that $D \sim 1/\sqrt{t}$, and therefore $D \to 0$ as $t \to \infty$. That is, such a diffusion process is so slow that the long-time diffusivity of the system is *zero*. This slowness is of course caused by the highly heterogeneous structure of a percolation network. Note that, even if the percolation network is above its percolation threshold, as long as $R < \xi_p$, diffusion is not Fickian. This also explains why $\tilde{W}_e$ has to depend on ω. Had we used a constant W_e in Eq. (47), Eq. (60) would have predicted that we *always* have normal (Fickian) diffusion regardless of the value of p, whereas percolation theory tells us that for $p \leq p_c$ the diffusivity is zero, since no sample-spanning cluster of open bonds exists. This non-Fickian diffusion is called *anomalous diffusion* (Gefen *et al.*, 1983), or *fractal diffusion* (Sahimi *et al.*, 1983b). We introduce a *random walk fractal dimension* D_w by

$$\langle R^2(t) \rangle \sim t^{2/D_w} , \qquad (8.64)$$

so that normal or Fickian diffusion corresponds to $D_w = 2$. It is then obvious that EMA and Eq. (63) predict a *universal value* $D_w = 4$ (independent of the details of the system) for diffusion in percolation networks at p_c (or above p_c for $R < \xi_p$). Equation (63) is valid as long as h_{-1}, or more generally f_{-1}, is finite. However, D_w will *not* be universal if h_{-1} (or f_{-1}) is divergent. For example, if $h(W) = (1-a)W^{-a}$, where $0 < a < 1$ (for which h_{-1} is divergent), EMA predicts (Sahimi *et al.*, 1983b) that at $p = p_c$

$$\langle R^2(t) \rangle \sim t^{(1-a)/(2-a)} , \qquad (8.65)$$

which reduces to Eq. (63) in the limit $a = 0$. Equation (65) implies that D_w depends heavily on the details of the distribution function. We should point out that such power-law distributions are not uncommon. In fact, the conductance distribution of the conducting channels in the Swiss-Cheese model discussed in Chapters 3 and 6 has this form (Halperin *et al.*, 1985), and therefore such distributions are of practical importance.

At steady state, i.e., in the limit $t \to \infty$ (or, equivalently, in the limit $\omega \to 0$), we have $\gamma_{\alpha\alpha} = -2/Z$ and Eq. (58) reduces to

$$\int f(W) \frac{W - W_e}{W + (Z/2 - 1)W_e} dW = 0 \quad . \qquad (8.66)$$

In this limit $f(W)$ is simply the distribution of the bond (pore) conductance W [since in this limit W can be interpreted as the bond conductance; consider Eq. (44) at steady state], and therefore $\sigma = W_e$ is the effective conductance of the system. Equation (66) was first derived by Kirkpatrick (1971, 1973). The same equation had been derived much earlier by Bruggeman (1935) and Landauer (1952) for a continuum in which the factor $Z/2$ had been replaced by d, the dimensionality of the system. In the original Bruggeman-Landauer theory, in order to derive the EMA a sphere was embedded in the effective medium, instead of a single bond in the network model, and an equation similar to (53) was solved. To treat anisotropic media, Hori and Yonezawa (1977), Thorpe and Sen (1985), and Xia and Thorpe (1988) extended the Bruggeman-Landauer theory to the case in which instead of a sphere an ellipse of a given aspect ratio was used. Moreover, Stroud (1975) showed that the conductivity of the sphere need not be isotropic, but can be tensorial.

Using Eq. (66), we can calculate the permeability and diffusivity of a porous medium. Consider an effective medium network where each bond or pore has a conductance $\sigma = W_e$, since in an EMA treatment of the problem all pores or bonds have the same conductance σ. We fix the pressures at two opposite faces of the network so as to produce an average pressure gradient $\langle \nabla P \rangle$. The total fluid flux Q crossing any plane perpendicular to $\langle \nabla P \rangle$ is the sum of the individual fluxes in the bonds intersecting the plane. Each pore flux is the pressure difference across it times σ/η. If we approximate the local pressure difference as the projection of the average pressure gradient along the bond length l, we find

$$Q = \sum \frac{\sigma}{\eta} \langle \nabla P \rangle \cdot \mathbf{l} \quad . \tag{8.67}$$

If we divide Q by the area S of the plane, we obtain an average velocity which, when compared with Darcy's law, yields an estimate of the permeability

$$K = \sigma \langle \frac{1}{S} \Sigma \mathbf{l} \cdot \mathbf{n} \rangle \quad , \tag{8.68}$$

where $\mathbf{n}$ is a unit vector along the pressure gradient. But, if the medium is statistically homogeneous and isotropic, any unit vector can be used. Equation (68) shows clearly why for random pore networks K and σ obey the same scaling law near the percolation threshold, as mentioned in Chapter 3. It should be pointed out that in this kind of approach, the pressure drop in a pore body (site) is neglected, and it is assumed that all of the pressure drop occurs in the pore throats (bonds) of the network. Koplik (1982) treated the case in which the pressure drops in both pore bodies and throats are taken into account. The analysis in this case is complex and is ignored here. So long as pore bodies are large and compact and pore throats are long and narrow, the above approximation is valid. To calculate the effective diffusivity D we use the following formula

$$D = \frac{1}{\pi d} \frac{\sigma l}{\langle r^2 \rangle} \phi \, , \tag{8.69}$$

where $\sigma = W_e$ is the EMA prediction of the conductance of the network, d is its dimensionality, l is the length of a pore throat, ϕ is the porosity of the network, r is a pore throat radius, and $\langle r^2 \rangle$ is an average with respect to the pore size distribution. It has been assumed that all the pore throats (bonds) have the same length.

An important test for any theory of transport in porous media is its ability to predict the percolation threshold of the system. Despite its simplicity and lack of detailed information about the microstructure of the porous medium (or the network that represents it) in its structure, EMA can predict the existence of a non-trivial p_c. As mentioned above, EMA predicts that

$$p_{cb} = \frac{2}{Z} \, , \tag{8.70}$$

for the bond percolation threshold of a network. This formula can be obtained easily if we use Eq. (59) in (66), and find that value of p for which $W_e = 0$, since at p_c the effective transport properties vanish. This prediction is accurate for two-dimensional networks, but not so for three-dimensional ones [compare predictions of Eq. (70) with the results given in Tables 3.1 and 3.2]. Moreover, EMA predicts that the critical exponents μ and e for the conductivity and permeability of the system near p_c are both equal one in *all* dimensions, which is an incorrect result (see Table 3.3). This can be seen easily if we use $f(W) = (1-p)\delta_+(W) + p\delta(W - W_0)$ in Eq. (66). We obtain

$$\frac{W_e}{W_0} = \frac{p - 2/Z}{1 - 2/Z} , \tag{8.71}$$

and thus W varies linearly with $p - p_c = p - 2/Z$. Even if we replace $\delta(W - W_0)$ with the more general distribution $h(W)$, we still would obtain $\mu = 1$ since, except in special cases discussed in Chapter 3, the critical exponents are independent of $h(W)$ or any other detail of the system. In general, as Koplik (1981) showed, EMA is very accurate if the system is not close to its p_c, regardless of the structure of $h(W)$. Its predictions are also more accurate for two-dimensional networks than for three-dimensional ones. The performance of EMA near p_c can be improved systematically (Blackman, 1976; Turban, 1978; Ahmed and Blackman, 1979; Sheng, 1980; Sahimi, 1984). For example, a cluster of several bonds with random conductances is embedded in the effective-medium network, instead of a single bond used in the above single-bond EMA, and Eq. (53) is solved to calculate W_e. The most accurate results are obtained with those clusters that preserve the symmetry of the network. Erdös and Haley (1976) showed how different averaging schemes can affect the performance of EMA, and suggested an averaging scheme that would improve EMA's performance.

One can also develop an EMA for the case in which there is a first-order chemical reaction (Sahimi, 1988b). It turns out that if the system is at steady state and the reaction rate coefficient is k_r, all one has to do is to replace ω everywhere with k_r (in this limit W_e is of course independent of time). If the system is not at steady state, we simply replace ω with $\omega + k_r$. Moreover, an EMA has also been extended to anisotropic networks (Bernasconi, 1974), which can be used for calculating the permeability tensor of fracture networks which are usually anisotropic (Harris, 1990, 1992). This is discussed in Chapter 10. An EMA can be developed for site percolation in random networks (Watson and Leath, 1974; Butcher, 1975; Bernasconi and Wiesmann, 1976; Joy and Strieder, 1978, 1979; Sahimi *et al.*, 1984). Finally, EMA has been modified to include the effect of a short-range correlation (Hori and Yonezawa, 1977; Hilfer, 1991a).

How does EMA perform if it is used for analyzing experimental data? Koplik *et al.* (1984) analyzed in detail a Massilon sandstone, used a serial sectioning method to determine an equivalent random network to its pore space and, using this information, and employed Eqs. (66) and (68) to calculate permeability and conductivity of the pore space. They found that the predicted Ks differ

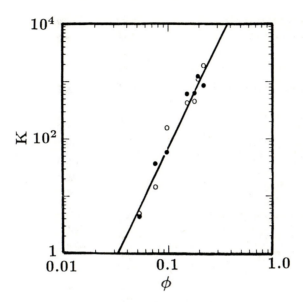

Figure 8.6: The logarithmic plot of dependence of the permeability K (in milli-darcy) of Fontainebleau sandstone on the porosity. The slope of the straight line is 3.8 (after Doyen, 1988).

from the data by about one order of magnitude, while the predicted σs differ by a factor of about 2. They attributed the difference to the fact that most sedimentary porous media, such as Massilon sandstone that they considered, are highly heterogeneous and anisotropic, properties that are not taken into account by Eq. (66). On the other hand, Doyen (1988) analyzed transport properties of Fontainebleau sandstones and used Eqs. (66) and (68) to predict them, and found that their K and σ could be predicted to within a factor of 3. Figure 8.6 compares the predictions of EMA with his data. Many other researchers have used EMA for predicting transport properties of disordered media, a list of whom is too long to be given here.

Before we end this section, we should mention here the work of Hori and Yonezawa (1977) who developed a cumulant expansion method for determining transport properties of disordered media. Their method is more accurate than EMA for three-dimensional media, while EMA is a better approximation for two-dimensional systems. In particular, the bond percolation threshold predicted by the cumulant expansion is given by

$$p_{cb} = 1 - \exp(-2/Z) \quad , \tag{8.72}$$

so that, for example, for the simple-cubic network Eq. (72) predicts, $p_{cb} \simeq 0.283$, which should be compared with the EMA's prediction, $p_{cb} = 1/3$, and the accepted value $p_{cb} \simeq 0.2488$ (see Table 3.2). This more accurate prediction is due to the fact that near p_c clustering of open (conducting) bonds, correlations between different clusters (remember that near the percolation threshold the correlation length ξ_p is very large), and fluctuations among the conductances of various clusters play essential roles. While EMA completely neglects such effects (since the whole system is represented by a uniform medium), the cumulant expansion does, to some extent, take such effects into account.

8.3.4 Position-space renormalization group methods

Our discussions so far should have made it clear to the reader that the main assumption behind EMA is that fluctuations in the potential field are small, since in deriving EMA we require the average of the fluctuations be zero. However, if the fluctuations are large as in, e.g., a fractal porous medium, or one that is near its percolation threshold, or in a megascopic porous medium with a broad distribution of the permeabilities, EMA breaks down and loses its accuracy. In such cases, a position-space renormalization group (PSRG) method is more appropriate, because in averaging the properties of the system this method takes into account the properties of the *pre-averaged* medium. It can also predict power-law scaling laws for the transport properties near the percolation threshold (see Chapter 3), a distinct advantage over EMA which always predicts the critical exponents of the conductivity and permeability to be unity. We describe the PSRG method for a random network model of a pore space, after which its generalization to other more complex systems will be clear.

Consider, for example, a square or a cubic network in which each bond is conducting, or has a non-zero permeability with probability p. The idea in any PSRG method is that, since our network is so large that we cannot calculate its properties exactly, we partition it into $b \times b$ or $b \times b \times b$ cells, where b is the number of bonds in any direction, and calculate their properties, which are hopefully representative of the properties of the original network. The shape of the cell can be selected arbitrarily, but it should be chosen in such a way that it preserves, as much as possible, the properties of the original network. For example, an important topological property of a square network is that it is *self-dual*. The dual of a two-dimensional network is obtained by connecting the centers of the neighboring polygons that constitute the network. For example, if we connect the centers of the hexagons in a hexagonal network, we obtain the triangular network. Thus these networks are dual of each other. However, if we connect the centers of the squares in a square network, we again obtain a square network, and thus this network is self-dual. This self-duality plays an important role in the percolation properties of the square network, and thus it is desirable to partition the network into self-dual cells. Figure 8.7 shows an example of such $b = 2$ cells for the square and cubic network, where the two-dimensional cell is self-dual.

The next step in a PSRG method is to replace each cell with one bond in

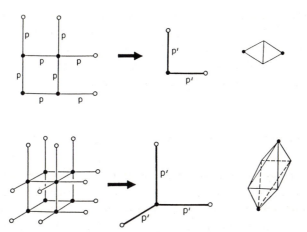

Figure 8.7: RG cells (left), their renormalized configurations (middle), and their equivalent electrical circuits (right).

each principal direction. If in the original network each bond is conducting with probability p, then the bonds that replace the cells would be conducting with probability $p' = R(p)$; this is also shown in Fig. 8.7. $R(p)$ is called the *renormalization group transformation*, and is the *sum* of the probabilities of *all* conducting configurations of the RG cell, which is obtained as follows. Since we are interested in percolation and transport in our network, and since the RG cell is supposed to represent our network, we solve the percolation and transport problem in each cell by applying a fixed potential difference across the cell in a given direction. For example, as far as percolation and transport are concerned, the 2×2 cell of Fig. 8.7 is equivalent to the circuit shown in there, usually called the Wheatstone bridge. Thus for this cell we only need to deal with 5 bonds, and for the $2 \times 2 \times 2$ cell we can construct an equivalent 12-bonds circuit, also shown in Fig. 8.7. To obtain $R(p)$, we find all conducting configurations of such circuits, with some bonds conducting (open) and some insulating (closed or missing). Thus for the 2×2 cell we obtain

$$p' = R(p) = p^5 + 5p^4q + 8p^3q^2 + 2p^2q^3 , \qquad (8.73)$$

where $q = 1 - p$. It is easy to see how this equation is obtained: There is only one conducting cell configuration with all five bonds conducting (probability p^5), 5 conducting configurations with 4 bonds conducting and one bond insulating (probability $5p^4q$), and so on.

As discussed in Chapter 3, the sample-spanning cluster at p_c is self-similar. This means that the RG transformation should remain invariant at p_c. The same thing should be true at $p = 1$ and $p = 0$, because under any reasonable transformation full and empty networks should be transformed to full and empty networks again. The points $p = 0, 1$, and p_c are thus called the *fixed points* of the transformation, and are denoted by p^*. Since the RG transformation should not change anything at these points, the implication is that at the fixed points the probability of having a conducting bond in the RG cell (p) and that of a bond in the renormalized cell $[p' = R(p)]$ should be the same. Thus the fixed points should be the solution of the polynomial equation

$$p^* = R(p^*) \,, \tag{8.74}$$

and indeed this equation usually has three roots which are $p^* = 0$, $p^* = 1$, and $p = p^*$, where p^* is the RG transformation prediction for p_c. For the RG cells of Fig. 8.7 we obtain $p^* = 1/2$ for both 2×2 and 3×3 cells, which is an exact result. In fact, it can be shown (Bernasconi, 1978) that the RG transformation for such two-dimensional cells for *any* b always predicts $p_c = 1/2$. For the $2 \times 2 \times 2$ cell we obtain $p^* \simeq 0.208$, which should be compared with the numerical estimate $p_c \simeq 0.249$.

So far, everything seems nice, simple, and correct. In fact, PSRG methods seem so simple that small-cell calculations can be done on the "back of a pocket", and the results of such calculations are reasonably accurate. The reader may guess that one should calculate $p_c = p^*$ for several cell sizes b, and then use finite-size scaling discussed in Chapter 3 according to which

$$p_c - p_c(b) \sim b^{-1/\nu_p} \,, \tag{8.75}$$

where p_c is the true percolation threshold of the network, and ν_p is the critical exponent of percolation correlation length ξ_p, to extrapolate the results to $b \rightarrow \infty$ and estimate p_c. Indeed, Reynolds *et al.* (1980) used this method and obtained very accurate estimates of site and bond percolation thresholds of the square network. However, as always life is more complex than we imagine! Ziff (1992) showed that as $b \rightarrow \infty$, the probability $R(p_c)$ approaches a *universal* value of $1/2$ for *all* two-dimensional networks. This implies that *asymptotically* Eq. (74) is *wrong*, because p_c is *not* universal and depends on the structure of the network, whereas $R(p_c)$ *is* universal (as $b \rightarrow \infty$). Moreover, Ziff (1992) showed that in two dimensions, Eq. (75) should be modified to

$$p_c - p_c(b) \sim b^{-1-1/\nu_p} \,, \tag{8.76}$$

so that in two dimensions one reaches the asymptotic regime *faster* than had been thought before.

To check this remarkable discovery, Stauffer *et al.* (1994) calculated $R(p_c)$ for bond percolation on a simple-cubic network for a sequence of increasing b's or linear size L of the network. Table 8.1, taken from Stauffer *et al.* (1994) and D. Stauffer (private communications, June 1994) shows their results. As this Table indicates, with increasing L the value of $R(p_c)$ approaches a constant value of about 0.42. However, observe that one needs large networks to obtain this value, although simulations with relatively small networks already indicate the approach to 0.42.

Table 8.1

Dependence of $R(p_c)$ on the linear size L of a simple-cubic network for bond percolation.

L	$R(p_c)$
3	0.353600
9	0.388622
17	0.391816
31	0.409680
41	0.396606
79	0.401197
101	0.4024950
131	0.403592
173	0.413333
367	0.410678

To check that this value is indeed universal, similar calculations were carried out for site percolation on the simple-cubic lattice, the results of which are given below.

Table 8.2

Dependence of $R(p_c)$ on the linear size L of a simple-cubic network for site percolation (D. Stauffer, private communication, June 1994).

L	$R(p_c)$
3	0.353600
9	0.388622
17	0.391816
31	0.391405
41	0.389223
79	0.403700
101	0.409200
131	0.423200
173	0.423000

Stauffer *et al.* (1994) also showed that Eq. (75) is still valid in three dimensions, and thus the applicability of Eq. (76) is limited to two dimensions. Thus, one can use a PSRG method to estimate p_c in two different ways. One is the "old" way of calculating $R(p_c)$ and using Eq. (74) to estimate p_c. However, this can be done only with small cell sizes (since for large b $R(p_c)$ will approach its universal value), and thus the estimated p_c is only an approximation. In this case, the critical exponent ν_p of the correlation length can be estimated from

$$\nu_p = \frac{\ln b}{\ln \lambda_p} , \qquad (8.77)$$

where $\lambda_p = dR(p)/dp$, evaluated at $p = p^*$. Thus for the 2×2 and 3×3 cells we obtain $\nu_p \simeq 1.43$ and 1.38, respectively, which should be compared with the exact value $\nu_p = 4/3$. For the $2 \times 2 \times 2$ cell we obtain $\nu_p \simeq 1.03$, which should be compared with the numerical estimate $\nu_p \simeq 0.88$. The second method is based on the discovery of Ziff (1992) and Stauffer *et al.* (1994). One varies p until $R(p)$ achieves its universal value of $1/2$ in two dimensions, or about 0.42 in three dimensions. The exponent ν_p is then calculated from Eq. (75) (in three dimensions) or (76) (in two dimensions). However, as Tables 8.1 and 8.2 indicate, only when the cell or network size is large, does the value of $R(p_c)$ approach its universal value.

We now discuss the PSRG approach for calculating the conductivity or permeability of a random network. In a PSRG approach for calculating the conductivity of a random network, one starts with a probability distribution $f_0(g)$ for the bond conductances g of the RG cell and replaces it with a new distribution $f_1(g)$, the probability distribution for the conductance of a bond in the renormalized cell. $f_1(g)$ is calculated by determining the equivalent conductance of the RG cell. Thus one obtains a recursion relation relating $f_1(g)$ to $f_0(g)$

$$f_1(g) = \int f_0(g_1)dg_1 f_0(g_2)dg_2 \cdots f_0(g_n)dg_n \delta(g_p - g') \,, \qquad (8.78)$$

where $g_1, \ldots, g_n$ are the conductances of the n bonds of the cell and g' is the equivalent conductance of the RG cell. For example, for the 5-bond cell of Fig. 8.7, one has

$$g' = \frac{g_1(g_2 g_3 + g_2 g_4 + g_3 g_4) + g_5(g_1 + g_2)(g_3 + g_4)}{(g_1 + g_4)(g_2 + g_3) + g_5(g_1 + g_2 + g_3 + g_4)} \,. \qquad (8.79)$$

For example, it is easy to show that if p

$$f_0(g) = (1 - p)\delta(g) + p\delta(g - g_0) \,, \qquad (8.80)$$

then for the 5-bond RG cell of Fig. 8.7, Eq. (78) yields

$$f_1(g) = [1 - R(p)]\delta(g) + 2p^3 q^2 \delta\left(g - \frac{1}{3}g_0\right) + 2p^2(1 + 2p)q^2 \delta\left(g - \frac{1}{2}g_0\right)$$
$$+ 4p^3 q \delta\left(g - \frac{3}{5}g_0\right) + p^4 \delta(g - g_0) \,, \qquad (8.81)$$

which is already more complex than Eq. (80). One iterates Eq. (78) again to obtain a new distribution $f_2(g)$ by substituting $f_1(g)$ into the right hand side of Eq. (78). The iteration process should continue until a distribution $f_\infty(g)$ is obtained, the shape of which does not change under further iterations. This is called the *fixed-point distribution* and the conductance of the original network is simply an average of this distribution. However, it is difficult to analytically iterate Eq. (78) many times. The common practice is to replace the distribution after iteration i by an *optimized* distribution $f_i^o(g)$, which mimics closely $f_i(g)$. The optimized $f_i^o(g)$ is usually taken to have the following form

$$f_i^o(g) = [1 - R(p)]\delta(g) + R(p)\delta[g - g^o(p)] , \qquad (8.82)$$

where $g^o(p)$ is an *optimized* conductance. In the past, various approximate schemes have been proposed for determining $g^o(p)$. One of the most accurate approximations was proposed by Bernasconi (1978). Suppose that after i iterations of Eq. (78) $f_i(g)$ is given by

$$f_i(g) = [1 - R(p)]\delta(g) + \sum_i a_i(p)\delta(g - g_i) , \qquad (8.83)$$

then $g^o(p)$ is approximated by

$$g^o(p) = \exp\left[\frac{1}{R(p)}\sum_i a_i(p)\ln g_i\right] . \qquad (8.84)$$

Once $g^o(p)$ is calculated, Eq. (78) is iterated again, a new distribution $f_{i+1}(g)$ and its optimized form $f_{i+1}^o(g)$ are determined, and so on. In practice, after a few iterations even an initially broad $f_0(g)$ converges quickly to a very stable distribution whose shape does not change under further rescaling. When this happens, $\lambda_c f_{n+1}(\lambda_c g) \simeq f_n(g)$, where λ_c is a constant. The critical exponent μ of the conductivity is then given by

$$\mu = \nu_p \frac{\ln \lambda_c}{\ln b} . \qquad (8.85)$$

In the case of the 2×2 RG cell of Fig. 8.7 we obtain, $\mu \simeq 1.32$, in excellent agreement with the accepted value $\mu \simeq 1.3$ (see Table 3.3). For the three-dimensional cell of Fig. 8.7 we obtain $\mu \simeq 2.14$, only 7% larger than $\mu \simeq 2.0$ for three-dimensional percolation networks. Moreover, the predicted conductivities in two dimensions are in good agreement with simulation results, but they are not as accurate in three dimensions, since the fixed point of the transformation, $p^* \simeq 0.208$, is significantly lower than than $p_c \simeq 0.2488$. Note that, since in a PSRG approach for the conductivity or permeability one does not use large RG cells (since computations become extremely complicated as b increases), for now we do not have to worry about the eventual universality of $R(p_c)$.

Position-space renormalization methods can also be used for estimating the permeability of a macroscopic porous medium modeled by a random network. The PSRG method for calculating the effective permeability of megascopic porous media is also similar to the above procedure, except that in this case one partitions the system into RG cells of equal sizes, to each one of which an equivalent permeability selected from a distribution is assigned. Then, similar to the above procedure, a renormalized permeability distribution is constructed using Eq. (78). A Monte Carlo sampling is then used to select the permeability of each

cell from the joint probability distribution of the RG cells' permeabilities. The
sampling and the iteration process are continued until a satisfactory representation of the permeability distribution is obtained. King (1989) used this method
for calculating permeability of megascopic porous media.

Young and Stinchcombe (1975) were the first to use PSRG methods for calculating percolation properties of random networks, while Stinchcombe and Watson
(1976) were the first to use these methods for calculating the electrical conductivity of percolation networks. However, in a simultaneous and little noted paper Madden (1976) used the same ideas for calculating transport properties of
porous media modeled by random networks, although he did not call his method
a renormalization approach. Many authors have proposed various variants of
PSRG methods for calculating both percolation and transport properties of random networks and other disordered systems (Straley, 1977a; Payandeh, 1980,
Reynolds *et al.*, 1980; Tsallis *et al.*, 1983; Sahimi *et al.*, 1984; Sahimi, 1988a).
Stanley *et al.* (1982) and Family (1984) reviewed most of the literature on this
subject.

Renormalization methods are usually very accurate for two-dimensional systems, and are flexible enough to be used for anisotropic and time-dependent
systems, although the treatment of the latter case is considerably more complex.
However, they have two drawbacks for three-dimensional systems. The first is
that the results for a percolation network with *any* type of $b = 2$ cell are not accurate. Moreover, in the treatment of the conductivity or permeability problem,
even after the first iteration of Eq. (78) the renormalized conductance distribution $f_1(g)$ is very complex; if we start with distribution (80), $f_1(g)$ will have
seventy-three components of the form $\delta(g - g_i)$, with $i = 1 - 73$. Hence analytical
calculation of $f_2(g)$ is very difficult if not impossible. The second drawback is
that even for a $b = 3$ RG cell, the RG transformation cannot be calculated analytically, because the total number of possible configurations of the RG cell is
of the order 3×10^{11}! Thus, one has to resort to a Monte Carlo renormalization
group method (Reynolds *et al.*, 1980), which is not any simpler than the simple
Monte Carlo method itself.

8.3.5 Renormalized effective-medium approximation

To circumvent the difficulties that PSRG methods encounter for three-dimensional
networks, Sahimi *et al.* (1983c, 1984) proposed a new method that combines
EMA and PSRG methods, and is called the renormalized EMA (REMA). The
idea is that an EMA is very accurate away from the percolation threshold, while
when we rescale a RG cell, the renormalized RG cell is farther away from p_c
than the original cell (because in the renormalized cell the correlation length ξ_p
is reduced by a factor $1/b$). Thus one may use an EMA with the renormalized
cell (or network) instead of the original network. That is, the pore conductance

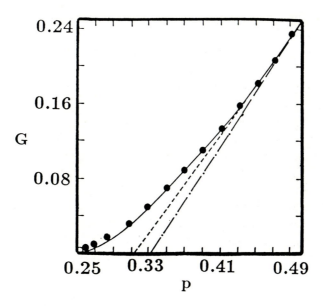

Figure 8.8: Comparison of the Monte Carlo data (circles) for the conductivity G of a simple-cubic network with the predictions of the REMA (solid curve), cluster EMA of Ahmed and Blackman (1979) (dashed curve), and the EMA (dashed-dotted curve) (after Sahimi *et al.*, 1983c).

distribution that one uses in Eq. (66) should be $f_1(g)$ [or $f_1(W)$] instead of $f_0(g)$. The new method is drastically more accurate than both EMA and PSRGs. For example, with the three-dimensional RG cell of Fig. 8.7, one obtains $p_{cb}^* \simeq 0.267$, only 7% larger thant the accepted value, $p_{cb} \simeq 0.2488$ for the cubic network. In general, if $R(p)$ is the RG transformation, REMA predicts that p_{cb} is the root of $R(p_{cb}) = 2/Z$ (instead of $p_{cb} = 2/Z$, which is the EMA's prediction, and $R(p_{cb}) = 1/2$ or 0.42, which is the PSRG's estimate). Using REMA, Sahimi *et al.* (1983c, 1984) and Sahimi (1988a) obtained very accurate predictions of transport properties of various two- and three-dimensional networks. Figure 8.8 compares the predictions of REMA for a simple-cubic network with those of EMA and the simulation results.

8.3.6 The Bethe lattice model

Most of the methods of estimating K and σ that have been reviewed so far are not exact and provide only approximate estimates of them. The only exception to this is the exact solution of electrical conduction on a Bethe lattice (Stinchcombe, 1974; Straley, 1977b). Consider a Bethe lattice of coordination number Z whose bond conductance distribution is $f(g)$. It has been established

(Straley, 1977b) that the conductivity σ of a Bethe lattice obeys the following scaling law near the percolation threshold

$$\sigma \sim (p - p_c)^3 , \qquad (8.86)$$

so that the critical exponent is $\mu = 3$, very different from $\mu \simeq 2$ for three-dimensional systems. Thus, the conductivity of a Bethe lattice is a poor approximation to that of three-dimensional networks. However, there is a happy coincidence here! Suppose that a function $\Upsilon(x)$ is the solution of the following nonlinear integral equation

$$\int_0^\infty \Upsilon(x) \exp(-\lambda x) dx = \left\{ \int_0^\infty \int_0^\infty \exp\left[-\frac{gx}{\lambda(g + x)} \right] f(g)\Upsilon(x) dg dx \right\}^{Z-1} .$$
$$(8.87)$$

It has not been possible so far to obtain an exact analytical solution for $\Upsilon(x)$, and thus Eq. (87) has to be solved numerically. A conductivity σ_m can be defined which is given by (Stinchcombe, 1974)

$$\sigma_m = Z \left[\int_0^\infty \int_0^\infty f(g)\Upsilon(x) \left(\frac{gx}{g + x} \right) dg dx \right]^{Z-2} . \qquad (8.88)$$

It can be shown (Stinchcombe, 1974) that near p_c

$$\sigma_m \sim (p - p_c)^2 , \qquad (8.89)$$

so that the critical exponent of σ_m is 2, identical with that of three-dimensional systems. The difference between σ and σ_m is as follows (Straley, 1977b). If we impose a unit potential gradient between any interior site of a Bethe lattice and its external boundary at infinity, then σ is the current *per bond* that flows through the lattice. Note that this is also the quantity that one calculates for three-dimensional networks. Thus, σ is the *macroscopic* conductivity of the network. On the other hand, σ_m is the *total* current that flows through the lattice in the same unit potential gradient, and thus σ_m is a *microscopic* conductivity.

Heiba *et al.* (1982) proposed σ_m as an accurate approximation to the conductivity of three-dimensional networks. The reason for this is clear. σ_m obeys the same scaling law near p_c as three-dimensional networks. In fact, it has been conjectured (Gingold and Lobb, 1990) that for three-dimensional systems μ is *exactly* 2. Moreover, as discussed in Chapter 5, the coordination number of a Bethe lattice can be adjusted so that its percolation threshold matches that of a three-dimensional network. For example, for the cubic network, $p_{cb} \simeq 0.2488$, and if we use a Bethe lattice of coordination number 5, then $p_{cb} = 1/(Z-1) = 1/(5-1) = 0.25$, which is less than 0.1% larger than that of the cubic network. Thus, a Bethe lattice of coordination number 5 can be used for estimating the conductivity of a cubic network. Figure 8.9 compares the conductivity σ of a cubic network obtained by Monte Carlo calculations and the microscopic conductivity σ_m of a

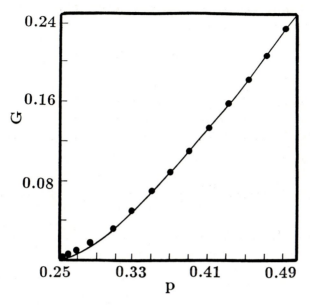

Figure 8.9: Comparison of the Monte Carlo data (circles) for the conductivity of a simple-cubic network with the predictions of a Bethe lattice of coordination number 5.

Bethe lattice of coordination number 5. It is evident that the difference between the two is so small that cannot be detected with a naked eye ! Thus, σ_m is an excellent estimator of the conductivity of three-dimensional networks, *regradless* of any particular form that $f(g)$ may have. The same method can be used for estimating permeability of three-dimensional networks. Sahimi (1988a) extended this to the case of unsteady-state diffusion and a first-order chemical reaction, while Sahimi (1993) extended the analysis to the case in which a non-Newtonian fluid flows through a disordered network.

8.3.7 Critical path analysis

A powerful idea for analyzing transport in heterogeneous media is the *critical path concept*. This concept was first proposed by Ambegaokar, Halperin and Langer (AHL) in 1971. These authors argued that transport in a disordered medium with a *broad* distribution of conductances (or any other kind of hetero-geneities) is dominated by those conductances whose magnitudes are larger than some characteristic value g_c, which is the smallest conductance such that the set of conductances $\{g|g > g_c\}$ forms a conducting sample-spanning cluster. This cluster is called the critical path. Therefore transport in a disordered medium with a broad conductance distribution reduces to a percolation problem with threshold value g_c. Shante (1977) and Kirkpatrick (1979) modified this idea by assigning all local conductances with values $g \geq g_c$ the value g_c, and setting all conductances with values $g < g_c$ to be zero (since the contribution of such bonds to the overall transport properties is very small). They then arrived at a trial solution for the sample conductance of the form

$$\sigma = ag_c[p(g_c) - p_c]^\mu , \qquad (8.90)$$

which is just the scaling law for the conductivity of the network discussed in Chapter 3. Here $p(g_c)$ denotes the probability that a given conductance is greater than or equal to g_c, and a is a constant. Equation (90) is now maximized with respect to g_c to obtain an estimate of g_c and thus σ. Computer simulations of Berman *et al.* (1986) for two-dimensional networks with various conductance distributions (Gaussian, log-normal, uniform, etc.) confirmed the quantitative accuracy of the AHL concept, even for relatively narrow conductance distributions. Therefore calculating effective transport properties of disordered media in which percolation may not seem to play any role can be reduced to determining the same properties for a percolation system. This indicates the broad applicability of percolation theory.

Katz and Thompson (1986, 1987) extended the ideas of AHL to estimate the permeability and electrical conductivity of porous media. In a porous medium the local hydraulic conductance is a function of the length l. Therefore the critical conductance g_c defines a characteristic length l_c. Since both flow and electrical conduction problems belong to the same class of percolation problems, the length that signals the percolation threshold in the flow problem also defines the threshold in the electrical conductivity problem. Thus we rewrite Eq. (90) as

$$\sigma = \phi g_c(l)[p(l) - p_c]^\mu , \qquad (8.91)$$

where the porosity ϕ ensures a proper normalization of the fluid or the electric-charge density. The function $g_c(l)$ is equal to $c_f l^3$ for the flow problem and $c_c l$ for the conduction problem. For appropriate choices of the function $p(l)$, the conductance $\sigma(l)$ achieves a maximum for some $l_{max} \leq l_c$. In general l_{max}^f for the flow problem is different from l_{max}^c for the conduction problem, because the transport paths have different weights for the two problems.

If $p(l)$ allows for a maximum in the conductance which occurs for $l_{max} \leq l_c$, then we can write

$$l_{max}^f = l_c - \Delta l_f = l_c \left[1 - \frac{\mu}{1 + \mu + l_c \mu p''(l_c)/p'(l_c)} \right] , \qquad (8.92)$$

$$l_{max}^c = l_c - \Delta l_c = l_c \left[1 - \frac{\mu}{3 + \mu + l_c \mu p''(l_c)/p'(l_c)} \right] . \qquad (8.93)$$

If the pore size distribution of the medium is very broad, then $l_c \mu p''(l_c)/p'(l_c) \ll$

1, and Eqs. (92) and (93) reduce to

$$l^f_{max} = l_c\left(1 - \frac{\mu}{1+\mu}\right) \simeq \frac{1}{3}l_c , \tag{8.94}$$

$$l^c_{max} = l_c\left(1 - \frac{\mu}{3+\mu}\right) \simeq \frac{3}{5}l_c , \tag{8.95}$$

if we use $\mu \simeq 2$ for three-dimensional percolation systems. Using these, we can establish a relation between σ and K. Writing

$$\sigma = a_1\phi[p(l^c_{max}) - p_c]^\mu \tag{8.96}$$

and

$$K = a_2\phi(l^f_{max})^2[p(l^f_{max}) - p_c]^\mu , \tag{8.97}$$

we obtain to first order in Δl_c or in Δl_f

$$p(l^{f,c}_{max}) - p_c = -\Delta l_{f,c}\, p'(l_c) . \tag{8.98}$$

To obtain the constants a_1 and a_2, Katz and Thompson (1986) assumed that at the local level the conductivity of the porous medium is σ_f, the conductivity of the fluid (usually brine) that saturates the pore space, and that the local pore geometry is cylindrical. These imply that $a_1 = \sigma_f$ and $a_2 = 1/32$. Therefore one obtains

$$K = a_3 l_c^2 \frac{\sigma}{\sigma_f} , \tag{8.99}$$

where $a_3 = 1/226$. A similar argument leads to (Katz and Thompson, 1987)

$$\frac{\sigma}{\sigma_f} = \frac{l^c_{max}}{l_c}\phi S(l^c_{max}) , \tag{8.100}$$

where $S(l^c_{max})$ is the volume fraction (saturation) of connected pore space involving pore widths of size l^c_{max} and larger.

Equations (99) and (100) involve no adjustable parameters. Every parameter is fixed and precisely defined. To obtain the characteristic length l_c, Katz and Thompson (1986, 1987) proposed the use of mercury porosimetry, discussed in Chapter 5. As we discussed there, mercury porosimetry is a percolation process. Hence consider a typical mercury porosimetry curve in which the pore volume of the injected mercury is measured as a function of the pressure; see Fig. 8.10. As can be seen, in the initial portion of the curve the curvature is positive. This portion is obtained before a sample-spanning cluster of pores, filled with mercury, has been formed. There is also an inflection point beyond which the pore volume increases rapidly with the pressure. This inflection point signals the formation of the sample-spanning cluster. Therefore from the Washburn equation (see Chapter 5) we must have $l \geq -4\sigma_{mv}\cos\theta/P_i$, where P_i is the pressure at the

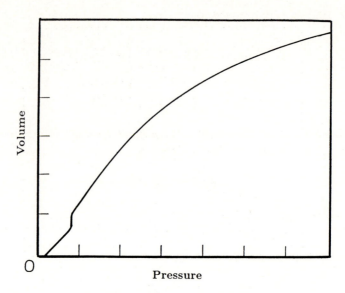

Figure 8.10: Schematic volume-pressure curve during mercury porosimetry.

inflection point, σ_{mv} is the surface tension between the mercury and the vaccum, and θ is the contact angle between the mercury and the pore surface. Hence

$$l_c = -\frac{4\sigma_{mv}\cos\theta}{P_i} \qquad (8.101)$$

defines the characteristic length l_c.

Figure 8.11 compares the logarithm of the permeability for a set of sandstones, calculated using Eq. (99), with the measured values. The dashed lines mark a factor of two. No adjustable parameter has been used, and the agreement between predictions and theory is very good. Note that, once l_c is determined from a mercury injection curve, the saturation $S(l_{max}^c)$ can also be determined immediately. Figure 8.12 compares the calculated and measured σ/σ_f. As in the case of permeability, no adjustable parameter has been used. Katz and Thompson (1986, 1987) contend that l_c can be estimated from mercury injection curves with an error of at most 15%. However, the error in the constant $1/226$ in Eq. (99) can be as large as a factor of 2. We emphasize that Eqs. (99) and (100) are *not* in general exact, but appear to provide very accurate estimates of K and σ. Note that if, as discussed in Chapter 2, the distribution of the heterogeneities (conductances or permeabilities) of a porous medium is very broad, the exponent μ can be non-universal, and therefore its non-universal value should be used in Eqs. (94) and (95).

Some related works should be mentioned here. Swanson (1981) had already recognized that during mercury injection (or flow of any non-wetting fluid in a porous medium) large pores dominate the flow paths, and that the inflection point in the pore volume-pressure curve signals the formation of a sample-

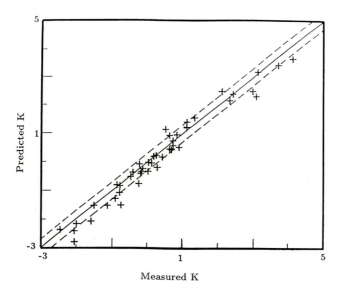

Figure 8.11: Logarithmic comparison of the predicted permeability K (in milli-darcy) with the measured K (after Katz and Thompson, 1986).

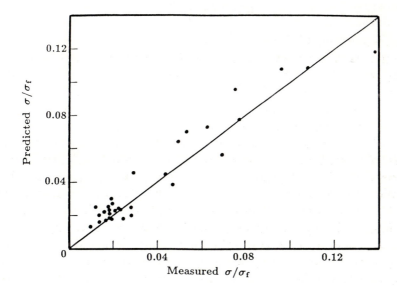

Figure 8.12: Logarithmic comparison of the predicted conductance σ with the measured values. σ_f is the conductivity of the fluid saturating the porous medium (after Katz and Thompson, 1987).

spanning cluster. Thus he postulated a relation between the permeability and the capillary pressure curve. He maximized the product of length and saturation, both of which can be estimated from the mercury-injection curve, and obtained

$$l \sim (S^2 l^2)_{max} \quad , \tag{8.102}$$

where l_{max} is a length scale very similar, both qualitatively and numerically, to the length scale l_c introduced by Katz and Thompson (1986, 1987). The agreement between Eq. (102) and the experimental data was very good, which is not entirely surprising given the similarity between Swanson's l_{max} (not to be confused with Katz and Thompsons's l_{max}) and l_c. However, it should be noted that the transport paths considered by Swanson were appropriate for electrical conduction not the fluid flow problem. Banavar and Johnson (1987) and Le Doussal (1989) calculated the coefficient a_3 in Eq. (99) slightly differently than Katz and Thompson (1986, 1987) did. For example, Banavar and Johnson (1987) gave, $a_3 \simeq 7.68 \times 10^{-3}$, which should be compared with Katz and Thompson's value, $a_3 = 1/226 \simeq 4.42 \times 10^{-3}$. However, Banavar and Johnson's predictions are still within the error bars of Fig. 8.11. Nyame and Ilbston (1980) used an empirical parameter similar to l_c to describe permeability in cement paste. Hagiwara (1984) replaced S^2 in Eq. (102) with σ to obtain $K \sim \sigma l^2$, which appears again to agree with data.

Finally, the ideas of AHL and Katz and Thompson can also be extended for calculating the permeability of fractured rocks. Indeed, Charlaix *et al.* (1987a) used arguments very similar to those of Katz and Thompson to calculate the permeability of fracture networks with a broad distribution of fracture apertures. These ideas can also be used for estimating the permeability of a porous medium saturated by a *non-Newtonian* fluid (Sahimi, 1993).

8.3.8 Transfer matrix method

So far we have discussed analytical methods of estimating K and σ. Now we discuss three numerical methods, the first of which is the transfer matrix method discussed in this section. In this method the conductivity or permeability of the network is calculated *exactly as the network is constructed*. This method was proposed by Derrida and Vannimenus (1982). We follow Derrida *et al.* (1984) and describe the method for a two-dimensional system, after which its generalization to three dimensions will be obvious. For the sake of efficiency, the computations are usually done for a strip, although the network can have any shape or structure.

Figure 8.13 shows the strip. It is built by adding one bond at a time to it, starting from the line on the left. At each stage of its construction, the left part of the strip is described by a $M \times M$ matrix **A**, where M is the number of sites in a section (column), including the sites of the first (top) and the last

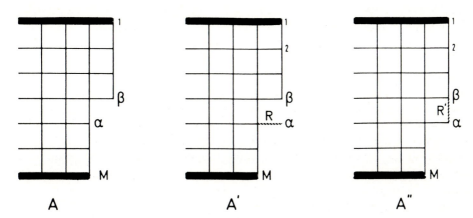

Figure 8.13: Construction of the random network in the transfer matrix method (after Derrida *et al.*, 1984).

(bottom) rows. To construct **A** we proceed as follows. If we want to estimate the conductivity of the left part of the strip, we attach to sites 1, 2, $\cdots$, M a wire which imposes a potential V_α on site α (see Fig. 8.13), as a result of which a current I_β flows in the wire attached to site β. Since the problem is linear, we must have, $I_\beta = \sum_\alpha A_{\alpha\beta} V_\alpha$. On the other hand, the currents depend only on the *difference* between the potentials, and thus we can always set $V_M = 0$. The only thing we now need to specify is how **A** is transformed as we add more bonds to the network (strip).

There are two kinds of bond that one can add to a two-dimensional strip. If we add a horizontal bond with resistance $R = 1/g$ (where g is selected according to its distribution) to site α (see Fig. 8.13), then **A** is transformed to $\mathbf{A}'$ whose entries A_{ij} are related to those of **A** by

$$A'_{ij} = A_{ij} - \frac{A_{i\alpha} A_{j\alpha} R}{1 + A_{\alpha\alpha} R}, \tag{8.103}$$

which can be proven easily by using the usual laws of connecting resistors together. Similarly, if we add a vertical bond with resistance R' between α and β, then

$$A''_{ij} = A'_{ij} + \frac{(\delta_{\alpha j} - \delta_{\beta j})(\delta_{\alpha i} - \delta_{\beta i})}{R'}, \tag{8.104}$$

where δ_{ij} is the usual Kronecker's symbol, namely, $\delta_{ij} = 1$ if $i = j$, and $\delta_{ij} = 0$ otherwise. Note that **A** is symmetric (as are all matrices that arise in the numerical solutions of conduction or diffusion), and this makes the computations more efficient. The conductivity of a strip of length $L \gg 1$ is then given by

$$\sigma = \lim_{L \to \infty} \frac{A(1,1)}{L}. \tag{8.105}$$

The same procedure can be used for calculating K.

Thus, calculation of σ or K becomes extremely simple. We do not store **A** for the entire $M \times L$ system, but only for the last two sections of it. For example, as soon as the construction of the strip reaches, e.g., the fourth section ($M = 4$), the first two sections are forgotten, and only the last two are kept. Moreover, the size of **A** is *independent* of L, so that one can use an extremely large L without worrying about the required computer memory. Derrida *et al.* (1984) give a computer program for transfer matrix calculation of the conductivity or permeability of a $M \times M \times L$ simple-cubic network. The calculations are similar to the two-dimensional case, except that M should be replaced by $M^2 + 2$, the number of sites in a section (plane). Equation (103) is then used for all bonds in the x-direction (the direction of macroscopic potential gradient), while Eq. (104) is used for those in the $y-$ and $z-$directions.

In its original form developed by Derrida and Vannimenus (1982), the transfer matrix method could be used to calculate only σ or K. It could provide no information about, e.g., the distribution of the potentials at the sites of the network, as **A** stores information only about the last two sections of the network. In its modified form developed by Duering and Bergman (1990), such distributions can also be calculated.

8.3.9 Random walk methods

As discussed above, from the time dependence of the mean square displacement of a random walker, Eq. (45), the effective diffusivity D (and σ) can be calculated. The first application of this idea for determining transport properties of composite or disordered systems appeared in a paper of Haji-Sheikh and Sparrow (1966), who studied heat conduction in a composite solid. Since this paper, many authors have used random walk methods to study transport in disordered media. In the context of percolation problems, Brandt (1975) appears to be the first to use this method to study diffusion of noble gases in glasses. But the method was popularized by de Gennes (1976), who made an analogy between the motion of a random walker in a disordered medium and that of an ant in a labyrinth. Mitescu and Roussenq (1976) followed de Gennes' idea and performed the first simulations on percolation clusters. Havlin and Ben-Avraham (1987), Haus and Kehr (1987) and Hughes (1994) provide extensive reviews of this subject. Bunde *et al.* (1985) considered a general two-component mixture, where each component constitutes a phase and both phases allow transport, and formulated a random-walk model for calculating the conductivity of such a mixture. Using vectorization on a supercomputer and multispin coding, one can develop a highly efficient computer algorithm for simulating random walks in a disordered medium; see Sahimi and Stauffer (1991) for details.

One can facilitate random-walk simulations by using *first-passage time* equations. The idea is that if a random walker moves in a homogeneous region of the system, there is no need to spend unnecessary time to simulate its detailed motion. Instead, the walker can take large steps to pass quickly through a homogeneous region and arrive at the interface between the two phases. The necessary time for taking large steps can be calculated *analytically*. Thus unlike conventional simulations in random networks in which the length of each step of the walk is only one lattice bond, and each time a step is taken the time is increased by one unit, in first-passage time simulations the walker takes long steps (providing its step does not take it outside of a phase), and the time is increased by an amount appropriate to that step. This basic idea was first used by Sahimi *et al.* (1982) for simulating hydrodynamic dispersion in a porous medium by a random-walk method; their work is discussed in Chapter 9. In the context of calculating the effective conductivity, diffusivity and reaction rate of a disordered medium, Zheng and Chiew (1989) and Kim and Torquato (1990, 1991) appear to be the first to have used this method. Let us now discuss briefly the first-passage time simulation for the conductivity and diffusivity.

Consider a multiphase system that consists of n-phases with conductivities $\sigma_1, \ldots, \sigma_n$ and volume fractions $\phi_1, \ldots, \phi_n$. In a first-passage time simulation one constructs the largest (imaginary) concentric sphere of radius R around a randomly chosen point in phase i, which just touches the multiphase interface. Suppose that the random walker is initially at the center of the imaginary sphere. The mean time t_m for the particle to reach a randomly-selected point on the surface of the sphere is $t_m(R) = R^2/(2d\sigma_i)$, where d is the dimensionality of the system. This time is recorded, and the process of constructing the sphere and calculating the time a point on its surface is reached is repeated, until the random walker comes within a very small distance of the multiphase interface. One then computes the mean time t_b for crossing the boundary and the probability of crossing the boundary, which is proportional to the ratio of the conductivities of the two phases. If the random walker crosses the interface and enters a new phase, it finds itself in a new homogeneous phase, and therefore the process of sphere construction is repeated. If the reference phase is taken to be phase 1, then, in the limit of long times, the effective conductivity of the system is given by (Kim and Torquato, 1990)

$$\frac{\sigma}{\sigma_1} = \frac{\left\langle \sum_i t_{1m}(R_i) + \sum_j t_{1m}(R_j) \right\rangle}{\left\langle \sum_i t_m(R_i) + \sum_j t_b(R_j) \right\rangle} \quad , \tag{8.106}$$

where $t_{1m}(R_i)$ is the mean first-passage (hitting) time for a walker in a homogeneous sphere of radius R_i, i denotes the number of phases, j is the number of paths crossing the interface, and $\langle \cdots \rangle$ denotes an average over all realizations of the disordered medium.

We now need the first-passage time equations which apply in a very small neighborhood of the interface between the two phases, say 1 and 2. These are given by

$$\nabla^2 p_1 = 0 \quad \text{in } \Omega,$$
$$p_1(\mathbf{x}) = 1 \quad \text{on } \partial\Omega_1,$$
$$p_1(\mathbf{x}) = 0 \quad \text{on } \partial\Omega_2,$$
$$p_1(\mathbf{x})|_1 = p_1(\mathbf{x})|_2 \quad \text{on } \Gamma,$$
$$\frac{\partial p_1}{\partial \mathbf{n}_1}\Big|_1 = \alpha \frac{\partial p_1}{\partial \mathbf{n}_1}\Big|_2 \quad \text{on } \Gamma,$$
$$p_2(\mathbf{x}) = 1 - p_1(\mathbf{x})$$
$$\sigma_i \nabla^2 t_b = -1 \quad \text{in } \Omega_i, \qquad (8.107)$$
$$t_b(\mathbf{x}) = 0 \quad \text{on } \partial\Omega,$$
$$t_b(\mathbf{x})|_1 = t_b(\mathbf{x})|_2 \quad \text{on } \Gamma,$$
$$\frac{\partial t_b}{\partial \mathbf{n}_1}\Big|_1 = \alpha \frac{\partial t_b}{\partial \mathbf{n}_1}\Big|_2 \quad \text{on } \Gamma$$

In these equations, p_1 (p_2) is the probability that the random walker, initially at $\mathbf{x}$ near the center of the imaginary sphere of radius R, hits $\partial\Omega_1$ ($\partial\Omega$) *for the first time* without hitting $\partial\Omega = \partial\Omega_1 + \partial\Omega_2$ *for the first time*, Γ is the interface surface, $\mathbf{n}_i$ is a unit vector outward normal from region Ω_i, $|_i$ denotes the approach to Γ from the region Ω_i and $\alpha = \sigma_2/\sigma_1$. In practice, p_1, p_2, and t_b are computed when the random walker comes within a prescribed small distance $Y\delta_i$ of the interface, where Y is the local radius of the curvature, and $\delta_i \ll 1$.

The solution to (107) is straightforward for flat interfaces (lines or planes) and can be derived analytically. The more interesting case is when the interface is curved. For example, in a random packing of spheres in which the spheres make the conducting phase and the channels between the non-overlapping spheres make up the insulating phase, or vice versa, the interface between the two phases is curved, and it is not possible to derive an exact analytical solution of Eqs. (107). For this case, Kim and Torquato (1990) derived an approximate, but very accurate, analytical solution to Eq. (107), which was used in their simulations.

The first-passage time simulation is particularly effective for simulating transport in continuum models of porous media discussed in Chapter 6. The efficiency of the method decreases, however, as the porosity of the pore space decreases since the search for the construction of the imaginary sphere becomes time consuming. Near the percolation threshold, the method is not efficient at all, and the vectorized version of the conventional random-walk simulation (Sahimi and Stauffer, 1991) should be used. Random-walk methods are particularly useful for estimating the electrical conductivity of porous media made of an insulating granular matrix saturated with a conducting pore fluid such as brine. A traditional method such as the finite-element technique is notoriously time consuming for such porous media since even if we use only 20^3 grains (a modest number), a very fine finite-element mesh with roughly 10^9 nodes would be required to solve the Laplace's equation accurately, a prospect which is totally impractical. For this reason alone, random-walk methods are the preferred technique for estimating diffusivity and conductivity of porous media. Evans *et al.* (1980), Abbasi *et al.* (1983), Nakamo and Evans (1983), Akanni *et al.* (1987), and Tomadakis and Sotirchos (1991) used random-walk methods to study both ordinary and Knudsen diffusion in a variety of porous media made of random distributions of penetrating or non-penetrating spheres (or disks in two dimensions). Likewise, Smith and Huizenga (1984) used the method to investigate Knudsen diffusion in a random assemblage of spheres. More recently, Schwartz and Banavar (1989) used random-walk simulations to calculate the electrical conductivity of the grain consolidation model of Roberts and Schwartz (1985), discussed in Chapter 4, with multisize particles. The results, shown in Fig. 8.14 in terms of the formation factor $F = \sigma_f/\sigma$, are in excellent agreement with the experimental data of Guyon *et al.* (1987) for sintered binary composites, and with other experimental data for similar systems (Oger *et al.*, 1986). In practice, the grain particles are not usually spherical, but in a random-walk simulation the particles can have any shape. Schwartz *et al.* (1989a) also calculated the electrical conductivities of a system that was originally a packing of spherical rubber grains, but was exposed to a uniaxial pressure that was applied to deform the particles. To model the system they constructed an unconsolidated sphere pack with a given particle distribution, and then compressed the system in one direction by a given α_c. This resulted in a system of spheroidal grains which were then allowed to grow along the three axes. Thus a sphere of radius R at (x_0, y_0, z_0) was replaced by the spheroid $\alpha_c^{2/3}(x - x_0)^2 + \alpha_c^{2/3}(y - y_0)^2 + \alpha_c^{-4/3}(z - z_0)^2 = R^2$. Note that the resulting system is no longer isotropic, and one has to calculate a conductivity tensor. The results of Schwartz *et al.* (1989a) for this system were in good agreement with the measurements of McLachlan *et al.* (1987). In order to speed up random-walk simulations in porous media which have *megascopic* inhomogeneities such as layering, Schwartz *et al.* (1989a) introduced a *weak bias* in their simulations. This causes the random walker to sample the pore space more efficiently, because in the direction of the bias the travelled distance is proportional to N_s rather than $N_s^{1/2}$, where N_s is the number of steps. Although the idea of biased diffusion for estimating conductivity of inhomogeneous systems is rather old (Miller and Abrahams, 1960), Schwartz *et al.* (1989a) appear to be the first to apply it to inhomogeneous porous media.

In contrast with diffusivity and conductivity, there is *no* random-walk method for estimating the permeability of a porous medium, because there is no general relation between K and σ, or K and D. This is not totally surprising, as σ and D are calculated from the solution to the Laplace's equation, which is a scalar equation, whereas K is calculated from the solution to Stokes' equation, a vector equation. Thus one cannot expect to have a general relation between these quantities. This lack of a general relation between a scalar and a vector system is well-known in statistical physics. Effective properties of scalar percolation (for example, D or σ) are not, in general, related to those of vector percolation (for

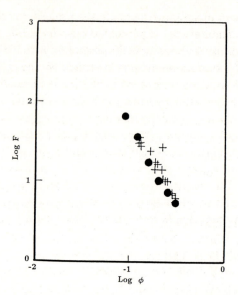

Figure 8.14: Comparison of the random walk predictions of the formation factor F of the grain consolidation model (circles) with the experimental data (+) (after Schwartz and Banavar, 1989).

example, the elastic moduli).

8.3.10 Network simulation

In the absence of a random-walk algorithm, one of the main numerical means of estimating the permeability of a disordered pore space has been computer simulations using a network model. Of course the same networks can also be used for estimating the diffusivity and conducitivity of the pore space. A pore throat (bond) shape (for example, cylindrical, channel-like, etc.) and a flow regime (for example, laminar) are assumed. The flow problem is then solved analytically for a single pore, from which an expression is obtained for the flow rate Q_{ij} in a pore throat $i - j$ in terms of the pressure drop along the pore, its length and effective radius, and the fluid viscosity. In most cases the pressure drop across a pore body, where pore throats meet, is ignored. One then writes down a mass balance for each node or each pore body i, which simply means that the net flow rate reaching it is zero, and thus

$$\sum_j Q_{ij} = 0 ,\qquad(8.108)$$

where the sum is over all pore bodies (sites) j that are connected to i by a pore throat. Writing this equation down for every interior node of the network results in a set of simultaneous equations for nodal pressures, from the solution of which the pressure field in the network, and thus the permeability can be calculated. The boundary conditions are usually an imposed flow rate, or a pressure gradient in one direction, and periodic boundary conditions in the other directions. One usually distributes the effective sizes of the pore throats according to a probability density function which represents the pore size distribution.

Various versions of this network simulation have been used in the last four decades. As already mentioned in Chapters 5 and 6, Owen (1952) appears to be the first who did relatively extensive computations with a three-dimensional network of large pore bodies and very narrow pore throats, and estimated the formation factor F. Fatt (1956) used a two-dimensional network of pore throats with distributed effective sizes, and calculated the permeability and relative permeabilities for two immiscible fluids (see Chapter 12). In the 1960s there were several works in which network models were used for calculating the permeability, conductivity and the formation factors, and to investigate the relationship between them (Fatt, 1960; Nicholson, 1968; Rink and Schopper, 1968; Greenberg and Brace, 1969; Weinbrandt and Fatt, 1969). This type of simulation was continued in the 1970s (Nicholson and Petropoulos, 1971, 1975, 1977; Shankland and Waff, 1974; Dullien, 1975), in the 1980s (Koplik, 1982; Koplik *et al.*, 1984; Seeburger and Nur, 1984; Doyen, 1988; Constantinides and Payatakes, 1989), and in the 1990s (David *et al.*, 1990; Mukhopadhyay and Sahimi, 1995). These works vary in the amount of detail included in the network model, the sizes of the networks that are used, and so on, but the essential idea behind all of them is the same as that discussed above. On the other hand, Bryant *et al.* (1993a,b) developed a network model based on Finney's random close packing of equal spheres (Finney, 1968; see Chapter 3), *without* invoking any major assumption regarding the microstructure of the network, and calculated the permeability of the network. Chu and Ng (1989) used a somewhat similar method. Bryant *et al.* (1993a,b) showed that the calculated permeability agrees well with the data, thus confirming the general validity of network models for calculating the permeability of a pore space. Finally, Adler (1985a,b,c,d) studied flow and transport in a variety of deterministic fractal networks.

The same method can be used for simulating flow in megascopic porous media. In this case, each bond of the network represents a region of the rock to which a permeability is assigned which is selected from a permeability distribution, which can contain long-range correlations, such as those induced by fractional Brownian motion discussed in Chapter 3; see Mukhopadhyay and Sahimi (1995).

8.4 Fractal transport and non-local formulation of diffusion

As discussed above, if a porous medium contains percolation disorder and is near its percolation threshold, or if it contains long-range correlations, then diffusion may be fractal. In this case, the mean square displacement $\langle R^2(t) \rangle$ of a diffusing particle provides valuable information about the behavior of the system. The mean square displacement $\langle R^2(t) \rangle$ of the particle at time t is related to the effective diffusivity D by Eq. (45), and the growth of $\langle R^2(t) \rangle$ with t is given by Eq. (64). Thus, in general there are three possibilities.

(i) $D_w = 2$, which means that $\langle R^2 \rangle$ grows linearly with t, and D is a constant. This is the familiar Fickian diffusion. In this case the average probability $P_d(r, t)$ of finding the diffusing particles at point $\mathbf{r}$ at time t, which is proportional to the concentration of the diffusing particles, is given by a Gaussian distribution

$$P_d(r, t) \sim t^{-d/2} \exp\left(-\frac{r^2}{4Dt}\right) , \qquad (8.109)$$

which obeys the classical diffusion equation, where $r = |\mathbf{r}|$ and the averaging is taken over all initial positions of the diffusants. As already emphasized, it is important to keep in mind the precise condition under which Eq. (109) is valid. If $R = \sqrt{\langle R^2(t) \rangle}$, then Eq. (109) is valid if R is much larger than the length scale over which the porous medium is homogeneous. For example, for percolation systems this length scale is the correlation length ξ_p, while for meagascopic porous media this length scale is the permeability correlation length ξ_k. Note that it follows from Eq. (109) that the probability $P_0 = P_d(0, t)$ of finding the diffusants at the origin at time t (or the concentration of the diffusants at the origin) is given by

$$P_0 \sim t^{-d/2} . \qquad (8.110)$$

(ii) Now suppose that $D_w > 2$, which implies that $\langle R^2(t) \rangle$ grows with t slower than linearly. This is the anomalous or fractal diffusion mentioned above. For this diffusion regime Eq. (45) tells us that D depends on t and vanishes as $t \to \infty$, i.e., diffusion is very slow and inefficient. Fractal diffusion takes place when R is much smaller than the length scale for macroscopic homogeneity of the medium. Therefore, one has fractal diffusion if $R \ll \xi_p$ for percolation systems, or if $R \ll \xi_k$ for megascopic porous media. In general, diffusion in *all* fractal media is anomalous, since such media are self-similar and do not possess a length scale for their macroscopic homogeneity. Note that at the percolation threshold p_c the correlation length is divergent, and therefore one always has fractal diffusion at p_c.

For diffusion in porous media with percolation disorder, we can relate the random walk fractal dimension D_w to the critical exponents of percolation defined in Chapter 3. Suppose that diffusion takes place only on the sample-spanning cluster of open pores. Near p_c we have $D \sim \xi_p^{-\theta_p}$, where $\theta_p = (\mu - \beta_p)/\nu_p$, and μ, β_p and ν_p are the percolation exponents defined in Chapter 3. Since fractal diffusion occurs only if $R \ll \xi_p$, we can replace ξ_p with R and write $D \sim \langle R^2 \rangle^{-\theta_p/2}$. On the other hand, Eq. (45) tells us that $D \sim d\langle R^2 \rangle/dt \sim \langle R^2 \rangle^{-\theta_p/2}$, which after integration yields, $\langle R^2 \rangle \sim t^{2/(2+\theta_p)}$, implying that

$$D_w = 2 + \frac{\mu - \beta_p}{\nu_p} = 2 + \theta_p \qquad (8.111)$$

This result was first derived by Gefen *et al.* (1983). Using the numerical values of the exponents given in Table 3.3, we find that $D_w(d = 2) \simeq 2.87$ and $D_w(d = 3) \simeq 3.8$, indicating fractal diffusion in both two and three dimensions. The three-dimensional value of D_w is only about 5% smaller than the EMA prediction $D_w = 4$, and thus this prediction of EMA is reasonably accurate. The crossover between fractal and nonfractal diffusion takes place at a crossover time t_{co} such that

$$t_{co} \sim \xi_p^{D_w} , \qquad (8.112)$$

so that for $t << t_{co}$ diffusion is fractal, whereas for $t >> t_{co}$ diffusion is Fickian. This implies that for percolation disorder in three dimensions, $t_{co} \sim \xi_p^{3.8} \sim (p - p_c)^{-3.34}$, whereas the EMA discussed above predicts that, $t_{co} \sim (p - p_c)^{-2}$. However, both Eq. (112) and EMA predict that as p_c is approached, the time scale t_{co} for reaching Fickian diffusion *diverges*, which is due to the fact that as p_c is approached the correlation length ξ_p also diverges, and thus the porous medium is a fractal object on a rapidly increasing length scale. This has important implications for interpreting experimental data for diffusivity of porous media, since one has to ensure that a Fickian regime has been reached before assigning a constant value to the diffusivity. Banavar *et al.* (1985) have proposed an experimental method for measuring D_w, based on nuclear magnetic resonance and magnetic-field gradient experiments.

(iii) Consider now the case $D_w < 2$, which means that $\langle R^2(t) \rangle$ grows with t *faster* than linearly. This is called *superdiffusion* (Sahimi, 1987). In general, it is *not* possible to have superdiffusion in fractal media. However, dispersion in heterogeneous porous media (to be discussed in Chapter 9) and turbulent diffusion (Shlesinger *et al.*, 1987) are two examples of transport processes that give rise to superdiffusion.

Neither fractal diffusion nor superdiffusion is described by the classical diffusion equation, and the exact form of $P_d(r, t)$ is not known. In fact, the exact form of $P_d(r, t)$ has been a controversial subject for several years. Many authors have studied this problem (see O'Shaughnessy and Procaccia, 1985; Guyer, 1985; Aharony and Harris, 1989; Cushman, 1991; to name but a few). In particular, Cushman (1991) introduced a wave-vector and frequency-dependent diffusion tensor $\hat{D}(\mathbf{k}, \omega)$, where the diffusive flux $\mathbf{j}(\mathbf{r}, t)$ is given by

$$\hat{\mathbf{j}}(\mathbf{k}, \omega) = \hat{D}(\mathbf{k}, \omega) \cdot ik\hat{P}_d(\mathbf{k}, \omega) , \qquad (8.113)$$

where a sign error in Cushman's Eq. (9) has been corrected, and $\mathbf{k}$ and ω denote the Fourier and Laplace transform variables conjugate to $\mathbf{r}$ and t, respectively. Thus, in real space and real time, Eq. (113) is equivalent to

$$\mathbf{j}(\mathbf{r}, t) = -\int_0^t dt' \int dV(\mathbf{r}') D(\mathbf{r}', t') \cdot \nabla P_d(\mathbf{r} - \mathbf{r}', t - t') . \qquad (8.114)$$

Classical diffusion corresponds to the special case

$$\mathbf{D}(\mathbf{r}, t) = \delta(\mathbf{r})\delta_+(t)\mathbf{D}_{classical} \; , \tag{8.115}$$

where $\mathbf{D}_{classical}$ is the constant diffusivity tensor when diffusion is described by the classical diffusion equation. Equation (114) tells that the relation between the flux and the concentration (which is proportional to P_d) is *non-local* in space and has a *memory*. If we now invoke the continuity equation

$$\frac{\partial P_d(\mathbf{r}, t)}{\partial t} + \nabla \cdot \mathbf{j}(\mathbf{r}, t) = 0 \; , \tag{8.116}$$

we obtain

$$\frac{\partial P_d(\mathbf{r}, t)}{\partial t} = \nabla \cdot \int_0^t dt' \int dV(\mathbf{r}')\mathbf{D}(\mathbf{r}', t') \cdot \nabla P_d(\mathbf{r} - \mathbf{r}', t - t') \; . \tag{8.117}$$

If $P_d(\mathbf{r}, 0) = \delta(\mathbf{r})$, then Fourier-Laplace transforming Eq. (116) and using the expression for the flux yields

$$\hat{P}_d(\mathbf{k}, \omega) = [\omega + \mathbf{k} \cdot \hat{\mathbf{D}}(\mathbf{k}, \omega) \cdot \mathbf{k}]^{-1} \; . \tag{8.118}$$

The Laplace transform of the mean square displacement can easily be shown to be

$$\langle \tilde{R}^2(\omega) \rangle = \frac{2}{\omega^2} \text{trace}[\hat{\mathbf{D}}(\mathbf{0}, \omega)] \; , \tag{8.119}$$

[compare this with Eq. (64)] so that if we use Eq. (64), we obtain

$$\text{trace}[\hat{\mathbf{D}}(\mathbf{0}, \omega)] \sim \omega^{1-2/D_w} \; , \tag{8.120}$$

which is the generalization to anisotropic media of Cushman's result, except that his formula inconsistently carries an $O(|\mathbf{k}|^2)$ term.

At this point we still have not obtained an explicit form for the probability density $P_d(r, t)$, and although Eq. (117) is exact, its solution is not easy to find. In the literature, two different equations have been proposed for $P_d(r, t)$. The first one was suggested by O'Shaughnessy and Procaccia (1985). They wrote down the flux as

$$\mathbf{j}(\mathbf{r}, t) = -D(\mathbf{r})\nabla P_d(\mathbf{r}, t) \; , \tag{8.121}$$

with (see Chapter 3)

$$D(\mathbf{r}) = a|\mathbf{r}|^{-\theta} \; , \tag{8.122}$$

where a is a contant. Given this form for the flux, it is not possible to Fourier transform Eq. (116). However, O'Shaughnessy and Procaccia (1985) used the radial form of the Laplace operator given by

$$\nabla^2 = \frac{d^2}{dr^2} + \frac{D_f - 1}{r}\frac{d}{dr} \; , \tag{8.123}$$

in an attempt to embody the restriction of the diffusion to a subspace of fractal dimension D_f. Given this, the solution to the generalized diffusion equation can be obtained

$$P_d(r,t) = \frac{D_w}{D_f \Gamma(D_f/D_w)} \left(\frac{1}{aD_w^2 t}\right)^{D_f/D_w} \exp\left(-\frac{r^{D_w}}{aD_w^2 t}\right), \qquad (8.124)$$

where $\Gamma(x)$ is the gamma function. From Eq. (124) it is easy to obtain Eq. (64). However, Eq. (124) does not satisfy Eq. (116), which is not surprising given the assumptions of O'Shaughnessy and Procaccia. In general, as the exact treatment of the problem in the previous sections indicated, an equation which is local and does not contain memory, is not expected to describe the long-time behavior of diffusion in heterogeneous and fractal porous media. However, Eq. (124) has been found to be relatively accurate at *short* times, which is also surprising since one expects the memory and non-locality effects to be particularly important at short times.

Another equation for $P_d(r,t)$, proposed by Guyer (1985), is given by

$$P_d(r,t) \sim t^{-D_f/D_w} \exp\left[-\left(\frac{r}{t^{1/D_w}}\right)^{\nu}\right], \qquad (8.125)$$

where $\nu = D_w/(D_w - 1)$. This equation has been found to be accurate at long times, although it also does not satisfy Eq. (116). Using Eq. (118), one can show that

$$\hat{P}_d(\mathbf{k},\omega) = \omega^{-1}(1 + b\omega^{-2/D_w}|\mathbf{k}|^2 + \cdots), \qquad (8.126)$$

where b is a constant, and that both Eq. (124) and (125) are in some sense consistent with (126), so that one should investigate higher order terms in the above expansion in order to understand the difference between Eqs. (124) and (125). The conclusion is that this problem remains unsolved. Note, however, that both Eq. (124) and Eq. (125) predict that for fractal porous media

$$P_0 \sim t^{-D_f/D_w}. \qquad (8.127)$$

Alexander and Orbach (1982) observed that for *both* two- and three-dimensional percolating networks the quantity D_f/D_w is nearly 2/3. The reason for this remarkable result is not yet understood.

8.5 Derivation of Archie's law

A useful empiricism for sedimentary rocks is Archie's law

$$\sigma = \sigma_f \phi^m, \qquad (8.128)$$

where σ_f is the electrical conductivity of the fluid saturating the porous media. The exponent m has been found to vary anywhere between 1.3 and 4, depending upon consolidation and other factors. Archie's law has been found to hold even for igneous rocks (Brace *et al.*, 1968; Brace and Orange, 1968). However, Archie's law may take a more complex form for clayey or shaly porous media because, e.g., the clays, which are capable of ion exchange, can complicate conduction mechanisms. Note that Archie's law implies that the fluid phase remains connected at all saturations, i.e., its percolation threshold is zero.

There have been many attempts to derive Archie's law in order to understand its origin. Here, we briefly review these works since most of them used a variation of EMA. Sen (1981, 1984), Sen *et al.* (1981), Mendelson and Cohen (1982), and Yonezawa and Cohen (1983) showed that a modification of EMA can be used to derive Archie's law. This version of EMA was called the *self-similar* EMA because of the assumption that a rock grain is coated with the fluid which includes coated rock grains, the coating at each level consisting of other coated grains. First consider the standard EMA, Eq. (66). With a binary distribution, $f(W) = p\delta(W - \sigma_f) + (1 - p)\delta(W - \sigma_r)$, and $\sigma = W_e$, Eq. (66) becomes

$$\sigma = \frac{-A \pm \sqrt{A^2 + 4(\gamma^{-1} - 1)\sigma_f\sigma_r}}{2(\gamma^{-1} - 1)} \quad , \tag{8.129}$$

where $\gamma^{-1} = Z/2$ for discrete EMA and $\gamma^{-1} = d$ for a d-dimensional continuum model, and

$$A = [\sigma_f - (\gamma^{-1} - 1)\sigma_r - \gamma^{-1}(\sigma_f - \sigma_r)p] \quad ,$$

where σ_r is the rock conductivity. Equation (129) has two solutions, but only one of them is physically meaningful. In the limits $p \to 0$ and $p \to 1$, we get

$$\sigma_{(1)} = \sigma_f \left[1 + (1 - p)\gamma^{-1}\frac{\sigma_r - \sigma_f}{\sigma_r + (\gamma^{-1} - 1)\sigma_f} \right] \quad , \tag{8.130}$$

$$\sigma_{(2)} = \sigma_r \left[1 + p\gamma^{-1}\frac{\sigma_f - \sigma_r}{\sigma_f + (\gamma^{-1} - 1)\sigma_r} \right] \quad . \tag{8.131}$$

The basic idea behind the self-similar EMA is as follows. One starts with the pure fluid system, replaces small portions of it by pieces of rock step by step, and applies EMA at each step. Assume that $\sigma^{(i)}$ is the conductivity of the mixture at a given step i, and replace a small volume Δq_i of the medium by grains of rock. Equation (66) yields

$$\frac{\sigma^{(i)} - \sigma^{(i+1)}}{\sigma^{(i)} + (\gamma^{-1} - 1)\sigma^{(i+1)}}(1 - \Delta q_i) + \frac{\sigma_r - \sigma^{(i+1)}}{\sigma_r + (\gamma^{-1} - 1)\sigma^{(i+1)}}\Delta q_i = 0 \quad . \tag{8.132}$$

If Δq_i is small enough, we obtain from Eq. (132)

$$\sigma^{(i+1)} = \sigma^{(i)} \left[1 + \gamma^{-1}\frac{\sigma_r - \sigma^{(i)}}{\sigma_r + (\gamma^{-1} - 1)\sigma^{(i)}}\Delta q_i \right] \quad . \tag{8.133}$$

Note that by denoting the volume fraction of the fluid at the ith stage by p_i we have, $\Delta p_i q_i = p_i - p_{i+1}$. If we use Eq. (133) repeatedly, we obtain for the case of $\sigma_r = 0$

$$\sigma^{(n+1)} = \sum_{i=1}^{n} \left[1 + \frac{\gamma^{-1}}{\gamma^{-1} - 1} \frac{p_{i+1} - p_i}{p_i} \right] \sigma^{(0)} \quad , \tag{8.134}$$

where $\sigma^{(0)} = \sigma_f$ and $p_0 = 1$ (where $p_0 = 1$ corresponds to pure fluid, the system we started with). If we take the limit $\Delta q_i \to 0$, then Eq. (132) becomes a differential equation

$$\frac{d\sigma^{(i)}}{\sigma^{(i)}} = -\frac{dp_i}{p_i} \frac{\sigma_f - \sigma^{(i)}}{\sigma_f + (\gamma^{-1} - 1)\sigma^{(i)}} \gamma^{-1} \quad , \tag{8.135}$$

which, after integrating and using the boundary condition $\sigma^{(i)} = \sigma_f$ for $p_i = 1$, yields Eq. (128) with

$$m = \frac{\gamma^{-1}}{\gamma^{-1} - 1} \quad . \tag{8.136}$$

Thus, if we interpret p as the porosity of the medium, then EMA produces Archie's law. Equation (136) also shows that, consistent with experimental data, m is not universal but depends on the connectivity of the system.

What is the geometrical interpretation of this result? For a network model of a pore space, Yonezawa and Cohen (1983) presented a nice interpretation which can be summarized as follows. At the ith stage every bond has the conductivity $\sigma^{(i)}$, and then a small fraction Δq_i of the bonds is replaced by a resistor with conductivity σ_r and use EMA to estimate the conductivity of the new system. This is equivalent to putting a resistor parallel to the original one on each bond. The conductivity $\sigma^{(ia)}$ of an added resistor should be

$$\sigma^{(ia)} = \gamma^{-1} \sigma^{(i)} \frac{\sigma_2 - \sigma^{(i)}}{\sigma_2 + (\gamma^{-1} - 1)\sigma^{(i)}} \Delta q_i \quad , \tag{8.137}$$

if the original resistor belongs to the host medium, otherwise it is given by Eq. (133). The implication is that even when $\sigma_r = 0$, the application of EMA at each stage makes the link between the nodes conducting, because of adding a parallel conducting resistor, and thus there is always a sample-spanning cluster of conducting bonds. Translating this for the rock-fluid system, it implies that this procedure guarantees the continuity of the fluid phase and the granularity of the rock grains. Note that in the original derivation of Sen *et al.* (1981) m was found to be $3/2$, which corresponds to $d = 3$ in the continuous EMA (or $Z = 6$ in the discrete EMA), which corresponds to spherical grains. For non-spherical particles $m > 3/2$, but under certain circumstances one can even have $m < 3/2$. Mendelson and Cohen (1982) give $m = \sum_i (1 - \mathcal{P}_i)^{-1}/3$, where $\mathcal{P}_i$'s are the depolarization factors and $\sum_i \mathcal{P}_i = 1$. Bussian (1983) generalized the self-similar EMA to include finite rock conductivity σ_r, and fitted the resulting formula to data, treating m and σ_r as adjustable parameters. He found $m \geq 3/2$ in almost all cases he considered, and argued that this is because clay gives a finite value to σ_r, and since clay particles are usually flat, they increase m (Mendelson and Cohen, 1982).

One major drawback of the above derivation of Archie's law is that it pertains only to a microstructure whose solid component is disjoint. This difficulty was circumvented by Sheng (1990) who generalized the self-similar EMA to a *three-component* system consisting of fluid, solid, and *cement material*. Component one, the starting phase, is composed of a mixture of fluid and cement material. Sheng (1990) showed that the self-similar EMA with three components reproduces Archie's law with

$$m = \frac{5 - 3\mathcal{P}}{3(1 - \mathcal{P}^2)} \, , \tag{8.138}$$

where $\mathcal{P}$ is the depolarization coefficient of the grains, but with the added feature that the solid grains also remain connected.

Although the self-similar EMA is successful in providing a derivation of Archie's law, its use for understanding various properties of rocks is not without conceptual difficulty. Generally speaking, rocks have porosities less than 40%. This is far from the dilute limit in which the assumptions of the models can be justified [recall that Eq. (133) is valid only in the dilute limit]. If the porosity is low, then the grains are in close contact with one another and the interaction between them is important. Such interactions cannot be taken into account correctly by an EMA. In fact, Milton (1984) showed that the self-similar EMA accounts for the interactions correctly only in the special case in which grains of any given size are surrounded by much smaller grains, and grains of the same size are far separated from each other. This is hardly the case in natural rocks. Moreover, rocks with very similar grains can have very different values of m, and rocks with very dissimilar grains can have very similar values of m. These cannot be explained with the self-similar EMA. Hilfer (1991b) presented an alternative derivation of Archie's law based on a percolation model. His result $m = 1 + \mu$, where μ is the conductivity exponent, indicates that m is universal, in contradiction with the data, unless μ is non-universal, which is the case only in certain model porous media.

Wong *et al.* (1984) showed that their shrinking tube model, discussed in Chapter 4, can reproduce Archie's law such that m would depend on the skewness of the pore size distribution. Their experiments with fused-glass beads and real rocks indicated that m is larger in porous media with a wider fluctuation of pore sizes, which their model also correctly predicted. Moreover, they showed that their model predicts that K is related to the porosity by

$$K \sim \phi^{m'} \, , \tag{8.139}$$

where $m' = 2m$. Equation (139) is consistent with the empirical Kozeny-Carman correlation, Eq. (27). Thus, Wong *et al.*'s model allows a unified derivation of two well-established and widely-used empirical laws. Note that if, e.g., $m = 3$, we obtain $m' = 6$, consistent with the experimental observations (Wyllie and Rose, 1950; Timur, 1968) that if K is to be related to ϕ by a power law, the exponent has to be large.

8.6 Relation between permeability and electrical conductivity

As discussed above, an exact relation between the permeability and electrical conductivity of porous media does not exist. Although many empirical and semi-empirical relations between K and σ have been proposed in the past, almost all of them "work" for only certain classes of porous media, and not for other porous media. For example, although Wong *et al.* (1984) found that $\sigma \sim \phi^m$ and $K \sim \phi^{m'}$, the relation $m' = 2m$ is restricted to their model, and is not expected to hold for a general porous medium. Johnson *et al.* (1986), however, introduced a well-defined parameter Λ, defined by

$$\Lambda = 2\frac{\int |E(\mathbf{r})|^2 dV_p}{\int |E(\mathbf{r})|^2 dS_p} \neq \frac{V_p}{S_p} \quad , \tag{8.140}$$

where $E(\mathbf{r})$ is the potential in the electrical conduction problem, and V_p and S_p are, respectively, the pore volume and the pore surface area. Note that V_p/S_p is a geometrical parameter that can be measured and is *independent* of any transport process. On the other hand, Λ is a dynamical property, defined for the specific problem of electrical conduction, and cannot be measured by geometrical analysis alone. Since $E(\mathbf{r})$ vanishes in certain regions of the pore space (for example, in an isolated region), Λ is roughly a measure of *dynamically connected* pores of the medium. Johnson *et al.* (1986) proposed that for three-dimensional porous media

$$K \simeq a_4\frac{\Lambda^2}{8F} = a_4\frac{\Lambda^2}{8}\frac{\sigma}{\sigma_f} \quad , \tag{8.141}$$

where $a_4 = O(1)$. It should be said at the outset that, for the reasons discussed above, this relation [and Eqs. (99) and (100)] cannot *in general* be exact, although, because of the physical meaning of Λ, it is certainly appealing. If we compare Eq. (141) with Katz and Thompson's (1986) relation, Eq. (99), we obtain

$$a_3 l_c^2 = a_4\frac{\Lambda^2}{8} \quad . \tag{8.142}$$

This is immediately indicative of the possibility that Λ can be measured, since l_c is obtained from a mercury-injection curve. Various authors have tested the validity of Eq. (141) (Banavar and Johnson, 1987; Straley *et al.*, 1987; Banavar *et al.*, 1988; Schwartz and Banavar, 1989; Saeger *et al.*, 1991; Kostek *et al.*, 1992; Schwartz *et al.*, 1993), using a variety of numerical and analytical methods, as well as experimental data for well-defined systems. Avellaneda and Torquato (1991) investigated the relation between K and σ, and derived the conditions under which an approximate relation between K and σ may be expected. It now appears that (Kostek *et al.*, 1992), unless the system contains two widely different relevant length scales, Eq. (141) would be very accurate.

8.7 Relation between permeability and nuclear magnetic resonance

Many years ago, Timur (1969) suggested that nuclear magnetic resonance (NMR) may be used as a way of measuring K. This may seem to be impractical, since most of the fluid in a porous medium is stored in the pore bodies, whereas K and other transport properties are controlled by the pore throats. Various authors (for example, de Gennes, 1982; Banavar *et al.*, 1985) have investigated this issue. In particular, Kenyon *et al.* (1988) studied the relation between NMR and the permeability of 56 water-saturated sandstones. They found that the decay of proton magnetization $M_z(t)$ is described by a stretched exponential

$$M_z(t) = m_0 \exp\left[-\left(\frac{t}{T_1}\right)^{\delta_1}\right] , \qquad (8.143)$$

where T_1 is a relaxation time discussed below, and that $\log K$ shows a very strong correlation with $\log(\phi^4 T_1^2)$, where ϕ is the porosity of the system. Figure 8.15 presents their data plotted in this fashion, from which we obtain

$$K \sim (\phi^4 T_1^2)^{\delta_2} , \qquad (8.144)$$

where $\delta_2 \simeq 0.7$. Billardo *et al.* (1991) carried out NMR experiments on, and measured the permeabilities of, 44 different sandstones and found that Eq. (144) fits their data very well. Banavar and Schwartz (1987) investigated the same problem in the grain consolidation model of Roberts and Schwartz (1985) and the shrinking-tube model of Wong *et al.* (1984) (see Chapter 4) and reached the same conclusion. These results may seem surprising until we ask ourselves: What is the meaning of the relaxation time T_1? The protons in the hydrogen of water molecules carry nuclear magnetic moment, which enables them to align themselves with an externally applied field. But because water molecules at room temperature are thermally agitated, only a few of the protons actually align themselves with the external field, which are, however, detectable. If the external field is removed, the system will go back to its equilibrium configuration. The time that the system needs to do this is also the same as the time that it needs to build up its external magnetization after the external field has been applied. This time is usually denoted by T_1, and is called spin-lattice relaxation time for protons. Why should T_1 be related to K? Experiments show that if one measures T_1 for water saturating a porous medium, one finds T_1 to be much smaller than it is for the same water in the bulk. This is because T_1 for water is affected strongly by surface relaxation mechanisms, and thus it is expected to be sensitive to the microstructure of the porous medium and provide insight into

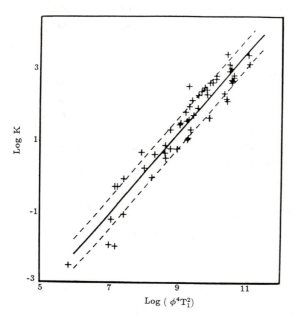

Figure 8.15: Correlation of the permeability K of water-saturated sandstones with the porosity ϕ and the NMR relaxation time T_1 (after Kenyon *et al.*, 1988).

the structure of the pore space. Thus it appears likely that NMR can serve as a means of measuring K.

Thompson *et al.* (1989) used deuterium NMR to study sandstones, carbonate rocks, and synthetic porous samples. The reason for using deuterium instead of proton is that (Williams and Fung, 1982) it has a much smaller magnetic moment than protons. They measured T_{1p}, the longitudinal relaxation time in the rotating frame, found that the magnetization $M_z(t)$ obeys a stretched-exponential similar to Eq. (143), with T_1 replaced by τ_{SE}, the corresponding stretched-exponential relaxation time. They argued that the parameter ℓ_c [see Eqs. (99) and (100)] is proportional to a time scale τ_{NMR}, the NMR relaxation time. Their experimental data appear to be in complete agreement with this argument. Moreover, they showed that their data is fully consistent with

$$K \sim (T_{1p}\phi)^2 \sim (\tau_{NMR}\phi)^2 \quad , \tag{8.145}$$

which, in some sense, is consistent with Eq. (144), except that the exponents of T_{1p} (or τ_{NMR}) and T_1 in these equations are not the same, although Thompson *et al.* (1989) also mentioned that their data could also be fitted with an equation similar to (144), but the quality of the resulting fit would be slightly worse than that provided by (145). Thompson *et al.* (1989) also showed that T_{1p} is proportional to the *width* of the pore size distribution, and also the water saturation in a partially-saturated porous medium. Since NMR measurements can be made at depth in the earth with well-logging tools, they provide one of the few means of estimating *in situ* permeabilities and saturations.

8.8 Dynamic permeability

So far, we have discussed the static permeability of a porous medium. But, what can we say about its dynamic permeability? If we measure dynamic permeability, can it provide extra information about the structure of the porous medium? Dynamic permeability $\tilde{K}(\omega)$ is defined by a generalized Darcy's law

$$\langle \mathbf{v}(\omega) \rangle = -\frac{\tilde{K}(\omega)}{\eta} \boldsymbol{\nabla} P(\omega) \quad , \tag{8.146}$$

where ω is the frequency, and $\boldsymbol{\nabla} P(\omega) = \boldsymbol{\nabla} Pe^{-i\omega t}$. This problem has been investigated by several groups. Auriault *et al.* (1985) made measurements of $\tilde{K}(\omega)$ to test the validity of Eq. (146). Charlaix *et al.* (1988b) also measured $\tilde{K}(\omega)$ for a variety of fused-glass beads and crushed glass in a frequency range of 0.1 Hz to 1kHz. Sheng and Zhou (1988), Zhou and Sheng (1989), Johnson *et al.* (1987), Chapman and Higdon (1992), and Knackstedt *et al.* (1993) studied the problem theoretically and numerically. [Some errors in the studies of Sheng and Zhou were corrected by Chapman and Higdon (1992).] Let us summarize some of the more important results, most of which were derived by Johnson *et al.* (1987).

We obviously have to use Eq. (146) together with

$$T_p(\omega)\rho_f \frac{\partial \langle \mathbf{v} \rangle}{\partial t} = -\boldsymbol{\nabla} P \quad , \tag{8.147}$$

as the starting point, where $T_p(\omega)$ is the frequency-dependent tortuosity (a dimensionless quantity), and ρ_f the density of the fluid. $T_p(\omega)$ and $\tilde{K}(\omega)$ are related to each other:

$$\tilde{K}(\omega) = \frac{i\eta\phi}{T_p(\omega)\omega\rho_f} \quad , \tag{8.148}$$

where $i = \sqrt{-1}$. As $\omega \to 0$, we obviously have $\tilde{K}(\omega) \to K$, and

$$\lim_{\omega \to 0} T_p(\omega) = \frac{i\eta\phi}{K\omega\rho_f} \quad . \tag{8.149}$$

Johnson *et al.* (1987) showed that in the high-frequency limit

$$\lim_{\omega \to \infty} T_p(\omega) = T_{p\infty} + \frac{C_\infty}{\sqrt{-i\omega}} \quad , \tag{8.150}$$

where $C_\infty = KB/(\eta\phi)$, B being the bulk modulus of the fluid. Moreover, one has

$$T_{p\infty} = \phi F \quad , \tag{8.151}$$

for *any* porous medium, a result that is generally attributed to Lord Rayleigh. Johnson *et al.* (1987) also showed that in the high-frequency limit, one has

$$\lim_{\omega \to \infty} T_p(\omega) = T_{p\infty} \left[1 + \sqrt{\frac{i\eta}{\omega \rho_f}} \frac{2}{\Lambda} \right] \quad , \tag{8.152}$$

$$\lim_{\omega \to \infty} \tilde{K}(\omega) = \frac{i\eta\phi}{T_{p\infty}\omega\rho_f} \left[1 - \sqrt{\frac{i\eta}{\omega \rho_f}} \frac{2}{\Lambda} \right] \quad , \tag{8.153}$$

[compare Eqs. (152) and (150)] so that the *high-frequency* limit provides information on Λ, and thus on K through Eq. (141). Moreover, Johnson *et al.* (1987) proposed that

$$T_p(\omega) = T_{p\infty} + \frac{i\eta\phi}{\omega K \rho_f} \sqrt{1 - \frac{4iT_{p\infty}^2 K^2 \omega \rho_f}{\eta \Lambda^2 \phi^2}} \quad , \tag{8.154}$$

$$\tilde{K}(\omega) = K \left(\sqrt{1 - \frac{4iT_{p\infty}^2 K^2 \omega \rho_f}{\eta \Lambda^2 \phi^2}} - \frac{iT_{p\infty} K \omega \rho_f}{\eta \phi} \right)^{-1} \quad . \tag{8.155}$$

These results, together with those of Sheng and Zhou, and the experimental data of Charlaix *et al.* (1988) also indicate that, in general, $\tilde{K}(\omega)$ obeys the following scaling equation

$$\frac{\tilde{K}(\omega)}{K} = f(\omega/\omega_c) \quad , \tag{8.156}$$

where $\omega_c = \eta\phi/(T_\infty K \rho_f)$ is a characteristic frequency at which a crossover in $\tilde{K}(\omega)$ from a viscous flow regime to an inertial one takes place. The function $f(\omega/\omega_c)$ is found to be universal, independent of the microstructure of the porous medium. Johnson (1989) showed that Eq. (155) works extremely well when compared with numerical simulations, and it also obeys Eq. (156). Therefore, although Eq. (156) tells us that we cannot gain much microstructural information about the porous medium by just measuring $\tilde{K}(\omega)$ [since $f(\omega/\omega_c)$ is universal], Eq. (153) can be used to obtain K, the static permeability of the porous medium (via Λ). Experimental data of Charlaix *et al.* (1988b), as well as the analytical and numerical calculations of Sheng and Zhou (1990), Chapman and Higdon (1992), and Knackstedt *et al.* (1993) seem to support the validity of Eq. (156).

8.9 Conclusions

Several methods of estimating transport properties of porous media are now available. Some of these methods, such as the effective-medium approximations, can provide an order of magnitude estimates of the transport properties, while others, such as the random-walk methods or network simulations, are even more accurate and are capable of providing precise predictions for the transport properties. However, when the heterogeneities are broadly distributed, some of the methods break down and are no longer accurate. Interesting relations have been found between various transport properties on one hand, and between them and experimental data that can be measured by certain techniques such as NMR on the other hand. Such relations have openned up the possibility of gaining a much deeper understanding of porous media and their static and dynamical properties.

Chapter 9

Dispersion in Porous Media

9.0 Introduction

So far we have discussed mainly flow and transport processes that involve only one fluid and one fluid phase. Beginning with this Chapter, we discuss those phenomena that are at the next level of complexity, namely, those that involve at least two fluids and one or two fluid phases. One of the most important of such phenomena is hydrodynamic dispersion which is discussed in this Chapter. Miscible displacement processes, which are generalizations of dispersion phenomena in which the viscosities of the two fluids are *not* the same, are discussed in Chapter 11, while two-phase flows are discussed in Chapter 12.

9.1 The phenomenon of dispersion

When two miscible fluids are brought into contact, with an initially sharp front separating them, a transition zone develops across the initial front, the two fluids slowly diffuse into one another, and after some time develop a diffused mixed zone. If one assumes that the volumes of the two fluids do not change upon this mixing, then the net transport of one of the fluids across any arbitrary plane can be represented by the Fick's second law of diffusion

$$\frac{\partial C}{\partial t} = D_m \nabla^2 C \quad .$$

(9.1)

Here C is solute the concentration, t is the time, and D_m is the molecular diffusivity. This mixing process is independent of whether or not there is a convective current through the medium. However, if the two fluids are also flowing, then there will be some additional mixing of a different sort: convective mixing. This mixing is caused by a non-uniform velocity field, which in turn may be caused by the morphology of the medium, the fluid flow condition, and chemical or physical interactions with the solid surface of the medium , and is called *hydrodynamic dispersion*. Dispersion is important to a wide variety of processes such as miscible displacements in enhanced oil recovery, salt water intrusion in coastal aquifers, where fresh and salt waters mix by a dispersion process, in-situ study of the characteristics of an aquifer, where a classical method of determining such characteristics is injecting fluid tracers in it and measuring their travel times, and the pollution of the surface waters because of industrial and nuclear wastes. Dispersion phenomena also occur in flow and reaction in packed-bed chemical reactors; these have been studied extensively by chemical engineers for a long time (see, for example, Bernard and Whilhelm, 1950, as one of the earliest papers on this subject).

9.2 Mechanisms of dispersion processes

In steady flow through a disordered porous medium, the *transit-time*, or *first-passage time*, of a fluid particle between entrance and exit planes depends on the path, or streamline, that it follows through the pore space. A population of particles passing the entrance plane at the same instant will arrive at the exit plane by a set of streamlines with a *distribution* of transit times. Thus a solute concentration front will spread in the mean-flow direction as it passes through the medium. The resulting first-passage time distribution (FPTD) is a measure of *longitudinal dispersion* in a porous medium.

Likewise, a population of particles passing simultaneously through a restricted area of the entrance plane will not follow entirely the mean flow to the exit plane, but will be dispersed in the transverse directions as well, that is, the population and the set of streamlines travelled will have a wider distribution of exit locations than of entrance locations. Thus, a concentration front will also spread laterally on the way to the exit plane. The distribution of the first-passage times for crossing the system at a given transverse plane is a measure of *transverse dispersion* in a porous medium.

Two basic mechanisms drive dispersion in macroscopically homogeneous, microscopically disordered porous media, and arise in the pore-level velocity field forced on the flowing fluid by the irregularity of the pore space. The first mechanism is *kinematic*: streamtubes divide and rejoin repeatedly at the junctions of flow passages in the highly interconnected pore space. The consequent tangling and divergence of streamlines is accentuated by the widely varying orientations of flow passages and coordination numbers of the pore space. The result is a wide variation in the lengths of the streamlines and their downstream transverse separations. The second mechanism is *dynamic*: the speed with which a given flow passage is traversed depends on the flow resistance or hydraulic conductance of the passage, its orientation, and the local pressure field. The two mechanisms conspire to produce broad FPTDs between entrance and exit plane. These two mechanisms suggest two possible geometrical aspects of dispersion processes, defined with respect to the mean-velocity direction, i.e., a longitudinal effect due to the difference between the velocity components in the direction of mean flow and a transverse effect due to the differences between local velocity components orthogonal to the direction of the mean flow.

These two mechanisms of dispersion do *not* depend on molecular diffusion. However, diffusion modifies the effects of the two basic mechanisms by moving material from one streamline to another, and also by the usually weaker streamwise diffusion of material relative to the average velocity. The solid matrix of a porous medium of course acts locally as a separator of streamlines and thus as a barrier to diffusion, and therefore the modification of dispersion by diffusion depends on pore space morphology and how it in turn affects local flow and concentration fields. The effect of molecular diffusion is usually important only at the pore level, where it acts to transfer the tracer particles out of slow or stagnant regions of the pore space.

9.3 The convective-diffusion equation

Dispersion processes in microscopically disordered and macroscopically isotropic and homogeneous porous media are usually modelled based on the convective-diffusion equation (CDE) (see Chapter 2)

$$
\frac{\partial C}{\partial t} + \langle \mathbf{v} \rangle \cdot \boldsymbol{\nabla} C = D_L \frac{\partial^2 C}{\partial x^2} + D_T \nabla_T^2 C \ , \tag{9.2}
$$

where $\langle \mathbf{v} \rangle$ is the macroscopic mean velocity, C is the mean concentration of the solute, and ∇_T^2 is the Laplacian in transverse directions. For the sake of simplicity we delete $\langle \cdot \rangle$ and denote the magnitude of the average fluid velocity vector by v. Thus the basic idea is to model dispersion processes as anisotropic diffusional spreading of concentration, the diffusivities being the longitudinal dispersion coefficient D_L and the transverse dispersion coefficient D_T. One important goal of any study of dispersion is to investigate the conditions under which dispersion processes in a given environment *cannot* be represented by the CDE.

Dispersion is said to be *diffusive* or *Gaussian* if it obeys a CDE. If a population of the solute particles is injected into the medium at $\mathbf{r}_0 = (x_0, y_0, z_0)$ at $t = 0$ [i.e., $C(x_0, y_0, z_0, 0) = C_0$], for diffusive dispersion the probability density $P(\mathbf{r}, t)$ obeys the Gaussian distribution

$$
P(\mathbf{r}, t) = (8\pi^3 D_L D_T^2 t)^{-3/2} \exp\left[-\frac{(x - x_0 - vt)^2}{4D_L t} - \frac{(y - y_0)^2}{4D_T t} - \frac{(z - z_0)^2}{4D_T t} \right] \ , \tag{9.3}
$$

where $P(\mathbf{r}, t)d\mathbf{r}$ is the probability that a solute particle is in a plane between $\mathbf{r}$ and $\mathbf{r} + d\mathbf{r}$ at time t, and $\mathbf{r} = (x, y, z)$. $P(\mathbf{r}, t)$ is proportional to C/C_0, and therefore Eq. (3) represents a solution of Eq. (2). If one defines $\mathcal{Q}(\zeta - \zeta_0, t)dt$ as the probability that a solute particle, beginning in the plane at ζ_0, will cross, *for the first time*, a plane at ζ between t and $t + dt$, then from Eq. (3) one can easily obtain the FPTD in a given direction, since $\mathcal{Q}$ and P are related

$$
P(\zeta - \zeta_0, t) = \int_0^t P(\zeta - \zeta_1, t - \tau)\mathcal{Q}(\zeta_1 - \zeta_0, \tau)d\tau \ , \tag{9.4}
$$

and therefore

$$
\mathcal{Q}(\zeta - \zeta_0, t) = |\zeta - \zeta_0|(4\pi D_\zeta t^3)^{-1/2} \exp\left[-\frac{(\zeta - \zeta_0 - v_\zeta t)^2}{4D_\zeta t} \right] \ , \tag{9.5}
$$

where D_ζ and v_ζ are the dispersion coefficient and the mean flow velocity in the ζ-direction, respectively. Various moments of Q yield information about the flow field and the dispersion processes. For example, for the longitudinal direction we have

$$\langle t \rangle = \frac{L}{v} \quad , \tag{9.6}$$

and

$$\langle t^2 \rangle = \langle t \rangle^2 \left(1 + \frac{2D_L}{Lv} \right) \quad , \tag{9.7}$$

where $L = \zeta - \zeta_0$. In general, one can easily show that for large L and to the leading order one has $\langle t^n \rangle \sim \langle t \rangle^n$, where $n > 1$ is any integer number. Of course, this is true if the description of dispersion by a CDE is appropriate, and therefore one way of showing that a CDE cannot describe a dispersion process in a certain medium is to show that $\langle t^n \rangle / \langle t \rangle^n$ $(n > 1)$ is *not* a constant (i.e., this ratio depends on t), and one needs more information to describe various moments of the FPTD. This is discussed later in this Chapter.

9.4 Measurement of dispersion coefficients

Since measurement of D_L and D_T is not as straightforward as that of the permeability K or the diffusivity D, we briefly discuss it here. There are two different ways of observing a dispersion phenomenon and measuring its properties. One can observe the variations of the solute concentration either as a function of the distance from the entrance to the system *at a fixed time*, or as a function of the time *at a fixed distance* from the entrance to the system. We describe methods for both cases.

We assume that the system is one dimensional, and that the fluid velocity v is constant. The initial and boundary conditions are: $C(x \geq 0, t = 0) = 0$, $C(x = 0, t > 0) = C_0$, and $C(x \to \infty, t \geq 0) = 0$. Then the solution to the one-dimensional version of Eq. (2) can be obtained by the Laplace transform technique. Assume that the length of the porous medium is L, and define the dimensionless quantities

$$\alpha_\pm = \frac{x \pm vt}{(4D_L t)^{1/2}} \quad , \tag{9.8}$$

then the solution of the CDE, subject to the above initial and boundary conditions, is given by

$$\frac{C}{C_0} = \frac{1}{2}\mathrm{erfc}\,(\alpha_-) + \frac{1}{2}\exp\left(\frac{xv}{D_L}\right)\mathrm{erfc}\,(\alpha_+) \tag{9.9}$$

where $\mathrm{erfc}(z)$ is the complementary error function. The second term of Eq. (9) is usually very small compared with the first term, and can be safely neglected, in which case

$$\frac{C}{C_0} = \frac{1}{2}\mathrm{erfc}\,(\alpha_-) = \frac{1}{\sqrt{\pi}}\int_{\alpha_-}^{\infty} \exp\,(-\kappa^2)d\kappa \quad , \tag{9.10}$$

where we used the definition of the complementary error function. Equation (9) is rewritten as

$$\frac{C}{C_0} = \frac{1}{\sqrt{2\pi}} \int_{\alpha'}^{\infty} \exp\left(-\frac{\kappa^2}{2}\right) d\kappa \ , \tag{9.11}$$

where $\alpha' = 2\alpha_-/\sqrt{2}$. Figure 9.1 shows concentration profiles C/C_0 versus $x_d = x/L$ and $t_d = \int_0^t v dt/(L\phi)$, which is a dimensionless time. Equation (11) tells us that, *at a fixed time*, the solution is a normal distribution function $1 - N[(x - \langle x \rangle)/s]$ with the average $\langle x \rangle = vt$, and the standard deviation $s = \sqrt{2D_L t}$. Using well-known properties of a normal distribution, we can write

$$N(1) \simeq 0.84 \ , \quad N(-1) \simeq 0.1587 \tag{9.12}$$

These two properties of a normal distribution allow us to measure D_L. The width w of the transition zone, the zone between a region with pure solute and a region with pure solvent, is usually defined as the difference between values of x at which $C/C_0 = 0.16$ and $C/C_0 = 0.84$. Thus

$$w = 2s = 2\sqrt{2D_L t} = x_{0.16} - x_{0.84} \ , \tag{9.13}$$

from which we obtain

$$D_L = \frac{(x_{0.16} - x_{0.84})^2}{8t} \ . \tag{9.14}$$

Thus, if we obtain a graph of C versus x at a fixed time t, we can easily determine D_L. In many cases it is easier to fix x (for example, at the exit of the system) and measure the concentration as a function of t. In this case, Eqs. (11) and (12) tell us that

$$D_L = \frac{1}{8} \left(\frac{x - v t_{0.16}}{\sqrt{t_{0.16}}} - \frac{x - v t_{0.84}}{\sqrt{t_{0.84}}} \right)^2 \ , \tag{9.15}$$

where $t_{0.16}$ is the time at which $C/C_0 = 0.16$.

It is often easier to do the measurements for an unconsolidated porous medium. Here we describe methods for measuring D_L and D_T in such porous media, but similar methods can also be used for consolidated porous media. One saturates a packed column with one fluid (the solvent), displaces it with another miscible fluid (the solute), and measures the fluid composition at the exit end of the column as a function of displacement. Brigham *et al.* (1961) developed a convenient method for determining D_L from data of this type. In this method one plots $\lambda_p = (V/V_c - 1)/\sqrt{V/V_c}$ versus the percent of the solute on an arithmetic probability paper, where V is the volume of the solute injected into the medium, and V_c is the volume of the column. Then D_L is given by

$$D_L = v L \left(\frac{\lambda_{p90} - \lambda_{p10}}{3.625} \right)^2 \ , \tag{9.16}$$

where L is the length of the column, and λ_{p90} is the value of λ_p when the solvent contains 90% displacing fluid. Equation (16) is obtained by an argument similar

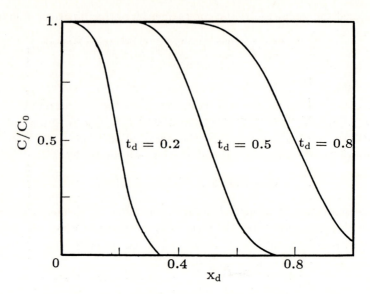

Figure 9.1: Normalized concentration profile C/C_0 versus the dimensionless distance x_d at various dimensionless times t_d.

to the one used for deriving Eq. (14), except that the mixing zone is defined as the region between values of x at which $C/C_0 = 0.1$ and 0.9. Figure 9.2 shows a schematic representation of this method.

Measuring D_T is more difficult. If the packed column is arranged as in Fig. 9.3, then a mixed zone develops in the transverse direction. If a concentration profile is made along a line perpendicular to the direction of the flow (longitudinal direction), a typical S-shaped profile will be observed (see Fig. 9.3). Then D_T is determined by plotting percent composition versus the distance from 50% composition on an arithmetic probability paper, and is estimated by the following formula

$$D_T = \frac{\text{v}}{L} \left(\frac{z_{0.9} - z_{0.1}}{3.625} \right)^2 , \qquad (9.17)$$

where $z_{0.9}$ is the transverse distance between 90% composition and 50% composition. The reader can guess the argument for deriving Eq. (17).

9.5 Dispersion in simple systems

The simplest system in which dispersion can be studied is laminar flow through a single capillary tube. This problem has been studied extensively, and in what follows we summarize the main results. As we show later in this Chapter, some of the results for dispersion in a capillary tube are surprisingly

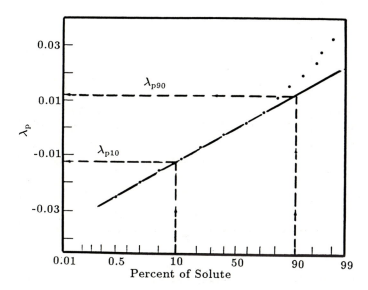

Figure 9.2: A typical solute-composition plot for determining the dispersion coefficient D_L, on arithmetic-probability paper. Circles are the data (after Brigham *et al.*, 1961).

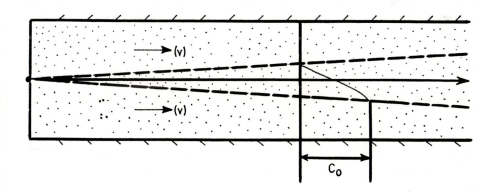

Figure 9.3: Schematic view of a transverse dispersion experiment. The area between the dashed lines is the dispersion zone (after Harleman and Rumer, 1963).

similar to those in porous media.

9.5.1 Dispersion in a capillary tube

Historically, Griffiths (1911) was the first to report some experimental results that demonstrated the essence of the dispersion process in a tube with diffusional effects present, but without mathematical treatment. He observed that a tracer fluid injected into a system of water spreads out symmetrically about a plane in the cross section which moves with the speed of flow. He commented that: "*It is obvious that the movement of the center of the column of the tracers must measure the mean speed of flow.*" It turned out that this was not as obvious as Griffiths had thought! Forty two years later, Taylor (1953) pointed out that this is a rather startling result for two reasons. First, because the water at the center of the tube moves with twice the mean speed of the flow (the Hagen-Poiseuille flow), the water at (or near) the center must approach the column of tracer, absorb the tracer as it passes through the column, and then reject the tracer as it leaves on the other side of the column. Secondly, although the velocity is asymmetrical about the plane moving at the mean speed, the column of tracer spreads out symmetrically.

Taylor (1953) and Aris (1956) studied dispersion in a cylindrical capillary tube of radius R. Starting from a CDE for a tube (see Chapter 2)

$$\frac{\partial C}{\partial t} + 2v_m \left[1 - \left(\frac{r}{R}\right)^2\right] \frac{\partial C}{\partial x} = D_m \left(\frac{\partial^2 C}{\partial r^2} + \frac{1}{r}\frac{\partial C}{\partial r} + \frac{\partial^2 C}{\partial x^2}\right) \quad , \qquad (9.18)$$

where v_m is the mean flow velocity in the tube, and defining a *mean* concentration C_m by

$$C_m = \frac{\int_0^{2\pi}\int_0^R C(r,x)r\,dr\,d\theta}{\int_0^{2\pi}\int_0^R r\,dr\,d\theta} = \frac{2}{R^2}\int_0^R Cr\,dr \quad , \qquad (9.19)$$

they showed that in the limit of long times

$$\frac{\partial C_m}{\partial t} = D_L \frac{\partial^2 C_m}{\partial x_1^2} \quad , \qquad (9.20)$$

where $x_1 = x - v_m t$ is the moving coordinate with respect to the mean-flow velocity, and

$$D_L = D_x + \frac{R^2 v_m^2}{48 D_r} \quad , \qquad (9.21)$$

where the subscripts x and r signify the fact that D_x and D_r are the contributions of axial and radial molecular diffusion, respectively. That is, if in Eq. (18) we delete $\partial^2 C/\partial x^2$ (i.e., neglect axial diffusion), D_x will also be deleted from Eq. (21); of course, $D_x = D_r = D_m$. Note that in Taylor-Aris dispersion D_L

depends *quadratically* on v_m. We define a Péclet number Pe by $Pe = Rv_m/D_m = t_{rd}/t_c$, where $t_{rd} = R^2/D_m$ is a radial diffusion time scale, and $t_c = R/v_m$ is the convection time scale. Thus, Pe is simply a measure of the competition between diffusion and convection. Then Eq. (21) is rewritten as

$$\frac{D_L}{D_m} = 1 + \frac{1}{48}Pe^2 \quad . \tag{9.22}$$

Aris (1956) also showed that for a cylindrical tube with a cross section of *any* shape, one has

$$D_L = D_m + \delta_s \frac{l_s^2 v_m^2}{D_m} \quad , \tag{9.23}$$

where l_s is a length scale of the tube, and δ_s is a shape factor that depends on the shape of the cross section. For example, for an elliptical cross section where the major and minor semi-axes are a and b, respectively, one has $l_s = a$ and

$$\delta_s = \frac{1}{48} \frac{24 - 24e^2 + 5e^4}{24 - 12e^2} \quad , \tag{9.24}$$

where $e = \sqrt{1 - b^2/a^2}$. For a circular cross section, $b = a$, $e = 0$ and $\delta = 1/48$, as expected. For dispersion in a duct of parallel plates with fully-developed laminar flow, $l_s = h$ and $\delta_s = 2/105$, where h is the half-width of the channel. Thus the quadratic dependence of D_L on v_m is independent of the shape of the cross section: it is the result of the competition between the equally-strong molecular diffusion and convection.

Aris (1956) conjectured that *any* initial distribution of concentration will ultimately approach a Gaussian distribution. Chatwin, in a series of papers (see, Chatwin, 1977, for earlier references to his work), proved this. Further important work on dispersion in tubes was done by Horn (1971) and Brenner (1980). In particular, Brenner (1980) generalized Taylor-Aris dispersion significantly and employed local and global spaces (for example, in the tube problem r is the local space and x is the global one). One can also exploit the equivalence between Langevin and Fokker-Planck equations and derive the Taylor-Aris results (Van den Broeck, 1982).

9.5.2 Dispersion in spatially-periodic models of porous media

At the next level of complexity are spatially-periodic models of porous media. We already discussed the geometrical structure of such models in Chapter 6. Diffusion, electrical conduction, and flow and dispersion in such models of porous media are discussed in Chapter 13 where we consider flow phenomena in unconsolidated porous media. For now it suffices to mention that Brenner (1980), Brenner and Adler (1982), Eidsath *et al.* (1983), Koch *et al.* (1989) and Salles *et al.* (1993) examined theoretically dispersion in spatially-periodic porous

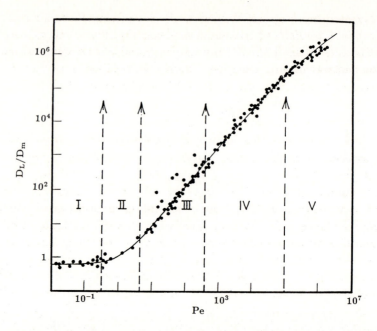

Figure 9.4: Dependence of the longitudinal dispersion coefficient D_L on the Péclet number Pe, and various dispersion regimes. D_m is the molecular diffusivity (after Fried and Combarnous, 1971).

media, and Gunn and Pryce (1969) measured D_L for flow parallel to one of the axes of a simple cubic lattice of spherical particles. In particular, Koch *et al.* (1989) showed that for a square array of cylinders or a cubic array of spheres, and in the limit $Pe \to \infty$, D_L depends quadratically on Pe, and that D_T approaches a constant value. These are in contradiction with dispersion in *disordered* porous media, for which both D_L and D_T have a much weaker dependence on Pe and, in particular, D_T does not reach a constant value (see below).

9.6 Dependence of dispersion coefficients on the Péclet number

Many researchers have carried out experimental studies of dispersion, almost exclusively, in beadpacks, unconsolidated sandpacks, and sandstones. Some of the results, mainly for unconsolidated sands, were compiled by Fried and Combarnous (1971). Work on sandstones, compiled by Perkins and Johnston (1963) and Legaski and Katz (1967) shows, however, that dispersion in consolidated porous media is similar to that in unconsolidated media. Figure 9.4 collects experimental data for D_L/D_m for sandpacks, which shows that there are five different regimes of dispersion. Figure 9.5 represents typical experimental data for D_T/D_m. For unconsolidated porous media the Péclet number is defined as

$Pe = d_g v/D_m$, where d_g is frequently taken to be the average diameter of a grain or bead. The five dispersion regimes, shown in Fig. 9.4, are as follows.

(i) $Pe < 0.3$. This is the *diffusion* regime in which convection is so slow that diffusion controls dispersion almost completely. In this regime, we have *isotropic* dispersion such that (Brigham *et al.*, 1961; Koplik *et al.*, 1988b)

$$\frac{D_L}{D_m} = \frac{D_T}{D_m} = \frac{1}{F\phi} \quad , \tag{9.25}$$

where, as usual, F is the formation factor and ϕ is the porosity of the medium. The quantity $1/(F\phi)$ varies commonly between 0.15 and 0.7, depending on the porous medium. Because of this isotropy, a concentrated sphere of solute will remain a sphere [rather than developing into an ellipsoid as indicated by Eq. (3)], but will increase in size as dispersion progresses. Although Eq. (25) has been quoted widely in the literature, a proof of it, even by a heuristic argument, is not usually given. In particular, the presence of ϕ is not obvious. However, if we consider the limit $Pe = 0$, then we can derive Eq. (25) by the following argument. As discussed in Chapter 3, Einstein's relation relates the electrical conductivity σ_f of a fluid to the molecular diffusivity D_m by $\sigma_f = ne^2 D_m/(kT)$, where n is the density of the charge carriers, e is the charge, k is the Boltzmann's constant, and T is the temperature of the system. For a porous medium, the same equation can be used, except that σ_f should be replaced by σ, the electrical conductivity of the medium, D_m by D, the effective diffusivity, and n by $n\phi$, the density of charge carriers in the medium, and thus $\sigma = n\phi e^2 D/(kT)$. If we take the quotient, we obtain $D/D_m = \sigma/(\phi\sigma_f) = 1/(F\phi)$, with $F = \sigma_f/\sigma$. This equation is equivalent to (25), since in the limit $Pe \to 0$ we have $D_L = D_T = D$.

(ii) $0.3 < Pe < 5$. This is the *transition* regime in which convection contributes to dispersion, but the effect of diffusion is still quite strong. D_L/D_m appears to increase with Pe, although it is difficult to say how!

(iii) $5 < Pe < 300$. This is the *power-law* regime. Convection dominates dispersion, but the effect of diffusion cannot be neglected, and one can write

$$\frac{D_L}{D_m} = \frac{1}{F\phi} + a_L Pe^{\beta_L} \quad , \tag{9.26}$$

$$\frac{D_T}{D_m} = \frac{1}{F\phi} + a_T Pe^{\beta_T} \quad . \tag{9.27}$$

The *average* values of β_L and β_T from all the available experimental data are $\beta_L \simeq 1.2$ and $\beta_L \simeq 0.9$. We call this regime the *boundary-layer* dispersion after Koch and Brady (1985), since, as we show below, this regime is consistent with the diffusive boundary layers near the solid surface, first found by Saffman (1959), where diffusion transfers materials from the very slow regions near the solid walls to faster streamlines. In practical applications β_L and β_T often are taken to be unity. The coefficients a_L and a_T depend on the heterogeneities of the pore space, and their typical values are $a_L \simeq 0.5$ and $a_T \simeq 0.01 - 0.05$.

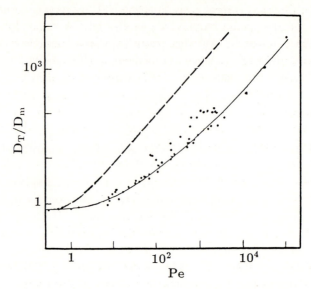

Figure 9.5: Dependence of the transverse dispersion coefficient D_T on the Péclet number Pe. For comparison, the $D_L - Pe$ curve (dashed) is also sketched (after Fried and Combarnous, 1971).

(iv) $300 < Pe < 10^5$. This is the regime of *pure convection*, and simple dimensional analysis indicates that (ignoring the $1/(F\phi)$ term which is small)

$$\frac{D_L}{D_m} \sim Pe \ , \tag{9.28}$$

$$\frac{D_T}{D_m} \sim Pe \ . \tag{9.29}$$

This is usually called *mechanical dispersion*. In this case dispersion is simply the result of a stochastic velocity field induced by the randomly distributed pore boundaries.

(v) $Pe > 10^5$. This is the *turbulent* dispersion regime. The Péclet number is no longer the only correlating parameter, and the Reynolds number should also be used. However, for flow through porous media this regime is not of interest.

(vi) There is a sixth dispersion regime that is not evident in Fig. 9.5. This is the so-called *holdup dispersion* (Koch and Brady, 1985), first studied by Carberry and Bretton (1958), Turner (1959), and Aris (1959). In this case, the solute is trapped in a dead-end region or inside the solid grains, from which it can escape only by molecular diffusion. One has (ignoring the $1/(F\phi)$ term)

$$\frac{D_L}{D_m} \sim Pe^2 \ , \tag{9.30}$$

$$\frac{D_T}{D_m} \sim Pe^2 \ , \tag{9.31}$$

which indicate a rather strong dependence of D_L and D_T on Pe. In a porous medium near its percolation threshold p_c, there are many dead-end pores, and therefore this regime may be relevant to such a porous medium.

Bacri *et al.* (1987) used an acoustic technique (Bacri *et al.*, 1984) and measured D_L for three different porous media, namely, an unconsolidated pack of glass beads, a fireproof brick, and a mill sandstone. In an acoustic technique the solute concentration is derived from the velocity variations of a sound propagating in several cross sections of the porous medium. They showed that pore-level disorder strongly affects D_L and its dependence on Pe. They also were able to observe power laws (26)-(29), depending on the breadth of the pore size distribution and connectivity of the pore space.

9.7 Models of dispersion in macroscopic porous media

Having gained a qualitative understanding of dispersion in porous media and what we may expect, let us now review and discuss various models of dispersion in macrosopic porous media.

9.7.1 Continuum models: The volume-averaging method

We already discussed this method in Chapter 8, where we considered single-phase flow problems. The works of Whitaker (1967), Bachmat (1969, 1972), Gray (1975), Carbonell and Whitaker (1983), Eidsath *et al.* (1983), Koch and Brady (1985), and Plumb and Whitaker (1988a) fall in this class of methods. As an example, we consider the work of Plumb and Whitaker (1988a). In their method one starts from a CDE for the liquid or pore (β) phase

$$\frac{\partial C}{\partial t} + \boldsymbol{\nabla} \cdot (C\mathbf{v}_\beta) = \boldsymbol{\nabla} \cdot (D_m \boldsymbol{\nabla} C) \ , \tag{9.32}$$

with the boundary conditions

$$-\mathbf{n}_{\beta\sigma} \cdot D_m \boldsymbol{\nabla} C = 0 \quad \text{at} \quad S_{\beta\sigma} \ , \tag{9.33}$$

$$C = \mathcal{F}(\mathbf{r}, t) \quad \text{at} \quad S_{\beta\sigma} \ , \tag{9.34}$$

and the initial condition, $C = C_0(\mathbf{r})$. If one defines the phase average concentration

$$\langle C \rangle = \frac{1}{V} \int_{V_\beta} C d\mathbf{V} \ , \tag{9.35}$$

and uses the averaging theorem, Eq. (8.4), *twice*, one can write

$$\langle \boldsymbol{\nabla} \cdot (D_m \boldsymbol{\nabla} C) \rangle = \boldsymbol{\nabla} \cdot \langle D_m \boldsymbol{\nabla} C \rangle + \frac{1}{V} \int_{A_{\beta\sigma}} \mathbf{n}_{\beta\sigma} \cdot D_m \boldsymbol{\nabla} C d\mathbf{A} = \boldsymbol{\nabla} \cdot \langle D_m \boldsymbol{\nabla} C \rangle \ , \tag{9.36}$$

which means that the average of Eq. (32) is given by

$$\frac{\partial \langle C \rangle}{\partial t} + \boldsymbol{\nabla} \cdot (C \mathbf{v}_\beta) = \boldsymbol{\nabla} \cdot \left[D_m \left(\boldsymbol{\nabla} \langle C \rangle + \frac{1}{V} \int_{A_{\beta\sigma}} \mathbf{n} \cdot C d\mathbf{A} \right) \right] . \qquad (9.37)$$

One now writes (see Chapter 8) $C = \langle C \rangle^\beta + \tilde{C}$, and $\mathbf{v}_\beta = \langle \mathbf{v}_\beta \rangle^\beta + \tilde{\mathbf{v}}_\beta$, and substitutes these into Eq. (36), to obtain

$$\epsilon_\beta \frac{\partial \langle C \rangle^\beta}{\partial t} + \boldsymbol{\nabla} \cdot (\epsilon_\beta \langle \mathbf{v}_\beta \rangle^\beta \langle C \rangle^\beta) + \boldsymbol{\nabla} \cdot \langle \tilde{C} \tilde{\mathbf{v}}_\beta \rangle =$$
$$\boldsymbol{\nabla} \cdot \left[\epsilon_\beta D_m \left(\boldsymbol{\nabla} \langle C \rangle^\beta + \frac{1}{V_\beta} \int_{A_{beta\sigma}} \mathbf{n}_{\beta\sigma} \cdot \tilde{C} d\mathbf{A} \right) \right] . \qquad (9.38)$$

Plumb and Whitaker (1988a) then go on with an order-of-magnitude analysis of various terms in Eq. (38) in order to delete some terms. Similar to the derivation of Darcy's law discussed in Chapter 8, a trial solution of the following form is then assumed

$$\tilde{C} = \mathbf{f} \cdot \boldsymbol{\nabla} \langle C \rangle^\beta , \qquad (9.39)$$

which reduces Eq. (38) to

$$\epsilon_\beta \frac{\partial \langle C \rangle^\beta}{\partial t} + \boldsymbol{\nabla} \cdot (\epsilon_\beta \langle \mathbf{v}_\beta \rangle^\beta \langle C \rangle^\beta) = \boldsymbol{\nabla} \cdot (\epsilon_\beta \mathbf{D}^* \cdot \boldsymbol{\nabla} \langle C \rangle^\beta) , \qquad (9.40)$$

where $\mathbf{D}^*$ is the dispersion tensor given by

$$\mathbf{D}^* = D_m \left[\mathbf{I} + \frac{1}{V_\beta} \int_{A_{\beta\sigma}} \mathbf{n}_{\beta\sigma} \cdot \mathbf{f} d\mathbf{A} \right] - \langle \tilde{\mathbf{v}}_\beta \mathbf{f} \rangle^\beta . \qquad (9.41)$$

This analysis shows that on a large enough length scale, such that the porous medium is homogeneous, a CDE for the *average* concentration, Eq. (40), holds. Note that $\mathbf{D}^*$ contains two terms. One is the contribution of molecular diffusion, while the other is due to hydrodynamic transport. However, the contribution of molecular diffusion appears only as D_m, not as $D_m/(F\phi)$, since the volume-averaging method of Plumb and Whitaker cannot take into account the effect of the tortuosity of a porous medium. As in the case of single-phase flow discussed in Chapter 8, there is an unknown function $\mathbf{f}$ which has to be determined. In practice, $\mathbf{f}$, which is the solution of the following boundary value problem

$$\tilde{\mathbf{v}}_\beta + \mathbf{v}_\beta \cdot \boldsymbol{\nabla} \mathbf{f} = D_m \nabla^2 \mathbf{f} , \qquad (9.42)$$

$$- \mathbf{n}_{\beta\sigma} \cdot \boldsymbol{\nabla} \mathbf{f} = \mathbf{n}_{\beta\sigma} \text{ at } A_{\beta\sigma} , \qquad (9.43)$$

provided that $D_m t / l_\beta^2 \gg 1$ (where l_β is the length scale associated with the β-phase), can be determined if a model of pore space is specified. However, if the pore space is disordered, then the numerical calculation of $\mathbf{f}$ is no easier than any numerical method that may be used to solve Eq. (32) directly.

A more sophisticated version of this method was developed by Koch and Brady (1985). These authors considered dispersion in unconsolidated porous media, such as a packed bed, in which the volume fraction of the porous particles is ϕ_p. They first formally related the average concentration field to the probability distribution of solid material, and then derived the effective dispersion coefficients in the *high porosity limit*, including the relevant proportionality constants. Their results for relatively high Péclet numbers can be summarized as follows

$$\frac{D_L}{D_m} = 1 + \frac{3}{4}Pe + \frac{\pi^2}{6}\phi_p Pe \ln Pe + \frac{D_m(1+\chi)^2}{15\kappa D_{mp}}\phi_p Pe^2 , \qquad (9.44)$$

$$\frac{D_T}{D_m} = 1 + \frac{63\sqrt{2}}{320}\sqrt{\phi_p}Pe . \qquad (9.45)$$

Here κ is the partition coefficient for the solute (i.e., partitioning of the solute between the fluid flowing through the bed and the fluid in the void space inside the particles), D_{mp} is the molecular diffusivity of the solute inside the particles, and χ is defined by

$$\chi = \frac{\phi_p(1 - \kappa^{-1})}{1 - \phi_p(1 - \kappa^{-1})} . \qquad (9.46)$$

An appealing aspect of Koch and Brady's work is that the fluid mechanical aspects of the problem are treated *without* any approximations. It can also be extended to the case where dispersion is not Gaussian and does not obey a CDE (see below). A major disadvantage of the method is the assumption of high porosity that has to be made in order to make the method numerically predictive.

9.7.2 Continuum models: Statistical-kinetic approach

These models are mathematical formulations either of the motion of a solute molecule, as was done by Beran (1968) and Todorovic (1970), or of an abstract entity, e.g., a "point." Various versions of this model were developed by Scheidegger (1965), Todorovic (1971, 1975, 1982) and Chaudhari and Scheidegger (1965). The Chaudhari-Scheidegger approach assumes that the solute concentration in a "cell" of a porous medium is a Markovian variable and, in this sense, it is somewhat different from the rest.

Beran (1968) treated a one-dimensional system; he assumed that the velocity field is an asymptotically stationary stochastic process and that two velocities separated by a large but finite time are uncorrelated. He invoked a formal analogy between a simple random walk (see Chapter 8) and the position of the solute molecules and then invoked a central-limit theorem to assert that the probability $P(x,t)$ [the one-dimensional analog of Eq. (3)] is Gaussian. By invoking an ergodic theorem, Beran asserted that $P(x,t)$ is proportional to $C(x,t)$, the solute concentration, but did not calculate the probability density for the velocities $v_1, v_2, \ldots, v_n$ of the solute particle after $1, 2, \ldots, n$ steps. The construction of this probability density is perhaps the most fundamental problem in the development of a model of dispersion. Todorovic (1970) also developed a theory of longitudinal dispersion and extended his treatment to transverse dispersion (Todorovic, 1971) and to the situation wherein a time-dependent injection of solute particles at a boundary is specified (Todorovic, 1975). Using a Markov process, he argued that the displacement $x(t)$ of the solute particle at time t is a Brownian (diffusive) process, and therefore its probability density is Gaussian. Bear (1972) and Chaudhari and Scheidegger (1965) also suggested that the theory of Markov processes can be used for modelling dispersion processes. The problem with these approaches is that, (i) a Gaussian distribution for the concentration profile is guaranteed, whereas in reality this is not always the case, and (ii) they do not provide a method of actually calculating the dispersion coefficients; therefore they are purely phenomenological and formal and have no practical use.

9.7.3 Continuum models: Fluid-mechanical models

These models are based on three basic attributes: (i) a *Lagrangian* description of the motion of solute-containing fluid through a single pore; (ii) specific assumptions about the medium, e.g., homogeneity and isotropy; and (iii) calculation of quantities of interest as statistical averages. In a Lagrangian approach the motion of a solute particle is followed, and the average velocity and the dispersion coefficients are defined as the time-rate of change of the mean and mean square positions of the particle, respectively. The works of Scheidegger (1954), Day (1956), de Josselin de Jong (1958), Saffman (1959, 1960), Haring and Greenkorn (1970), and Bear (1972) are in this group. Saffman's work is the most general of these, and hence is discussed briefly. His model consisted of a network of randomly oriented and distributed straight capillaries, in each one of which the flow was uniform. The path of solute particles was regarded as a random walk (see Chapter 8) in which the length, direction, and duration of each step were random variables.

Saffman was careful to introduce a dynamical basis for his model founded explicitly on fluid-mechanical ideas. He assumed that all the pores have an equal circular cross-section of radius R, and that flow was laminar in all tubes. Saffman distinguished five cases in his first (1959) paper:

(i) $t_c \ll t_{rd}$, where t_c is the convective time spent by a solute particle in the pore and t_{rd} the time required for appreciable radial diffusion of this particle, $t_{rd} = R^2/(8D_m)$ (i.e., the time that the solute particle spends to jump a distance $R/2$ from one streamline to another), where R is the pore radius. Thus radial diffusion is negligible and the duration of a step is $t = t_c = l/v_m$, where l is the pore length.

(ii) The solute particle is on a streamline close to the pore wall, i.e., its speed is small and molecular diffusion occurs. The duration of a step is then $t = t_{rd} + l/v_m$, i.e., the particle makes *one jump* from the streamline close to the pore wall to another one whose speed is v_m, and then is convected out of the pore.

(iii) $t_{rd} < t \ll t_{ad}$, where t_{ad} is the time for appreciable *axial* diffusion; $t_{ad} = l^2/(2D_m)$. The effect of axial diffusion is negligible and $t = t_{rd} + l/v_m$.

(iv) $t_{rd} < t \leq t_{ad}$, which means that the pore is very narrow and $t = l/v_m$.

(v) $t_{ad} \ll t$. Thus the duration of a step is $t = t_{ad}$.

Saffman found that in all cases D_T is given by

$$D_T = \frac{3}{16} lv \quad . \tag{9.47}$$

However, D_L was found to depend on the regime considered. If

$$D_L = \frac{1}{2} v l s^2 \quad , \tag{9.48}$$

then

$$s^2 = \frac{1}{3} \ln \frac{3vt_{ad}}{l} + \frac{1}{12} \left(\ln \frac{6vt_{rd}}{l} \right)^2 - \frac{1}{4} \ln \frac{6vt_{rd}}{l} + \frac{19}{24} \quad , \tag{9.49}$$

if

$$\frac{vt_{ad}/l}{\sqrt{n_s \ln(3vt_{ad})/l}} \ll 1 \quad ; \tag{9.50}$$

$$s^2 = \frac{1}{6} \ln \frac{27vt_m}{2l} + \frac{1}{12} \left(\ln \frac{6vt_{rd}}{l} \right)^2 - \frac{1}{4} \ln \frac{6vt_{rd}}{l} + \frac{19}{24} \quad , \tag{9.51}$$

if $\ln n_s \gg 2$, and

$$\frac{3vt_{rd}/l}{n_s^{1/2}(\ln n_s^{1/2})} \ll 1 \quad , \quad \frac{3vt_{ad}/l}{n_s^{1/2}(\ln n_s^{1/2})} \gg 1 \quad ;$$

$$s^2 = \frac{1}{48} \left(\ln \frac{54vt_m}{l} \right)^2 \quad , \tag{9.52}$$

if $\ln n_s \gg 2$, and

$$\frac{4vt_{rd}/l}{n_s^{1/2} \ln n_s^{1/2}} \gg 1 \quad , \quad \frac{4vt_{ad}/l}{n_s^{1/2} \ln n_s^{1/2}} \gg 1 \quad ,$$

where n_s is the mean number of steps taken by the fluid particles after a large time t_m, at which D_L and D_T are measured, and is equal to $3\langle x \rangle/(2l)$, with $\langle x \rangle$ being the mean longitudinal position at time T_m. If we neglect the constant $19/24$ in Eqs. (49) and (51), which is usually much smaller than the other terms in these equations, Saffman's results can be summarized as

$$\frac{D_L}{D_m} \sim Pe(\ln Pe)^\alpha \quad . \tag{9.53}$$

where $\alpha = 1$ or 2. Equation (53) can now be compared with Eq. (26). If we take $\alpha = 1$ and fit the experimental data to this equation, the resulting fit would be as accurate as that provided by Eq. (26) if $\beta_L \simeq 1.25$. On the other hand, if we take $\alpha = 2$, the resulting fit would be compatible with Eq. (26) if $\beta_L \simeq 1.15$. This explains two interesting features of all experimental data: (i) The data indicate that β_L is either about 1.13-1.16 [obtained by Legaski and Katz (1967) for Bandera sandstone, by Salter and Mohanty (1982) for Berea sandstone, and by Blackwell *et al.* (1959) for packed unconsolidated sands], or about 1.24 - 1.30 [reported by Brigham *et al.* (1961) and Pakula and Greenkorn (1971) for glass beads, and by Legaski and Katz (1967) for Boise and Nordosaria sandstones and for Dolomites], with an overall average of about 1.2, as mentioned above. (ii) β_L is probably not universal; it depends on the strength of competition between molecular diffusion and convection, which in turn depends on the pore shapes. On the other hand, Eq. (47) is *not* completely compatible with Eq. (27), since most data [see, for example, Blackwell (1962)] indicate that $\beta_T \simeq 0.9$, as mentioned above. Similar to β_L, it is likely that β_T is *not* universal.

Saffman also found that dispersion *cannot* be described by a CDE, unless T_m is sufficiently large. Saffman's analysis is clearly indicative of the significance of molecular diffusion to dispersion in microscopically disordered porous media, no matter how small it may be, as long as it is not exactly zero. In the absence of molecular diffusion, a solute particle, which is travelling along a streamline very close to a pore wall, will need a huge amount of time to escape from this region, and $D_L = 0$. The logarithmic terms in Eqs. (49)-(52) are exactly due to such singularities in the dispersion process. However, diffusion intervenes and transfers the fluid particle to a much a faster streamline.

Saffman's results are presumably valid if Pe is large but finite. In his second paper (Saffman, 1960), he considered the case where Pe is "less than some large value," and found that both D_L and D_T depend quadratically on Pe. The agreement between Saffman's results and various experimental data ranges from reasonable to good. In this author's opinion, Saffman's work is the most detailed and careful analytical analysis of dispersion in microscopically disordered porous media which, however, has not been fully appreciated. However, his work also has its shortcomings. Saffman did not allow the possibility of a pore size distribution (all pores were assumed to have the same radius). Haring and Greenkorn (1970) rederived some of Saffman's results assuming a pore size distribution. Moreover, in Saffman's work the flow field is represented by a sort of mean-field approximation, and there are no correlations between successive steps of the walk. This restriction can also be removed by Monte Carlo calculations of network models of pore space, as was first done by Sahimi *et al.* (1982), which is discussed below. Finally, it is worth mentioning that the logarithmic singularities found by Saffman were rediscovered by Aronovitz and Nelson (1984) in what they called "diffusion in steady flow" through a porous medium, which is nothing but hydrodynamic dispersion discussed here!

9.7.4 Network models

   These models belong to the class of fluid-mechanical models that we already discussed, except that the mean-field nature of the flow field and the absence of disorder and heterogeneity are explicitly lifted. As already mentioned in Chapter 3, Torelli and Scheidegger (1972) appear to be the first to propose a random network model for studying dispersion processes in porous media, although they did not report any result. Torelli (1972) did simulate dispersion processes in flow through a random network, but his results pertain to a sort of dispersion not related to what we are interested in here.

   Sahimi *et al.* (1982) were the first who used random network models of porous media to simulate dispersion. In their model, one first determines the flow field in the network by the method discussed in Chapter 8, where we discussed network models for calculating the permeability of a porous medium. Then solute particles are injected into the network at random at the upstream plane $x = 0$. Each particle selects a streamline at random. The convective travel time for a given pore is given by $t = l/v_p$, where l is the length of the pore, and v_p is the pore flow velocity. Complete mixing at the nodes is assumed, and therefore the probability that a pore is selected, once a particle has arrived at a node, is proportional to the flow rate in that pore. The FPTD $\mathcal{Q}$ for the particles are computed by fixing the longitudinal or lateral positions and measuring the time at which the particles arrive at these positions for the first time. It is easy to see that D_ζ, the dispersion coefficient in the ζ-direction, is given by

$$D_\zeta = \int_0^\infty \mathcal{Q}(\zeta - \zeta_0, t)\frac{S_\zeta^2}{2t}dt \ , \tag{9.54}$$

where $\zeta_0(x_0)$ is the starting position of the particles, $s_x^2 = (x - x_0 - vt)^2$, and, $s_\zeta^2 = (\zeta - \zeta_0)^2$ for $\zeta = y$ or z.

   However, this random-walk method is appropriate for mechanical dispersion, since pore-level molecular diffusion has been ignored, and only the effect of a stochastic velocity field throughout the network has been taken into account. To include the effect of molecular diffusion and simulate the boundary-layer dispersion, the following method was adopted (Sahimi and Imdakm, 1988). The convective time t_c for travelling along a streamline in a pore is first calculated. If $t_c \gg t_{rd}$, where t_{rd} is the radial diffusion time scale discussed above, then one sets $t = t_c + t_{rd}$, since the tracer has enough time to diffuse to a faster streamline. To simulate holdup dispersion, the tracer particles are allowed to diffuse into the dead-end pores of the network. Transport in such pores is only by molecular diffusion. In a series of papers, Sahimi *et al.* (1982, 1983a, 1986a,b) and Sahimi and Imdakm (1988) showed that such network models can reporduce and simulate all the regimes of dispersion discussed above. In particular, Eqs. (26)-(31) can all be predicted by these models.

de Arcangelis *et al.* (1986a) proposed another model which they called the *probability propagation algorithm*. In this model a one-dimensional CDE is assumed to hold for each pore of the network,

$$\frac{\partial C}{\partial t} + v_m \frac{\partial C}{\partial x} = D_m \frac{\partial^2 C}{\partial x^2} \quad . \tag{9.55}$$

Consider a network of capillary tubes $\{ij\}$. The concentration C_{ij} in each tube obeys Eq. (55), with the initial condition $C_{ij}(x_{ij}, 0) = 0$, and three boundary conditions: (i) a unit pulse of input flux at node i at $t = 0$,

$$\sum_{\{j\}} S_{ij} \left(v_{mij} C_{ij} - D_m \frac{\partial C_{ij}}{\partial x_{ij}} \right)_{x_{ij}=0} = \delta(t) \quad , \tag{9.56}$$

where S_{ij} is the cross-section area of tube $i - j$, and v_{mij} the mean-flow velocity in that tube; (ii) a common concentration $C_i(t)$ at the starting junction, $C_{ij}(0, t) = C_i(t)$ for all j; and (iii) a sink at each tube end, $C_{ij}(l, t) = 0$, for all j, corresponding to the fact that a tracer reaching the end acts as a source for the junction problem at the new node. The first-passage time probability is given by $q_{ij}(t) = -S_{ij} D_m \partial C_{ij}(l, t)/\partial x_{ij}$. Equation (55) is easily solved in the Laplace transform space. The solution is given by

$$\hat{C}_{ij}(x, \lambda) = A_{ij} \exp(\alpha_{ij} x) + B_{ij} \exp(\beta_{ij} x) \quad , \tag{9.57}$$

$$\alpha_{ij} \, , \, \beta_{ij} = \frac{v_{ij} \pm \sqrt{v_{ij}^2 + 4 D_m \lambda}}{2 D_m} \quad , \tag{9.58}$$

where A_{ij} and B_{ij} are determined from the above boundary conditions, and λ is the Laplace transform variable conjugate to t. Then it is easy to see that

$$\hat{q}_{ij}(\lambda) = \hat{C}_{ij}(\lambda) S_{ij} \frac{\alpha_{ij} - \beta_{ij}}{\exp(-\beta_{ij} l) - \exp(-\alpha_{ij} l)} \quad . \tag{9.59}$$

Having determined $\hat{q}_{ij}(\lambda)$, we obtain the FPTD $\hat{Q}(L, \lambda)$ for the *entire* network

$$\hat{Q}(L, \lambda) = \sum_{\Gamma} \prod_{i,j \in \Gamma} \hat{q}_{ij}(\lambda) \quad , \tag{9.60}$$

where the sum is over *all* paths Γ from the inlet to the outlet of the network.

To compute this sum efficiently, de Arcangelis *et al.* (1986a) ordered the nodes of the network in decreasing pressure, starting with the inlet and finishing with the outlet. At each node i, a quantity $\hat{Q}_i(\lambda)$ is introduced that is a partial sum of Eq. (60), over paths running from the inlet to site i. For a delta-function input of tracer, one initially has $Q_I = 1$ at the inlet I and $Q_i = 0$ elsewhere. One then proceeds recursively through the pressure-ordered node list, propagating the quantity $\mathcal{Q}_i$ from each node i to its network neighbors j according to the rule $\hat{\mathcal{Q}}_j(\lambda) \rightarrow \hat{\mathcal{Q}}_j(\lambda) + \hat{\mathcal{Q}}_i(\lambda)\hat{q}_{ij}(\lambda)$, $\hat{\mathcal{Q}}_i(\lambda) \rightarrow 0$. After all the internal nodes have been propagated once in this way, the quantity $\hat{\mathcal{Q}}_0(\lambda)$ at the outlet contains all terms of Eq. (60) corresponding to purely downstream paths. However, because molecular diffusion is present, the solute motion includes upstream paths as well. Hence, after one sweep through the network, one has $\hat{\mathcal{Q}}_n \neq 0$ for internal nodes n. By repeated sweeps through the network, the contributions of paths with progressively more upstream steps are included. Once $\hat{\mathcal{Q}}(L, \lambda)$ is determined, it is inverted to the time domain and D_L is calculated using Eq. (54). Note that this model, in the mechanical dispersion regime (i.e., with no boundary layer diffusion), is equivalent to the random-walk model of Sahimi and co-workers. de Arcangelis *et al.* (1986a) showed that this method can reproduce the results for both mechanical and boundary-layer dispersion. The method is very efficient as long as the network is well-connected. For percolation networks near the percolation threshold the method is very inefficient because calculating the sum in Eq. (60) becomes very time-consuming.

In a later paper, Koplik *et al.* (1988b) used another method for studying dispersion in random networks. In this method, one first calculates the flow field throughout the network by the method already discussed. Assuming that dispersion in each pore obeys a CDE, Eq. (56) with its right-hand side being zero (which is simply a statement of the continuity of mass at each node), is written for all interior nodes of the network. The resulting set of linear equations for nodal concentrations is solved (in the Laplace transform space), from which D_L is calculated. Roux *et al.* (1986) used the same method, except that they used a transfer matrix method discussed in Chapter 7. Sahimi and Jue (1989) and Sahimi (1992b) used a similar method to study convection and diffusion of *large* molecules in porous media, i.e., when the hydrodynamic radius of the molecules is comparable to the pore sizes.

9.8 Long-time tails: Dead-end pores versus disorder

As already discussed, molecular diffusion transfers fluid particles into and out of stagnant, dead-end or low-velocity regions of the pore space. Many experimental measurements of the concentration distribution during dispersion indicate the presence of a long-time tail in the concentration profiles, and diffusion into and out of the stagnant regions is often used to explain such long-time tails. Such a phenomenon has, in fact, been of great interest for a long time. Carberry and Bretton (1958), Aris (1959), and Turner (1959) were probably the first who studied dispersion in systems with stagnant regions. In particular, Aris (1959) showed that $D_L/D_m \sim Pe^2$, a result that was rediscovered by Koch and Brady (1985). In the early 1960s there were several studies of the relation between the observed long-time tails and the effect of dead-end pores. Deans (1963) and

Coats and Smith (1964) attributed the long-time tails to the presence of dead-end pores, which can cause long delays in the travel times, and hence long tails in the concentration profiles. They developed a semi-empirical model to account for this, which is discussed below. Brigham (1974) and Baker (1977) found that trapping in the dead-end pores is needed to describe dispersion in carbonate rocks but not in sandstones. They proposed that the origin of stagnant regions in carbonate rocks is either regular or bimodal porosity. This was recently disputed by Gist *et al.* (1990), who measured dispersion coefficients in a variety of sandstones and carbonate rocks. Their mercury capillary-pressure data for Austin Chalk and Indiana limestone indicated the presence of bimodal porosity, yet no long-time tails were observed in the measured concentration profiles.

Deans (1963), Coats and Smith (1964), Passioura (1971), Baker (1977), Rao *et al.* (1980), and Salter and Mohanty (1982) all investigated the effect of the long-time tails and dead-end pores. In Baker's model, which is the most sophisticated of these works, it is assumed that a fraction ϕ_f of the pore volume is available for flow, while $1 - \phi_f$ is the stagnant or dead-end fraction. A one-dimensional CDE is assumed (thus ignoring transverse dispersion), modified to account for the effect of the stagnant regions:

$$\phi_f \frac{\partial C_f}{\partial t} + (1 - \phi_f) \frac{\partial C_s}{\partial t} + \mathrm{v} \frac{\partial C_f}{\partial x} = D_L \frac{\partial^2 C_f}{\partial x^2} \qquad (9.61)$$

where C_f and C_s are, respectively, the concentrations of the solute in the flowing and stagnant regions. This equation is augmented by a mass balance between the stagnant and flowing fluids

$$(1 - \phi_f) \frac{\partial C_s}{\partial t} = k_c (C_f - C_s) \quad , \qquad (9.62)$$

where k_c is the mass transfer coefficient. k_c^{-1} can be interpreted as the time that the solute particles spend in the stagnant regions. Equations (61) and (62), with the appropriate initial and boundary conditions, are then solved. Normally, ϕ_f and k_c are not known *a priori* and are treated as adjustable parameters.

Bacri *et al.* (1990a), who used an acoustic technique to measure the concentration and velocity profiles in dispersion in unsaturated porous media (i.e., dispersion in one fluid phase while another immiscible fluid is also present), Charlaix *et al.* (1987b), who measured dispersion coefficients and concentration profiles in sintered-glass bead packs, and Gist *et al.* (1990), who did the same in a variety of sandstones and carbonate rocks, all used the Coats-Smith-Baker model to fit their data and found very good fits. However, while Bacri *et al.* (1990a) attributed the long-time tails in their data to the fact that the length of their medium was too short to allow for the development of Gaussian dispersion (see Fig. 9.6), Charlaix *et al.* (1987b) and Gist *et al.* (1990) attributed this to the heterogeneous nature of their porous medium. Thus it is important to understand why the Coats-Smith-Baker model is able to provide such good fits to the data (see below).

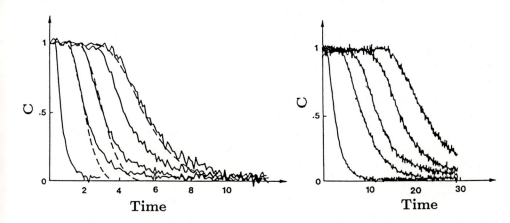

Figure 9.6: Unsaturated concentration profiles C at different cross sections of the porous medium, with mean flow velocity v = 3.6 cm/hr (left) and v = 0.9 cm/hr (right). The dashed curves correspond to Gaussian profiles (after Bacri *et al.*, 1990a).

From their studies of dispersion in consolidated porous media, Gist *et al.* (1990) identified two cases in which long-time tails can occur. The first case is that of a heterogeneous porous medium, when the permeability contrast between various regions is strong enough. This gives rise to a long-time tail in the concentration profile. The second case is that of a *narrow* pore-size distribution, in which the permeability heterogeneities are due to defects in the packing density. If the long-time tails are in fact due to permeability heterogeneities, the implications for the scale up of laboratory results to field conditions can be important. For example, the Coats-Smith-Baker model predicts that the long-time tails will disappear if k_c^{-1} is much smaller than the total travel time of the fluid particles [this is easily seen by inspecting Eqs. (61) and (62)], whereas long-time tails will persist if there are strong permeability heterogeneities at any length scale. This is also consistent with studies and measurements of tracer dispersion in groundwater flow in heterogeneous aquifers (Pickens and Grisak, 1981). Before we go on and explain this complex phenomenon, let us study first dispersion in short porous media, a closely related subject.

9.9 Dispersion in short porous media

As discussed above, Bacri *et al.* (1990a) attributed the long-time tails of their concentration profiles to the small size of their sample. Thus dispersion in short porous media can be important, because then the mixing zone will be large compared with the medium's length. Brenner (1962) and Brigham (1974) were among the first to investigate this issue, with Brigham (1974) making a comprehensive and definitive analysis of this problem.

We already gave the solution of a CDE for one-dimensional dispersion in a porous medium of length L and a step change in the inlet concentration at time $t = 0$; see Eqs. (8) and (9). We define

$$\beta_{\pm} = \frac{L \pm v(t - V_{inj}/Q)}{\sqrt{4D_L(t - V_{inj}/Q)}} \quad , \tag{9.63}$$

where Q is the volumetric flow rate and V_{inj} is the total volume of injected tracer solution. Then the solution for a pulse input of total volume V_{inj} is found by superimposing two step-change solutions. The outlet solute concentration in this case is given by

$$\frac{C}{C_0} = \frac{1}{2}\text{erfc}\,(\alpha_-) + \frac{1}{2}\exp\left(\frac{Lv}{D_L}\right)\text{erfc}\,(\alpha_+) - \frac{1}{2}\text{erfc}\,(\beta_-) - \frac{1}{2}\exp\left(\frac{Lv}{D_L}\right)\text{erfc}\,(\beta_+)\,,$$
$$\tag{9.64}$$

where $\alpha_{\pm}$ is defined by Eq. (8). If one is interested only in observation times $t \gg V_{inj}/Q$, then the above solution can be simplified to

$$\frac{C}{C_0} = \left(\frac{L}{2\pi}\right)\left\{[D_L(t - V_{inj}/Q)]^{-1/2} - (D_L t)^{-1/2}\right\}\left[\exp(-\beta_-^2) + \exp(-\alpha_-^2)\right]$$
$$\tag{9.65}$$

Gist *et al.* (1990) used Eq. (65) to fit their concentration profiles and found that the resulting fits are as accurate as those provided by the Coats-Smith-Baker model.

Brigham (1974) showed that if the Coats-Smith-Baker model is adjusted at the effluent boundary to account for the difference between in-situ and flowing concentrations, then Eqs. (9), (64) and (65) and the Coats-Smith-Baker model will essentially provide identical fits to the data. This explains why Bacri *et al.* (1990a) could fit their data for a short porous medium with the Coats-Smith-Baker model.

Koch and Brady (1987) also considered dispersion in porous media of short to moderate lengths. They derived an expression for the Fourier transform of the concentration and the effective dispersion coefficients. They showed that the characteristic time τ_{KB} for reaching a diffusive transport described by a CDE is related to a Péclet number Pe_1 by

$$\tau_{KB} \sim Pe_1^{-2/3} \tag{9.66}$$

where $Pe_1 = dv/D_m$, and d is the typical grain size *before* the grains are fused to produce a consolidated porous medium. They found qualitative agreement between Eq. (66) and the data of Charlaix *et al.* (1987b). Bacri *et al.* (1990a) also used the Koch-Brady expression for the concentration profile, but found only qualitative agreement between the predictions and their data, whereas the Coats-Smith-Baker model provided an accurate fit to their data. Perhaps the reason for this discrepancy is that the Koch-Brady results are valid in the limit of high porosities, whereas the data of Bacri *et al.* (1990a) and Charlaix *et al.* (1987b) are both for porosities that are beyond the region of validity of the Koch-Brady results. Koch and Brady (1987) also proposed that τ_{KB} can be taken to be the same as k_c^{-1} in the Coats-Smith-Baker model.

9.10 Dispersion in porous media with percolation disorder

In this section we discuss dispersion in percolation networks. As Katz and Thompson (1986, 1987) showed (see Chapter 8), flow in a porous medium with a broad pore size or permeability distribution may be mapped onto an equivalent percolation problem. The same must be true about dispersion, since a broad pore size distribution gives rise to a broad distribution of pore flow velocities which in turn affects dispersion. There are two features of percolation networks that can influence dispersion. One is the fact that there are a large number of dead-end pores near the percolation threshold p_c, and thus holdup dispersion can be important. The second is the fact that for length scales shorter than the percolation correlation length ξ_p, the sample-spanning cluster and its backbone are fractal objects, and thus dispersion is not expected to be described by a CDE. We call this regime fractal dispersion. Two important characteristic quantities are the *dispersivities* $\alpha_L = D_L/v$, and $\alpha_T = D_T/v$ (which are proportional to each other, but α_L is usually larger than α_T). Physically, the dispersivities represent the length scale over which a CDE can describe dispersion, and thus in some sense, they are similar to ξ_p because dispersion in a percolation system can be described by a CDE if the dominant length scale of the system is larger than ξ_p.

In their simulations of dispersion in percolation networks, Sahimi *et al.* (1982, 1983a, 1986a,b) found that as p_c is approached, the dispersivities and dispersion coefficients also increase dramatically. This can be attributed to the fact that near p_c the transport paths are very tortuous, resulting in broad FPTDs and hence large dispersive mixing of the two fluids. Figure 9.7 shows their results for dispersion in a percolating square network. The increase in dispersivities and the dispersion coefficients near p_c was confirmed by the experiments of Charlaix *et al.* (1987b,1988a) and Hulin *et al.* (1988a), who studied dispersion in model porous media and measured D_L. Charlaix *et al.* (1988a) constructed two-dimensional hexagonal networks of pores whose effective diameters were of the order of millimeters. They found that as the fraction of open pores decreased, D_L increased sharply, and that Eq. (26) was satisfied. But even when dispersion coefficients were measured quite close to p_c, the quadratic dependence of D_L/D_m on Pe [Eq. (30)] was *not* observed (although the fraction of dead-end pores is quite large near p_c), presumably because the exchange time between the flowing fluids and the dead-end regions was so long that could not be detected during their experiment. Hulin *et al.* (1988a) measured D_L in bidispersed sintered glass

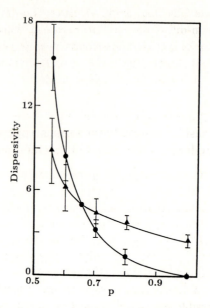

Figure 9.7: Longitudinal (circles) and transverse (triangles) dispersivities in a percolating square network, in which a fraction p of the bonds carry flow. The dispersivities are measured in units of the length of a bond (pore). At $p = 1$ there is no disorder, and thus the dispersion coefficients and the dispersivities are both zero (after Sahimi *et al.*, 1986b).

materials prepared from mixtures of two sizes of beads. They observed that when the porosity was decreased from 30% to 12%, D_L increased by a factor of 30. The results of these studies also indicated that dispersion is more sensitive to *large-scale* inhomogeneities of a porous medium than to its detailed *local* structure. Somewhat similar results were obtained by Charlaix *et al.* (1987b).

We should mention here a paper of de Gennes (1983a) in which he studied dispersion near p_c. de Gennes (1983a) presented a very long discussion and derivation to show that, in calculating D_L, the average flow velocity that one must use has to be based on the *total* travel time of the solute particles in the sample-spanning cluster, rather than the travel time along the backbone *alone*. Intuitively, this is clear and even in experimental measurements of the concentration profiles and D_L, there is no way to measure the travel time along the backbone alone, and what is routinely measured is the total travel or transit time.

In Chapter 3 we discussed universal scaling laws for various properties of percolation networks near the percolation threshold. What are the scaling laws for D_L and D_T near p_c? From our discussions so far, it must be clear that D_L and D_T are sensitive to the structure of the porous medium. Similar to fractal diffusion discussed in Chapter 8, we may define a crossover time τ_{co} such that for $t \gg \tau_{co}$ dispersion is Gaussian or diffusive and follows a CDE, whereas for $t \ll \tau_{co}$ dispersion is non-diffusive, with the crossover between the two regimes taking place at about $t \simeq \tau_{co}$. For dispersion near p_c, this time scale can be estimated from

$$\tau_{co} \sim \frac{\xi_p^2}{D_L} , \qquad (9.67)$$

since the dominant length scale of the system is ξ_p. To derive the scaling laws for D_L and D_T, we must consider separately the various dispersion regimes discussed above. Let us first introduce two random-walk fractal dimensions by

$$\langle \Delta x^2 \rangle \sim t^{2/D_w^l} , \qquad (9.68)$$

$$\langle y^2 \rangle \sim \langle z^2 \rangle \sim t^{2/D_w^t} , \qquad (9.69)$$

where $\langle \Delta x^2 \rangle = \langle (x - \langle x \rangle)^2 \rangle = \langle x^2 \rangle - \langle x \rangle^2$. These two equations are defined for length scales $L \ll \xi_p$. Two average flow velocities can also be defined. One is an average velocity v_c, defined in terms of the travel time in the sample-spanning cluster. Then, $v_c \sim K/X^A$, where K is the permeability of the system, and X^A is the percolation accessible fraction discussed in Chapter 3. Thus, near p_c

$$v_c \sim (p - p_c)^{e - \beta_p} \sim \xi_p^{-\theta_c} \qquad (9.70)$$

where, $\theta_c = (e - \beta_p)/\nu_p$, and e, ν_p and β_p are, respectively, the critical exponents of the permeability, the correlation length, and the accessible fraction which were defined in Chapter 3. On the other hand, if an average particle velocity v_B is defined in terms of the travel times along the backbone, then $v_B \sim K/X^B$, where X^B is the percolation backbone fraction defined in Chapter 3. Thus, near p_c

$$v_B \sim (p - p_c)^{e - \beta_B} \sim \xi_p^{-\theta_B} \qquad (9.71)$$

where, $\theta_B = (e - \beta_B)/\nu_p$, and β_B is the critical exponent of X^B defined in Chapter 3. For length scales $L \ll \xi_p$, we should replace ξ_p in Eqs. (70) and (71) by L, and therefore $v_c \sim L^{-\theta_c}$ and $v_B \sim L^{-\theta_B}$, respectively. We also define a *macroscopic* Péclet number

$$Pe_m = \frac{v \xi_p}{D} , \qquad (9.72)$$

where D is the effective diffusivity of the system, and v can be either v_c or v_B. For $L \ll \xi_p$, we replace ξ_p in Eq. (72) by L. Having defined these quantities, we can now investigate the scaling of the dispersion coefficients and τ_{co} near p_c.

   (i) Let us first consider the small Péclet number regime discussed above. In this case convection has no effect and, $D_L \sim D_T \sim D \sim (p - p_c)^{\mu - \beta_p}$, as found in Chapter 3, where μ is the critical exponent of the conductivity of the system. For $L \ll \xi_p$ we have fractal diffusion and from our discussion in Chapter 7 we can immediately write

$$D_w^l = D_w^t = 2 + \theta_p . \qquad (9.73)$$

where $\theta_p = (\mu - \beta_p)/\nu_p$, and D_w is the fractal dimension of the random walk defined in Chapter 8. Moreover, Eq. (67) tells us that

$$\tau_{co} \sim (p - p_c)^{-\mu - 2\nu_p + \beta_p} \sim \xi_p^{2+\theta_p} \ , \tag{9.74}$$

so that $\tau_{co} \sim L^{2+\theta_p}$ for $L \ll \xi_p$. These equations are valid for dispersion in the entire sample-spanning cluster in the limit $Pe_m \to 0$. For dispersion in the backbone θ_p in Eqs. (73) and (74) should be replaced with $(\mu - \beta_B)/\nu_p$. Moreover, for the nth moment of the FPTD we have $\langle t^n \rangle \sim \langle t \rangle^n$, where $n > 1$ is an integer, and therefore for $L \ll \xi_p$ we have, $\langle t^n \rangle \sim (L^2/D)^n$, so that

$$\langle t^n \rangle \sim L^{n(2+\theta_p)} \ . \tag{9.75}$$

(ii) Suppose now that dispersion takes place only in the backbone of the network, and that Pe_m is relatively large. Although any porous medium has a large number of dead-end pores near its p_c, as the experiments of Charlaix *et al.* (1988a) indiated, the medium has to be extremely close to p_c if the effect of the dead-end pores is to be seen, so that dispersion along the backbone has practical importance. For dispersion in the backbone, we have $D_L/D \sim Pe_m$, and $D_T/D \sim Pe_m$ (mechanical dispersion), since the logarithmic correction indicated by Eq. (53) is neglected in scaling analyses (because $\ln x$ grows with x slower than any power of x, and therefore it is equivalent to a zero critical exponent), which means that $D_L \sim D_T \sim \xi_p v_B \sim \xi_p^{1-\theta_B} \sim (p - p_c)^{e - \beta_B - \nu_p}$. Using the numerical values of e, β_B, and ν_p given in Table 3.3, we obtain $D_L \sim D_T \sim (p - p_c)^{-0.56}$ in two dimensions, and $D_L \sim D_T \sim (p - p_c)^{0.04}$, so that D_L and D_T *diverge* in two dimensions but *vanish* very weakly in three dimensions. This demonstrates the strong effect of the backbone structure on dispersion processes. As discussed in Chapter 3, the backbone can be approximated by nodes, links, and blobs. Links are the bonds or pores that connect the blobs and the remaining multiply-connected bonds aggregate together in the blobs. The blobs are very dense in two dimensions, providing a wide variety of paths for the solute particles with broad FPTDs. As a result, D_L and D_T diverge as p_c is approached. On the other hand, the blobs are not very dense in three dimensions, which means that FPTDs are not broad enough to give rise to divergent D_L and D_T. In a hypothetical porous medium modeled by a Bethe lattice (for which $e = 3$, $\beta_B = 1$ and $\nu_p = 1/2$), $D_L \sim D_T \sim (p - p_c)^{1/2}$, which indicates the strong effect of closed loops of a network (which are absent in Bethe lattices) on D_L and D_T. For $L \ll \xi_p$ we replace ξ_p with L, and thus

$$D_L \sim L^{1-\theta_B} \tag{9.76}$$

with a similar scaling law for D_T. As all length scales of the system must be proportional to each other (and to L), Eq. (76) can be rewritten as $D_L \sim \langle \Delta x^2 \rangle^{(1-\theta_B)/2}$. Using a fundamental property of random-walk processes, $D_L \sim d\langle \Delta x^2 \rangle/dt$, we obtain $d\langle \Delta x^2 \rangle/dt \sim \langle \Delta x^2 \rangle^{(1-\theta_B)/2}$, which after integration yields

$$\langle \Delta x^2 \rangle \sim t^{2/(1+\theta_B)} \ , \tag{9.77}$$

with a similar equation for $\langle y^2 \rangle$ and $\langle z^2 \rangle$. Equation (77) implies that

$$D_L \sim t^{(1-\theta_B)/(1+\theta_B)} , \qquad (9.78)$$

and that

$$D_w^l = D_w^t = 1 + \theta_B . \qquad (9.79)$$

Equation (77) implies that in two dimensions $\langle \Delta x^2 \rangle \sim \langle y^2 \rangle \sim t^{1.26}$, and in three dimensions $\langle \Delta x^2 \rangle \sim \langle y^2 \rangle \sim \langle z^2 \rangle \sim t^{0.97}$. That is, dispersion is *superdiffusive* in two dimension, so that the mean square displacements of the solute particles grow with time faster than linearly, whereas dispersion is *fractal* or *subdiffusive* in three dimensions, so that the mean square displacements grow slower than linearly with time.

On the other hand, Eq. (71) tells that for $L << \xi_p$ we have $v_B \sim L^{-\theta_B} \sim \langle x \rangle^{-\theta_B}$, and since $v_B \sim d\langle x \rangle/dt$, we obtain, after integration

$$v_B \sim t^{-\theta_B/(1+\theta_B)} , \qquad (9.80)$$

which is in sharp contrast with diffusive dispersion for which v_B is *constant*. Finally, since $\alpha_L = D_L/v_B$, we obtain

$$\alpha_L \sim t^{1/(1+\theta_B)} , \qquad (9.81)$$

which means that for non-diffusive dispersion, the dispersivity depends on t. Equations (78) and (80) are manifestations of *dynamical fractals* discussed in Chapter 2, where dynamical properties of a system depend on a non-integer power of t. The time scale τ_{co} is given by

$$\tau_{co} \sim (p - p_c)^{-e+\beta_B-\nu_p} \sim \xi_p^{1+\theta_B} , \qquad (9.82)$$

and for $L << \xi_p$ we have $\tau_{co} \sim L^{1+\theta_B}$. Equation (82) should be compared with Eq. (74). It is easy to show that $\langle t^n \rangle \sim (p - p_c)^{-\nu_p-n(e-\beta_B)} \sim \xi_p^{1+n\theta_B}$, and $\langle t \rangle \sim \xi_p^{1+\theta_B}$, so that $\langle t^n \rangle/\langle t \rangle^n \sim \xi_p^{1-n}$. That is, from the scaling of $\langle t \rangle$ alone one *cannot* obtain the scaling of $\langle t^n \rangle$ for $n > 1$. For $\xi_p \gg L$, we have

$$\langle t^n \rangle \sim L^{1+n\theta_B} , \qquad (9.83)$$

which should be compared with Eq. (75) if we replace θ_p by θ_B.

(iii) Now consider holdup dispersion discussed above. We have, $D_L \sim (v_c\xi_p)^2/D$, which is the same as Eq. (30) in which the length scale is ξ_p and the molecular diffusivity D_m has been replaced by D, as suggested by de Gennes (1983a). We can therefore write, $D_L \sim \xi_p^{2-2\theta_c+\theta_p} \sim (p - p_c)^{-2\nu_p-\mu-\beta_p+2e}$, with a similar result for D_T, and thus $D_L \sim (p - p_c)^{-1.5}$, and $D_L \sim (p - p_c)^{-0.17}$ in two and three dimensions, respectively. That is, as p_c is approached dispersion coefficients *diverge*, which undoubtedly is due to the contribution of dead-end pores and the long times that the solute particles spend there. For $L \ll \xi_p$, we have

$$D_L \sim L^{2-2\theta_c+\theta_p} . \qquad (9.84)$$

Using the same technique as before, we find that

$$\langle \Delta x^2 \rangle \sim t^{2/(2\theta_c - \theta_p)} , \tag{9.85}$$

with similar scalings for $\langle y^2 \rangle$ and $\langle z^2 \rangle$. Thus

$$D_L \sim t^{(2-2\theta_c+\theta_p)/(2\theta_c-\theta_p)} , \tag{9.86}$$

and

$$D_w^l = D_w^t = 2\theta_c - \theta_p , \tag{9.87}$$

and therefore $\langle \Delta x^2 \rangle \sim t^{2.3}$ and $\langle \Delta x^2 \rangle \sim t^{1.1}$ in two and three dimensions, respectively. That is, one *always* has superdiffusive dispersion when holdup dispersion, i.e., trapping of the solute particles in the stagnant regions of the pore space, is dominant. It is now straightforward to show that

$$v_c \sim t^{-\theta_c/(1+\theta_c)} , \tag{9.88}$$

and therefore

$$\alpha_L \sim t^{(2+\theta_p)/[(2\theta_c-\theta_p)(1+\theta_c)]} , \tag{9.89}$$

which is more complex than Eq. (81). The time scale τ_{co} is given by

$$\tau_{co} \sim (p - p_c)^{-2e+\mu+\beta_p} \sim \xi_p^{2\theta_c-\theta_p} , \tag{9.90}$$

and $\tau_{co} \sim L^{2\theta_c-\theta_p}$ for $L \ll \xi_p$. It is now straightforward to show that $\langle t^n \rangle \sim (p - p_c)^{-n\nu_p(\theta_p+2)+\nu_p} \sim \xi_p^{n(\theta_p+2)-1}$, so that $\langle t^n \rangle/\langle t \rangle^n \sim \xi_p^{n-1}$, and therefore the scaling of $\langle t \rangle$ alone is *not* enough for obtaining that of $\langle t^n \rangle$ for any $n > 1$. In the $L \ll \xi_p$ regime we have

$$\langle t^n \rangle \sim L^{n(\theta_p+2)-1} . \tag{9.91}$$

Aside from Eqs. (73)-(75), all of the above equations were derived by Sahimi (1987) and were confirmed by Monte Carlo simulations of Sahimi and Imdakm (1988) and Koplik *et al.* (1988b). The face that $\langle t^n \rangle/\langle t \rangle^n$ depends on n means that *there is no unique time scale for characterizing non-Gaussian dispersion.* Koplik *et al.* (1988b) also proposed an elegant and general scaling form for the FPTD given by

$$Q(t) = \frac{1}{t_d} F_s \left(\frac{t}{t_d}, \frac{t_c}{t_d} \right) , \tag{9.92}$$

where t_d and t_c are the diffusion and convective time scales, respectively, and $t_d \sim \xi_p^{2+\theta_p}$, as Eq. (74) indicates, which is also the largest time that the solute particles can spend in the dead-end pores, since the length of the longest dead-end branches is of the order ξ_p. The scaling function F_s has the following limiting behavior

$$F_s(x, y) \rightarrow \begin{cases} F_1(x) & \text{as } y \rightarrow \infty , \\ y F_2(x) & \text{as } y \rightarrow 0 . \end{cases} \tag{9.93}$$

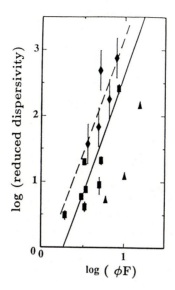

Figure 9.8: Dependence of the reduced longitudinal dispersivity α_L/d_g on the porosity ϕ and the formation factor F in sandstones (squares), epoxies (triangles), and carbonates (diamonds). The solid and dashed lines are the best fit of the sandstones and carbonates data, respectively (after Gist *et al.*, 1990).

The case of pure diffusion corresponds to $y = t_c/t_d \rightarrow \infty$, while the convective limit corresponds to $y \rightarrow 0$. Numerical simulations support this scaling representation of $\mathcal{Q}(t)$.

Gist *et al.* (1990) studied dispersion in a variety of sandstones and carbonate rocks, and used the percolation ideas of Katz and Thompson (1986, 1987), discussed in Chapter 8, to quantify their results. Following Sahimi *et al.* (1982, 1983a, 1986a,b) they argued that the fundamental quantity to be considered is the ratio ξ_p/d_g, where d_g is the mean grain size. Since $\xi_p/d \sim (X^A)^{-\nu_p/\beta_p}$, and as $\nu_p/\beta_p \simeq 2$ (see Table 3.2), one can write $\xi_p/d_g \sim (X^A)^{-2}$. Because X^A is roughly proportional to the fluid saturation S, we can write

$$\frac{\xi_p}{d_g} \sim S^{-2} \quad . \tag{9.94}$$

Gist *et al.* (1990) derived a relation between $\alpha_r = \alpha_L/d_g$ and ξ_p/d_g using the percolation model of Sahimi *et al.* Their final result is

$$\alpha_r \sim \left(\frac{\xi_p}{d_g}\right)^{2.2} \sim S^{-4.4} \quad . \tag{9.95}$$

Their data for sandstones, epoxies and carbonates supported this relation; see Fig. 9.8. This confirms the relevance of percolation to dispersion in porous media and the scaling laws discussed above.

The last question to be addressed in this section is: What is the equation for the probability density function $P(\mathbf{r}, t)$ (or the solute concentration) for non-Gaussian dispersion? For example, as our discussion of dispersion in percolation networks near p_c indicates, dispersion in a fractal porous medium is not expected to be Gaussian, for which $P(\mathbf{r}, t)$ is given by Eq. (3). In Chapter 8 we discussed the appropriate form of $P(\mathbf{r}, t)$ for fractal diffusion. For non-Gaussian dispersion in a porous medium with a fractal dimension D_f, Sahimi (1987) proposed the following equation for $P(\mathbf{r}, t)$, in the limit of long times

$$P(\mathbf{r}, t) \sim t^{-d_s} \exp\left[-\alpha_1 \left(\frac{|x - \langle x \rangle|}{t^{1/D_w^l}}\right)^{\nu_l} - \alpha_2 \left(\frac{|y|}{t^{1/D_w^t}}\right)^{\nu_t} - \alpha_3 \left(\frac{|z|}{t^{1/D_w^t}}\right)^{\nu_t}\right], \quad (9.96)$$

where αs are constant, and as we already showed, $D_w^l = D_w^t$ for most cases. Here $d_s = D_f/D_w^l$, and $\nu_l = D_w^l(D_w^l - 1)^{-1}$, and $\nu_t = D_w^t(D_w^t - 1)^{-1}$. Equation (96) reduces to Eq. (3) when $D_f = d$ and $D_w^l = D_w^t = 2$. Monte Carlo simulations of Sahimi and Imdakm (1988) seem to support the validity of Eq. (96), but no rigorous derivation of it is yet available, and the matter is still an open question.

9.11 Dispersion in megascopic porous media

Dispersion in megascopic porous media has attracted considerable attention by *both* hydrologists and *politicians* in the past two decades, as a result of growing concerns about pollution and water quality. Because of intensifying exploitation of groundwater, and the increase in solute concentrations in aquifers due to salt-water intrusion, leaking repositories, and use of fertilizers, dispersion in megascopic porous media has been a main topic of research. Moreover, dispersion in miscible displacement processes is an important phenomenon during oil recovery processes, and depending on the magnitudes of D_L and D_T and other physical parameters of the process, dispersion can help or hurt a miscible displacement process and its efficiency.

As discussed in Chapter 1, in studying transport in heterogeneous porous media, one should define clearly what is meant by heterogeneous. In Chapter 1 we discussed four different scales of heterogeneities which are microscopic, macroscopic, megascopic, and gigascopic scales. Bhattacharya and Gupta (1983) discussed a variety of length scales ranging from kinetic and Taylorian to the Darcy scales, while Dagan (1986) considered length scales ranging from pore to laboratory to formation to regional levels. Cushman (1984) provided a brief review of the general problem of the development of N-scale transport equations. A complete treatment of the problem at all these length scales is not currently available.

We remind the reader that dispersion in megascopic porous media is purely mechanical, arising as a result of large-scale spatial variations of the permeability of the medium and the resulting random velocity field. Thus Eqs. (28) and (29)

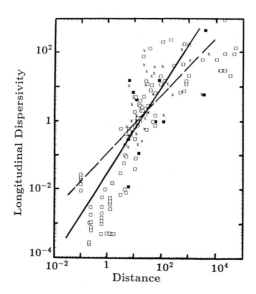

Figure 9.9: Field-scale longitudinal dispersivity as a function of the distance from the source (both are in meters). Solid line represents the best fit to about 75% of the data, while the dashed line represents Gaussian dispersion (after Arya *et al.*, 1988).

are generally expected to hold, in which the length scale used in Pe may be the permeability correlation length ξ_k. In this sense, dispersion in a megascopic porous medium is somewhat *simpler* than that in a macroscopic medium. The main problem to be solved for dispersion in a megascopic porous medium is the relation between the implied prefactors in Eqs. (28) and (29) and the distribution of large-scale heterogeneities of the medium.

In the early years of investigating dispersion in megascopic porous media with large-scale permeability and porosity variations, a CDE served as the starting point for analyzing field data and the treatment of the problem was completely deterministic. However, a considerable amount of data have indicated unequivocally that D_L and D_T measured in a field are larger by several orders of magnitude than those measured in a laboratory, and an entirely deterministic approach cannot provide a completely satisfactory explanation for such data. Moreover, field experiments of Sudicky and Cherry (1979), Pickens and Grisak (1981), Sudicky and Frind (1982), Sudicky *et al.* (1985), and Molz *et al.* (1983) (for a review see Gelhar *et al.*, 1992) indicate that dispersion coefficients and dispersivities are often scale-dependent, and that the apparent dispersivities α_L and α_T seem to increase with the transit times of the solute particles, similar to those in percolation systems for length scales $L \ll \xi_p$. Figure 9.9, taken from Arya *et al.* (1988), demonstrates this phenomenon very clearly. As the distance from the source increases, so does the dispersivity and no asymptotic limit in which α_L is

constant is apparent. Warren and Price (1961) seem to be the first to investigate dispersion in megascopic porous media, taking into account the effect of the
permeability distribution. They used a Monte Carlo method which is discussed
in the next section.

Despite their many shortcomings, deterministic approaches are still advocated and used by many. Although they provide valuable insights into the
problem, deterministic approaches fail to provide any quantitative predictions
for dispersion coefficients and the solute concentration profile. Typical of such
approaches is the large-scale volume averaging method of Plumb and Whitaker
(1988a,b), which we now discuss.

9.11.1 Continuum models: Large-scale volume-averaging

Plumb and Whitaker (1988a,b) (see also Tompson and Gray, 1986) considered
a *two-scale* problem and developed a large-scale averaging technique for determining the macroscopic transport equation. The starting point of their analysis
was Eq. (40), which has to be averaged over regions in which the permeability
and porosity vary spatially. To do this one writes

$$\epsilon_\beta = \{\epsilon_\beta\} + \tilde{\epsilon}_\beta \,, \tag{9.97}$$

$$\langle \mathbf{v} \rangle^\beta = \{\langle \mathbf{v} \rangle^\beta\} + \tilde{\mathbf{v}}_\beta \,, \tag{9.98}$$

$$\langle C \rangle^\beta = \{\langle C \rangle^\beta\} + \tilde{C} \,, \tag{9.99}$$

$$\mathbf{D}^* = \{\mathbf{D}^*\} + \tilde{\mathbf{D}} \,, \tag{9.100}$$

where $\{\cdot\}$ denotes the large-scale quantity. Substituting these expressions in
Eq. (4) and averaging the results over regions in which the permeability and
porosity vary appreciably, Plumb and Whitaker (1988a,b) derived a large-scale
averaged equation for the average concentration which contained such terms
as $\nabla\nabla\nabla\{\langle C \rangle^\beta\}$ and $\nabla\nabla\partial\{\langle C \rangle^\beta\}/\partial t$, indicating that dispersion in megascopic
porous media does not, in general, obey a CDE. As our discussion of dispersion
in percolation systems in the non-Gaussian regime indicates, when dispersion
does not obey a CDE, one has to deal with time- and scale-dependent dispersion
coefficients. The work of Plumb and Whitaker is valuable in that it demonstrates
clearly the deviations of dispersion in megascopic porous media from a conventional CDE description. The main difficulty with the approach of Plumb and
Whitaker (1988a,b) is that, except for very simplified models of a pore space,
the numerical solution of their averaged equation is extremely difficult to obtain.
The problem has to be solved first at the local level in order to use the solution
as the starting point for determining the solution of the averaged megascopic
equation. Moreover, as discussed above, the main contribution to dispersion in a
megascopic porous medium comes from large-scale variations of the permeability
and porosity of the medium, and local or pore-level phenomena such as diffusion

do not play an important role. It is, however, not clear how such large scale-permeability and porosity variations can be incorporated into the simple models that one has to work with.

Koch and Brady (1988) also studied dispersion in megascopic porous media using volume-averaging techniques. They were able to show that if the correlation length ξ_k for the permeability fluctuations is finite, dispersion is diffusive and obeys a CDE. However, if ξ_k is divergent, then superdiffusive dispersion occurs in which the mean square displacements grow with time faster than linearly, completely similar to superdiffusive dispersion near and at the percolation threshold for length scales $L \ll \xi_p$. Thus, one may conclude that ξ_k plays a role very similar to percolation correlation length ξ_p. Moreover, Koch and Brady (1988) showed that, in the superdiffusive dispersion regime the space-time evolution of the concentration is *universal*, and is *uniquely* related to the covariance of the permeability field. This is again completely similar to superdiffusive dispersion in percolation systems at length scales $L << \xi_p$.

9.11.2 Continuum models: Stochastic-spectral methods

This method has been popular with geologists and hydrologists, and has been used extensively. The main motivation for using such methods is that the complex geohydrological structures of aquifers, the nonuniformity and unsteadiness of flow, and other influencing factors make dispersion a very complex phenomenon. Field measurements of dispersivities are often costly and time consuming. For example, one needs to drill many observation wells to monitor the spread of the solute, and the spreading itself is often very time consuming and slow, and a few *years* may be needed for completing the investigations. The level of uncertainties in all of the operations and measurements is quite high, and therefore stochastic methods have been advocated so that the concepts of randomness, uncertainty and errors can be introduced into the models and analyses. Early analytical studies of this problem, using stochastic concepts, were carried out by Mercado (1967) and Buyevich *et al.* (1969), but not all the ingredients were known at that time, and it has been only in the last decade or so that a more comprehensive analysis of this problem has become possible. The assumption of ergodicity is implicit in the stochastic approaches. That is, one assumes that dispersion of a solute in an ensemble of porous media with the assigned statistical properties mimics the situation in a real field, which is a single realization of a megascopic porous medium with large-scale variations of the permeability, porosity, and other properties. This assumption is valid if the scale of the flow system is large compared with the correlation length of the system. Thus, if the permeability, porosity, or other properties are fractally distributed, which implies that the correlation length is as large as the linear size of the system, the stochastic-spectral approach breaks down.

To give the reader some ideas about stochastic-spectral models of dispersion in megascopic porous media, let us discuss briefly the work of Gelhar *et al.* (1979), which is representative of this class of models. The starting point is a CDE at the local level

$$\frac{\partial C}{\partial t} + \frac{\partial}{\partial x}(vC) = \frac{\partial}{\partial x}\left(D_L \frac{\partial C}{\partial x}\right) + \frac{\partial}{\partial z}\left(D_T \frac{\partial C}{\partial z}\right) \quad , \tag{9.101}$$

where it is assumed that C, D_L and D_T, which are *local* properties, are random processes with

$$C(x, z, t) = C_m(x, t) + c(x, z, t) \quad , \tag{9.102}$$

$$v = v_m + u \quad , \tag{9.103}$$

$$D_L = D_{Lm} + d_L \quad , \tag{9.104}$$

$$D_T = D_{Tm} + d_T \tag{9.105}$$

where subscript m denotes a mean value, e.g., $C_m(x, t) = \langle C(x, z, t)\rangle$, with the averaging being taken with respect to the vertical depth z. $c(x, z, t)$, d_L and d_T are *fluctuations* such that their mean values are zero. If we substitute Eqs. (102)-(105) into Eq. (101) and take the average of both sides, we find that

$$\frac{\partial C_m}{\partial t} + \frac{\partial}{\partial x}(v_m C_m) + \frac{\partial}{\partial x}\langle uc\rangle = \frac{\partial}{\partial x}\left(D_{Lm}\frac{\partial C_m}{\partial x}\right) + \frac{\partial}{\partial x}\left\langle d_L \frac{\partial c}{\partial x}\right\rangle + \frac{\partial}{\partial x}\left\langle d_T \frac{\partial c}{\partial z}\right\rangle \quad , \tag{9.106}$$

where we have now used the fact that the mean quantities are independent of z. We subtract Eq. (106) from Eq. (101) and use the coordinate $\zeta = x - v_m t$ to obtain

$$\underset{1}{\frac{\partial C}{\partial t}} \quad \underset{2}{+u\frac{\partial C_m}{\partial \zeta}} \quad \underset{3}{+\frac{\partial}{\partial \zeta}(uc - \langle uc\rangle)} \quad = \quad \underset{4}{D_{Tm}\frac{\partial^2 c}{\partial z^2}} \quad \underset{5}{+\frac{\partial}{\partial \zeta}\left(d_L \frac{\partial c}{\partial \zeta}\right)} \quad \underset{6}{+D_{Lm}\frac{\partial^2 c}{\partial \zeta^2}}$$

$$\underset{7}{\frac{\partial}{\partial z}\left(d_T \frac{\partial c}{\partial z} - \langle d_T \frac{\partial c}{\partial z}\rangle\right) +} \quad \underset{8}{\frac{\partial}{\partial \zeta}\left(d_L \frac{\partial c}{\partial \zeta} - \langle d_T \frac{\partial c}{\partial \zeta}\rangle\right)} \quad . \tag{9.107}$$

If the perturbation u is small, then the second order terms (numbered 3, 7 and 8) can be neglected and one obtains an approximate equation of the form

$$\frac{\partial c}{\partial t} + u\frac{\partial C_m}{\partial \zeta} = D_{Tm}\frac{\partial^2 c}{\partial z^2} + d_L \frac{\partial^2 c}{\partial \zeta^2} + D_{Lm}\frac{\partial^2 c}{\partial \zeta^2} \quad , \tag{9.108}$$

so that, even at this level of approximation, one already has the additional term $d_L \partial^2 c/\partial\zeta^2$. Equation (108) is solved by assuming that the permeability is a statistically-homogeneous random process. To solve this equation, one introduces the spectral representation of the random variables. If the field permeability K is written as, $K = K_m + k$, where $K_m = \langle K\rangle$, and $\langle k\rangle = 0$, then

$$k = \int_{-\infty}^{\infty} e^{i\omega z}dZ_k(\omega) \quad , \tag{9.109}$$

where ω is the wave number and $Z_k(\omega)$ is a complex stochastic process with orthogonal increments. The random processes u, c, d_L, and d_T also have similar spectral representations. Based on the experimental results of Harleman and Rumer (1963) that α_L seems to be proportional to $\sqrt{K}$ (this may be intuitively clear, as $\sqrt{K}$ is some sort of a length scale, as is α_L), it is easy to show that $d_L/D_{Lm} = 3k/(2K_m)$. If we introduce a spectral representation for c

$$c = \int_{-\infty}^{\infty} e^{i\omega z}\, dZ_c(\omega) \quad , \tag{9.110}$$

then Eq. (108) becomes

$$\frac{\partial y}{\partial t} + a_T v_m \omega^2 y - D_{Lm}\frac{\partial^2 y}{\partial \zeta^2} = V_s(\zeta, t) \tag{9.111}$$

with

$$\begin{aligned}
y &= dZ_c \quad , \\
V_s &= v_m \frac{dZ_k}{k_m} G \quad , \\
G &= -\frac{\partial C_m}{\partial \zeta} + \frac{3}{2}a_L\frac{\partial^2 C_m}{\partial \zeta^2} \quad ,
\end{aligned}$$

and $a_L = D_{Lm}/v_m$ and $a_T = D_{Tm}/v_m$. The cross spectrum of u and c, $S_{uc}(\omega)$, and the spectrum of k, $S_{kk}(\omega)$, can be represented by (see, for example, Lumley and Panofsky, 1974)

$$\begin{aligned}
S_{uc}(\omega) &= E(dZ_u dZ_c^*) \quad , \tag{9.112}\\
S_{kk}(\omega) &= E(dZ_k dZ_k^*) \quad , \tag{9.113}
\end{aligned}$$

where E denotes the expected value, and $*$ denotes the complex conjugate. Since, $dZ_u/v_m = dZ_k/k$, we obtain

$$S_{uc}(\omega) = \frac{S_{kk}(\omega)^2}{k_m^2}\left\{ v_m G\frac{1 - e^{-\beta t}}{a_T\omega^2} - \left(\frac{\partial G}{\partial t} - D_{Lm}\frac{\partial^2 G}{\partial \zeta^2}\right)\left[\frac{1 - (1 + \beta t)e^{-\beta t}}{a_T^2\omega^4}\right]\right\} \quad ,$$
$$\tag{9.114}$$

where $\beta = a_T v_m \omega^2$. Similarly, $\langle uc \rangle$ is given by

$$\langle uc \rangle = A v_m G - B\left(\frac{\partial G}{\partial t} - D_{Lm}\frac{\partial^2 G}{\partial \zeta^2}\right) \quad , \tag{9.115}$$

and

$$A = \int_{-\infty}^{\infty} \frac{S_{kk}(\omega)}{k_m^2}\frac{1 - e^{-\beta t}}{a_T\omega^2}\,d\omega \quad ,$$

$$B = \int_{-\infty}^{\infty} \frac{S_{kk}(\omega)}{k_m^2}\frac{1 - e^{-\beta t}}{a_T^2\omega^4}\,d\omega \quad .$$

Using all of these results, Eq. (107) is rewritten as

$$
\frac{\partial C_m}{\partial t} = (A + a_L)\mathrm{v}_m \frac{\partial^2 C_m}{\partial \zeta^2} - B \frac{\partial^3 C_m}{\partial \zeta^2 \partial t} - 3 a_L A \mathrm{v}_m \frac{\partial^3 C_m}{\partial \zeta^3} + 3 a_L B \frac{\partial^4 C_m}{\partial \zeta^3 \partial t}
$$

$$
+ \left(a_L B \mathrm{v}_m + \frac{9}{4} a_L^2 A \mathrm{v}_m \right) \frac{\partial^4 C_m}{\partial \zeta^4} + \cdots \tag{9.116}
$$

Equation (116) indicates that the average concentration C_m does not obey a CDE, a result similar to that obtained by Plumb and Whitaker (1988b). The rest of the analysis is clear: a spectrum $S_{kk}(\omega)$ is assumed and the quantities A and B are calculated. Having determined A and B, one can proceed to analyze Eq. (116). Gelhar and Axness (1983) extended this analysis to three-dimensional heterogeneous media and found that the dispersion coefficients depend linearly on the average velocity, which is not surprising [see Eqs. (28) and (29)].

The above method assumes that all of the randomness is due to the permeability field. An alternative approach relies on a stochastic representation of the velocity (Tang *et al.*, 1982), and develops an ensemble average equation containing coupling between the velocity and concentration fluctuations (which is similar to that found by Gelhar *et al.*, 1979), that leads to a coefficient in the stochastic transport equation which is similar to D_L in a conventional CDE. This term, the *ensemble dispersion coefficient*, depends upon the variance-covariance structure of the velocity field. If neighboring velocities are uncorrelated, the ensemble dispersion coefficients *increase* as a function of travel distance from the source. If the covariance of the velocity field is an exponentially-decaying function, then ensemble dispersion coefficients reach a constant value. This analysis shows clearly the effect of correlations in the stochastic treatment of dispersion. Stochastic models of the type that we discussed here can be useful if an adequate representation of the velocity or permeability fields is available. The interested reader should consult Dagan (1986, 1987) and Haldorsen and Damsleth (1990) for more details and references on stochastic modelling of transport in megascopic porous media.

9.11.3 Discrete models and Monte Carlo methods

We already discussed in Chapter 6 the model of megascopic porous media that was developed by Warren and Price (1961), Warren and Skiba (1964), and Heller (1972), and its improvement by Smith and Freeze (1979) and Smith and Schwartz (1980, 1981a,b) who incorporated short-range correlations between the neighboring blocks. Smith and Schwartz used this model for studying dispersion in megascopic porous media. They used an algorithm for the motion of solute particles that included both deterministic and random displacements. In their simulations, a solute particle is released in the flow field. For each time step a velocity is calculated by linearly interpolating the value from four surrounding values (in a two-dimensional system). The particle is then moved a distance that is fixed by the magnitude of the time step and the velocity. This is the deterministic portion of the displacement. Relocation from the deterministic position is accomplished first by moving the particle a distance d_x in a direction that coincides with the flow vector, and second a distance d_y in a direction normal to it. The random displacements d_x and d_y are calculated from

$$d_x = (0.5 - [R])\sqrt{24 D_{Ll} \Delta t} \ , \tag{9.117}$$

$$d_y = (0.5 - [R])\sqrt{24 D_{Tl} \Delta t} \ , \tag{9.118}$$

where Δt is the time step, $[R]$ a random number uniformly distributed in $(0, 1)$, and D_{Ll} and D_{Tl} are *local* dispersion coefficients. Using this model, Smith and Schwartz investigated many aspects of dispersion in heterogeneous media and showed that strong permeability heterogeneities give rise to non-Gaussian dispersion. They also showed that when the permeability correlation length ξ_k is of the order of the system length, a unique dispersion coefficient may not be possible to define, in agreement with the previous results discussed above.

9.11.4 Fractal models

In our discussion of models of megascopic porous media in Chapter 5, we mentioned Hewetts's (1986) work on modelling transport in such systems. Hewett analyzed vertical porosity logs of megascopic porous media and found that their distributions often obey fractal statistics. More precisely, they obey a fractional Gaussian noise (fGn), discussed in Chapter 3, with $H > 0.5$. Vertical porosity logs analyzed by Hewett (1986) produced values $H \simeq 0.7 - 0.8$, indicating long-range positive correlations. Typical data are shown in Fig. 9.10. The exponent H, so obtained, was subsequently used for generating fractal distributions of the permeability of the porous medium with long-range correlations. A fractional Brownian motion (fBm), which was introduced and discussed in Chapter 3, was used for generating the long-range correlations.

Arya *et al.* (1988) analyzed over 130 greatly-varying dispersivities, collected on length scales up to 100 km. The data collected by them, which are shown in Fig. 9.9, show large scatter, but their analysis indicated that about 75% of them follow the following scaling law

$$\alpha_L \sim L^\delta \ , \tag{9.119}$$

where L is the length scale of measurements, or the distance from the source where the solute is injected into the solvent in the porous medium. Arya *et al.* (1988) suggested that $\delta \simeq 0.75$. Neuman (1990) presented a different analysis of these and other data, and proposed that there are in fact *two* distinct regimes. The first regime was proposed to be for $L << 100$m for which $\delta \simeq 1.5$, while the

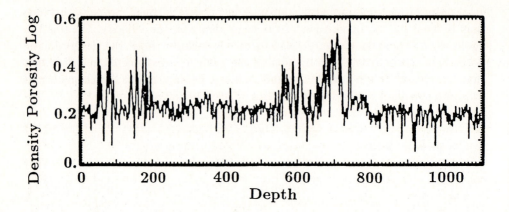

Figure 9.10: A typical porosity log (after Hewett, 1986).

second regime was for $L >> 100$m, for which δ was found to be close to unity. Equation (119) is reminiscent of dispersion in percolation networks discussed above. As our discussion of dispersion in percolation networks indicates, scale-dependence of α_L also implies that it is time dependent. In general, we write

$$\alpha_L \sim t^v ,   \qquad (9.120)$$

which is again similar to our results for percolation systems at length scales $L << \xi_p$, Eqs. (81) and (89). A *non-universal* v has been found to provide reasonably accurate fits of various field data, including those shown in Fig. 8.10, with $v \simeq 0.5 - 0.6$. Of course, scale- and time-dependence of α_L implies the same for v and D_L, also in agreement with our results for percolation networks.

In Hewett's work, the dispersivity α_L was implicitly assumed to obey Eq. (120) with

$$v = 2H - 1 ,   \qquad (9.121)$$

so that with $H \simeq 0.75$ one obtains $\alpha_L \sim t^{0.5}$. A similar result was obtained by Philip (1986), and Ababou and Gelhar (1990), and is also implicitly assumed by Arya *et al.* (1988). Philip's work also predicted that at short times, $\alpha_L \sim t$, and the constraint $0.5 < H < 1$ was also proposed, consistent with Hewett's analysis of porosity logs. This equation was then used in numerical simulation of dispersion. Using these results, transport processes in megascopic porous media were simulated in a series of papers by the Chevron group (Hewett and Behrens, 1988; Mathews *et al.*, 1989; Emanuel *et al.*, 1989). The phenomena that were simulated were dispersion, a miscible displacement (discussed in Chapter 11), and a waterflood (a process in which water is injected into the medium to displace the oil; see Chapter 12), which is an immiscible displacement process. For example, for simulating dispersion, one first generates the permeability and porosity fields, using fBm and fGn distributions. The simulations are conditional (see Chapter

6), because the permeability and porosity fields have to honor some actual data collected at certain places in the field. The flow field is then calculated using Darcy's law, which reduces the problem to calculation of the pressure field. Often a finite-difference approximation with a rectangular grid is used (see Chapter 14). The convective-diffusion equation is then discretized and solved with the resulting flow field. In practice, the discretized equations have to be solved for a very fine grid, so that the effect of permeability and porosity distributions with their long-range correlations can be captured. Comparison of the simulation results with the field data shows that such conditional simulations with fractal distributions of the permeability and porosity are far more realistic than the conventional simulations. Thus, the results for dispersion in percolation systems at length scales $L << \xi_p$, and the results of the Chevron group demonstrate most definitively the relevance of fractal statistics to modelling megascopic porous media and transport processes in such media.

A different approach to the application of fractal concepts to dispersion in megascopic porous media was proposed by Wheatcraft and Tyler (1988). To simulate dispersion in such porous media, they generated fractal and self-similar heterogeneities and, using a Lagrangian random walk model of dispersion, they showed that dispersion coefficients and dispersivities grow with the distance travelled. Although their model does seem to provide an explanation for the observed dependence of the dispersivities on time and length scales in terms of fractal concepts, their random walk algorithm is not adequate enough to provide quantitative information, because their random walk is *unbiased*, i.e., the random walker selects its steps randomly, whereas if the motion of the solute particles is to be modelled by a random walk, it has to be a biased walk, the bias being generated dynamically by the flow field.

9.11.5 Long-range correlated percolation model

So far the percolation models that we have used to discuss and explain various phenomena have been based on random percolation. Such percolation models give us the first theoretical justification for the existence of scale- and time-dependent dispersion coefficients and dispersivities in megascopic porous media. However, random percolation models predict *universal* scaling of α_L and D_L with L and t, whereas, as discussed above, field measurements and observations indicate non-universal scalings. We now use fBm to propose a percolation model with long-range correlations, and show that it can provide an explanation for the scale-dependence of the dispersivities and the dispersion coefficients of megascopic porous media discussed above. For more details see Sahimi (1994a).

We propose the following percolation model with long-range correlations. To each bond of a network we assign a number selected from an fBm, which is interpreted as the permeability of a portion of a megascopic porous medium

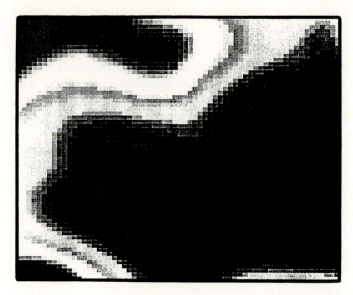

Figure 9.11: A square network in which the bonds permeabilities obey an fBm with $H = 0.8$, and 30% of the bonds with the lowest permeabilities have been removed. Darkest and lightest areas correspond, respectively, to the lowest and highest permeable regions.

over which it is homogeneous. These numbers are therefore infinitely correlated ($H > 0.5$) or anticorrelated ($H < 0.5$). In uncorrelated percolation the bonds of the network are removed randomly. However, to preserve the correlations between the bonds in the correlated model, we remove those bonds that have been assigned the *smallest* numbers. The idea is that in a porous medium with a broad distribution (such as the fBm) of the permeabilities, a finite fraction of the regions should have very small permeabilities, and therefore their contribution to the overall properties of the system would be negligible. Figure 9.11 shows a square network in which the bond permeabilities have been selected according to a fBm with $H = 0.8$, and 30% of the bonds with the smallest permeabilities have been removed. For comparison, we show in Fig. 9.12 the same network in which the same fraction of the bonds have been removed *at random*. The striking difference between the two systems can be understood by considering the nature of fBm. Since for $H > 0.5$ the correlations are positive, most bonds with large or small permeabilities are clustered together. As a result, removal of the low-permeability bonds does *not* generate much randomness in the system. Moreover, as we can see in Fig. 9.11, the percolation cluster generated by this model does not have many dead-end bonds and is close to its backbone. This assertion is confirmed by our numerical results discussed below. On the other hand, if we consider the percolation cluster for $H < 0.5$ (the anticorrelated case), it contains much more randomness.

Various properties of this percolation model have been investigated, using large scale-simulations (see Chapter 8) and finite-size scaling (see Chapter 3). In particular, p_c, ν_p, $\hat{e} = e/\nu_p$, $\hat{\mu} = \mu/\nu_p$, and the fractal dimensions D_p and D_{BB} of the sample-spanning cluster and its backbone have been calculated. It has been found that for $1/2 < H < 1$, the range of interest to us in this book, p_c decreases with increasing H, while the reverse is true for $0 < H < 1/2$. Moreover, ν_p and D_p essentially retain their value for random percolation, except when $H \simeq 1$, where $D_p \to 2$. To calculate $\hat{e}$ and $\hat{\mu}$, the numbers generated by fBm are taken as the permeability or conductance of the bonds. It has been found that for $H > 1/2$ D_{BB} increases with H and that $D_{BB} \to 2$ as $H \to 1$, i.e., the cluster becomes compact. Moreover, $D_p \simeq D_{BB}$, confirming the assertion that for $H > 1/2$ the cluster and its backbone are similar. Table 9.1 presents the results for $1/2 < H < 1$, which, unlike random percolation for which the critical exponents are universal, indicates a smooth dependence of the exponents on H.

Table 9.1

Values of the critical exponents for two-dimensional percolation with long-range correlation, as a function of the Hurst exponent H.

H	$\hat{e}$	$\hat{\mu}$	D_{BB}
0.50	0.98	0.98	1.64
0.60	0.91	0.95	1.82
0.75	0.86	0.80	1.85
0.90	0.82	0.50	1.89
0.98	0.76	0.32	1.96

We now show how our correlated percolation model can explain experimental data on hydrodynamic dispersion in megascopic porous media. As discussed above, if the permeabilities are distributed according to an fBm, then the pore space must contain regions of very low permeabilities, the elimination of which gives rise to a correlated percolation structure whose backbone is very close to the cluster itself, and in fact it becomes identical with it for large enough H (see Table 9.1). Therefore, since the fraction of the dead-end pores or stagnant regions in the system is very small, molecular diffusion that transfers the solute into and out of such regions plays no significant role. This means that dispersion is dominated by the stochastic velocity field imposed on the medium by the permeability distribution and, consistent with the field data, D_L depends on the average flow velocity v as

$$D_L \sim \xi v , \qquad (9.122)$$

where ξ is some appropriate length scale. Under such conditions, the role of diffusion is to transfer the solute out of the slow boundary layer zones near

Figure 9.12: The same network as in Fig. 11, except that 30% of the bonds have been removed at *random*.

the pore surfaces, and its effect appears only as a logarithmic correction to Eq. (122); see Eq. (53). Note that, had we not removed the low permeability zones, diffusion into and out of such zones would have been important, and Eq. (30) would have told us that we would have had $D_L \sim v^2$, contradicting the field data.

Because flow takes place only in the backbone of a percolation structure, we should consider dispersion in the backbone of a correlated percolation cluster. Since the permeabilities are infinitely correlated, their correlation length is larger than any other relevant length scale of the system, and therefore the only dominant length scale of the system is its linear size L, implying that the system is a backbone fractal for *any* L. Thus, Eqs. (78) and (81) can be used, implying that $v = 1/(1 + \theta_B)$, except that when calculating θ_B, we must use the numerical values of the relevant exponents given in Table 9.1. We argue that it is our *two-dimensional* correlated percolation that is relevant to the field-scale data, since such data are obtained at large distances from the source (up to several tens of kilometers), whereas the thickness of such porous media is at most a few hundred meters, and therefore such porous media are long and thin, and thus essentially two dimensional. Since the analysis of various field-scale permeability data by Hewett (1986) and others have indicated that $H > 1/2$ (and mostly $0.7 < H < 0.9$), from Table 9.1 we obtain a *nonuniversal* $v = 1/(1 + \theta_B) \simeq 0.53 - 0.62$ for this range, completely consistent with the experimental data discussed above. Note that, if we do not remove the low permeability zones, we would have to use Eq. (89), which would yield values for v that are not in agreement with the experimental data. For example, we would obtain $v(H = 0.6) \simeq 2$, completely inconsistent with the field data.

We thus propose that percolation with long-range correlation is relevant to flow phenomena in megascopic porous media and aquifers, as it provides a consistent explanation for the experimental data on dispersion coefficients and dispersivities collected in field experiments.

9.12 Dispersion in stratified porous media

The last topic to be discussed in this Chapter is dispersion in stratified porous media. Natural rocks are often stratified, made of various layers in which the structural and transport properties may vary greatly from stratum to stratum. For this reason, dispersion and transport in stratified porous media have always been of interest. The earliest studies on transport in stratified porous media appear to be those of Koonce and Blackwell (1965) and Goddin *et al.* (1966), who studied the displacement of oil by water or a solvent in a stratified porous medium. However, these works do not belong to this Chapter and are discussed in Chapter 12.

Marle *et al.* (1967) and Güven *et al.* (1984, 1985) applied Taylor-Aris dispersion theory discussed above to a system of N strata that communicate with one another. Marle *et al.* (1967) obtained a complex integral expression for the longitudinal dispersion coefficient involving porosity, velocity, and *local* transverse dispersion coefficient which were all functions of the distance perpendicular to the strata. Lake and Hirasaki (1981) and Van den Broeck and Mazo (1983, 1984) also considered Taylor-Aris dispersion in a stratified medium. In particular, Van den Broeck and Mazo (1983, 1984) derived several interesting results, including the FPTD and the longitudinal dispersion coefficient. Gelhar *et al.*'s (1979) work discussed above can be thought of as a method of studying dispersion in two-dimensional stratified porous media, since their equations represent averages over the vertical distance z, and the permeability field is assumed to depend on the distance perpendicular to the strata. Plumb and Whitaker (1988a,b) used their large-scale volume-averaging method discussed above to study dispersion in stratified porous media.

In a seminal paper, Matheron and de Marsily (1980) studied dispersion analytically in a two-dimensional stratified porous medium, using analytical methods and asymptotic expansions. The direction of the flow velocity was assumed to be parallel to the bedding and constant for a given stratum. It was further assumed that the component of the velocity along the direction of macroscopic flow field is a weakly-stationary stochastic process. The permeability was assumed to be an isotropic stochastic process and the medium was assumed to be of infinite extent in *both* directions. Matheron and de Marsily (1980) showed that under such conditions dispersion is *never* diffusive. The reason is that since the system is infinitely large and heterogeneous, a travelling solute particle always samples new regions and strata with new heterogeneities. As a result, a diffusive dispersion regime can never be reached. Matheron and de Marsily also showed that if dispersion is to be Gaussian, then the integral of the covariance of the velocity (or permeability) must be zero, as its Laplace transform must depend linearly on the Laplace transform variable λ near $\lambda = 0$. However, for most realistic situations, this will not be the case. On the other hand, if the macroscopic flow is *not* strictly parallel to the stratification (i.e., a small but finite perpendicular flow component is added), then dispersion will asymptotically be Gaussian if this integral is finite.

Bouchaud *et al.* (1990) extended Matheron and de Marsily's work by studying a random walk in a two-dimensional stratified medium, containing random velocity fields. If the velocities in the x-direction, the macroscopic flow direction, are a function of the vertical distance, then Bouchaud *et al.* showed that

$$\langle \Delta x^2 \rangle \sim t^{3/2} \ , \tag{9.123}$$

i.e., one has superdiffusive dispersion. They also showed that there are large sample-to-sample fluctuations. The probability density $P(x,t)$, when averaged over various environments (realizations of the medium), was found to be non-Gaussian and is approximately given by

$$\langle P(x,t) \rangle \sim t^{-3/4} f(x/t^{3/4}) \ , \tag{9.124}$$

where $f(u)$ is a scaling function with the properties that $f(u) \sim \exp(-u^\delta)$ for $u \gg 1$, with $\delta = 4/3$. These results show clearly the non-Gaussian nature of dispersion in megascopic and stratified heterogeneous media, and the inadequacy of a CDE for describing it.

9.13 Conclusions

From our discussion of dispersion processes in porous media, it is clear that these phenomena are very sensitive to the spatial heterogeneities of the media. For this reason, dispersion has been advocated as a sensitive probe of the structure of heterogeneous porous media. Moreover, it is clear that superdiffusive dispersion is a generic property of transport in heterogeneous porous media with long-range correlations and, as such, it is very different from pure diffusion in such media which is usually fractal and very slow.

Chapter 10

Flow and Dispersion in Fractured Rock

10.0 Introduction

In Chapters 8 and 9 we described flow and dispersion processes in disordered porous media. However, the porous media considered there are characterized by a *single* family of transport paths, namely, those that are made exclusively of the pores. Thus, transport in such pore systems is chacracterized by a single transport equation. In a great many cases, the description of a porous system by a single family of transport paths, or a single porosity contributed by the family, is totally inadequate, and transport *cannot* be described by a single transport equation. As already discussed in Chapters 5 and 7, most natural rock masses consist of interconnected and intertwined networks of fractures and pores, which imply the existence of *two* distinct porosities, one of which is contributed by the fractures, while the other one is contributed by the pores. In some cases, e.g., carbonate rocks, one may need *three* degrees of porosity for characterizing rock. Although most of the porosity is contributed by the pores, the fractures provide the most effective transport paths. Moreover, transport processes along the pores and fractures are very different. For example, if the fracture network is sample-spanning, it may be thought of as the backbone of the system in which transport occurs, while the porous matrix may act as a capacitor which is charged by exchange with adjacent fractures.

In addition to fractured rock that is the focus of this Chapter, there are several other systems of scientific and industrial importance in which transport takes place through two or more distinct families of transport paths. An example is provided by porous catalyst particles which usually contain very large pores (*macropores*) and very small pores (*micropores*). The existence of two distinct types of pores gives rise to considerable complications in modelling transport in porous catalysts. For example, transport in the micropores is hindered (restricted) in comparison with that in the macropores because the size of the transporting molecules is often comparable with the size of the micropores. Moreover, while it is easy for the molecules to enter the macropores from the micropores, the reverse is not true, implying that the rates of exchange between the two types of pores are not equal.

Metals and polycrystals provide another example. In these materials, transport often proceeds simultaneously through *two* distinct families of paths, the *bulk* and the *grain boundaries, dislocations and internal cracks*. Finally, coalbed methane reservoirs provide another important example of a system with distinct families of transport paths. Such reservoirs are made of large fractures and very small pores. It is widely accepted that production of methane gas, originally adsorbed on the coalbed matrix, occurs by desorption from the matrix and its subsequent diffusion towards fractures. However, the pores are so small that they do not allow the influx of water from the fractures, and their sizes are also comparable to the size of the methane molecules. Thus, molecular transport in the fractures of coalbed methane reservoirs is very different from that in the pores.

In general, if there are N distinct families of transport paths, transport in the system must be modelled by N coupled transport equations, with the coupling needed to account for the *exchange* between the various families of transport paths. In this Chapter we study flow and dispersion in fractured rock, which corresponds to the $N = 2$ case, although the network model that we develop is quite general and applies to a system with an arbitrary number of distinct families of transport paths. As usual, we consider both the continuum and discrete models and discuss the progress that has been made so far.

10.1 Flow in a single fracture: Continuum and discrete models

If the fracture surface were perfectly smooth, then flow in a single fracture could be modelled as flow between parallel flat plates, for which volumetric flow rate Q is given by, $Q = (wb^3\Delta P)/(12\eta L)$, where w is the width of the fracture, b is its aperture, $\Delta P/L$ is the pressure gradient applied to the fracture, and η is the viscosity of the fluid. Thus, according to this model Q depends on the third power of b. As discussed in Chapter 5, the surface of fractures is very rough, and often the roughness obeys fractal statistics. There is theoretical as well as experimental evidence (Tsang and Witherspoon, 1981; Brown, 1987a,b) for deviations from the cubic law, $Q \sim b^3$, which has been attributed to the surface roughness. For this reason, flow in a single fracture with a realistic shape cannot be solved analytically, and only numerical solutions of the problem can be obtained, which is why this problem has received considerable attention. The numerical methods used so far have been mostly based on discretizing the governing equations, the usual continuity and momentum equations (see Chapter 2), by a finite-difference or finite-element method (see Chapter 14) and solving the resulting equations. However, if the effect of surface roughness of a fracture is to be taken into account, the grid has to be very refined near the surface and this requires intensive computations. Moreover, because of the surface roughness, the fracture aperture varies from point to point. Thus, if the aperture could be measured at a large number of points in the void space of the fracture, then one could use the data with the discretized equations. However, this would also be very tedious and computationally intensive.

It is known that fracture aperture decreases under stress. It has also been observed that under stress the fluid flow through a single fracture is usually along certain channels. Figure 10.1, taken from Billaux (1990), shows the image of the

Figure 10.1: A binary image of a single fracture (left), its equivalent skeleton (middle) in which the dead-end parts and small loops have been removed, and its equivalent network (right). Dark areas represent the void space (after Billaux, 1990).

cross section of a single fracture with its contact and void areas under normal stress. Also shown is the same system after the dead-end channels and small loops of channels have been removed. This figure tells us that *each* fracture should be mapped onto an equivalent network of channels, an example of which is also shown in Fig. 10.1. Thus, flow calculations in a single channel can be reduced to that in a network. The structure of the network depends of course on the morphology of the fracture, and in particular on the distribution of its contact area. Data for such characteristic properties have to be obtained with careful (and often very tedious) measurements. Once the channel network is constructed, flow calculations in the network (that is, in the fracture) can be carried out, a subject already discussed in Chapter 8. This approach was developed by Moreno *et al.* (1988), Tsang *et al.* (1988) and Nolte *et al.* (1989). Figure 10.2, adopted from Moreno *et al.* (1988), shows the distribution of fluid fluxes in an idealized two-dimensional network model of a single fracture, in which the apertures were distributed according to a log-normal distribution (see Chapter 5). The apertures are correlated with each other, with a spatial covariance that decays exponentially. The thick lines show the channels through which most of the fluid flows, and are clearly indicative of channeling phenomenon discussed above. We should point out that mapping flow through a single fracture onto flow through a network also implies that percolation theory is relevant even to flow through a single fracture, as stress decreases the void space of the fracture and creates flow channels. Moreover, as Fig. 10.2 indicates, most of the fluid flows through a few channels, and thus the connectivity of the channels is important to fluid flow through a single fracture.

10.2 Flow in fractured rock

Flow in fractured rock has been studied by both a continuum approach and

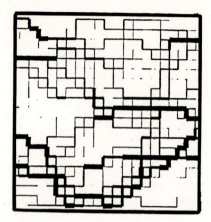

Figure 10.2: A two-dimensional network representation of a single fracture, and its distribution of flow rates. The thickness of the lines is proportional to the flow rates. Observe that the effective network has the structure of a percolation backbone shown in Fig. 3.8 (after Moreno *et al.*, 1988).

by network models. The continuum approach has relied heavily on the multi-porosity model, and the petroleum industry has made extensive use of this model. Network models can be studied by an effective-medium approximation, a generalization of what we discussed in Chapter 8 for porous media, and by Monte Carlo computer simulations. As usual, we discuss both approaches in an attempt to understand their strengths and shortcomings.

10.2.1 Continuum approach: The multi-porosity models

To see the basis for the multi-porosity models, consider first a single-porosity system, e.g., a porous medium. Suppose that the fluid is compressible with a compressibility factor c. Then, an equation of state for the fluid is usually written as (see, for example, Muskat, 1937)

$$\rho = \rho_0 \left(\frac{P}{P_0}\right)^m \exp\left[c(P - P_0)\right], \qquad (10.1)$$

where ρ is the density of the fluid, P is the pressure, and ρ_0 and P_0 denote some reference state. The value of the parameter m depends on the fluid. For liquids $m = 0$, while for gases, $m = 1$ for isothermal processes, and $m = C_v/C_p$ for adiabatic processes, where C_v and C_p denote, respectively, the specific heat at constant volume and constant pressure. We restrict our attention to liquids for which $m = 0$. If we assume that the liquid is only slightly compressible, we can expand the exponential term in Eq. (1) and keep only the first term to obtain

$$\rho \simeq \rho_0 + c\rho_0(P - P_0). \qquad (10.2)$$

On the other hand, the equation of continuity, using Darcy's law, is given by

$$\phi \frac{\partial \rho}{\partial t} = \nabla \cdot \left(\frac{\rho K}{\eta} \nabla P \right) , \qquad (10.3)$$

where ϕ is the porosity of the medium, and K is the permeability of the medium. Substituting Eq. (2) into (3) yields

$$\frac{K}{\eta} \nabla^2 P = c\phi \frac{\partial P}{\partial t} . \qquad (10.4)$$

The idea behind the double-porosity model is to generalize Eq. (4) to a system of fractures and pores. Thus, if subscripts 1 and 2 denote the pore and fracture systems, respectively, Barenblatt and Zheltov (1960) proposed the following set of equations for describing flow of a slightly compressible fluid through a system of fractures and pores

$$\frac{K_1}{\eta} \nabla^2 P_1 - \frac{\alpha}{\eta}(P_1 - P_2) = c_1 \phi_1 \frac{\partial P_1}{\partial t} , \qquad (10.5)$$

$$\frac{K_2}{\eta} \nabla^2 P_2 + \frac{\alpha}{\eta}(P_1 - P_2) = c_2 \phi_2 \frac{\partial P_2}{\partial t} , \qquad (10.6)$$

where α is a characteristic of the porous system, and has the units $1/(\text{length})^2$. The second terms on the left-hand sides of Eqs. (5) and (6) represent the *exchange* between the fracture and pore systems, if it is assumed that the fluid exchange is a pseudo-steady-state process. This assumption can be relaxed, and is discussed further below. Note that this formulation of the problem assumes that the fracture and pore systems can be characterized by the two permeabilities K_1 and K_2 and the two porosities ϕ_1 and ϕ_2. If there are large-scale variations of the permeabilities in the system, as in megascopic systems, then Eqs. (5) and (6) should be written in more general forms to allow for the spatial variations of the permeabilities. Thus Eqs. (5) and (6) are rewritten as [see Eq. (3)]

$$\nabla \cdot (K_1 \nabla P_1) - \alpha(P_1 - P_2) = c_1 \eta \phi_1 \frac{\partial P_1}{\partial t} , \qquad (10.7)$$

$$\nabla \cdot (K_2 \nabla P_2) + \alpha(P_1 - P_2) = c_2 \eta \phi_2 \frac{\partial P_2}{\partial t} , \qquad (10.8)$$

although the double-porosity model used in the petroleum industry is described by Eqs. (5) and (6).

If we assume that, compared with K_2, K_1 is negligible, and that the porosity ϕ_2 of the fracture system (i.e., its storage capacity) is also negligible compared with ϕ_1, the porosity of the pore system, then it is not very difficult to show that

$$\frac{\partial P_i}{\partial t} - \frac{K_2}{\alpha} \frac{\partial}{\partial t}[\nabla \cdot (\nabla P_i)] - \frac{K_2}{c_1 \phi_1 \eta} \nabla \cdot (\nabla P_i) = 0 , \quad i = 1, 2 . \qquad (10.9)$$

In the limit $\alpha \to \infty$ Eq. (9) reduces to the classical diffusion equation which describes flow of a slightly compressible fluid in a porous medium with K_2 and ϕ_1 as its permeability and porosity. Equation (9) was first given by Barenblatt *et al.* (1960). The celebrated Warren-Root model (Warren and Root, 1963) of naturally fractured reservoirs is just this limiting case, except that Warren and Root did not neglect the storage capacity of the fracture system. Thus their model is described by

$$\alpha(P_2 - P_1) = c_1\eta\phi_1\frac{\partial P_1}{\partial t} , \tag{10.10}$$

$$K_2\nabla \cdot (\nabla P_2) + \alpha(P_1 - P_2) = c_2\eta\phi_2\frac{\partial P_2}{\partial t} . \tag{10.11}$$

The solution of the double-porosity model usually is obtained for a configuration of the system in which one or more wells have been dug in the reservoir. Then the main goal of the model is to calculate the behavior of the pressure at the well(s). This is usually called *well-testing*, and is a standard method of gaining information about the structure of the reservoir, since the solutions of Eqs. (10) and (11), which contain characteristic parameters such as the permeability and porosity, can be fitted to experimental data, from which Ks and ϕs may be estimated. For example, the solution of the problem for an axisymmetric system (i.e., in cylindrical coordinates without angular or axial dependence; see Chapter 2) can be obtained. We introduce dimensionless quantities defined by

$$P_{di} = \frac{2\pi K_2 h(P_0 - P_i)}{\eta Q} , \quad i = 1,2 \tag{10.12}$$

where h is thickness of the reservoir, and Q is the flow rate of the well at which the response of the reservoir is measured,

$$r_d = \frac{r}{r_w} , \tag{10.13}$$

where r_w is the radius of the well, and

$$t_d = \frac{K_2 t}{\eta r_w^2(c_1\phi_1 + c_2\phi_2)} , \tag{10.14}$$

$$\Phi = \frac{c_2\phi_2}{c_1\phi_1 + c_2\phi_2} , \tag{10.15}$$

$$z = \frac{\alpha r_w^2}{K_2} . \tag{10.16}$$

Equations (10) and (11) in dimensionless forms are then given by

$$z(P_{d2} - P_{d1}) = (1 - \Phi)\frac{\partial P_{d1}}{\partial t_d} , \tag{10.17}$$

$$\frac{1}{r_d}\frac{\partial}{\partial r_d}\left(r_d\frac{\partial P_{d2}}{\partial r_d}\right) + z(P_{d1} - P_{d2}) = \Phi\frac{\partial P_{d2}}{\partial t_d} . \tag{10.18}$$

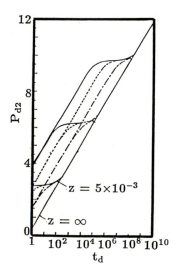

Figure 10.3: Typical solutions of the Warren-Root model for an infinite reservoir. The inclined curves are, from top to bottom, for $\Phi = 0.001$, 0.01, 0.1, and 1. The two upper horizontal curves are for $z = 5 \times 10^{-6}$ and 5×10^{-9} (after Warren and Root, 1963).

The initial and boundary conditions are $P_{d2}(\infty, t_d) = 0$, if the reservoir is of infinite extent, and

$$\frac{\partial P_{d2}}{\partial r_d}\Big|_{r_{de}} = 0 \qquad (10.19)$$

if the reservoir is of finite extent, where r_{de} is the external radius of the reservoir [hence Eq. (19) is simply the zero-flux condition on the boundary of the reservoir]. Moreover,

$$\frac{\partial P_{d2}}{\partial r_d}\Big|_{r_d=1} = -1 \qquad (10.20)$$

and $P_{d1}(r_d, 0) = P_{d2}(r_d, 0) = 0$. Warren and Root (1963) obtained an approximate solution of this boundary-value problem in the Laplace transform space. If λ is the Laplace transform variable conjugate to t, then the Warren-Root solution in the Laplace transform space is given by

$$\tilde{P}_{d2} = \frac{Y_0[r_d\sqrt{\lambda f(\lambda)}]}{\lambda\sqrt{\lambda f(\lambda)}Y_1[\sqrt{\lambda f(\lambda)}]} \ , \qquad (10.21)$$

with

$$f(\lambda) = \frac{\Phi(1-\Phi)\lambda + z}{(1-\Phi)\lambda + z} \ ,$$

where Y_i is the ith-order modified Bessel function of the second kind. Note that the solution of the Warren-Root model tells us that two parameters, namely,

Φ, which is the ratio of the storage capacity of the fracture system and the total storage capacity of the system, and z, which is related to the permeability of the fracture system, suffice to characterize the flow behavior in a naturally fractured system. Figure 10.3 displays typical Warren-Root solutions for an infinite reservoir and some values of Φ and z. Bourdet *et al.* (1983) proposed that in order to distinguish between various types of reservoir, one should look at a plot of $\partial P_d/\partial(\log t_d) = t_d \partial P_d/\partial t_d$ versus t_d. Typical behavior obtained in this way is shown in Fig. 10.4. For a homogeneous, single-porosity medium, the curve should become horizontal, and thus such a plot can be characteristic of a fractured reservoir.

Fluid exchange between the pore and fracture systems is not necessarily a pseudo-steady-state process. One may argue that the exchange between the two systems is a transient phenomenon, and thus the exchange terms should be proportional to the gradient of the pressure at the interface between the two systems, rather than the difference between the pressures in the bulk of the two systems, as assumed above. This idea was first proposed by Kazemi (1969), who also obtained the solution of the problem for a relatively-simple case.

As discussed in Chapter 7, one can generalize the double-porosity model to a triple-porosity model, or more generally to a multi-porosity model, in which one has three or more distinct porosities and permeabilities that characterize the system. The triple-porosity model was first proposed by Closmann (1975). In this model one has two types of porous system, one with good flow properties, and another one with poor properties. The third degree of porosity is contributed by the fracture system. It has been proposed that such a model may be appropriate for describing fractured carbonate reservoirs. If subscript 3 denotes the fracture system, and 1 and 2 denote the two porous systems, then a triple-porosity model is expressed by the following equations

$$K_1 \nabla^2 P_1 + \alpha_1(P_3 - P_1) = c_1 \eta \phi_1 \frac{\partial P_1}{\partial t} , \qquad (10.22)$$

$$K_2 \nabla^2 P_2 + \alpha_2(P_3 - P_2) = c_2 \eta \phi_2 \frac{\partial P_2}{\partial t} , \qquad (10.23)$$

$$K_3 \nabla^2 P_3 - \alpha_1(P_3 - P_1) - \alpha_2(P_3 - P_2) = c_3 \eta \phi_3 \frac{\partial P_3}{\partial t} . \qquad (10.24)$$

Closmann (1975) assumed that, compared with K_3, K_1 and K_2 are negligible. Abdassah and Ershaghi (1986) modified this model by assuming that the exchange terms are proportional to the pressure gradients between various systems, as was done by Kazemi (1969) for the double-porosity model. Such models can also be modified in order to take into account the effect of stratification in the reservoir.

As can be seen, multi-porosity models are basically the solution of a set of coupled transient diffusion equations, with the coupling terms representing the fluid exchange between the pore and fracture systems. Thus, depending on the

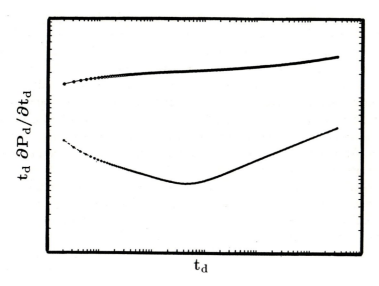

Figure 10.4: Typical dependence of the derivative of the pressure on the time for a double-porosity medium. For comparison the behavior of the pressure (upper curve) is also shown.

initial and boundary conditions, one may obtain a variety of solutions. It is not possible to give a list of the various analytical solutions that have been obtained so far, as the list is too long. The interested reader should consult Chen (1989) for a comprehensive list of references and a more complete discussion.

Although multi-porosity models have helped us gain a better understanding of flow in a fractured rock, they suffer from several shortcomings, the most serious of which is the fact that they do not include any information about the morphology of the pore and fracture systems. As we have emphasized throughout this book, the morphology of a porous system, the connectivity of its pores and/or fractures, the pore size and aperture distributions, the pore and fracture densities in various regions of the system, and the roughness of the fracture and pore surfaces, are all important parameters that have deep influence on the behavior of the system. Continuum multi-porosity models ignore all of these, and represent the pore and fracture systems by their effective permeabilities and porosities. Given that fracture network of many naturally fractured systems is a fractal object (see Chapter 4), such a representation is totally inadequate. To take such effects into account, one has to resort to network models of pore and fracture systems. This is discussed in the next section.

10.2.2 Network models: Exact formulation and EMAs

Exact formulation of a network model of a multi-porosity system was developed by Hughes and Sahimi (1993). The starting point of the analysis is Eqs. (7) and (8), the set of coupled diffusion equations in which the permeabilities vary spatially. This set is then discretized to obtain the network model equivalent of a multi-porosity system. The difference between network and continuum models of multi-porosity systems is that in the network model the topology of the fracture and pore systems, and the pore size and fracture aperture distributions are explicitly included. Thus, in analogy with our formulation of the problem in Chapter 8 for porous media, consider the following set of coupled discretized diffusion equations, which is the natural generalization of the familiar master equation on a d-dimensional network discussed in Chapter 8. To accommodate the existence of N distinct families of transport paths (and porosities) the set has to contain N coupled equations:

$$\frac{\partial \mathbf{P}_i(t)}{\partial t} = \sum_j \mathbf{W}_{ij}[\mathbf{P}_j(t) - \mathbf{P}_i(t)] + \mathbf{E}_i \mathbf{P}_i(t) \,. \tag{10.25}$$

Here $\mathbf{P}_i(t)$ is a column vector, the dimension of which corresponds to the number of distinct transport paths. Consistent with the probabilistic interpretation of a discretized diffusion equation given in Chapter 8, the sth component of $\mathbf{P}_i(t)$ gives the probability that at time t, a randomly-moving particle will be found in path s at network site i. The *transition matrix* $\mathbf{W}_{ij}$ governs the rate at which the bond joining sites i and j is crossed. The ss' entry of this matrix gives the rate at which particles in path s' at site j move to path s at site i, and is related to the permeabilities at sites or grid points i and j. The *exchange matrix* $\mathbf{E}_i$ gives the instantaneous rate of transition between transport paths at site i, or the rate of exchange between distinct transport paths.

    If the underlying network is translationally invariant, if the transition matrix $\mathbf{W}_{ij}$ is a function only of the relative position of sites i and j, and if the exchange matrix $\mathbf{E}_i$ is independent of position, Eq. (25) is easily solved by Fourier analysis. To illustrate the more general situation, consider a one-dimensional system with two paths for which transitions between sites are restricted to nearest-neighbor sites, so that for $j = i \pm 1$,

$$\mathbf{W}_{ij} = \begin{pmatrix} \alpha_{ij} & 0 \\ 0 & \beta_{ij} \end{pmatrix} \text{ and } \mathbf{E}_i = \begin{pmatrix} -\mu_i & \nu_i \\ \mu_i & -\nu_i \end{pmatrix} \,. \tag{10.26}$$

In this case, Eq. (25) is equivalent to two scalar equations which are spatially discretized versions of continuum equations of double diffusion in one space dimension discussed above.

    As the model of a disordered medium with N distinct families of transport paths, we make the following assumptions. (i) All network sites are topologically equivalent. The set of nearest neighbors of site i is denoted by $\{i\}$, so that $j \in \{i\}$ means that site j is a nearest neighbor of site i. The coordination number of the network is denoted by Z. (ii) The transition matrix $\mathbf{W}_{ij}$ is non-zero only when the sites i and j are nearest neighbors. (iii) The transition matrices $\mathbf{W}_{ij}$

are independent, identically distributed random variables. (iv) The exchange matrices $\mathbf{E}_i$ are independent, identically distributed random variables. (v) The $\mathbf{W}_{ij}$ and $\mathbf{E}_k$ matrices are mutually independent. It is not assumed that the for a given matrix $\mathbf{W}_{ij}$ or $\mathbf{E}_k$ the individual entries are independent. Assumptions (iii)-(v) may be relaxed, but are imposed here for simplicity.

The explicit solution of Eq. (25) for a given realization of the disorder is unlikely to be available (except in one dimension) and indeed is not of particular interest. We require a statistical characterization of the large-scale effective transport properties of the system. In one dimension, some exact results can be derived (see below), but apart from this case, exact solution of this problem is unlikely. We develop an exact but implicit solution for matching the disordered system to a uniform system with an effective transition matrix which is the same for all bonds and an effective exchange matrix which is the same for all sites. A discussion of the physical interpretation of the exact solution in the case of a single family of transport paths (a porous medium) was given in Chapter 8. In what follows the diagonal matrix with diagonal elements (in order) $\rho_1, \rho_2, \cdots,$ ρ_N is denoted by $\mathrm{diag}\{\rho_1, \rho_2, \cdots, \rho_N\}$ or, more briefly, by $\mathrm{diag}\{\rho_n\}$.

We attempt to match the disordered system to an "equivalent" uniform system. As discussed in Chapter 8, the matching procedure is performed in the Laplace transform space, in which case Eq. (25) becomes

$$\lambda \tilde{\mathbf{P}}_i - \mathbf{P}_i(0) = \sum_{j \in \{i\}} \mathbf{W}_{ij}[\tilde{\mathbf{P}}_j(t) - \tilde{\mathbf{P}}_i(t)] + \mathbf{E}_i \tilde{\mathbf{P}}_i . \qquad (10.27)$$

This equation is to be matched to the equation

$$\lambda \tilde{\mathbf{P}}_i^e - \mathbf{P}_i(0) = \sum_{j \in \{i\}} \tilde{\mathbf{W}}_e(\lambda)[\tilde{\mathbf{P}}_j^e(\lambda) - \tilde{\mathbf{P}}_i^e(\lambda)] + \tilde{\mathbf{E}}_e \tilde{\mathbf{P}}_i^e , \qquad (10.28)$$

for the uniform system. Similar to our analysis in Chapter 8, the effective transition matrix $\tilde{\mathbf{W}}_e$ and the effective exchange matrix $\tilde{\mathbf{E}}_e$ are functions of the Laplace transform variable λ. Hence, in the time domain, the disordered system is actually being matched to a *generalized master equation* with N families of transport paths:

$$\frac{\partial \mathbf{P}_i^e(t)}{\partial t} = \int_0^t \sum_{j \in \{i\}} \mathbf{W}_e(t - \tau)[\mathbf{P}_j^e(\tau) - \mathbf{P}_i^e(\tau)]d\tau$$

$$+ \int_0^t \mathbf{E}_e(t - \tau)\mathbf{P}_i^e(\tau)d\tau , \qquad (10.29)$$

where $\mathbf{W}_e(t)$ and $\mathbf{E}_e(t)$ are *memory kernels*. We know from our analyses of the problem with a single family of transport paths given in Chapter 7 that when transport is confined to a fractal subset of the network (for example, a percolation cluster at length scales smaller than the correlation length of percolation), the memory kernels can be slowly decaying, leading to fractal (subdiffusive) transport, in which the mean square displacement of a diffusing particle on the network grows more slowly than linearly with time. This is particularly important, since as our discussion in Chapter 5 indicated, fracture network of natural rock is often a fractal object, and at the largest length scales may have the characteristics of a percolation cluster at the percolation threshold. Only a formulation of the problem similar to the one developed here can take such important properties into account. The analog of Eq. (29) for a continuum is given by

$$\frac{\partial \mathbf{P}}{\partial t} = \int_0^t \mathbf{K}_e(t-\tau)\nabla^2\mathbf{P}(\mathbf{x},\tau)d\tau + \int_0^t \mathbf{E}_e(t-\tau)\mathbf{P}(\mathbf{x},\tau)d\tau , \qquad (10.30)$$

where $\mathbf{x}$ is a point in space. Equation (30) tells us that matching a heterogeneous continuum to a homogeneous one induces memory. Equation (30) should be compared with Eqs. (5) and (6), or with (22)-(24), which contain no memory.

Since most of our equations are derived in the Laplace transform space, we delete the tilde sign for the memory kernels and use $\mathbf{W}_e(\lambda)$ and $\mathbf{E}_e(\lambda)$. The initial condition can be removed by subtracting Eqs. (27) and (28), and after a little rearrangement we obtain

$$(Z\mathbf{I}+\mathbf{A})[\tilde{\mathbf{P}}_i(\lambda)-\tilde{\mathbf{P}}_i^e(\lambda)] - \sum_{j\in\{i\}}[\tilde{\mathbf{P}}_j(\lambda)-\tilde{\mathbf{P}}_j^e(\lambda)]$$

$$= -\sum_{j\in\{i\}}\mathbf{\Delta}_{ij}[\tilde{\mathbf{P}}_i(\lambda)-\tilde{\mathbf{P}}_j(\lambda)] + \mathbf{\Gamma}_i\tilde{\mathbf{P}}_i(\lambda) , \qquad (10.31)$$

where

$$\mathbf{A} = (\mathbf{W}_e)^{-1}(\lambda\mathbf{I}-\mathbf{E}_e) , \qquad (10.32)$$

$$\mathbf{\Delta}_{ij} = (\mathbf{W}_e)^{-1}\mathbf{W}_{ij}-\mathbf{I} , \qquad (10.33)$$

$$\mathbf{\Gamma}_i = (\mathbf{W}_e)^{-1}(\mathbf{E}_i-\mathbf{E}_e) , \qquad (10.34)$$

The matrix Green function $\mathbf{G}_{ij}(\mathbf{A})$ for the difference operator on the left-hand side of Eq. (31) satisfies

$$(Z\mathbf{I}+\mathbf{A})\mathbf{G}_{ik}(\mathbf{A}) - \sum_{j\in\{i\}}\mathbf{G}_{jk}(\mathbf{A}) = -\delta_{ik}\mathbf{I} . \qquad (10.35)$$

Equation (31) has the formal solution

$$\tilde{\mathbf{P}}_i(\lambda)-\tilde{\mathbf{P}}_i^e(\lambda) = \sum_j\sum_{k\in\{j\}}\mathbf{G}_{ij}(\mathbf{A})\mathbf{\Delta}_{jk}[\tilde{\mathbf{P}}_j(\lambda)-\tilde{\mathbf{P}}_k(\lambda)]$$

$$-\sum_j\mathbf{G}_{ij}(\mathbf{A})\mathbf{\Gamma}_j\tilde{\mathbf{P}}_j(\lambda) , \qquad (10.36)$$

which is exact, but only determines $\tilde{\mathbf{P}}_i(\lambda)$ implicitly. However, it expresses the fluctuations in the solution of the disordered system from the solution of the uniform one as a sum over the fluctuations in the transition matrices $\mathbf{W}_{ij}$ (via $\mathbf{\Delta}_{jk}$) and the exchange matrices $\mathbf{E}_i$ (via $\mathbf{\Gamma}_i$), and is a convenient starting point for the construction of effective-medium approximations (EMAs). We defer for the moment the discussion of the implementation of the EMA to assemble some properties of the uniform system described by Eq. (28). These properties are needed to translate the EMA predictions of $\mathbf{W}_e(\lambda)$ and $\mathbf{E}_e(\lambda)$ into observable quantities and will also prepare us for some subtleties which arise in the exact analysis in one space dimension (see below).

For d-dimensional regular networks, the site index i is replaced by a vector $\mathbf{r}$ with integer entries. The Green function can be constructed by discrete Fourier analysis, and we find in particular that the Green function at the origin, which we need for constructing an EMA, is $\mathbf{G}_{00}(\mathbf{A}) = -\mathbf{G}(\mathbf{A})$, where

$$\mathbf{G}(\mathbf{A}) = \frac{1}{(2\pi)^d} \int_{-\pi}^{\pi} \cdots \int_{-\pi}^{\pi} \{Z[1 - \Lambda(\mathbf{k})]\mathbf{I} + \mathbf{A}\}^{-1} d^d\mathbf{k} \,, \tag{10.37}$$

and

$$\Lambda(\mathbf{k}) = Z^{-1} \sum_{\mathbf{r}\in\{0\}} \exp(i\mathbf{k}\cdot\mathbf{r}) \,. \tag{10.38}$$

We introduce a matrix $\mathbf{H}$ to diagonalize $\mathbf{A}$, so that

$$\mathbf{H}^{-1}\mathbf{A}\mathbf{H} = \mathrm{diag}\{a_n\} \,. \tag{10.39}$$

One example of $\mathbf{H}$ is given below. Then

$$\mathbf{G}(\mathbf{A}) = \mathbf{H}\,\mathrm{diag}\{g(a_n)\}\,\mathbf{H}^{-1} \,, \tag{10.40}$$

where

$$g(a) = \frac{1}{(2\pi)^d} \int_{-\pi}^{\pi} \cdots \int_{-\pi}^{\pi} \frac{d^d\mathbf{k}}{Z[1 - \Lambda(\mathbf{k})] + a} \tag{10.41}$$

is the value at the origin of the (scalar) lattice Green function discussed in Chapter 8. In particular, in one dimension we have

$$g(a) = \frac{1}{\sqrt{a(4 + a)}} \,. \tag{10.42}$$

The Laplace transform $\langle \tilde{\mathbf{R}}^2(\lambda) \rangle$ of the column vector mean square displacement

$$\langle \mathbf{R}^2(t) \rangle = \sum_{\mathbf{r}} |\mathbf{r}|^2 \mathbf{P}_{\mathbf{r}}(t) \tag{10.43}$$

is $-\nabla^2_{\mathbf{k}}\hat{\mathbf{P}}(\mathbf{k}, \lambda)|_{\mathbf{k}=0}$, where

$$\hat{\mathbf{P}}(\mathbf{k}, \lambda) = \sum_{\mathbf{r}} \exp(i\mathbf{k}\cdot\mathbf{r})\tilde{\mathbf{P}}_{\mathbf{r}}(\lambda) \,, \tag{10.44}$$

is the discrete Fourier transform of $\tilde{\mathbf{P}}_{\mathbf{r}}(\lambda)$. For the initial condition

$$\mathbf{P}_{\mathbf{r}}(0) = \delta_{\mathbf{r}0}\mathbf{u} \,, \tag{10.45}$$

taking the discrete Fourier transform of Eq. (28) we deduce that

$$\hat{\mathbf{P}}(\mathbf{k}, \lambda) = \mathbf{F}(\mathbf{k}, \lambda)^{-1}\mathbf{u} , \tag{10.46}$$

where

$$\mathbf{F}(\mathbf{k}, \lambda) = \lambda\mathbf{I} + Z[1 - \Lambda(\mathbf{k})]\mathbf{W}_e(\lambda) - \mathbf{E}_e(\lambda) \tag{10.47}$$

and $\Lambda(\mathbf{k})$ is defined by Eq. (38). We differentiate the equation

$$\mathbf{F}(\mathbf{k}, \lambda)\mathbf{F}(\mathbf{k}, \lambda)^{-1} = \mathbf{I} \tag{10.48}$$

with respect to the components of $\mathbf{k}$, let $\mathbf{k} \to \mathbf{0}$, and note that $-\nabla_{\mathbf{k}}^2\Lambda(\mathbf{k})|_{\mathbf{k}=0} = 1$, to deduce that

$$\langle\tilde{\mathbf{R}}^2(\lambda)\rangle = Z[\lambda\mathbf{I} - \mathbf{E}_e(\lambda)]^{-1}\mathbf{W}_e(\lambda)[\lambda\mathbf{I} - \mathbf{E}_e(\lambda)]^{-1}\mathbf{u} . \tag{10.49}$$

As discussed in Chapter 8, the dominant large-t behavior of $\langle\mathbf{R}^2(t)\rangle$ can be deduced from the dominant small-λ behavior of $\langle\tilde{\mathbf{R}}^2(\lambda)\rangle$, although the latter has to be calculated carefully since $\mathbf{E}_e(\lambda \to 0)$ is a singular matrix.

In analysing site occupancy probabilities it is convenient to introduce the *propagator matrix* $\mathbf{M_r}(t)$, defined by

$$\mathbf{P_r}(t) = \mathbf{M_r}(t)\mathbf{u} , \tag{10.50}$$

so that

$$\tilde{\mathbf{M}}_{\mathbf{r}}(\lambda) = \frac{1}{(2\pi)^d}\int_{-\pi}^{\pi}\cdots\int_{-\pi}^{\pi}e^{i\mathbf{k}\cdot\mathbf{r}}\mathbf{F}(\mathbf{k}, \lambda)^{-1}d^d\mathbf{k} . \tag{10.51}$$

After a little algebra, one may show that the Laplace transform of the propagator from the origin is given by

$$\tilde{\mathbf{M}}_0(\lambda) = \mathbf{G}(\mathbf{A})\mathbf{A}(\lambda\mathbf{I} - \mathbf{E}_e)^{-1} = \mathbf{G}(\mathbf{A})(\mathbf{W}_e)^{-1} , \tag{10.52}$$

where as before $\mathbf{A} = (\mathbf{W}_e)^{-1}(\lambda\mathbf{I} - \mathbf{E}_e)$, so that $\mathbf{A}$ is a function of λ.

We are now ready to construct an EMA for the problem. We confine our attention to two cases, conveniently described as *transition disorder* and *exchange disorder*. The general case in which both transition and exchange disorder are present is much more difficult, and is not treated here.

Transition disorder: This type of disorder corresponds to, e.g., randomness in the shapes and sizes of the elements of the transport paths. For example, as we know by now, in rock masses the pores are usually characterized by a pore size distribution, and the fractures by a distribution of their hydraulic conductances or apertures. For laminar flow, the flow rate Q in a pore is proportional to the fourth power of its effective radius, whereas for a fracture Q is proportional to the nth power of its aperture, where n can vary anywhere from 3 to 6, which is why flow in the fracture network is very different from that in the porous matrix. Thus, we assume that the exchange matrices are the same for each site (so $\mathbf{\Gamma}_i \equiv \mathbf{0}$) and we write $\mathbf{E}_i \equiv \mathbf{E}_e$. We construct the *single-bond* EMA by allowing only one bond to have a transition matrix which differs from $\mathbf{W}_e$. If this bond joins sites 0 and 1, Eq. (36) reduces to

$$\tilde{\mathbf{P}}_i(\lambda) - \tilde{\mathbf{P}}_i^e(\lambda) = \mathbf{G}_{i0}(\mathbf{A})\mathbf{\Delta}_{01}[\tilde{\mathbf{P}}_0(\lambda) - \tilde{\mathbf{P}}_1(\lambda)]$$
$$+ \mathbf{G}_{i1}(\mathbf{A})\mathbf{\Delta}_{10}[\tilde{\mathbf{P}}_1(\lambda) - \tilde{\mathbf{P}}_0(\lambda)] \,. \tag{10.53}$$

If we note that $\mathbf{G}_{01}(\mathbf{A}) = \mathbf{G}_{10}(\mathbf{A})$ and let $\mathbf{G}_{00}(\mathbf{A}) = \mathbf{G}_{11}(\mathbf{A}) = -\mathbf{G}(\mathbf{A})$, we may deduce from Eq. (53) an explicit expression for the variation in $\tilde{\mathbf{P}}_i(\lambda)$ across the 0-1 bond:

$$\tilde{\mathbf{P}}_0 - \tilde{\mathbf{P}}_1 = \{\mathbf{I} + \frac{2}{Z}[\mathbf{I} - \mathbf{A}\mathbf{G}(\mathbf{A})]\mathbf{\Delta}_{01}\}^{-1}(\tilde{\mathbf{P}}_0^e - \tilde{\mathbf{P}}_1^e) \,. \tag{10.54}$$

To derive Eq. (54), we note from symmetry that Eq. (35) reduces for $i = k = 0$ to

$$(Z\mathbf{I} + \mathbf{A})\mathbf{G}_{00}(\mathbf{A}) - Z\mathbf{G}_{01}(\mathbf{A}) = -\mathbf{I} \,. \tag{10.55}$$

The EMA is constructed by requiring that

$$\langle \tilde{\mathbf{P}}_0(\lambda) - \tilde{\mathbf{P}}_1(\lambda) \rangle = \tilde{\mathbf{P}}_0^e(\lambda) - \tilde{\mathbf{P}}_1^e(\lambda) \,, \tag{10.56}$$

where the angle brackets denote the average over the disorder in the transition matrices. Hence we arrive at our working equation

$$\langle \{\mathbf{I} + \frac{2}{Z}[\mathbf{I} - \mathbf{A}\mathbf{G}(\mathbf{A})]\mathbf{\Delta}\}^{-1} \rangle = \mathbf{I} \,, \tag{10.57}$$

where $\mathbf{G}(\mathbf{A})$ is given by Eq. (40) and we have for brevity dropped the bond index 01 from $\mathbf{\Delta}_{01}$.

Exchange disorder: This type of disorder corresponds to an asymmetry in the exchange rates. For example, the rate of molecular transport from the micropores to the macropores of a porous catalyst is not the same as that from the macropores to the micropores, because molecular sizes are often comparable to those of the micropores. Since the effective sizes of the pores are often distributed quantities, the rate of exchange between the macropores and micropores varies in space. The same is true about the fractures and pores of rock. In this case we assume that the transition matrices are the same for each site (so $\mathbf{\Delta}_{ij} \equiv \mathbf{0}$) and we write $\mathbf{W}_{ij} \equiv \mathbf{W}_e$. We make the *single-site* EMA by allowing only one site (site 0, say) to have an exchange matrix which differs from $\mathbf{E}_e$ which reduces Eq. (36) to

$$\tilde{\mathbf{P}}_i(\lambda) - \tilde{\mathbf{P}}_i^e(\lambda) = -\mathbf{G}_{i0}(\mathbf{A})\mathbf{\Gamma}_0\tilde{\mathbf{P}}_0^e(\lambda) \,. \tag{10.58}$$

For $i = 0$ we deduce that

$$\tilde{\mathbf{P}}_0(\lambda) = [\mathbf{I} - \mathbf{G}(\mathbf{A})\mathbf{\Gamma}_0]^{-1}\tilde{\mathbf{P}}_0(\lambda) \,. \tag{10.59}$$

This gives the self-consistency condition

$$\langle [\mathbf{I} - \mathbf{G}(\mathbf{A})\mathbf{\Gamma}]^{-1} \rangle = \mathbf{I} \,, \tag{10.60}$$

where the averaging is over the disorder in the exchange matrices and we have for brevity dropped the site index 0 from Γ_0.

Having assembled the general effective medium formalism, we now examine some representative predictions of the approximation. Consider the case in which there is no exchange matrix (i.e., $\mathbf{E}_i = \mathbf{0}$), but allow the transition matrices $\mathbf{W}_{ij}$ to have off-diagonal terms. Then $\mathbf{A} = \lambda(\mathbf{W}_e)^{-1} \to \mathbf{0}$ as $\lambda \to 0$ and indeed $\mathbf{AG(A)} \to \mathbf{0}$. Note that $\mathbf{G(A)}$ has a finite limit as $\mathbf{A} \to \mathbf{0}$ if $d \geq 3$ and a modest divergence in lower dimensions. Thus Eq. (57) reduces to a simple equation for the $\lambda \to 0$ limiting form of the effective transition rate matrix:

$$\langle [(1 - \frac{2}{Z})\mathbf{I} + \frac{2}{Z}\mathbf{W}_e(0)^{-1}\mathbf{W}]^{-1} \rangle = \mathbf{I} . \tag{10.61}$$

with the average taken over the distribution of the random variable $\mathbf{W}$. In space dimension $d \geq 2$, we have N^2 simultaneous nonlinear equations to solve for the N^2 elements of the matrix $\mathbf{W}_e(0)$. However, if $d = 1$ (so that $Z = 2$), the equations greatly simplify and we find the prediction that

$$\mathbf{W}_e(0) = \langle \mathbf{W}^{-1} \rangle^{-1} . \tag{10.62}$$

From Eq. (52), the propagator from the origin is given by

$$\langle \tilde{\mathbf{M}}_0(\lambda) \rangle = \mathbf{G}[\lambda(\mathbf{W}_e)^{-1}](\mathbf{W}_e)^{-1} . \tag{10.63}$$

If we introduce a matrix $\mathbf{V}$ to diagonalize $(\mathbf{W}_e)^{-1}$, so that

$$(\mathbf{W}_e)^{-1} = \mathbf{V}\text{diag}\{\omega_n^2\}\mathbf{V}^{-1} , \tag{10.64}$$

we see that

$$\langle \tilde{\mathbf{M}}_0(\lambda) \rangle = \mathbf{V}\text{diag}\{\omega_n^2 g(\lambda\omega_n^2)\}\mathbf{V}^{-1} . \tag{10.65}$$

In one dimension, we have $\mathbf{W}_e(0)$ predicted explicitly by Eq. (62), while $g(a) \sim 1/(2\sqrt{a})$ as $a \to 0$, so that we predict that

$$\langle \tilde{\mathbf{M}}_0(\lambda) \rangle = \frac{1}{2\sqrt{\lambda}}\mathbf{V}\text{diag}\{\omega_n\}\mathbf{V}^{-1} . \tag{10.66}$$

It can be shown that this is an exact result.

We can derive some results for a network of arbitrary dimension d, with two families of transport paths, as in rock with fractures and pores, and diagonal exchange matrices, so that we have

$$\mathbf{W} = \begin{pmatrix} u & 0 \\ 0 & v \end{pmatrix}, \quad \mathbf{W}^0 = \begin{pmatrix} u^0 & 0 \\ 0 & v^0 \end{pmatrix}, \tag{10.67}$$

$$\mathbf{E} = \begin{pmatrix} -\mu & \nu \\ \mu & -\nu \end{pmatrix}, \quad \mathbf{E}^0 = \begin{pmatrix} -\mu^0 & \nu^0 \\ \mu^0 & -\nu^0 \end{pmatrix}, \tag{10.68}$$

$$\mathbf{A} = \begin{pmatrix} u^{-1}(\lambda + \mu) & -u^{-1}\nu \\ -v^{-1}\mu & v^{-1}(\lambda + \nu) \end{pmatrix} . \tag{10.69}$$

For brevity, we write

$$\kappa = \frac{\mu}{u} + \frac{\nu}{v} \quad \text{and} \quad \kappa^0 = \frac{\mu^0}{u^0} + \frac{\nu^0}{v^0} \,. \tag{10.70}$$

The eigenvalues of $\mathbf{A}$ are the solutions a_1 and a_2 of the equation

$$\left(\frac{\lambda + \mu^0}{u^0} - a\right)\left(\frac{\lambda + \nu^0}{v^0} - a\right) - \left(\frac{\mu^0}{u^0}\right)\left(\frac{\nu^0}{v^0}\right) = 0 \tag{10.71}$$

and so as $\lambda \to 0$ we have

$$a_1 \to 0 \quad \text{and} \quad a_2 \to \kappa^0 \,. \tag{10.72}$$

The corresponding eigenvectors for $\lambda = 0$ are $\begin{pmatrix} \nu^0 \\ \mu^0 \end{pmatrix}$ and $\begin{pmatrix} -\nu^0 \\ u^0 \end{pmatrix}$, and so the diagonalizing matrices are

$$\mathbf{H} = \begin{pmatrix} \nu^0 & -\nu^0 \\ \mu^0 & u^0 \end{pmatrix} \quad \text{and} \quad \mathbf{H}^{-1} = \frac{1}{\kappa^0 \mu^0 \nu^0} \begin{pmatrix} u^0 & \nu^0 \\ -\mu^0 & \nu^0 \end{pmatrix} \,. \tag{10.73}$$

Although $\mathbf{G}(\mathbf{A}) = \mathbf{H}\,\mathrm{diag}\{g(a_n)\}\mathbf{H}^{-1}$, it is not convenient to calculate $\mathbf{G}(\mathbf{A})$ this way, since $g(a_1)$ diverges as $\lambda \to 0$ for $d = 1$ and 2. However, we note that as $\mathbf{A}\mathbf{G}(\mathbf{A}) = \mathbf{G}(\mathbf{A})\mathbf{A} = \mathbf{H}\,\mathrm{diag}\{a_n g(a_n)\}\mathbf{H}^{-1}$, and $ag(a) \to 0$ as $\lambda \to 0$, in the limit $\lambda \to 0$ we obtain

$$\mathbf{A}\mathbf{G}(\mathbf{A}) = \mathbf{G}(\mathbf{A})\mathbf{A} = \frac{g(\kappa^0)}{u^0 v^0}\begin{pmatrix} \mu^0 \nu^0 & -\nu^0 \nu^0 \\ -\mu^0 u^0 & \nu^0 u^0 \end{pmatrix} \,. \tag{10.74}$$

Similarly in the limit $\lambda \to 0$,

$$\mathbf{G}(\mathbf{A})^{-1} = \frac{1}{g(\kappa^0)\kappa^0 u^0 v^0}\begin{pmatrix} \mu^0 \nu^0 & -\nu^0 \nu^0 \\ -\mu^0 u^0 & \nu^0 u^0 \end{pmatrix} \,. \tag{10.75}$$

Consider now *exchange disorder only*, so that $\mathbf{W} = \mathbf{W}_e$ and we may therefore take $u^0 = u$ and $v^0 = v$ as known constants. The analysis is simplified by noting that since

$$\Gamma = (\mathbf{W}_e)^{-1}(\mathbf{E} - \mathbf{E}_e) = \mathbf{A}(\lambda\mathbf{I} - \mathbf{E}_e)^{-1}(\mathbf{E} - \mathbf{E}_e) \,, \tag{10.76}$$

Eq. (60) is equivalent to the assertion that

$$\langle\{\mathbf{I} - \mathrm{diag}\{a_n g(a_n)\}\mathbf{H}^{-1}(\lambda\mathbf{I} - \mathbf{E}_e)^{-1}(\mathbf{E} - \mathbf{E}_e)\mathbf{H}\}^{-1}\rangle = \mathbf{I} \,. \tag{10.77}$$

Although $(\lambda\mathbf{I} - \mathbf{E}_e)^{-1}$ diverges as $\lambda \to 0$,

$$(\lambda\mathbf{I} - \mathbf{E}_e)^{-1}(\mathbf{E} - \mathbf{E}_e)$$

$$= \frac{1}{\mu^0 + \nu^0 + \lambda}\begin{pmatrix} -\mu + \mu^0 & \nu - \nu^0 \\ \mu - \mu^0 & -\nu + \nu^0 \end{pmatrix} \tag{10.78}$$

has a finite small-λ limit. After some algebra, the matrix self-consistency condition (77) reduces to four scalar equations:

$$\langle \frac{1}{D^0}(1 - \frac{\mu^0 - \mu}{u^0})\rangle = \langle \frac{1}{D^0}(1 - \frac{\nu^0 - \nu}{v^0})\rangle = 1 \qquad (10.79)$$

and

$$\langle \frac{\mu^0 - \mu}{D^0 u^0}\rangle = \langle \frac{\nu^0 - \nu}{D^0 v^0}\rangle = 0 , \qquad (10.80)$$

with $D^0 = 1 + g(\kappa^0)(\kappa - \kappa^0)$. Although it may appear that the problem is overdetermined, since

$$\langle \frac{1 + g(\kappa^0)(\kappa - \kappa^0)}{D^0}\rangle = 1 , \qquad (10.81)$$

it is easy to prove that if the pair of conditions (80) hold, then $\langle 1/D^0 \rangle = 1$ and Eqs. (79) are also satisfied. It follows that

$$\mu^0 = \langle \frac{\mu}{1 + g(\kappa^0)(\kappa - \kappa^0)}\rangle \qquad (10.82)$$

and

$$\nu^0 = \langle \frac{\nu}{1 + g(\kappa^0)(\kappa - \kappa^0)}\rangle , \qquad (10.83)$$

where κ^0 is determined from the equation

$$\langle \frac{1}{1 + g(\kappa^0)(\kappa - \kappa^0)}\rangle = 1 . \qquad (10.84)$$

Although we have not exhibited it explicitly here, we emphasize that the last three equations apply only in the limit $\lambda \to 0$ (i.e., the steady-state limit). The general case of positive λ (i.e., the transient problem) is more complicated, but can be studied by numerical simulation.

To illustrate the predictions of the EMA, consider the case of *binary disorder*, where (μ, ν) takes the values (μ^*, ν^*) and $(0, 0)$ with probabilities p and $1 - p$ respectively, and for brevity write $\kappa^* = \mu^*/u^0 + \nu^*/v^0$. Equation (84) becomes

$$(1 - p) = (1 - \kappa^0/\kappa^*)[1 - g(\kappa^0)\kappa^0] . \qquad (10.85)$$

In one dimension, this leads to a quadratic equation for κ^0, but in higher dimensions the scalar lattice Green function g is not an elementary function and the equation has to be solved numerically. The more general case in which (μ, ν) takes the values (μ_1, ν_1) and (μ_2, ν_2) with probabilities p and $1 - p$, respectively, where $\nu_1 << \mu_1$ and $\nu_2 << \mu_2$, is of practical interest since it may correspond to, e.g., transport in a porous catalyst with micropores and macropores, or rock with tight pores and large fractures. As discussed above, transport in the micropores is hindered and slow and, moreover, while it is easy for the molecules to enter the macropores from the micropores, the reverse is not true and, therefore, the rates of exchange between the two types of pores are not in general equal. The

EMA developed here should be particulary accurate for such problems, since the system is not critical and does not have a fractal structure.

Consider now two-path systems with transition disorder only. This case appears more subtle. If we attempt to write $\mu^0 = \mu = $ constant, we find that the self-consistency condition (57) reduces in the $\lambda = 0$ limit to *three* independent scalar equations for the *two* unknown functions $u^0(0)$ and $v^0(0)$, which in general admit no solution. This suggests that for transition disorder, the uniform system which represents the macroscopic transport properties of the disordered system must either possess off-diagonal terms in the matrix $\mathbf{W}_e(\lambda)$ or have an exchange matrix $\mathbf{E}_e(\lambda)$ which differs from the exchange matrix for each realization of the system. In other words, matching a disordered system with several distinct families of transport paths to a uniform system induces not only memory, but also *additional couplings* absent from the original system. This explain why simply coupled diffusion equations that are used in the continuum multi-porosity models discussed above, often are found to be poor models for transport in fractured rock. This is perhaps the most important insight provided by our network formulation of the problem.

Even the $d = 1$ limit of this problem with $N \geq 2$ is non-trivial and far more complex than the $N = 1$ case. For example, whereas percolation-like disorder divides a linear chain into finite segments and prohibits macroscopic transport, in our problem one can have global transport even if m ($m < N$) paths have been disrupted by percolation disorder. A more interesting case arises if transport in one family of paths is much faster than in the other(s). For example, transport in fracture network of rock is much faster than that in the pores, or diffusion in the macropores of a catalyst is much faster than that in the micropores. As we have already learned in Chapter 5, in natural rock it is often true that the fracture network has a fractal structure. Moreover, at the largest length scales (of order of a kilometer or more) the fracture network is found to have the structure of the sample-spanning percolation cluster at the percolation threshold (see Chapters 5 and 7), while the pore network is well-connected. In this case one may expect a variety of interesting results to emerge. In particular, the system can behave quite differently on various time scales which would depend on how the transport process is partitioned between the two networks, and whether these networks are macroscopically connected. Using the exact formulation and EMA developed here, one may study such issues, some of which have been addressed by Hughes and Sahimi (1993), to whom we refere the interested reader.

10.2.3 Anisotropic EMA and the permeability of fracture networks

As we discussed in Chapters 5 and 7, fracture networks of rock are often anisotropic, and therefore they are characterized by a permeability tensor rather than a single permeability. If we are interested only in this tensor, then we can

Figure 10.5: A two-dimensional fractured system with two different sets of fractures, each having its own orientation (after Harris, 1990).

use an anisotropic EMA to obtain it. To do this, one can use the anisotropic EMA, an extension of the EMAs discussed in Chapter 8. This extension was developed by Bernasconi (1974). Its use for fracture networks was discussed by Harris (1990, 1992).

We characterize a d-dimensional anisotropic network by d effective permeabilities K_i with $i = 1, \cdots, d$, one for each principal direction. The anisotropy of the network may be caused by several factors. For example, each principal direction may be characterized by its own permeability distribution, as most rocks are stratified and thus have different distributions of heterogeneities in different directions. Alternatively, the anisotropy may be caused by the orientation distribution of the fractures. For example, consider a two-dimensional fracture system that consists of two fracture sets, with the fractures in each set having the same orientation. An example is shown in Fig. 10.5. It is clear that this system is anisotropic and is characterized by two permeabilities, one for each of its principal directions.

We consider both types of anisotropies, and give some results. The interested reader should consult the original references for more detail. Consider first an anisotropic network in which each principal direction is characterized by its own conductance distribution $f_i(W)$ with $i = 1, \cdots, d$. For a simple-cubic network in d-dimensions, the effective conductances W_{ei} are predicted by the anisotropic EMA (AEMA) to be the solutions of the following equations (Bernasconi, 1974)

$$\int_0^\infty f_i(W) \frac{W - W_{ei}}{W + \mathcal{S}_i} dW = 0 \,, \quad i = 1, \cdots, d \,, \tag{10.86}$$

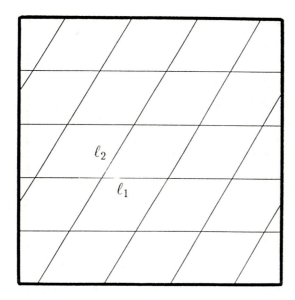

Figure 10.6: The equivalent network of the fractured system shown in Fig. 10.5 (after Harris, 1990).

which is very similar to Eq. (8.66) for isotropic media. For a square network

$$\mathcal{S}_1 = W_{e1} \frac{\arctan\sqrt{W_{e2}/W_{e1}}}{\mathrm{acrtan}\sqrt{W_{e1}/W_{e2}}} \ , \tag{10.87}$$

and $\mathcal{S}_2$ is obtained by interchanging W_{e1} and W_{e2}. For a simple-cubic network

$$\mathcal{S}_1 = W_{e1} \frac{\mathrm{acrtan}\sqrt{W_{e1}^{-1}(W_{e1}W_{e2} + W_{e1}W_{e3} + W_{e2}W_{e3})}}{\mathrm{acrtan}[W_{e1}(W_{e1}W_{e2} + W_{e1}W_{e3} + W_{e2}W_{e3})]^{-1/2}} \ , \tag{10.88}$$

and $\mathcal{S}_2$ and $\mathcal{S}_3$ are obtained by cyclic rotation of W_{e1}, W_{e2}, and W_{e3}. Observe that in the limit in which all the effective conductances are equal, these equations reduce to Eq. (8.66) for isotropic media, as they should. Once the effective conductances are calculated, the effective permeabilities are determined by the same method discussed in Chapter 8 for isotropic media.

Consider now the second type of anisotropy (Harris, 1990). For concreteness we consider the fracture network shown in Fig. 10.5. This network consists of two sets of fractures. One set consists of long, parallel, and equally spaced fractures, while the second set consists of those whose centers are randomly and uniformly distributed on lines parallel to the fracture profiles of the first set. This system can be mapped exactly onto the network shown in Fig. 10.6. Suppose that in the network shown in Fig. 10.6 p_i is the probability that a bond (fracture) of type i has non-zero (hydraulic) conductance. Suppose also that ζ is the number of set 2 fracture centers per unit length lying on a line parallel to and midway between a pair of adjacent set 1 fractures. Then

$$p_2 = \zeta \ell_2 , \qquad p_1 = 1 \tag{10.89}$$

where ℓ_2 is the length of the fractures in set 2 (see Fig. 10.6). The network shown in Fig. 10.6 is anisotropic, not only because of the orientations of the fractures, but also due to the fact that set 1 fractures and set 2 fractures have different conductance distributions, as their lengths are very different. However, since the network is still 4-coordinated, and the angles that the bonds make with each other are the same everywhere, Eqs. (86) and (87) can be used. The two conductance distributions are $f_1(W) = h_1(W)$, and $f_2(W) = (1 - p_2)\delta(W) + p_2 h_2(W)$, where h_1 and h_2 are two normalized probability distributions that describe the conductances of the fractures in set 1 and set 2. Note that, because $p_1 = 1$, $f_1(W)$ does not contain any $\delta(W)$. From Eqs. (86) and (87) it can be seen easily that

$$\tan^2[\frac{1}{2}\pi(1 - p_2)] \leq \frac{W_{e1}}{W_{e2}} \leq \tan^2(\frac{1}{2}\pi p_1) . \tag{10.90}$$

Thus, the idea is to map the fracture network onto an equivalent regular network, for which the Green functions that are necessary for constructing the EMA and AEMA (see Chapter 8) can be determined. The functions S_i that are defined above are in fact obtained from such Green functions. Harris (1990, 1992) has worked out several examples, and Mukhopadhyay and Sahimi (1995) have derived the general AEMA for an arbitrary regular network. The interested reader should consult these papers.

10.3 Dispersion in a single fracture

Dispersion in fractured rock is important to both oil recovery processes and groundwater flow, and has received considerable attention from researchers in both fields. However, as discussed in Chapter 9, dispersion is a far more complex phenomenon than the problem of flow in porous or fractured rock. Similar to the studies of flow in a single fracture and in fractured rock, previous studies of dispersion in such systems can also be divided into two groups. In the first group are studies that investigated transport in a single fracture. This group includes the works of Grisak and Pickens (1980), Neretnieks (1980), Tang *et al.* (1981), Noorishad and Mehran (1982), Novakowski *et al.* (1985), Raven *et al.* (1988), and Lowell (1989). These studies were concerned with the effect of the matrix on transport through a single fracture, and how experimental data should be interpreted. Some of these works were analytical, while others required numerical simulations. They are useful for modelling transport in a network of fractures, since a fracture network simulation requires the solution of the problem at the level of a single fracture.

Novakowski *et al.* (1985) and Raven *et al.* (1988) carried out dispersion experiments in a single natural fracture. In the work of the former authors, a discrete
fracture at a depth of approximately 100 m, located between two boreholes in
a relatively horizontal attitude over a distance of about 10 m was used. The
experimental technique involved injecting a slug of tracer particles in a steady
groundwater flow field, established between a pumping and recharging borehole
and monitoring the tracer breakthrough by sampling the withdrawal water directly. High injection and withdrawal flow rates were maintained in order to
minimize the residence time in both the injection and sampling intervals. Radioactive Bromine 82, ^{82}Br, and a fluorescent dye were used as the tracer (solute),
and their concentration at the withdrawal borehole was measured. The concentration profile of the tracer was then fitted to the solution of a one-dimensional
convective-diffusion equation (CDE) for a finite system, discussed in Chapter 9;
see Eq. (9.9). Figure 10.7 compares the predictions of the fitted model with the
experimental data. Similar experimental studies were carried out by Raven *et
al.* (1988). The tracer breakthrough curves were determined from samples of
the withdrawn groundwater and were interpreted using the first-passage distributions discussed in Chapter 9. However, to interpret their data, the authors
used a dispersion model similar to the Coats-Smith-Baker model discussed in
Chapter 9, namely, in addition to the usual CDE, they also used an equation to
account for the exchange of the solute between the mobile and stagnant fluids.
They found that such a model is a more accurate representation of their data
than Eq. (9.9).

10.4 Dispersion in fractured rock

Several authors have investigated dispersion in fracture network of rock. They
include Schwartz *et al.* (1983), Endo *et al.* (1984), Smith and Schwartz (1984),
Hull *et al.* (1987), Robinson and Gale (1990), Cacas *et al.* (1990a,b), and Tsang
et al. (1991). These authors studied dispersion in a discrete network of interconnected fractures, and are perhaps the most advanced studies of transport in
fractured rocks. Hull *et al.* (1987) and Robinson and Gale (1990) constructed
two-dimensional networks of fractures in which they carried out dispersion experiments. They also performed Monte Carlo simulation of dispersion in similar
networks in order to test the validity of the assumptions that they made in their
models.

Simulation of dispersion in a network of fractures has to be done at three
different levels. The most elementary level is a single fracture already discussed
above. In principle, one has to map each fracture onto an equivalent network,
and then construct a network of such networks to represent the overall fractured
rock. However, this would make the computations very intensive, and to our
knowledge has not been attempted extensively. Most authors represent each

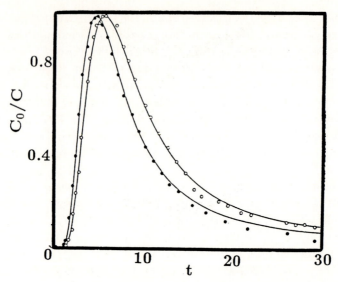

Figure 10.7: Comparison of the experimental data (circles) for normalized concentration C_0/C versus time (in hours) in a single fracture and the predictions of Eq. (9.9). The two curves correspond to two different flow rates, and a longitudinal dispersivity of 1.4 m was used in the equation (after Novakowski *et al.*, 1985).

fracture by a single channel to which effective properties, such as aperture and width, are assigned.

At the next level of complexity is dispersion at the intersection of two or more fractures. Most authors have assumed that the solute and solvent mix completely at the fracture intersections, similar to a class of model developed by Sahimi *et al.* for dispersion in pore networks which we discussed in Chapter 9. The assumption of complete mixing at the fracture intersections is partly based on the works of Castillo (1972) and Krizek *et al.* (1972), who used a plexiglass model of two intersecting fractures. Their work indicated that the assumption of complete mixing at the intersections was justified. However, Hull and Koslow (1986) pointed out that these works were not conclusive. On the other hand, Wilson and Witherspoon (1976) reported that there is little or no mixing at the fracture intersections. Thus, Endo *et al.* (1984) studied dispersion in a fracture network in which *no* mixing was assumed at the intersections. More recent laboratory work of Hull and Koslow (1986) and Robinson and Gale (1990), using discrete networks of fractures, indicated that the assumption of complete mixing at the fracture intersections is not justified. Although there is some mixing at the intersections, its intensity depends on the morphology of the fracture network and the flow regime in it. For example, if the flow field is slow enough, diffusive mixing may occur. Note that what happens at the intersection of the fractures has a strong influence on the distribution of the solute particles in the fracture

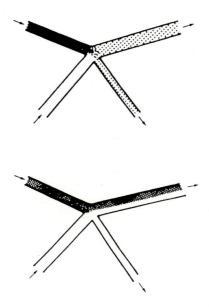

Figure 10.8: Two scenarios for dispersion at an intersection of fractures: complete mixing (top) and no mixing (bottom) (after Cacas *et al.*, 1990b).

network. Figure 10.8 shows two scenarios for dispersion in a fracture intersection. In one of them complete mixing of the solute and solvent is assumed, while in the other one no mixing is assumed. It is obvious that the two systems have very different distribution of the solute particles.

Once dispersion at the level of a single fracture and at the intersection of two or more fractures are understood, one can study dispersion at the level of a network of fractures. In most of fracture networks models that have been used so far, dispersion is simulated by a random walk method described in Chapter 9 for pore networks and for megascopic porous media. For example, in the work of Schwartz *et al.* (1983) and Smith and Schwartz (1984), fractured rock is represented by a two-dimensional network of interconnected discrete fractures. Such models were already discussed in Chapter 7. The fractures are oriented at various angles with respect to the direction of the macroscopic mean flow. After calculating the flow field throughout the network, tracer (solute) particles are released into the network and their motion, which is governed by a random walk biased by the macroscopic mean flow (see Chapter 9), is monitored. In most cases, it has been found that the most important controlling factors are the orientation of the fracture sets with respect to the macroscopic mean flow, and the density of the fractures in a given region of the system. Moreover, the distribution of the apertures can have a strong influence on the distribution of the solute particles throughout the fracture network. Thus, uncertainties in predicting the solute concentration profile is larger in networks with larger variability in their aperture distribution.

One of the most important results of such studies of dispersion in a network of fractures is that in most cases dispersion is *not* Gaussian or diffusive. A typical solute concentration profile that is obtained with such models is shown in Fig. 10.9, and it is clear that this profile cannot be represented by the soltution of a CDE with appropriate initial and boundary conditions. As already discussed in Chapter 6, several careful experimental studies have indicated that fracture networks of rock may be fractal objects. If so, dispersion in such networks should be characterized by dispersivities and dispersion coefficients that vary with time and length scales, similar to those discussed in Chapter 9 for megascopic porous media. Even though the networks used so far in the Monte Carlo simulations of dispersion in fracture networks have not been fractal, the fact that dispersion has not been found to be diffusive indicates that, dispersion in fractured rock is very complex. On the other hand, assuming that a fracture network is a fractal object, Ross (1986) provided some arguments about the variations of the dispersivities with the scale of observations and the distance traveled.

10.5 Conclusions

Flow and dispersion in fractured rock are complex phenomena, and although several continuum models have been developed for studying them, they are often inadequate and cannot provide quantitative, and in some cases even qualitative, predictions for the quantities of interest. The most important factors that contribute to this complexity are the structure of the pore and fracture networks, the exchange mechanism between the two networks, and the heterogeneities that exist at different scales. In our opinion, fractured rock represents a definitive example of a system for which network models are definitely more appropriate than the continuum models.

Chapter 11
Miscible Displacements

11.0 Introduction

In Chapters 9 and 10 we discussed dispersion processes in porous media and fractured rock. Dispersion involves two fluids and one fluid phase, and the viscosities and densities of the solute and the solvent are assumed to be equal. In this Chapter we discuss displacement of one fluid by another miscible fluid having a different viscosity and density. If we inject a fluid into a system saturated with another fluid, and if the two fluids mix in all proportions and their mixture remain a single phase, the two fluids are said to be *first-contact miscible*. Intermediate-molecular weight hydrocarbons such as propane and butane have this property. In other situations, the injected and in-place fluids may form two different phases, i.e., they are not first-contact miscible. However, mass transfer between the two phases and repeated contact between them can achieve miscibility; this is usually called *multiple contact* or *dynamic miscibility*. In the petroleum industry, the miscible injection fluids that achieve either first-contact or dynamic miscibility are usually called miscible solvents. In this Chapter we restrict ourselves to first-contact miscible problems, since modelling multiple contact miscibility involves thermodynamic phase equilibria calculations that are beyond the scope of our book.

Miscible displacement processes have received considerable attention since early 1950s. In the 1950s and early 1960s over 100 projects were undertaken to study the feasibility and economics of miscible displacement processes as effective tools of increasing oil production. In most of these projects hydrocarbon miscible flooding was used. However, most of the hydrocarbons used are normally less viscous than the oil and this viscosity contrast, together with the so-called gravity segregation, make a miscible displacement much less efficient than desired. For this reason, miscible displacement processes have not been used as widely as immiscible displacement processes, such as water flooding, in oil recovery processes. In most miscible displacement processes the injected fluid is either a hydrocarbon or, less frequently, flue gas, nitrogen, or CO_2. The temperature and pressure that one needs for miscibility of the oil and these agents are often so high that they limit the number of prospective reservoirs. For example, CO_2 and medium to heavy hydrocarbons become miscible only at high temperatures and pressures (Sahimi *et al.*, 1985; Sahimi and Taylor, 1991). Another negative factor is the cost of a miscible displacement process. It may happen that a miscible displacement is more efficient in terms of the amount of the recovered oil than an immiscible process, but the total cost of the miscible displacement (including the cost of transporting the fluid agent to the oil field from other locations) is so high that makes it unattractive from an economical point of view. Flue gas and nitrogen have only limited application as agents of a miscible displacement process in deep and high pressure reservoirs. In the United States, CO_2 has the greatest potential for miscible displacement and oil recovery, but its availability is questionable.

In a paper of Paterson (1984) a connection was suggested between miscible displacement processes and diffusion-limited aggregation (DLA) models, which are the basic models of non-equilibrium processes (for a review of DLA models see Meakin, 1988) and give rise to fractal structures. Since then there has been a great deal of interest in this connection, especially in the statistical physics community. We discuss and review in this Chapter a few important continuum models of miscible displacements and contrast them with the statistical and DLA-like models. We also discuss the stability of miscible displacement processes. This Chapter is by no means exhaustive. The interested reader can consult, for example, the monograph by Stalkup (1984) which provides a detailed discussion of these processes, the classical methods of studying them, and the experimental data.

11.1 Factors affecting miscible displacement processes

As discussed by Stalkup (1984), many factors contribute to the efficiency of a miscible displacement. Among the most important of these are the following.

(i) *Mobility Ratio*: The mobility λ_i of a fluid i is defined as the ratio of the effective permeability of the rock K_i and the fluid's viscosity η_i: $\lambda_i = K_i/\eta_i$. When one fluid displaces another, the mobility ratio M is defined as the ratio of the mobilities of the displacing and displaced fluids, and is one of the most important influencing factors in any displacement process. Normally, M is not constant because mixing changes the effective viscosities of the two fluids. In addition, the viscosity of the mixed zone also depends on those of the displacing and displaced fluids. In many cases the viscosity η_m of the mixed zone is estimated from the following empirical law due to Koval (1963)

$$\eta_m = \left(\frac{C_s}{\eta_s^{1/4}} + \frac{1 - C_s}{\eta_o^{1/4}} \right)^{-4}, \tag{11.1}$$

where C_s is the solvent concentration, and η_s and η_o are the viscosities of the solvent and oil (the displaced fluid), respectively. Sometimes, instead of C_s, the solvent volume fraction is used in Eq. (1). If, in addition to the solvent, another fluid such as water is also injected into the medium, as is often done in order to reduce the mobility of the solvent, then it is not completely clear how to define an effective value of M. In many miscible displacement processes, one has to deal with more than one displacing front. For example, in tertiary oil recovery processes (which are undertaken when a process such as water flooding is no longer effective) there are usually more than one displacing fronts. The problem of defining an effective M is even more complex in such a situation, and no completely satisfactory method has been developed yet.

(ii) *Dispersion*: Dispersive mixing can decrease the viscosity and density contrasts between the displacing and displaced fluids, and in many cases this can be very useful. As discussed in Chapter 9, two major contributing factors to dispersion are small- and large-scale variations of fluid velocities (or the permeabilities), and molecular diffusion, both of which help mixing of the two miscible fluids. Since the longitudinal dispersion coefficient D_L is usually larger than the transverse dispersion coefficient D_T, more dispersive mixing takes place in the direction of macroscopic flow. Even at the level of a single pore, it is difficult to study rigorously dispersive mixing of two fluids of unequal viscosities. In essence, we have to rework the Taylor-Aris dispersion theory discussed in Chapter 9, except that the viscosity of the mixing zone depends on the concentrations of the two fluids. To the best of my knowledge, no serious attempt has ever been made to tackle this problem. Since, as discussed in Chapter 9, pore space heterogeneities strongly affect D_L and D_T, this implies that such heterogeneities also affect miscible displacements.

Mobility ratio and gravity also affect dispersion. If $M > 1$, viscous instability develops (see below) in which case the displacement is no longer a simple process. However, if $M < 1$, the usual dispersion mechanisms discussed in Chapters 9 and 10 are operative. Moreover, since no instability develops, the effect of pore space heterogeneities is also supressed. On the other hand, if in a miscible displacement a less dense fluid displaces a denser fluid, then gravity supresses the effect of dispersive mixing (see below).

In some situations, longitudinal dispersion affects a miscible displacement more strongly than transverse dispersion, and vice versa. For example, when large fingers of the displacing fluids develop, which is often the case when $M > 1$, then there would be a large surface area on the sides of the fingers which allows for significant transverse dispersion to occur. This can help join the fingers, stabilize the displacement, and increase its efficiency. By contrast, longitudinal dispersion can only take place at the tip of the fingers, and therefore its effect is much weaker than that of transverse dispersion. For this reason, models that ignore transverse dispersion are usually not completely adequate for describing a miscible displacement.

(iii) *Porous Media Stratification*: We already discussed in Chapter 9 the effect of stratification on dispersion. Obviously, if dispersion is affected by stratification, so also are mixing and miscible displacement. It is clear that the displacing fluid preferentially chooses the strata with higher permeabilities. As a result, large amounts of the fluid to be displaced can be left behind in the strata with low permeabilities. If we try to displace this fluid by injecting more displacing agent into the low permeability strata, some of the agent will inevitably enter the high permeability strata and do "nothing," since such strata have already been swept by the displacing fluid, and this is not very efficient. The effect of stratification is even stronger when $M > 1$. Another phenomenon that affects miscible displacements in stratified media is the crossflow of displacing and displaced fluids between the strata. Depending on the direction of the displacement process, crossflows can help or hinder the efficiency of the process. Another factor that influences miscible displacements is dead-end pores. We already discussed in Chapter 9 the effect of dead-end pores on dispersion which, in turn, affects mixing and displacement.

11.2 Viscous fingering

If the displacing and displaced fluids are first-contact miscible, and if $M < 1$, then the displacement process is very simple and efficient. The displaced fluid moves ahead of the displacing fluid, the displacement front is stable, and there is a mixed zone between the pure displacing and displaced fluids regions. However, a real miscible displacement process is not as simple as this. If $M > 1$, the front is unstable and many fingers of the displacing fluid develop that penetrate the displaced fluid, leaving behind large amounts of the displaced fluid. These fingers have very irregular shapes, and their formation reduces strongly the efficiency of the displacement process. Figure 11.1 shows the effect of M on the formation and shape of the fingers. These experiments were carried out in a quarter of the so-called five-spot geometry made of consolidated sand, i.e., a geometry in which the injection point (or well in a field application) is at the center of the system and four production points (wells) are at the four corners of the system. The porous medium used in these experiments was essentially two dimensional. It is clear that as M increases the amount of swept oil decreases, and thin and long fingers of the displacing fluid are formed.

Why do viscous fingers form? Collins (1961) gives a simple but very clear illustration to explain this phenomenon. Suppose that a porous medium is saturated with oil, and a displacing or solvent fluid is injected into the medium to displace it. Assume that dispersion is negligible. The displacing fluid displaces the oil linearly, i.e., if the porous medium is homogeneous, the front will remain a flat plane throughout the process. Suppose now that the displacing fluid encounters a small region of higher permeability. Then the front will travel faster in that region and produce a bump that is a distance ϵ ahead of the rest of the front. If K, ϕ, and ΔP are, respectively, the permeability, porosity, and pressure difference along the medium, then using Darcy's law we can write

$$\frac{dx_f}{dt} = \frac{K \Delta P}{\eta_s \phi [ML + (1 - M)x_f]} , \qquad (11.2)$$

when x_f is the position of the front, L the medium's length, and $M = \eta_0/\eta_s$.

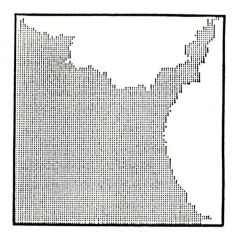 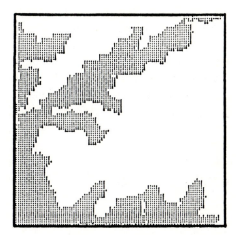

Figure 11.1: Patterns of viscous fingers for $M = 1$ (left) and $M = 10$ (right). Dark area represents the displacing fluid.

Similarly, we can write

$$\frac{d(x_f + \epsilon)}{dt} = \frac{K\Delta P}{\eta_s \phi[ML + (1 - M)(x_f + \epsilon)]} \quad , \tag{11.3}$$

and therefore

$$\frac{d\epsilon}{dt} = \frac{-K\Delta P(1 - M)}{\eta_s \phi[ML + (1 - M)x_f]^2}\epsilon = c\epsilon \quad , \tag{11.4}$$

if $\epsilon \ll x_f$, and hence $\epsilon = e^{ct}$. However, $c > 0$ if $M > 1$, which means that ϵ grows exponentially with t, and a long thin finger is formed. But the finger is not formed, or if it does it dies out, if $M < 1$. Although this example is too simple, it does illustrate the effect of M in the formation of viscous fingers. We should remark that finger formation is not restricted to miscible displacements, and can also occur in immiscible displacements which are discussed in Chapter 12. Because of its significance, viscous fingering has been studied for a long time, and several review papers have discussed this subject. Among these are those of Wooding and Morel-Seytoux (1976), who reviewed viscous fingering in a porous medium, Bensimon *et al.* (1986) and Saffman (1986) who discussed immiscible viscous fingering only in a Hele-Shaw cell, and Homsy (1987) who considered the problems in both Hele-Shaw cells and porous media. A Hele-Shaw cell (Hele-Shaw, 1898) is an essentially two-dimensional system confined between two flat plates of length L that are separated by a small distance b. Flow in this cell can be rectilinear, i.e., the fluid is injected into the system at one face of it and produced at the opposite face, or it can be radial in which the fluid is injected into the system at its center, and the system is like a cylinder of a large radius and small height. Viscous fingering also belongs to the general class of pattern-selection problems such as crystal growth phenomena, for which Kessler *et al.* (1988) have provided a comprehensive review.

There have also been many experimental studies of viscous fingering, both in miscible and immiscible displacements. Some of these include those of Blackwell *et al.* (1959), Benham and Olson (1963), Slobod and Thomas (1963), Greenkorn *et al.* (1965), Kyle and Perrine (1965), Perkins *et al.* (1965), Mahaffey *et al.* (1966), Perkins and Johnston (1969), and more recently, those of Paterson (1981, 1983, 1985), Paterson *et al.* (1982), Park *et al.* (1984), Chen and Wilkinson (1985), Måløy *et al.* (1985, 1987), Nittman *et al.* (1985), Ben-Jacob *et al.* (1986), Daccord *et al.* (1986), Lenormand *et al.* (1988), Tabeling and Libchaber (1986), and Bacri *et al.* (1991). Most of these experiments were carried out in either a Hele-Shaw cell, or a porous medium with microscopic disorder. Flow in the Hele-Shaw cells was either rectilinear or radial. The fluids were mostly oil and water, in the case of immiscible displacements, or oil and a miscible fluid. Most of the experimental works were also accompanied by a linear stability analysis (see below), and most authors were interested in the qualitative aspects of the displacement process, e.g., the shapes of the fingers and the effect of M on them. Some of these experiments will be compared with the theoretical predictions. In addition, viscous fingering and pattern formation in viscoelastic media (Van Dame *et al.*, 1987a,b), and with smectic and nematic liquid crystals (Buka *et al.*, 1986; Horváth *et al.*, 1987a) have also been studied. Finally, Horváth *et al.* (1987b) studied viscous fingering in a Hele-Shaw cell with a set of parallel grooves on one of the plates to investigate the effect of an uniaxial anisotropy on the patterns of displacement formed in the cell.

Before we go on with the discussion of miscible displacements, we would like to remind the reader that our policy throughout this book has always been to mention, and give credit to, the first important work on any subject we discuss. Hence, we should mention that Hill (1952) appears to be the first who published experimental and simple analytical results on viscous fingering and its stability. Later, Saffman and Taylor (1958) and Chouke *et al.* (1959) did the first rigorous analysis of the problem (although they considered immiscible viscous fingers discussed in Chapter 12). Homsy (1987) suggested that one must call this phenomenon the "Hill instability" problem, instead of the now-popular "Saffman-Taylor instability." We take a middle-of-the-road approach and refer to this as the "Hill-Saffman-Taylor instability."

11.3 Continuum models of miscible displacements in Hele-Shaw cells

We first consider miscible displacements in a Hele-Shaw cell of length L in which the two parallel plates are separated by a small distance b. For single-phase flow, in the absence of dispersion, and in the limits of small Reynolds number and small b/L, the governing equations are the continuity equation, $\nabla \cdot \mathbf{v} = 0$, and the Darcy's law. Because flow is confined between the two parallel plates, one can define an equivalent permeability for the cell given by $K = b^2/12$. These equations are the same as those of two-dimensional incompressible fluids in a porous medium. As long as dispersive mixing is absent, the same analogy is also valid between viscous fingers in Hele-Shaw cells and a two-dimensional porous medium, which is why the study of miscible displacements in Hele-Shaw cells has been popular. If dispersion is present, the analogy breaks down because, because as we discussed in Chapters 9 and 10, transverse dispersion is always present in a porous medium, whereas a Hele-Shaw cell with its thin gap between the two plates cannot support significant transverse dispersion in the third direction. With dispersion present, one has to add a convective-diffusion equation (CDE) (see Chapter 9) to the set of governing equations, and keep in mind that the viscosity η_m of the mixed zone and the effective viscosities of the two fluids are now dependent upon the concentration C_s of the solvent (and thus that of the fluid to be displaced). Such concentration-dependent viscosities also make the governing equations non-linear. The solution to this set of equations depends on the value of the Péclet number, which is the ratio of diffusion and convection time scales and was already discussed in Chapter 9, and on two other dimensionless groups which are

$$A \;=\; \frac{\eta_1 - \eta_2}{\eta_1 + \eta_2} = \frac{M - 1}{M + 1} \;, \tag{11.5}$$

$$B \;=\; \frac{gK(\rho_1 - \rho_2)}{v(\eta_1 + \eta_2)} \;, \tag{11.6}$$

where g is the gravity, and $M = \eta_1/\eta_2$. The most important aspects of a miscible displacement is its stability and efficiency. Thus, let us first outline the general approach to stability analysis of miscible displacements, as we refer to it frequently in the subsequent sections. Later in this Chapter we present a more quantitative description of the stability analysis, but for now we restrict ourselves to a qualitative description.

The standard method of stability analysis of miscible (and immiscible) displacments can be summarized as follows. In the first step the governing equations are introduced which represent the initial state of the system before it is perturbed. These are called the *base-state* equations. Next, one introduces perturbations into the dependent variables of the model. These perturbations can be caused by many factors, including the heterogeneities of the pore space, or the viscosity contrast between the two fluids. Thus, for example, the fluid velocity is written as $\mathbf{v} = \bar{\mathbf{v}} + \mathbf{v}'$, the pressure as $P = \bar{P} + P'$, and the concentration as $C = \bar{C} + C'$, where the bars indicate the unperturbed (base-state) or mean values, and the primes represent the perturbations. This substitution results in a set of equations which, when subtracted from the unperturbed equations, yields the governing equations for the perturbations. In the next step, the governing equations for the perturbations are solved either analytically or numerically, from the solution of which stability criteria are obtained. Clearly, if the perturbations grow with time, then the displacement is unstable. At first, this may seem to be a straightforward exercise which can be carried out with essentially no difficulty.

However, the governing equations for the perturbations are usually non-linear and difficult to solve, even numerically. In that case, the equations are linearized in order to make the computations more tractable. This results in a *linear stability criterion* which is useful for the *onset* (short-time behavior) of instability, but cannot be used for predicting the long time behavior of the displacement process. One common feature of linear stability theory is that one can decompose the initial perturbations into separate Fourier components, so that the stability of each component can be investigated separately. This often simplifies the problem considerably. This also suggests that Fourier analysis (also known as spectral analysis) is a very convenient way to solve the perturbation equations. It also introduces the terminology of wave theory, and thus the stability criterion may be expressed in terms of perturbations that have wavelengths greater or less than a critical value. Of course, if the critical wavelengths can be measured, then the stability criterion can be much more quantitative.

In this section, we present a mostly qualitative discussion of the available results. In the next section, where we discuss the results for porous media, we give a more quantitative discussion, some of which is also applicable to Hele-Shaw cells. Homsy (1987) provides an excellent and comprehensive review of the subject, and what follows in this section is essentially a summary of his discussion, plus our own remarks and comments and a discussion of the works since the publication of his paper.

Dispersion limits the wavelengths or frequencies of the Fourier components that can be unstable, which is why it usually has a positive effect on a miscible displacement. In the presence of dispersion the problem is always time dependent, transport coefficients such as the viscosity and dispersion coefficients are concentration (and hence time) dependent, and one cannot find a physical steady-state solution. Chouke was the first who analyzed the effect of dispersion, but his results appeared almost 30 years after their derivation in an Appendix to the paper of Gardner and Ypma (1982). For now we assume that the disturbances are of the form of normal modes proportional to $\exp(\omega t + i\alpha y)$. Chouke considered the case of a jump in viscosity, i.e., a base-case solution in which longitudinal dispersion is absent, and the concentration field can be divided into two distinct regions, one of which contains pure solvent or the displacing fluid, while the other contains pure displaced fluid. However, Chouke did allow both longitudinal and transverse dispersion to act on the disturbances. His result for ω, which essentially measures the rate of growth of the disturbances, is given by

$$\omega = \frac{1}{2}(A\alpha - \alpha^3 Pe^{-2} - \alpha\sqrt{\alpha^2 Pe^{-2} + 2A\alpha Pe^{-1}}) \ , \qquad (11.7)$$

where the cutoff wave number is given by

$$\alpha_c = \frac{APe}{4} \ , \qquad (11.8)$$

and ω is maximum when

$$\alpha_m = \frac{(2\sqrt{5} - 4)APe}{4} . \tag{11.9}$$

Thus the Péclet number $Pe = LV/D_m$ provides a physical mechanism for the introduction of a cutoff length scale. Equation (7) should be compared with the result in the absence of dispersion, i.e., when $Pe \to \infty$, $\omega = (A + B)\alpha$, which is unphysical because it implies that smaller wavelengths are even more unstable than the larger ones. An equation such as (7), which relates ω to α, is called a *dispersion relation*.

In the early 1960s several attempts were made to investigate the stability of miscible displacements in Hele-Shaw cells and in what were claimed to be porous media. However, the model porous medium that was used in these studies was in fact similar to a Hele-Shaw cell, since only an effective permeability was used for characterizing the medium, and no other morphological information about the structure of the medium was included in these studies. The approach of introducing small perturbations that we discussed above was taken by Perrine (1961) and Wooding (1962). In particular, Wooding (1962) treated the stability of a time-dependent base solution (i.e., the solution of the governing equations *without* the perturbations), and considered buoyancy-driven fingering. By expressing the disturbance quantities as an expansion in Hermite polynomials

$$H_n(x) = (-1)^n e^{x^2} \frac{d^n}{dx^n} e^{-x^2} , \tag{11.10}$$

and truncating the expansion beyond the first term, he obtained a dispersion relation (i.e., a relation between ω and α) which was qualitatively similar to Eq. (7). He also argued that all disturbances will die out if dispersion is given enough time to act, but did not specify what constitutes "enough". Another approach was based on a macrostatistical method, first used by Scheidegger and Johnson (1961), in which the fingers were treated statistically. Thus only the average cross-sectional areas occupied by the fingers were taken into account, and the shape and size of the individual fingers were neglected. Dougherty (1963), Koval (1963), and Perrine (1963) used this approach and took into account the effect of dispersion. Koval's model has been used widely in the petroleum industry, and is discussed in the next section.

Schowalter (1965) studied fingering in which both density and viscosity variations were present. Although, as discussed above, the governing equations do not allow a steady base solution, he assumed one anyway and obtained a dispersion relation and a cutoff wave number. Heller (1966) considered horizontal miscible displacement in a rectangular system, and studied the early growth or decay of perturbations using a Fourier analysis. He also included the effect of dispersion and approximated the profiles by straight-line segments (or "ramp-shaped" profiles as he called them), and obtained the dependence of the growth exponent

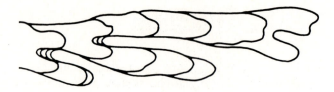

Figure 11.2: Tip splitting in viscous fingering in a Hele-Shaw cell (after Liang, 1986).

on the wave number. However, the assumption of straight-line segments for the profiles makes his results of limited applicability.

Peters *et al.* (1984) considered a miscible displacement process, took into account the effect of dispersion, and performed a Fourier analysis to derive a linear stability criterion. If the perturbations are proportional to $\exp(\omega t + \alpha_x x + \alpha_y y)$, then they proposed that the displacement is stable if

$$\frac{1}{8\pi^2}(\alpha_x^2 + \alpha_y^2) < \frac{1}{4\pi}\sqrt{\frac{K g d\rho/dC - v d\eta/dC}{\phi \delta_t \bar{\eta} D_T}} \qquad (11.11)$$

where δ_t is the length of the mixing zone. The right-hand side of (11) defines a critical frequency or wave number ω_c, which should be compared with Eq. (7).

Tan and Homsy (1986) considered this problem for the case of no density difference, and a situation in which the concentration dependence of the viscosity is given by, $\eta_m(C) = \exp(-C \ln M)$. The flow was rectilinear and the domain was unbounded. Both isotropic ($D_L = D_T$) and anisotropic dispersion were considered, and a quasi-steady state assumption was made. They showed that, for the isotropic case, Chouke's result is essentially correct in the sense of predicting correctly the magnitude of the growth rates and the preferred wave numbers. However, they also found that at longer times dispersion causes a shift to larger wave lengths, and stabilizes the flow to some extent. Tan and Homsy (1986) also showed that transverse dispersion causes a shift to smaller length scales, as expected. Hickernell and Yortsos (1986) considered the case in which the amount of the solvent injected into the system varied with time. This results in a spatially-varying mobility profile. They ignored the effect of dispersion and showed that a finite thickness of the zone of viscosity variations provides a cutoff scale. Chang and Slattery (1986) showed that, when there is a steep change in concentration and $M > 1$, the displacement can be unstable at the injection boundary. But, if the concentration is changed sufficiently slowly with time at the entrance to the system, the displacement will be stable to small perturbations, *regardless of the value of M*.

In some situations, dispersion can play a role similar to that of surface tension in immiscible displacements. As we discuss in Chapter 12, surface tension is

Figure 11.3: Shielding, in which one finger spreads out and shields other fingers (after Zimmerman and Homsy, 1991).

responsible for *tip splitting* in the fingers, which is the main mechanism of pattern formation in many phonemena such as crystal growth, and dispersion can cause a similar phenomenon in miscible displacements. In tip splitting, the tip of a finger becomes unstable and splits into two branches that compete with each other for further growth. An example is shown in Fig. 11.2. The experiments of Wooding (1969) in a Hele-Shaw cell, in the presence of buoyancy forces, are indicative of this phenomenon. In Wooding's experiments, transverse dispersion causes lateral *spreading* of the fingers, as expected. However, because of this spreading, the tips of the fingers can become unstable since their typical breadth exceeds the cutoff length scale, which is also set by transverse dispersion. Because of this instability, tip splitting can occur if Pe exceeds a critical value which, according to Tan and Homsy (1987), depends on both A and M. For example, at $M = e^3$, $250 < (Pe/A_r)_{critical} < 300$, if dispersion is isotropic (i.e., $D_L = D_T$), where A_r is the aspect ratio of the cell. Note that if transverse dispersion is weak, then tip splitting may not happen at all, because before it can start one must have spreading of the fingers, which can happen only if transverse dispersion is strong enough.

Numerical simulations of miscible displacements in rectilinear flows (such as those in Hele-Shaw cells) with weak dispersion (high values of Pe) are particularly difficult. The first attempt in this direction was made by Peaceman and Rachford (1962), using finite-difference methods, which was not successful because of large numerical errors that dominated the solution. Since their pioneering attempt, a lot of effort has been dedicated to this problem. The main reason for this difficulty is that for large values of Pe, viscous fingering can occur on many scales, and there is no unique power of Pe with which one can rescale all lengths. Zimmerman and Homsy (1991) proposed a method which appears to tackle this problem to a large extent. They used a two-dimensional (discrete) Hartley transform, which for an arbitrary function $g(x,y)$ is defined by

$$G(\alpha_x, \alpha_y) = \frac{1}{\sqrt{N_x N_y}} \sum_x \sum_y \text{cas}\left(\frac{2\pi x \alpha_x}{N_x} + \frac{2\pi y \alpha_y}{N_y}\right) g(x,y) \ , \qquad (11.12)$$

where N_x and N_y are the number of collocation points (the points at which a discrete Hartley transform is computed), α_x and α_y are the wave numbers in the longitudinal and transverse directions, respectively, and cas(x) is the so-called "cosine and sine" function, formed by summing the cosine and sine of its argument. The advantage of using the Hartley transform is that it can be easily inverted, because it is its own inverse. An efficient method for computing a two-dimensional discrete transform via $N_x + N_y$ discrete fast Hartley transforms has been devised by Bracewell *et al.* (1986). The Hartley transformation recasts the system of partial differential equations governing flow and the concentration into an ordinary differential equation for dC/dt (which is similar to Fourier transforming of a partial differential equation). The resulting equation is then integrated and transformed back into the (x, y) space. Zimmerman and Homsy (1991) used the Taylor-Aris result, Eq. (9.23), to relate D_L to Pe, in which they used $\delta_s = 2/105$, the appropriate value for flow between two parallel plates (Hele-Shaw cell), and assumed that $D_T = D_m$ (i.e., a constant, velocity-independent D_T), where D_m is the molecular diffusion. As such, the *results* are valid only for Hele-Shaw-like systems, although the authors claimed that the results are also valid for porous media. Over a wide range of values of Pe, they observed a variety of phenomena, including spreading, tip splitting, shielding, and pairing. In a shielding process, one finger gets ahead of its neighbors and grows explosively. Eventually, the tip of the finger spreads out and *shields* the neighboring fingers. An example is shown in Fig. 11.3. Pairing is a phenomenon by which pairs of fingers join and form a larger finger. This phenomenon has been observed both with isotropic dispersion ($D_L = D_T$) and anisotropic dispersion, although as far as porous media flows are concerned, isotropic dispersion is not relevant. It can also be found in immiscible fingering as was shown by Tryggvason and Aref (1985) and Kessler and Levine (1986). According to Tan and Homsy (1987), pairing occurs because of unequal cross-flow about neighboring fingers. This allows a finger to spread, which shields the growth of the neighboring finger and results in its eventual collapse. If tip splitting is absent, then pairing eventually results in the reduction of the number of fingers to one or two. The results of Zimmerman and Homsy (1991) also indicated great sensitivity in the complex two-dimensional fingering patterns that evolve on the size of the initial noise. However, when an averaging was performed over the cross-section area, it was found that the one-dimensional average concentration profiles were similar with large Pe and D_L/D_m. This means that it may be possible to describe nonlinear viscous fingers by a one-dimensional model, which is invariant with respect to Pe and D_m/D_L, for large Pe and small D_m/D_L. In a sense, this is the same as the idea of Scheidegger and Johnson (1961) already mentioned above. It is also similar to the work of Fayer's (1988) who constructed an approximate one-dimensional model with adjustable parameters that had clear physical meaning. Koval (1963) and Todd and Longstaff (1972) also constructed empirical one-dimensional models which could predict the evolution of the average concentration profile well. Their models are discussed below. Finally, Christie

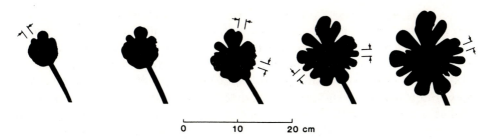

Figure 11.4: Viscous fingers in a radial Hele-Shaw cell. The gap between the plates is $b = 0.3$ cm, and the exposures are, from left to right, at times $t = 12$, 17, 21, and 31 sec, respectively. The indicated distance is $4b$ (after Paterson, 1985).

and Bond (1987) developed a numerical method for the evolution of both linear and non-linear fingers. They used a finite-difference method (see Chapter 14) for the longitudinal direction and Fourier decomposition in the transverse direction.

The effect of the cell geometry is significant. Wilson (1975) and Paterson (1985) studied miscible viscous fingering in a radial Hele-Shaw cell both theoretically and experimentally. Paterson (1985) ignored dispersion and obtained the following relation for ω

$$\omega = Am - 1 \quad , \tag{11.13}$$

where m is a discrete azimuthal wave number. This equation indicates that there is no cutoff azimuthal wave number, which is not surprising. Thus Paterson (1985) suggested that in a radial Hele-Shaw cell, and in the absence of dispersion, fingers can form on *all* length scales, down to the size of the gap between the plates. Based on a heuristic argument, Paterson (1985) estimated that the cutoff wavelength is approximately $4b$. Figure 11.4 shows the pattern of miscible viscous fingers in a radial Hele-Shaw cell. While his experimental results appear to agree with his rough estimate of the cutoff wavelength, Paterson's argument in the general case in which dispersion is present is not expected to hold, because, as discussed above, transverse dispersion (however small in a Hele-Shaw cell) helps the fingers to spread which leads to tip splitting. Thus fingers cannot form down to the scale b. This has been discussed in detail by Homsy (1987).

11.4 Continuum models of miscible displacements in porous media

Since 1950s there have been many experimental studies of miscible displacements in porous media, most of which were made of sand or other unconsolidated materials. In particular, Slobod and Thomas (1963) and Perkins *et al.* (1965) used X-ray techniques to make visualization studies of viscous fingers. Normally, one can observe viscous fingers over length scales that are several times larger than the typical pore size. Since in these experiments the transverse Pe (i.e., the Péclet number based on D_T instead of D_m) is relatively low, finger growth is probably due to the spreading phenomenon discussed above. A careful examination of the pictures taken during the experiments of Slobod and Thomas (1963) and Perkins *et al.* (1965) shows no tip splitting. However, Habermann's (1960) experiments (which are similar to those shown in Fig. 11.1) and those of Mahaffey *et al.*'s (1966) do show tip splitting due to shielding, similar to those in a Hele-Shaw cell already discussed. The dominant length scale in all of these experiments is much larger than a typical pore size. However, if one carries out experiments in a porous medium in which Pe increases with decreasing length scales, one can no longer claim that the dominant length scale is much larger than the typical pore size. An example is the work of Paterson *et al.* (1982) who studied miscible viscous fingers in packed beds. The viscous finger patterns that they obtained are similar to DLAs, i.e., they have a very open structure with many thin branches (see below). This indicates that, for a given pore space there is a crossover between the continuum and discrete description. An important and interesting question is the shape of viscous fingers for the case in which there is no mechanism for the creation of a cutoff scale, e.g., in the absence of dispersion. These is discussed below where we present discrete models of viscous fingering.

How can we describe miscible displacements and viscous fingers in a porous medium? One method is based on averaged continuum equations, most of which were discussed above and the rest are reviewed below. This method can then describe any instability that is smooth on the length scale of the continuum equations. Thus all of the above discussions regarding miscible displacements in Hele-Shaw cells and similar geometries are equally applicable to porous media, provided that dispersion phenomena discussed in Chapter 9 are properly taken into account. For example, the approach of Zimmerman and Homsy (1991) can be used for miscible displacement in porous media if, instead of what they used, we use the appropriate equations for the velocity (or Pe) dependence of D_L and D_T discussed in Chapter 9. In the strict absence of dispersion (i.e., the limit $Pe \rightarrow \infty$, where Pe is based on D_L or D_T and not D_m), viscous fingers will occur on all scales, with growth rates that increase with decreasing scale. This means that a scale will be reached at which a continuum description is no longer appropriate, and one has to develop a pore-level (discrete) model. Such models are discussed below. The initial-value problem in this case is ill-posed, but one can seek solutions that contain discontinuities. Since dispersion is absent in this case, there will be a step jump in the viscosity profiles (from the displacing fluid to the displaced fluid). As a result, the pressure obeys the Laplace equation, and the pressure and fluid fluxes are continuous across the front separating the displaced and displacing fluids. One may have all sorts of singularities in the solution, and different non-uniformities can appear in different boundary-value problems. Shraiman and Bensimon (1984) and Howison (1986) investigated such phenomena, but we do not consider them here.

Let us now describe briefly two one-dimensional continuum (and empirical) models of miscible displacements that have been popular in the petroleum industry. The first model is due to Koval (1963). In his model a given displacing fluid is assumed to travel at a constant, characteristic velocity v. If flow is linear, then

$$v = \left(\frac{dx}{dt}\right)_S = \frac{Q}{A\phi}\left(\frac{df_s}{dS}\right) , \tag{11.14}$$

where S is the saturation of the displacing fluid, Q is the volumetric flow rate, $\mathcal{A}$ is the cross-sectional area, and f_s is the volume fraction of the solvent (the displacing fluid) in the flowing fluid. To make further progress, Koval (1963) made an analogy between miscible and immiscible displacements. For an immiscible displacement of oil by water, the Buckley-Leverett equation (see Chapter 12), if gravity and capillary pressure are negligible, relates f_w to the permeabilities K_w and K_o of the water and oil phases by the following equation

$$f_w = \frac{1}{1 + \dfrac{K_o}{K_w}\dfrac{\eta_w}{\eta_o}} . \tag{11.15}$$

Koval argued that permeability to either oil or the displacing fluid can be expressed as the total permeability K multiplied by the average saturation of each fluid. Thus, he argued that if viscous fingering is the dominant phenomenon, then in analogy with Eq. (15) one can write

$$f_s = \frac{1}{1 + \dfrac{1}{H_i}\dfrac{\eta_{es}}{\eta_{eo}}\dfrac{1-S}{S}} . \tag{11.16}$$

In this equation η_{es} and η_{eo} are the *effective* viscosities of the solvent and oil (the displaced fluid), respectively, and H_i is called the *heterogeneity index* that characterizes the heterogeneity of a porous medium. As discussed above, because of mixing that occurs during a miscible displacement, the effective viscosities of the two fluids are not the same as those of the pure fluids. Based on experimental data, Koval (1963) suggested the following expression

$$M_e = \frac{\eta_{eo}}{\eta_{es}} = \left[0.78 + 0.22\left(\frac{\eta_o}{\eta_s}\right)^{1/4}\right]^4 . \tag{11.17}$$

To correlate H_i with some measurable quantity, Koval defined a homogeneous porous medium as one in which the oil recovery, after 1 pore volume of the solvent has been injected into the medium, is 99%. For a homogeneous porous medium $H_i = 1$, while any other porous medium for which the recovery is less than 99% is characterized by a $H_i > 1$. Therefore this is an empirical method of characterizing the heterogeneities of the medium. Figure 11.5 presents the correlation between H_i and the recovery.

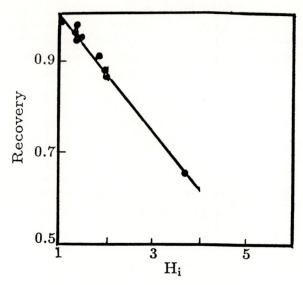

Figure 11.5: Dependence of the heterogeneity index H_i on the recovery, based on matched viscosity experiments (after Koval, 1963).

The solution to Eqs. (14) and (16) at solvent breakthrough (i.e., at the point where the solvent forms a sample-spanning cluster) is

$$V_{iBT} = \frac{1}{M_e H_i} \qquad (11.18)$$

where V_{iBT} is the injected pore volume of the solvent at the breakthrough. The fractional pore volume of the recovered oil f_o is predicted to be

$$f_o = \frac{2\sqrt{M_e H_i V_{iBT}} - V_{iBT} - 1}{M_e H_i - 1}, \qquad (11.19)$$

and the length ℓ_f of the region in which fingering has occured is given by

$$\ell_f = \left(M_e H_i - \frac{1}{M_e H_i} \right)_{x_m}, \qquad (11.20)$$

where x_m is the mean displacement distance. Figure 11.6 compares the predictions of Koval's model with the experimental data of Blackwell *et al.* (1959). The agreement is very good. It can be shown that Koval's model can also be used for predicting miscible displacements and their recovery efficiency in Hele-Shaw cells. Despite its apparent success, Koval's model suffers from two fundamental shortcomings. One is the empirical nature of Eq. (17). It is not clear at all why this equation should be an adequate representation of the viscosity ratio, which in effect measures the effect of mixing during a miscible displacement. The second shortcoming of the model is the inadequacy of the heterogeneity index H_i for

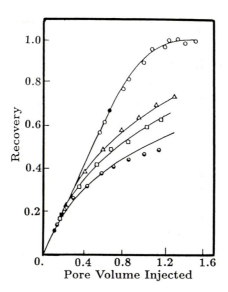

Figure 11.6: Comparison of the predictions of Koval's model (curves) with the experimental data of Blackwell *et al.* (1959). The results are, from top to bottom, for $M = 5$, 86, 150, and 375 (after Koval, 1963).

characterizing the structure of a porous medium. The correlation shown in Fig. 11.5 is based on laboratory data with macroscopic porous media, whereas field applications of miscible displacements involve megascopic porous media that are characterized by long-range correlations, the effect of which cannot possibly be included in the correlation shown in Fig. 11.5.

The second model is due to Todd and Longstaff (1972). In their model the average concentration $\bar{C}_s$ of the solvent is described by

$$\frac{\partial \bar{C}_s}{\partial t_D} + \frac{\partial \bar{f}_s}{\partial x_D} = 0 \ , \tag{11.21}$$

where $\bar{f}_s$ represents the average of f_s and is a function of $\bar{C}_s$, and x_D and t_D are dimensionless distance and time, respectively. This equation is of course a limiting case of a CDE in which the dispersion term has been neglected. Todd and Longstaff assumed that

$$\eta_{eo} = \eta_o^{1-\zeta}\eta_m^\zeta \ , \tag{11.22}$$

with a similar expression for η_{es}, where η_m is given by Eq. (1) in which $\bar{C}_s$ instead of C_s is used. $\bar{f}_s$ is given by

$$\bar{f}_s = \frac{\bar{C}_s}{\bar{C}_s + M_e^{-1}(1 - \bar{C}_s)} \ . \tag{11.23}$$

If Eqs. (21) and (23) are solved, one finds that the average concentration $\bar{C}_s$ moves with a speed $d\bar{f}_s/d\bar{C}_s$. Therefore, the leading edge of the finger, where

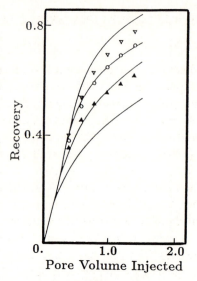

Figure 11.7: Comparison of the predictions of Todd-Longstaff model (curves) with the experimental data for a Hele-Shaw cell (solid triangles), bead pack (circles), and consolidated sand pack (open triangles). The results are, from top to bottom, for $\zeta = 1, 2/3, 1/3$, and 0, and $M = 10$ (after Todd and Longstaff, 1972).

$\bar{C}_s = 0$, moves at a speed $d\bar{f}_s/d\bar{C}_s = M_e$, whereas the trailing edge, where $\bar{C}_s = 0$, moves at a speed $d\bar{f}_s/d\bar{C}_s = 1/M_e$. Todd and Longstaff proposed that $\zeta = 2/3$ should be a reasonable estimate. To investigate the effect of the heterogeneities on the performance of the model, the predictions of their model were fitted to the experimental data for three different systems, namely, a Hele-Shaw cell, a bead pack (an unconsolidated porous medium), and a consolidated sand pack, using ζ as the adjustable parameter. The Hele-Shaw system does not contain any heterogeneities, while the other two systems contain very different heterogeneities. Figure 11.7 compares the fitted predictions with the data, and it is clear that it is impossible to predict all the three sets of data with a single value of ζ. Note also the difference between the Hele-Shaw results and those of the consolidated sand pack. It has been suggested that a good fit of the model's predictions to the experimental data of Blackwell *et al.* (1959) is obtained if one sets $\zeta = 1 - \log M_e / \log M$, where $M = \eta_o/\eta_s$. Thus, it should be clear to the reader that the model of Todd and Longstaff (1972) suffers from the same shortcomings as those of Koval's model discussed above. Because of their simplicity, and despite their shortcomings, these two models have been used heavily in the petroleum industry. A more recent model proposed by Fayer (1988) is in the same spirit as these two models, and thus enjoys the same degree of success and also suffers from the same shortcomings. More discussion of such models and their generalization are given by Blunt *et al.* (1994).

11.4.1 Numerical simulation of miscible displacements in porous media

In the petroleum industry, a standard approach to numerical simulation of miscible displacements in porous media, with or without the effect of dispersion, is based on solving the governing equations (those of flow and the concentration) with finite-difference techniques (see Chapter 14). The earliest of such methods was that of Peaceman and Rachford (1962) mentioned above, while the works of Giordano and Salter (1984), Giordano *et al.* (1985), and Christie and Bond (1987) represent some of the latest of such attempts. There are two major problems with such methods. The first is that the resolution of the results can be poor, making it very difficult to study viscous fingers. The second problem is that the initiation of instabilities in such numerical simulations requires variations of the permeability of the porous medium. However, the results are often very sensitive to the initial permeability variations that are used in the simulations. Although supercomputer technology has made it possible to use very fine finite-difference grids, the simulations require intensive computations and a very large and detailed grid. For a discussion of such finite-difference methods see Christie (1988).

An alternative approach is based on the method of weighted residuals (Finlayson, 1980), proposed by Tan and Homsy (1987) and Hatziavramidis (1988). Fourier expansions (a spectral method) and Chebyshev polynomials

$$T_n(x) = \cos(n \arccos x) \,, \tag{11.24}$$

are used in this method to obtain the solutions of the governing equations, which are accurate and relatively efficient. To obtain numerical solutions of the resulting equations, collocation methods (Finlayson, 1980) are used which means that these equations are solved exactly at the collocation points. One can also use a Fast Fourier Transform (FFT) to refine the grids. Even the system with the refined grid can be solved more efficiently than those with the standard finite-difference method. However, the method can become complex if simulations for long times are needed. Moreover, FFTs can be used if the number of grid points is an integral power of two, which means that most of the advantage gained from the use of FFTs can be lost. Finally, one usually has to use periodic boundary conditions in order to avoid the so-called Gibbs phenomenon, which is characterized by wiggly outlet concentration curves. Other methods of numerical simulation of the continuum models of miscible displacements were discussed above where we considered these phenomena in Hele-Shaw cells.

11.4.2 Stability analysis of miscible displacements

We already gave a qualitative discussion of how a stability analysis of miscible displacements is performed. In this section we provide a more quantitative discussion of this analysis. The main goals of this section are to illustrate (i) how a stability analysis is actually carried out, and (ii) how far such an analysis can take one.

As already discussed in Chapters 9 and 10, dispersion coefficients depend on the mean flow velocities, and the precise form of their dependence depends on the value of the Pe. However, for simplicity, we assume that D_L and D_T are related to the flow velocities v_x and v_y through the following relations (Bear, 1972; see Chapter 9):

$$D_L = D_m + \alpha_L |v| + \frac{(\alpha_L - \alpha_T)v_x^2}{|v|} \ , \tag{11.25}$$

$$D_T = D_m + \alpha_T |v| + \frac{(\alpha_L - \alpha_T)v_y^2}{|v|} \ , \tag{11.26}$$

which describe longitudinal and transverse dispersion coefficients in terms of the corresponding dispersivities α_L and α_T, and the molecular diffusion coefficient D_m. For convenience, we shall follow the notation in Yortsos and Zeybek (1988). The governing equations are the CDE and the continuity equation coupled with Darcy's law:

$$\phi \frac{\partial C}{\partial t} + v_x \frac{\partial C}{\partial x} + v_y \frac{\partial C}{\partial y} = \frac{\partial}{\partial x}\left(D_L \frac{\partial C}{\partial x}\right) + \frac{\partial}{\partial y}\left(D_T \frac{\partial C}{\partial y}\right) \ , \tag{11.27}$$

$$\frac{\partial v_x}{\partial x} + \frac{\partial v_y}{\partial y} = 0 \ , \tag{11.28}$$

$$v_x = -\frac{k}{\eta}\frac{\partial P}{\partial x} \ , \tag{11.29}$$

$$v_y = -\frac{k}{\eta}\frac{\partial P}{\partial y} \ . \tag{11.30}$$

Implicit in the above continuum description is the assumption that the local Péclet number vl/D_m is small.

The base-state solution (i.e., the solution to unperturbed or mean quantities), corresponding to a constant injection rate in a rectilinear flow geometry, is given by the well-known diffusive profile

$$\bar{C} = \frac{1}{2}\text{erfc}\left[\frac{\xi}{2\sqrt{t}}\right] \tag{11.31}$$

$$\frac{\partial \bar{P}}{\partial \xi} = -\frac{1}{\lambda(\bar{C})} \tag{11.32}$$

where $\xi = (x - vt)/L$ is a moving coordinate, λ denotes a normalized mobility (inverse of viscosity), and all lengths are scaled with D_{L0}/v, where D_{L0} is the base-state dispersion coefficient. Writing $C = \bar{C} + C'$, $\mathbf{v} = \bar{\mathbf{v}} + \mathbf{v}'$, and $P = \bar{P} + P'$, and using normal modes for concentration and flow rate, respectively

$$(C', v'_x) = (\Sigma, \Phi) \exp(\omega t + i\alpha y) \ , \qquad (11.33)$$

the following equations are obtained

$$\Sigma_{\xi\xi} - (\epsilon\alpha^2 + \omega)\Sigma = \Phi C_\xi - L_c(\Phi C_\xi)\xi \ , \qquad (11.34)$$

$$\lambda(\lambda^{-1}\Phi_{xi})_\xi - \alpha^2\Phi = -\alpha^2 R\Sigma \ , \qquad (11.35)$$

where the concentration dependence of the mobility is taken to be (Tan and Homsy, 1986)

$$\lambda(C) = \exp(RC) \ , \qquad (11.36)$$

and $R = \ln M$. The subscripts denote derivatives with respect to the variables. Two important terms, ϵ, and L_c

$$\epsilon = \frac{D_m + \alpha_T \mathrm{v}}{D_m + \alpha_L \mathrm{v}} \ , \qquad (11.37)$$

$$L_c = \frac{\alpha_L \mathrm{v}}{D_m + \alpha_L \mathrm{v}} \ , \qquad (11.38)$$

appear in Eq. (34). One is ϵ which is a measure of flow-induced anisotropic dispersion, which is characteristic of porous media, and the other one is L_c which is a measure of the contribution of mechanical dispersion to total dispersion. It must be stressed that in the works of Chouke, and of Tan and Homsy (1986), the term containing L_c in (34) is absent, thereby restricting their conclusions to essentially constant (although still anisotropic) dispersion. Zimmerman and Homsy (1991) used velocity-dependent dispersion coefficients, although, as mentioned above, their functional forms were different from (25) and (26).

Based on our discussion in the previous sections, one expects that the onset of instability and related features would be dictated by the sharpest mobility contrast, namely, those associated with a step concentration profile, which also allow for an analytical solution given by (Yortsos and Zeybek, 1988)

$$\alpha R \left[1 + L_c \gamma_0 \tanh \left(\frac{R}{2} \right) \right] = 2\gamma_0 (\alpha + \gamma_0) \ , \qquad (11.39)$$

where

$$\gamma_0 = \sqrt{\epsilon\alpha^2 + \omega} > 0 \ . \qquad (11.40)$$

In general, the solution of (34) leads to parabolic-like profiles, examples of which are shown in Fig. 11.8. The case $L_c = 0$ corresponds to the result of Tan and Homsy (1986): for an unfavorable mobility ratio ($R > 0$) large wavelengths are

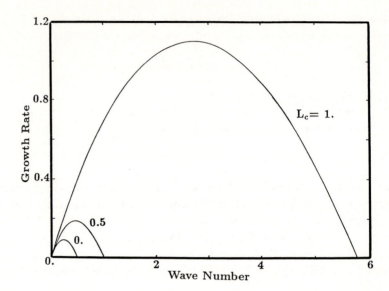

Figure 11.8: Step-profile results for the growth rate (after Yortsos and Zeybek, 1988).

unstable, while a strong stabilization due to transverse dispersion is exerted at smaller wavelengths. A cutoff wave number can be identified

$$\alpha_c = \frac{R}{2(\epsilon + \sqrt{\epsilon})} \ . \tag{11.41}$$

As expected, α_c increases with increasingly unfavorable mobility, and with an increase in the ratio of longitudinal to transverse dispersion $(1/\epsilon)$. However, the limits of the continuum description should be kept in mind. The size of the most unstable disturbance scales with the characteristic length, which for large enough flow rates becomes equal to the dispersivity α_L, which is normally a multiple of the typical pore size (or the length scale of the heterogeneities). It is apparent that a possible conflict may develop between the above result and the continuum description, precluding meaningful predictions over scales of the order of the microscale.

While the case $L_c = 0$ leads to expected results, a distinct sensitivity develops as L_c takes non-zero values (see Fig. 11.8). This effect is present only because of the velocity dependence of the dispersion coefficients, and it can be best quantified in terms of the following combination

$$B_c = \frac{RL_c}{2} \tanh\left(\frac{R}{2}\right) - \sqrt{\epsilon} - 1 \ . \tag{11.42}$$

The following results may then be shown (Yortsos and Zeybek, 1988):
(i) When $B_c < 0$ (which is *always* the case if $L_c = 0$), at small enough viscosity

ratio and for $L_c \neq 0$, the cutoff wave number is finite

$$\alpha_c = \frac{R}{2\sqrt{\epsilon}}(-B_c) \quad , \tag{11.43}$$

although it increases as L_c or B_c does.
(ii) On the other hand, when $B_c > 0$, a finite cutoff does not exist, with the rate of growth increasing indefinitely at large α as

$$\omega \sim B_c(B_c + 2\sqrt{\epsilon})\alpha^2 > 0 \quad . \tag{11.44}$$

Clearly, such is the case for a sufficiently high (but finite) mobility contrast, as long as $L_c \neq 0$, as shown in Fig. 11.8. This unexpected and rather remarkable result was obtained on the basis of a step base state, which is subject to a singular behavior in the large (as well as in the small) α region. To better understand the proper dependence, Yortsos and Zeybek (1988) attempted a more rigorous asymptotic analysis valid for base states near a step profile, namely, for

$$\bar{C} = \frac{1}{2}\text{erfc}(c\xi) \quad , \tag{11.45}$$

where $c \gg 1$. Their results showed that the step profile prediction, inequality (44), is invalid at large α when $B_c > 0$, and in fact a cutoff wave number does exist. However, the latter was found to increase monotonically and without bound as c increases, namely, as the profile becomes step-like, provided that $B_c > 0$. Thus the essential predictions that qualitatively different instability behavior is obtained by changing the sign of B_c, remains intact. Most of the above results were confirmed by the experimental study of Bacri *et al.* (1991).

The implications of these results are straightforward. Because of the dependence of the dispersion coefficients on the flow velocity, an essential feature of hydrodynamic dispersion in porous media, and for mobility ratios that exceed a critical value dictated by the given process conditions, a miscible displacement is predicted to be unstable *at all wavelengths*. Under such conditions, there is no finite preferred mode, and in fact the above continuum description is ill-posed and breaks down. This remarkable result raises serious doubts about our ability to describe the conditions for the onset of instability in miscible displacements. Recall that this result is the outcome of several hypotheses, including the validity of a continuum description, with dispersion formulated by a CDE and the dispersion coefficients represented by Eqs. (25) and (26). If these predictions are to be taken seriously, the breakdown of the continuum hypothesis beyond a finite M calls for an alternative description (to the present CDE-based description), and vice versa. One concludes that at least for large enough mobility ratios, our present description of miscible displacements is inadequate, particularly at the early, and the most important, stage of the process. Thus although Zimmerman and Homsy (1991) and others used a CDE, one needs to establish that their formulation is actually applicable to the early stages of the growth of the fingers.

Figure 11.9: A typical two-dimensional diffusion-limited aggregate.

The above analysis demonstrates clearly the significance of anisotropic dispersion (i.e., the inequality of D_L and D_T). Of course, all of the above results are limited to linear fingers, and non-linear fingers require full numerical simulations of the above equations, provided that the proper forms of velocity-dependence of the dispersion coefficients are used.

One problem with all of the above continuum models and analyses of stability of miscible displacements is that it is difficult to incorporate the effect of rock heterogeneity into the analysis. This is especially so if rock is strongly disordered with correlations over many length scales, in which case not only the above analyses cannot account for such effects, a continuum description may not even be valid to begin with. This is certainly the case if the pore space is fractal. Continuum models can incorporate the effect of long-range correlations, if for their numerical simulation a very fine grid is used (see Chapters 8 and 14). For this reason, discrete models, which are flexible enough to include the effect of pore space disorder, offer an alternative to the continuum formulations. Such an alternative is very useful for obtaining insight into the effect of disorder on viscous fingering. We now describe and discuss the discrete models.

11.5 Discrete models of miscible displacements

In this section we discuss discrete models of miscible displacements, and compare their performance with the experimental data and the predictions of continuum models.

11.5.1 Diffusion-limited aggregation models

In the DLA model one starts with an occupied site (the "seed") of a lattice, located either at the center of the lattice or on its edges. Random walkers are released, one at a time, far from the seed particle and are allowed to move randomly on the lattice (for simulation of random walks see Chapter 8). If they visit an empty site adjacent to an occupied one, the aggregate of the occupied sites advances by one site and occupies the last site visited by the walker. The walker is removed, another one is released and so on. After a large number of particles have joined the aggregate, it takes on a random structure with many branches, an example of which is shown in Fig. 11.9. Now, imagine that the aggregate represents a displacing fluid, and the empty sites represent the displaced or the defending fluid. Thus the original seed particle represents the point at which the displacing fluid is injected into the system. Since the particles perform their random walks on the empty sites, the probability $P(\mathbf{r})$ of finding the particle at a position $\mathbf{r}$ in this region obeys Laplace's equation, $\nabla^2 P = 0$. Because the walkers never move into the aggregate, the probability of finding them there is zero, $P = 0$. If the walkers are reflected at the "walls" of the system, one has $(\nabla P)_n = 0$, where n denotes normal to the wall. Finally, the mean speed at which the front between the displacing and displaced fluids (between the aggregate and the empty sites) advances is proportional to the probability on the side of the displaced fluid next to the front, i.e., $\mathbf{v} = (\nabla P)_n$. But these are essentially the same equations for the displacement of a viscous fluid by a miscible inviscid one, *in the absence of dispersion*. This can be seen by realizing that in this limit the continuity equation, $\nabla \cdot \mathbf{v} = 0$, together with Darcy's law, $\mathbf{v} \sim \nabla P$, yield $\nabla^2 P = 0$ for the displaced fluid, and $P \simeq$ constant for the displacing fluid. But because in the absence of dispersion the problem is linear, the constant can be arbitrary and thus it can be taken to be zero. In other words, a miscible displacement in the limits $M = \infty$ and no dispersion can be simulated by this algorithm. The DLA model was originally invented by Witten and Sander (1981, 1983) to simulate aggregation of small particles, and Meakin and many others have studied it extensively (see Meakin, 1988, for a review of the subject). As mentioned earlier, Paterson (1984) was the first to note the above analogy between the DLA model and miscible displacements.

If, in an unstable miscible displacement, we reverse the direction of the pressure gradient and allow the more viscous fluid to displace the less viscous (or inviscid) one, the displacement will be stable. This can be simulated by allowing the displaced fluid to advance each time the n_v visits occur. This was suggested by Paterson (1984) and Tang (1985), and may be called *anti*-DLA (since it essentially represents the reverse of the DLA process). Paterson (1984) compared the results of such simulations with the experimental data of Habermann (1960) and found reasonable agreement. Tang (1985) compared his results with the exact steady-state solution of Saffman and Taylor (1958), and the exact unsteady-state

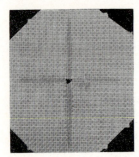

Figure 11.10: Viscous fingers developed by displacement of glycerine by oil in three etched-glass networks. The network on the left is totally ordered, whereas the one on the right is completely disordered (after Chen and Wilkinson, 1985).

solution of Shraiman and Bensimon (1984), and found excellent agreement.

There are, however, several problems with the analogy between the DLA model and miscible displacements. Recall that a miscible displacement can be unstable if $M > 1$. The DLA algorithm produces an aggregate that (i) has a fractal structure with a fractal dimension of about $D_f \simeq 1.7$ and 2.45, in two and three dimensions, respectively, and (ii) contains a large number of very tiny fingers in which tip splitting occurs at all times (see Fig. 11.9). Chen and Wilkinson (1985) displaced glycerine by oil in an etched-glass network and showed that, if the system is perfectly homogeneous (if all the pores have the same effective radius), then the fingers form ordered (dendrites) patterns (see Fig. 11.10), in which growth occurs mostly along the coordinates of the system, whereas the DLA algorithm would generate a random fractal for exactly the same situation. The reason is that the surface of a DLA structure is dominated by noise: random walkers taking *random trajectories* that start far from the surface of the aggregate arrive at the vicinity of the surface one at a time. These random trajectories make the surface very rough. Thus one has to somehow reduce the noise and randomness of these trajectories. Two algorithms were introduced to accomplish this.

In the first algorithm, one introduces a "sticking probability," the probability p_s that the front or the aggregate advances by one unit once a random walker is in an empty site next to it. The DLA case corresponds to $p_s = 1$. However, if we let p_s to be very small, then the surface of the aggregate becomes very smooth, because a random walker will encounter the front many times (on average roughly $1/p_s$ times) before the front actually advances. This method was discussed by Meakin (1986) and does result in a smooth surface. In the second algorithm, that was proposed by Tang (1985) and Szép *et al.* (1985), each time an empty site next to the front is visited by a random walker, a counter registers the event. The front does not advance to an empty site in its neighborhood unless it has been visited at least n_v times, so that $n_v = 1$ corresponds to the DLA case, and

$n_v \to \infty$ represents the noiseless limit of the DLA model. Kertész and Vicsek (1986) showed that this modification of the DLA algorithm can reproduce the patterns obtained by Chen and Wilkinson (see also Siddiqui and Sahimi, 1990a).

Several other aspects of DLA-based simulation of miscible displacements deserve careful considerations. (i) The first question that arises is whether this type of simulation can be generalized to the case of finite values of M (the DLA case corresponds to $M = \infty$). (ii) The second question is whether this type of simulation can be a *quantitative* tool for simulating miscible displacements in a *disordered* porous medium, at least in the limit $M = \infty$. (iii) One should remember the fact that this type of simulation shows a sensitive dependence on the lattice size, and therefore it is highly important to establish links between the parameters of the simulations and the physical parameters that can be measured. (iv) One needs to generalize the DLA model for a porous medium with a pore size distribution (as in a macroscopic porous medium), or for one with a permeability distribution (as in a megascopic porous medium). (v) Another important question is whether it is possible to generalize a DLA-based simulation to include the effect of dispersion since, as we discussed above, dispersion plays a fundamental role in miscible displacements, whereas in most of the published papers using DLA-based simulation, this effect has completely been ignored. These questions are taken up below.

We should mention the experiments of Nittman *et al.* (1985) and Daccord *et al.* (1986). These authors used a viscoelastic fluid to study fingering in both radial and rectilinear Hele-Shaw cells. Their viscous fingers show very localized tip splitting and shieldings which are also much stronger than those obtained with a Newtonian fluid. Their viscous fingers also have a fractal structure with a fractal dimension close to that of DLAs. These results contradict those obtained with Newtonian fluids discussed above, and remain unexplained.

Before closing this section, we would like to mention that the DLA algorithm has been generalized to include the effect of surface tension, as in an immiscible displacement in a Hele-Shaw cell, which is the classical Hill-Saffman-Taylor problem. This model was suggested by Szép *et al.* (1985) and Kadanoff (1985). The problem is to force the probability $P(\mathbf{r})$ at the interface between the two fluids to have a value of the form

$$P(\mathbf{r}) = a_1 \kappa + a_2 \quad , \tag{11.46}$$

where κ is the curvature of the interface, a_1 is proportional to the surface tension, and a_2 is a constant. According to the model proposed by Kadanoff (1985) and Szép *et al.* (1985), one should allow the walkers to leave the interface and walk through the displacing fluid until they finally reach the interface again. One allows the walkers to leave the displacing fluid with a probability proportional to $P(\mathbf{r})$ given by Eq. (46), which measures the net flux of walkers through each interface bond, and then moves the displaced or displacing fluid forward whenever the flux is $-n_v$ or $+n_v$. In this manner one has a complete and meaningful model for simulating immiscible displacements in a Hele-Shaw cell. Various versions of this basic model were used by Sahimi and Yortsos (1985), Meakin *et al.* (1987), Sarkar and Jensen (1987), and Tao *et al.* (1988) to investigate different aspects of the problem. Vicsek (1984, 1985) suggested a variant of the DLA in which the particles stick to the aggregate with a probability proportional to $P(\mathbf{r})$ given by Eq. (46), *without* the reshuffling of the interface suggested by Kadanoff (1985) and Szép *et al.* (1985). Sarkar (1985) showed that Vicsek's model simulates the *early stages* of the Hill-Saffman-Taylor problem. Finally, Liang (1986) applied the Kadanoff-Szép *et al.* model very successfully to the study of immiscible displacements in a Hele-Shaw cell. He compared his results with those obtained by other methods, such as conformal mapping and boundary integral methods (see Bensimon *et al.*, 1986, for a review), and experimental data, and found good agreement.

11.5.2 The dielectric breakdown model

This model was proposed by Niemeyer *et al.* (1984) to study dielectric breakdown phenomena in disordered solids, but it can also be used for simulating viscous fingers in the absence of dispersion, since in this limit the equations for the two phenomena are similar. A discrete version of the Laplace's equation is solved. This means that, for example, on a square lattice the unknown function F_i at lattice site i is given by $4F_i = F_{i+1} + F_{i-1} + F_{i+L} + F_{i-L}$, where L is the linear size of the lattice, and $(i + 1, i - 1, i + L, i - L)$ are the nearest-neighbor sites of i. Similar equations can be written down for other lattices. For example, for the simple-cubic lattice we have $6F_i = F_{i+1} + F_{i-1} + F_{i+L} + F_{i-L} + F_{i+L^2} + F_{i-L^2}$. As the boundary condition, the unknown function is specified at the boundary. In the context of a miscible displacement, the unknown function is the pressure, so the simulation is appropriate only for the case of a viscous fluid displaced by an inviscid fluid ($M = \infty$). At each step of the simulation, one site on the boundary between the two regions (between the displaced and displacing fluids) is selected for advancement of the front between the two regions, with the probability of selection being proportional to a power of the gradient of the unknown function, e.g., the pressure in the miscible displacement problem. This algorithm is usually called the dielectric breakdown (DB) model. The same acronym was used by Ben-Jacob *et al.* (1985) to denote dense branching morphology formed in various fluid displacements. The DB model can also be simulated by a random walk technique. The method would be similar to that of DLA, except that in the DB model the front advances to a nearest-neighbor empty site if this site is visited by a random walker *which also crosses the front* (whereas in the DLA model the front advances as soon as the empty site is visited by the walker). Thus the boundary conditions at the front are *not* the same for the DB and DLA models. However, although the difference between the DLA and DM models seems minor, the results are very different, indicating the sensitivity of the two models to the boundary conditions.

11.5.3 The gradient-governed growth model

DeGregoria (1985) and Sherwood and Nittman (1986) introduced an algorithm for simulating miscible displacements at finite values of M, and in the absence of dispersion. The model is usually called the gradient-governed growth (GGG) model. This is essentially an extension of the DB model to the case when both fluids have a finite viscosity. A discrete version of the Laplace's equation is solved in the region occupied by each fluid to yield the pressure field (see above). The front between the two fluids is advanced at each time step, with the selection probability being proportional to the local pressure gradient between a point on the front and the point to which the front is to advance. The model is *wrong* in a microscopic sense, because the velocity field is determined assuming the entire front is moving instantaneously, yet only one bond at the front is moved at a time. DeGregoria (1985) and Sherwood and Nittmann (1986) used this model to simulate miscible displacement at finite values of M, in the absence of dispersion. Using a 100×100 square network, DeGregoria (1985) obtained reasonable agreement with Habermann's (1960) data. Sherwood (1986) used the same model for investigating the size distribution of the islands of the displaced fluid that are formed when the displacing fluid completely surrounds a portion of the displaced fluid.

11.5.4 The two-walker model

This model was proposed by Siddiqui and Sahimi (1990a) for simulating miscible displacements at finite values of M, in the absence of dispersion, *using only random walkers*. Since the pressure in each fluid region obeys the Laplace's equation, one random walker for each fluid region (displacing and displaced) is used. Because in the absence of dispersion and surface tension the front always advances forward, both random walkers advance the front, upon contacting it, with a probability proportional to $p_1 = (M + 1)^{-1}$, for the particle in the displacing fluid region, and $p_2 = M/(M + 1)$, for the particle in the displaced fluid region. As such, this model can be thought of as the random walk version of the GGG model. Because of different mobilities of the two fluids, the lengths of each step of the random walkers are *different* in each fluid region. The results of this model are in complete agreement with those obtained with DB and GGG models. Siddiqui and Sahimi (1990a) also generalized the model to the case in which there is a pore size or pore permeability distribution. In this case, the random walkers take each step with a probability proportional to the (pore or region) permeabilities (see also Meakin, 1987; Blumberg Selinger *et al.*, 1989). A somewhat similar model was proposed by Leclerc and Neale (1988), although the precise relation between these two models is not clear to us.

Among the discrete models described so far, the DLA model and its two-walker generalization are the only ones that use *only* random walkers, and thus, from a computational point of view, they can be very efficient. Therefore, the relevant question is whether such random walk models can provide quantitative predictions for miscible displacements in which dispersion plays no important role. At first the answer may seem affirmative, since the governing equations for both phenomena are identical. However, this only guarantees that the universal properties of the two processes to be the same, but not necessarily the equality of non-universal quantities of interest. For example, as far as a petroleum or chemical engineer is concerned, the most important quantity to predict is the volume fraction of the displaced fluid at the breakthrough point, i.e., at the point at which a sample-spanning cluster of the displacing fluid is formed, since this is a quantitative measure of the efficiency of the displacement process. Murat and Aharony (1986) and Meakin *et al.* (1989) used the DLA and DB models to simulate viscous fingers in the absence of dispersion. They found that the two models are not always the same, but their results were not conclusive. As explained by Sahimi and Siddiqui (1987), the proper comparison is between DLA-like models that use only random walkers and a deterministic model in the absence of dispersion, because in an actual experiment the front does not advance one pore at a time, as in the DLA, DB and GGG models and their generalizations, but advances in several pores simultaneously, which is the basis of a deterministic model. This is discussed below after such a deterministic model is described.

11.5.5 Probabilistic models

The discrete models discussed so far do not explicitly take into account the effect of dispersion. We now discuss three models that can accomplish this, two of which use probabilistic arguments and are discussed in this section, while the third one is completely deterministic and is discussed in the next section.

The first of the two probabilistic models is due to King and Scher (1987, 1990), whose details are as follows. Consider first the case of miscible displacement *without* dispersion. For point injection of fluids the governing equations are

$$\frac{\partial C}{\partial t} + \mathbf{u} \cdot \boldsymbol{\nabla} C = \delta^2(\mathbf{x}) \quad , \tag{11.47}$$

$$\mathbf{u} = \frac{\mathbf{v}}{\hat{Q}} = \boldsymbol{\nabla} \times (\psi \hat{\mathbf{z}}) \quad , \tag{11.48}$$

where $\hat{Q}$ is the injection rate (volume per unit thickness per unit time), ψ is the stream function, and $\delta^2(\mathbf{x})$ is the two-dimensional Dirac delta function. The injected volume of fluid provides a natural time variable

$$\tau = \frac{\int_0^t \hat{Q}(t')dt'}{\psi} \quad . \tag{11.49}$$

If there is no dispersion, then the solution is simple: A concentration bank $C = 1$ (i.e., the pure fluid) displacing $C = 0$, which is also what the DLA-like models predict. However, in general, the concentrations need not form a bank, since dispersion intervenes and develops a mixed zone. To develop a probabilistic model that takes this effect into account, King and Scher (1987, 1990) interpreted $(\partial C/\partial \tau)d^2x$ as a *two-dimensional* probability density function for concentration evolution. Equations (47) and (48) tell us that we can write

$$\frac{\partial C}{\partial \tau}d^2x = -\frac{\partial C}{\partial \xi_1}\frac{\partial \psi}{\partial \xi_2}d\xi_1 d\xi_2 = -dC\,d\psi \quad , \tag{11.50}$$

where ξ_1 and ξ_2 are local tangential and normal coordinates on the front. Equation (50) can now be given a probabilistic interpretation. The probability of concentration evolution at $\mathbf{x}$ [i.e., $(\partial C/\partial \tau)d^2x$] is the product of the probability $d\psi$ of fluid flow through $\mathbf{x}$ and the probability dC of a concentration gradient moving through $\mathbf{x}$. For a given *realization*, one samples the cumulants C and ψ, i.e., determines the flux contour $C = r_1$, and the streamline $\psi = r_2$, and calculates their intersection in the plane, which is also the point at which concentration is modified, where r_1 and r_2 are two random numbers uniformly distributed in $(0,1)$.

In such a simulation, one has to employ a probabilistic interpretation of the finite-difference version of Eq. (47). We integrate this equation over a rectangular spatial region A_{ij} and time interval $\Delta\tau$ to obtain δC_{ij}, the change in the average concentration C_{ij}. This is given by

$$\delta C_{ij} = -\frac{\Delta\tau}{\Delta x \Delta y}\oint_{\partial A_{ij}}\bar{C}d\psi \quad . \tag{11.51}$$

Obviously, if $\delta C_{ij}/\Delta C$ is properly normalized, then it can be interpreted as the growth probability at site ij (i.e., as the probability that the displacing fluid and the front between the two fluids advance). One now has to fix ΔC. If we choose $\Delta C = 1$, we have a situation similar to the DLA model (i.e., a cluster of $C = 1$ respresenting the aggregate of the displacing fluid, facing a cluster of $C = 0$ representing the displaced fluid). According to Eq. (51), the growth probability is non-zero only when the boundary integral overlaps the cluster edge. This method has the advantage that finite values of M can be used in the simulations. But it also implies that very large clusters would be very difficult to generate with this method, since the simulations would be excessively time consuming.

We can now add the effect of dispersion. Consider first the static case $\mathbf{v} = 0$. The evolution equation is simply the diffusion equation, $\partial C/\partial T - D_L\nabla^2 C = 0$, where $T = D_L t$ (King and Scher assumed the equality of D_L and D_T, which is not justified). In discrete form

$$\delta C_{ij} = \frac{\Delta T}{\Delta x \Delta y} \oint_{\partial A_{ij}} \hat{\mathbf{n}} \cdot \boldsymbol{\nabla} \bar{C} dl \quad , \tag{11.52}$$

which should be compared with Eq. (51). If the time step ΔT satisfies $\Delta T/(\Delta x \Delta y) \leq \Delta C$, then

$$\frac{\delta C_{ij}}{\Delta C} = -\sum_{faces} \oint_{\partial A_{ij}} \left(\frac{\Delta T}{\Delta x \Delta y \Delta C} \right) \left(-\hat{\mathbf{n}} \cdot \boldsymbol{\nabla} \bar{C} \right) dl \quad . \tag{11.53}$$

For the full problem (convection and diffusion) we can split the time evolution as

$$\phi \frac{\partial C}{\partial t} = \hat{Q} \frac{\partial C}{\partial \tau} + \phi D_L \frac{\partial C}{\partial T} \quad . \tag{11.54}$$

In this equation $\partial/\partial t$ and $\partial/\partial T$ represent separate convection and dispersion processes. Convective evolution is described by Eq. (52), while dispersion evolution is represented by Eq. (54). Thus we have a sequential finger evolution in which convection initiates the growth, which is then modified or moderated by dispersion. If we fix the time interval δt, then $\delta \tau = (\hat{Q}/\psi)\delta t$, and $\delta T = D_L \delta t$, which then define the Péclet number $Pe = \delta \tau/\delta T$ (see also Chapter 9). In practice, $\delta \tau$ is set by $\delta \tau = \Delta \tau = \Delta x \Delta y \Delta C$, implying that $\Delta T = \Delta \tau/Pe$. In most practical cases, $Pe > 1$, and Eq. (54) is properly normalized. However, if $Pe < 1$, δT is subdivided into n_D intervals to obtain $\Delta T = (\delta \tau/Pe)n_D$, where $n_D > Pe^{-1}$. King and Scher (1990) simulated miscible displacements with dispersion for various values of M and obtained reasonable agreement with the data of Habermann (1960). It is interesting to note that, although in the GGG and the two-walker models dispersion is not present *explicitly*, the numerical algorithms seem to contain implicitly the effect of dispersion, since the predictions of the models agree with Habermann's data, as in an actual experiment dispersion is always present.

The second probabilistic model is due to Araktingi and Orr (1988). In their model the porous medium is represented by a three- or two-dimensional system of cubic (square) grid blocks. At the beginning of each time step the pressure field is calculated, given the distribution of the permeability and the current distribution of fluid viscosities. Tracer particles that carry a finite concentration of displacing fluid are injected into the system and are moved with velocities based on the pressure field. The velocities are calculated at the midpoint between grid nodes. Velocities for particles that are not on such nodes are obtained by linear interpolation. After moving the particles by convection to their current position, the effect of dispersion is simulated by random perturbations of particle positions in the longitudinal and transverse directions. As discussed in Chapter 9, for diffusive dispersion the standard deviations of the motion of the particles are given by, $\sigma_x = \sqrt{2D_L t}$ and $\sigma_y = \sqrt{2D_T t}$. Thus the distribution of the particles about a mean position can be simulated by multiplying these standard

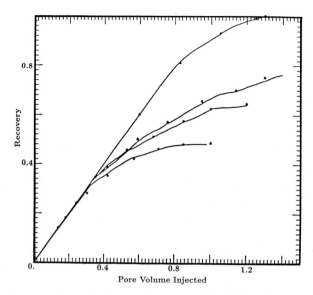

Figure 11.11: Comparison of the predictions of Araktingi-Orr model (curves) with the experimental data (symbols) of Blackwell *et al.* (1959). The results are, from top to bottom, for $M = 5, 86, 150,$ and 375 (after Araktingi and Orr, 1988).

deviations by a number between -6 and $+6$. This number is obtained by adding a sequence of 12 random numbers, distributed normally with a zero mean and unit standard deviation, to -6. The values $+6$ and -6 were selected because, on a practical basis, the probability of a particle moving beyond 6 standard deviations on either side of the mean is less than 1%. After the particles arrive at their new position, the current time step is determined. To avoid having particles travel a distance greater than a grid block, the time step is chosen to allow movement equal to half of a grid block length (or width), travelled at the highest existing velocity. The new pressure field is calculated, and a new position for each particle is determined. This procedure is repeated many times. It should be noted that both this method and that of King and Scher require *a priori* estimates of D_L and D_T.

Araktingi and Orr (1988) compared their results with the experimental data of Blackwell *et al.* (1959), using $D_L = 1.6 \times 10^{-3}$ cm^2/sec, and $D_T = 6.5 \times 10^{-5}$ cm^2/sec. Figure 11.11 compares their results with the data and the agreement is very good (compare this figure with Fig. 11.6). Although this model suffers from fluctuations due to the limited number of particles used in the simulation, the ensemble-average properties over several realizations compare well with the data of Blackwell *et al.* (1959). Note that the probabilistic method suggested by Araktingi and Orr (1988) is very similar to the Monte Carlo method of Smith and Schwartz (1980, 1981a,b) for studying dispersion in megascopic porous media discussed in Chapter 9. The only difference is that Araktingi and Orr (1988) treated the case in which the viscosities of the two fluids were *not* the same. The main disadvantage of these models is their long simulation times. This is especially true when the effects of gravity and transverse dispersion are to be included in the simulations.

11.5.6 Deterministic network models

These models were invented for investigating viscous fingers in pore networks. Nobles and Janzen (1958) had already used analog resistor networks to study the effect of M on miscible displacements. Random networks and deterministic flow models were originally proposed by Simon and Kelsey (1971, 1972), but their model was too simple and the networks used were too small. There are two such models, one of which is applicable to miscible displacements in the absence of dispersion, while the other can also take into account the effect of dispersion.

Let us first discuss the case in which dispersion in neglected. The porous medium is represented by a network of interconnected pores, usually assumed to be cylindrical tubes with distributed radii. Consider a tube of length l_{ij} and radius R_{ij} which connects nodes i and j, of which a portion x_{ij} is occupied by the displacing fluid and the rest by the displaced fluid. The pressure difference $P_i - P_j$ along the tube is given by

$$P_i - P_j = \frac{8\eta_2(x_{ij}/M + l_{ij} - x_{ij})Q_{ij}}{\pi R_{ij}^4} = \frac{Q_{ij}}{g_{ij}} \quad, \tag{11.55}$$

where η_2 is the viscosity of the displaced fluid, Q_{ij} is the flow rate in the tube between i and j, and g_{ij} is the hydraulic conductance of the tube. At each node i of the network we have conservation of fluid fluxes, $\Sigma_j Q_{ij} = 0$, which, when written for every interior node of the network, yields a set of linear equations for the nodal pressures. The boundary conditions are such that at the injection point $P = 1$, while at the production point $P = 0$. After determining the nodal pressures, we move the front a distance

$$\Delta x_{ij} = \frac{Q_{ij}}{\pi R_{ij}^2}\Delta t \quad, \tag{11.56}$$

into one of the pores adjacent to the interface. We choose the time step Δt to be the time necessary to exactly move the front to reach a node through the fastest tube. We then update all other fronts (i.e., we move them into the slower pores adjacent to the front and *partially* fill such pores) and, for the new configurations of the regions of displacing and displaced fluids, calculate the pressure field and repeat the entire process. This method was used by Chen and Wilkinson (1985), King (1987b), Siddiqui and Sahimi (1990a), and Ferer *et al.* (1992) to investigate miscible displacements without dispersion. The main

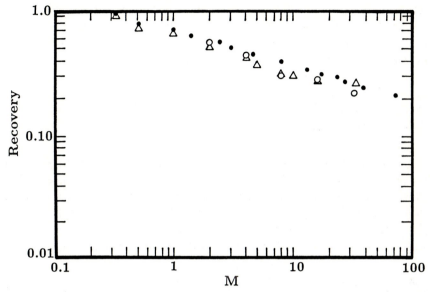

Figure 11.12: Comparison of the predictions of the two-walker model (triangles) and the deterministic network model (open circles) with the experimental data of Habermann (solid circles) (after Siddiqui and Sahimi, 1990a).

advantage of this method is that it allows one to investigate the effect of pore-level or large-scale heterogeneities on the displacement process. It is also free of the type of noise that is generated by the DLA-like models. Figure 11.12 compares the predictions of this model with Habermann's data, and the agreement is good.

As discussed above, an important question is whether non-universal properties of miscible displacements, such as their sweep efficiency, can be predicted by DLA-like models. Answering this question requires a careful comparison between the predictions of the deterministic model in the absence of dispersion and those of DLA-like models. Chan *et al.* (1986, 1988), Sahimi and Siddiqui (1987), and Siddiqui and Sahimi (1990a) studied this problem in detail. Sahimi and Siddiqui (1987) computed the sweep efficiency using both models and found them to be generally different. Moreover, they found that although the fractal dimensions of the two models are identical if the pore system is well connected, the two models are characterized by different fractal dimensions if the pore system is poorly connected, e.g., if the pore space is a percolation cluster near the percolation threshold. Chan *et al.* (1986, 1988) argued essentially along the same lines as those of Sahimi and Siddiqui, except that their model of pore space was different from that of Sahimi and Siddiqui. In Chan *et al.*'s model, the porous medium is represented by a network of interconnected *tubes* and *chambers*. The tubes have small diameter, and thus contribute most of the resistance to fluid flow. These tubes connect the grid points of the network a distance l apart at which the chambers are located. The chambers have volumes much greater than those of the tubes, thus making a negligible contribution to the hydrodynamic resistance. The fluid capacity of each chamber, i.e., its volume per specified volume l^3, is a randomly distributed quantity. Chan *et al.* showed that, unless the distribution of the fluid capacity of the chambers is exponential, the predicted values of the sweep efficiency by their model and the DLA model will be different. The conclusion is that although the DLA-like models often provide a good description of the universal properties of miscible displacements without dispersion, the mapping between the two problems is not exact but approximate, although a very good one in many situations.

The second deterministic network model is capable of taking into account the effect of dispersion. As in the first model, we first calculate the pressure field throughout the network. There is, however, one difference between this case and the first case: the portion x_{ij} is occupied by a *mixture* of displacing and displaced fluid with an effective viscosity η_m given by, e.g., Eq. (1). Thus M should be calculated and used in Eq. (55) based on the effective viscosity η_m and the viscosity η_2 of the displaced fluid, i.e., $M = \eta_2/\eta_m$. We then assume that within each pore a one-dimensional CDE governs the concentration distribution, whose solution is given by Eqs. (9.57) and (9.58). Using the fact that C is a conserved quantity, and employing the distribution of pore velocities, we can determine the concentration distribution in the entire network, since for every node i, we can write $\sum_j S_{ij} J_{ij} = 0$, where $J_{ij} = v_{ij} C_i - D_m \partial C_i/\partial x$ is the total flux (convective plus diffusive) in the pore that connects sites i and j, and S_{ij} is the cross-section area of pore ij. This equation, when written for every node of the network, yields a set of equations for nodal concentrations C'_i, which is solved numerically. Once the concentration distribution is determined, we proceed as in the first case except that the distance Δx_{ij} that the front moves in a pore is given by, $\Delta x_{ij} = J_{ij} \Delta t/(\pi R_{ij}^2)$. The advantage of this method is that one does not need to provide D_L and D_T as phenomenological inputs to the model, as is done in the models of King and Scher (1987, 1990) and Araktingi and Orr (1988). It also allows one to investigate the effect of rock heterogeneities on miscible displacements in the presence of dispersion.

11.6 Crossover from fractal to compact displacement

Since the universal properties of viscous fingering, in the limit $M \rightarrow \infty$ and without dispersion, and those of DLAs are the same, one may conclude that viscous fingers are fractal objects in the limit $M = \infty$. But what about the case of viscous fingers at finite values of M? We already know that viscous fingers may be unstable if $M > 1$, which might mean that they are fractal objects for *any* $M > 1$, with the instability manifesting itself in the fractal structure. However, this implies that the density of the displacing fluid would vanish as the displacement proceeds, as this is a general property of any fractal object.

This means that very thin sections of fluids would have to support a vast and
tenuous network of the fingers at the tip. In fact, if the displaced fluid has a
finite mobility, one would expect the fingers to become *thicker*, as opposed to the
thinner fingers with the fractal behavior. However, for $M > 1$ the displacement
can be unstable, if dispersion effects are absent, and we cannot expect it to be
smooth. The cluster of the displacing fluid is also not compact, since there is
no intermediate length scale between the size of the system and that of a pore.
Thus, as argued by King (1987b), only the *surface* of the viscous fingers (i.e., the
front between the two fluids) can have a fractal-like character, and this can be
interpreted as the manifestation of the instability of the process. This argument
is supported by the simulations of King (1987b), Blunt and King (1988), Siddiqui
and Sahimi (1990a), and Ferer *et al.* (1992) who used the deterministic model,
and by those of Siddiqui and Sahimi (1990b) who used the two-walker method.
Even the surface roughness and its fractal character should gradually disappear
as the size of the system increases. This was demonstrated by King and Scher
(1990) who argued that if the linear size of the system exceeds a crossover size
L_{co}, then there would be a crossover from a fractal-like behavior to a compact
displacement, where L_{co} is given by

$$L_{co} \sim \left(\frac{M-1}{2}\right)^{1/D_f} \, , \qquad (11.57)$$

where D_f is the fractal dimension of DLAs. King and Scher (1990) called this
phenomenon *viscous relaxation*. Earlier Monte Carlo simulations of DeGregoria
(1986) using the GGG model had already provided evidence for this phenomenon.
DeGregoria observed that, for finite values of M, the volume fraction of displaced
fluid increases as the size of the system increases, which implies increased stability. Note that Eq. (57) implies that no such crossover can occur if $M \rightarrow \infty$.
Using renormalization group methods (see Chapter 8), Lee *et al.* (1990) also
reached the same conclusion. They showed that in a renormalization group
treatment of this problem there are *two* fixed points. One is called the Eden
fixed point which corresponds to a compact, non-fractal structure such as what
one obtains with $M = 1$, while the other one is the DLA point, corresponding to
a fractal structure, which is what one obtains in the $M = \infty$ limit, which they
showed to be a saddle point. This implies that for *any* finite M viscous fingers
"eventually" approach a Euclidean limit, in the sense of taking up a compact
shape, perhaps with a fractal or rough surface, where "eventually" means either
long times, or very large length scales, or a fine enough grid. Furthermore, Lee
et al. (1990) proposed the following crossover scaling law

$$m(R_g) \sim R_g^d F\left(\frac{R_g^\delta}{M}\right) \, , \qquad (11.58)$$

where m is the "mass" of the cluster of displacing fluid, R_g the radius of gyration
of the cluster, d the dimensionality of the system, and δ a crossover exponent.

$F(x)$ is a scaling function with the properties that $F(x) \sim 1$ for $x \ll 1$, and $F(x) \sim x^a$ for $x \gg 1$, with $d + a\delta = 2$.

The crossover between fractal and non-fractal miscible displacements has two practical implications. The first is that the scale up of numerical or laboratory experiments of unstable displacements ought to be done with caution since, as suggested by Eq. (57), there will always be a crossover to a stable displacement if one waits long enough, or if the scale up is done for a large enough scale. The other implication is that heterogeneities of a scale of the order of the crossover length L_{co} or larger dominate viscous fingering (see also below).

11.7 Miscible displacements in megascopic porous media

Miscible displacements in megascopic porous media have also been studied by various methods, although these studies are not as extensive as those for macroscopic porous media. For example, Tan and Homsy (1992) used their continuum model to study miscible displacements in megascopic porous media, in which the heterogeneities were modelled as stationary random functions of space. The permeability correlation length ξ_k was finite. They found that the fingers grow linearly in time in a manner similar to that of macroscopic media discussed above. This result is not surprising since ξ_k was assumed to be finite. As an example of discrete simulations, we mention Araktingi and Orr's (1988) simulations in which they assumed that the mean of the permeability distribution is independent of the location, and that the spatial correlation between any two regions depends only on the distance between them. A heterogeneity index H_i was used to characterize a permeability field, which they defined it by

$$H_i = \sigma_{\ln K}^2 \hat{\xi}_K \quad , \tag{11.59}$$

where $\sigma_{\ln K}$ is the standard deviation of log permeability and $\hat{\xi}_K$ is the dimensionless correlation length of permeability $\hat{\xi}_K = \xi_K / L$, L being the system length. Note that this definition of H_i is somewhat different from that proposed by Koval. As discussed earlier, H_i combines the variability (as measured by $\sigma_{\ln K}$) and the spatial correlations of the permeability field. Figure 11.13 compares their simulation results for two values of M and two values of H_i. It is clear that for low values of H_i, i.e., a more homogeneous porous medium, the effect of M is strong, which is expected. However, as H_i increases the effect of M diminishes, and the shape of the fingers is dominated by the permeability heterogeneities. For example, for $H_i = 0.77$, the shapes of the fingers are almost identical for $M = 1$ and 20. If this is the case, then a simple generalization of a DLA-like model (or any other random-walk method) should suffice for simulating miscible displacements in megascopic reservoirs, since as discussed in Chapter 9, pore-level dispersion is not important in megascopic porous media, and the velocity field fluctuations, induced by the permeability field, is the most important factor, and this effect is easily captured by a DLA-like model.

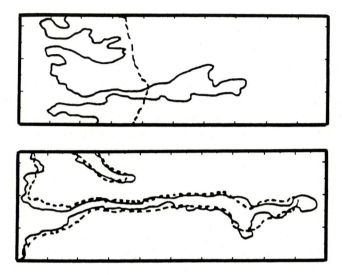

Figure 11.13: Comparison of viscous fingers for the heterogeneity index $H_i = 0.25$ (top) and $H_i = 0.77$ (bottom), and viscosity ratio $M = 1$ (dashed curves) and $M = 20$ (solid curves) (after Araktingi and Orr, 1988).

One of the factors that distinguishes a megascopic porous medium from a macroscopic one is the extent of the spatial correlations in the permeability distribution. If there are no spatial correlations, or they are of finite extent, we do not expect a miscible displacement in the megascopic medium to be very different from that in a macroscopic one. One way of incorporating the effect of long-range permeability correlations is of course through the use of fractal statistics that we already discussed in Chapters 6, 8 and 9. Emanuel *et al.* (1989) and Mathews *et al.* (1989) used such statistics in their simulations of displacement processes. They showed that using such statistics leads to significant improvement in the predictive ability of their models. These authors used finite-difference techniques for solving their equations, which may limit the size of the system and the extent of correlations that they can simulate (since one usually needs a very fine grid structure with finite-difference techniques in order to achieve reliable accuracy). On the other hand, Sahimi and Knackstedt (1994) used a random-walk model to study miscible displacements in megascopic porous media, in which the permeabilities obeyed the statistics of fractional Brownian motion discussed in Chapters 3, 6 and 8. They showed that M does not play any important role, and the permeability distribution controls the efficiency of the displacement.

11.8 Conclusions

At the end of this Chapter we should again emphasize the role of dispersion and permeability heterogeneities in any miscible displacement process. Since 1984 there has been an explosion of papers on miscible displacements and their relation with fractal phenomena such as DLAs. These studies have added to our understanding of miscible displacements, and in particular have pointed out the relation between miscible displacements and fractal structures, and the crossover between fractal and non-fractal displacements. However, most of these papers have not taken into account the effect of dispersion, as a result of which the predicted sweep efficiencies are not in quantitative agreement with the data. More work, both numerical and analytical, is necessary before we can claim with some confidence that we understand miscible displacements and their stability.

Chapter 12

Immiscible Displacements and Multiphase Flows

12.0 Introduction

In this Chapter we turn our attention to immiscible displacements and multiphase flows in porous media and fractured rock. A large number of factors can affect this class of phenomena, among which are capillary, viscous, and gravity forces, the viscosities of the two fluids and the interfacial tension separating them, chemical and physical properties of the surface of the pores or the fractures, i.e., whether or not there are surface active agents, or whether the surface is fractal, the morphologies of the pore space and the fracture network, and the wettability of the fluids. Obviously, multiphase flows and immiscible displacements in a porous system involve a set of complex phenomena, and to date no model has been developed that can take into account the effect of all of these factors. We start our discussion by studying wettability, as it has a strong influence on the distribution of two or more immiscible fluids in a porous system.

12.1 Wettability, contact angles and their measurement

In Chapter 5 we discussed briefly wettability and its effect on fluid flow. Generally speaking, the rock-fluid interactions are what we call wettability. It has a strong effect on the flow of two immiscible fluids in a porous medium, conventional and enhanced oil recovery processes, and many other phenomena (for example, coating operations), and has been studied for a long time by petroleum engineers (Bartell and Miller, 1928; Owens and Archer, 1971; Salathiel, 1973; McCaffery and Benion, 1974; Batycky *et al.*, 1981), and others. It is also known that oil recovery process itself can alter reservoir wettability (Wagner and Leach, 1959; Reed and Healy, 1977).

Consider, as an example, an experiment in which a drop of water is placed on a surface immersed in oil. Then a contact angle is formed that can vary anywhere from 0° to 180° (see Fig. 5.1). As can be seen, there are three different surface tensions are related by the Young-Dupŕe equation

$$\sigma_{ow} \cos \theta = \sigma_{os} - \sigma_{ws} \quad , \tag{12.1}$$

where σ_{ow} is the interfacial tension between oil and water, and σ_{os} and σ_{ws} are the surface tensions between oil and the solid surface, and water and the solid surface, respectively. Normally, the contact angle θ is measured through the water phase. Strictly speaking, if $\theta < 90°$, the surface can be considered as water-wet. However, in practice, $\theta < 65°$ for water-wet systems, while $105° < \theta < 180°$ for oil-wet systems. If $65° < \theta < 105°$, the system is said to be *intermediately-wet*, and has no strong preference for any of the two fluids. Another important case is *mixed wettability*, in which the wettability of the surface changes from pore to pore or from one portion of the surface to another. This is caused by *chemical* heterogeneity of the surface, and is actually the situation one has to deal with in most oil reservoirs.

The study of moving contact lines and contact angles goes back to Washburn (1921) who proposed Eq. (5.9) for a cylindrical tube. This equation is invalid if the length of the tube (for which the equation was intended) is much longer than its diameter, and if the diameter is small enough that gravity cannot have a significant effect on the shape of the moving contact line or meniscus. Fisher and Lark (1979) accumulated experimental evidence in support of this equation. However, Eq. (5.9) neglects the details of the surface and its effect on the moving contact line: it provides only an overall picture of what happens. For example, in almost all cases of practical interest, one has to deal with the no-slip boundary condition. Huh and Scriven (1971) were the first to analyze moving contact angles and lines and claimed that, with no-slip boundary condition, the stresses at the contact line become *divergent* (or the total dissipation diverges). They attributed this to the existence of, among other things, discontinuous processes around the contact line. As pointed out by Dussan V. and Davis (1974), this anomaly is due to the fact that Huh and Scriven (1971) had worked with a *planar* interface, and that their equations failed to satisfy the normal stress boundary condition at the interface between the two fluids. It is now well-known that there are two ways of removing the singularity. If the advancing fluid wets the solid surface perfectly, then a thin precursor film forms ahead of the contact line and the dissipation divergence (which is logarithmic) has a cutoff at the film thickness. On the other hand, if the advancing fluid does not wet the surface completely, slip can occur within a length l_θ from the contact line. This length can act as a cutoff and prevent the divergence of the total dissipation. These matters have been discussed by Dussan V. (1979).

On a moving contact line the interface exerts a force $\sigma_{fs} \cos \theta_D$, where θ_D is the apparent *dynamic* contact angle (as opposed to a static contact angle θ_S which is well-defined on a homogeneous surface), and σ_{fs} is the surface tension between the fluid and the solid. There is also an additional viscous force F_v on the contact line. If the capillary number $Ca = \eta v / \sigma_{ow}$ is small, F_v would also be small compared with the capillary forces, except within a distance l_θ from the contact line. This gives us a method of measuring F_v from θ_D, measured far from the contact line, since $F_v = \sigma_{fs} (\cos \theta_D - \cos \theta_S)$. Roughness, chemical heterogeneity and other factors can make this picture more complex, which is discussed below. The dynamic contact angle θ_D can be measured by optical methods (see, for example, Hoffman, 1975).

Many methods have been developed for measuring the wettability of a system, and Anderson (1986) has given a thorough discussion of them. Here, we restrict ourselves to three quantitative methods. More details are given by Anderson (1986).

12.1.1 Sessile drop method

There are several methods for measuring the contact angles. These include the tilting plate method, capillary rise method, tensiometric methods, vertical rod method, cylinder method, and sessile drops or bubbles method, and Adamson (1990) discusses most of them. However, the petroleum industry does not use most of these methods, because they work best when one deals with pure fluids, and clean, artificial cores which are rarely encountered in practice. Perhaps the most popular method in the petroleum industry is the sessile drop method (see, for example, McCaffery, 1972), and its modification by Treiber *et al.* (1972). In the latter method the mineral to be tested is put in a test cell which is made of inert material. This prevents contamination of the surface which can alter the true contact angle. Two flat and polished mineral crystals, that are usually either quartz or calcite crystals (sedimentary rocks are composed of such crystals; see Chapter 4), are mounted parallel to one another. The apparatus has to be completely clean so that the true contact angle can be measured. The cell containing the mineral crystals is then filled with de-oxygenated synthetic formation brine. It usually takes a few days for the oil-crystal interface to be clearly established. This process is called *aging*. Then the two crystals are displaced parallel to each other to shift the oil drop. Thus the brine can move over a portion of the surface that is covered with oil. In this way an *advancing* contact angle θ_A is measured. Usually, it takes a day or two before θ_A reaches its equilibrium value. The surface is aged again, the water is advanced again, and so on. The sessile drop method is similar to this procedure, except that only one flat crystal is used. A drop of oil is formed at the end of a fine capillary tube and brought into contact with the flat surface (see Fig. 5.1).

If the oil contains surface-active agents, θ_A increases with aging until an adsorption equilibrium is reached. This may take a long time. Usually, the measured contact angles show hysteresis, that is, the contact angle θ_A of an interface that was recently advanced differs from the apparent contact angle θ_R that recently receded. This hysteresis is presumably due to the existence of many metastable positions of the contact line. The difference $\theta_A - \theta_R$ can be as large as 60°. According to Adamson (1990) there are at least three reasons for this hysteresis which are (i) surface roughness, (2) surface heterogeneity, and (3) surface immobility on a macromolecular scale. As pointed out by Morrow (1970, 1976), surface roughness and pore geometry can affect the contact line between the two fluids and the surface, and thus change the apparent contact angle. If

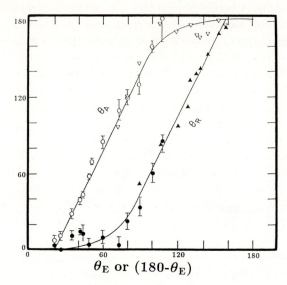

$$\theta_{\mathrm{E}} \text{ or } (180 - \theta_{\mathrm{E}})$$

Figure 12.1: Recently advanced and recently receded contact angles (in degrees) θ_A and θ_R, measured on roughened Teflon, versus intrinsic contact angle θ_E measured on smooth Teflon. Open circles (triangles) are for advancing θ_E (180-θ_E), while solid circles (triangles) are for receding angles (after Morrow, 1976).

the surface is smooth, then θ is fixed. However, in many natural porous media such as oil reservoirs there are sharp edges which give rise to a wide range of contact angles. Figure 12.1 demonstrates this clearly. The contact angles θ_A and θ_R were measured on a roughened teflon surface, while the intrinsic angle θ_E was measured on a smooth teflon surface; the contact angles were changed by varying the salinity. We shall return to this phenomenon later. Moreover, compositional heterogeneity of a surface gives rise to θ_A and θ_R that can change from pore to pore, and one problem with contact angle measurements is that they cannot take such effects into account. In addition, no information can be gained about the absence or presence of various coatings on the pore surface of a reservoir rock.

12.1.2 Amott method

In this method (Amott, 1959) a core is prepared by centrifuging under brine until the residual oil saturation (ROS) is reached. This is that volume fraction of oil in the pore space, residing in isolated finite clusters of pores, that can no longer be displaced by the brine in the centrifuge. Then four steps are taken for measuring the wettability of the core. (i) The core is immersed in oil, and the volume of water displaced by free or spontaneous imbibition of oil into the core is measured, in a period of time that may be as long as one or two weeks. (ii) The core in oil is centrifuged until the irreducible water saturation (IWS), i.e., the water saturation that can no longer be reduced by centrifuging, is reached and the total amount of displaced water is measured. (iii) The core is immersed in brine, and the volume of oil displaced by water is measured. (iv) Finally, the core is centrifuged in brine until ROS is reached, and the total amount of displaced oil is measured. To decide whether the core is water or oil wet, two quantities are calculated. One is $R_o = V_{wi}/V_{wt}$, where V_{wi} is the water volume displaced by free imbibition of oil, and V_{wt} is the total volume of water displaced by free and forced (centrifugal) displacements. The second quantity is $R_w = V_{oi}/V_{ot}$, where the notations have similar meanings. Now, if the porous medium is water wet, then $R_w > 0$ and $R_o = 0$, because in a water-wet porous medium there can be no free imbibition of oil (the oil to be forced into the medium), and vice versa. In a sense, the method measures the average wettability of the system. Sometimes, an index $I_W = R_w - R_o$ is measured which can vary anywhere from $+1$ (completely water-wet porous media) to -1 (completely oil-wet porous media). According to Cuiec (1984), for $0.3 \leq I \leq 1$ the medium must be considered as water-wet, for $-0.3 \leq I \leq 0.3$ it is intermediately-wet, and for $-1 \leq I \leq -0.3$ it is oil-wet. The main problem is that if a porous medium is close to being intermediately-wet, then this method is not very sensitive or accurate, simply because free imbibition of either fluid cannot take place in significant amounts.

12.1.3 U.S. Bureau of Mines method

This method was developed by Donaldson *et al.* (1969) and, similar to the Amott method, also measures the average wettability of the system. It compares the work necessary for one fluid to displace the other. The wetting fluid requires less work to displace the non-wetting fluid from the core than the opposite. It can be shown (see, for example, Morrow, 1970) that the required work is proportional to the area under the capillary pressure curve (in a capillary pressure-water saturation plot; see Chapter 5). If the porous medium is strongly water-wet, most of the water will imbibe freely into the core, and the area under the curve for water will be very small. Thus the capillary pressure curves for the two displacements are measured, and a wettability index $I_W = \log(A_o/A_w)$ is calculated, where A_o and A_w are the areas under the oil- and water-drive curves, respectively. Obviously, if $I_W > 0$, then the porous medium is water-wet, and if $I_W < 0$, it is oil-wet. If, $I_W \simeq 0$, then the porous medium is close to being intermediately-wet. For most natural porous media such as oil reservoirs $-1.5 \leq I_W \leq 1$.

12.2 The effect of surface roughness on contact angles

We already mentioned the experimental work of Morrow (1970, 1976) who investigated the effect of surface roughness on contact angles. There have also been a few recent theoretical studies of the effect of a heterogeneous surface on moving contact angles and contact lines (Joanny and de Gennes, 1984; Pomeau and Vannimenus, 1985; Robbins and Joanny, 1987; Joanny and Robbins, 1990). For example, Joanny and Robbins (1990) studied the motion of a contact line on a surface with periodic heterogeneities. Although such heterogeneities do not usually occur on natural surfaces, their study can provide clues as to how such problems may be studied on a real heterogeneous surface. They considered the case in which the contact line is advanced at a constant force F or a constant velocity v. In the first case, motion starts if $F > F_{C1}$, where F_{C1} is some threshold force which is related to the static θ_A or θ_R defined above. For smooth heterogeneity, they found that, $F - F_{C1} \sim v^2$. If the contact line is moved at a constant velocity, the results are somewhat different: one has two regimes, namely, the weak and strong pinning regimes. By pinning we mean a phenomenon in which the interface is pinned to the solid surface and does not move. In the strong pinning regime, which is more interesting and relevant, there is another threshold F_{C2} which approaches F_{C1} as the surface becomes more heterogeneous. For smooth heterogeneities, they obtained $F - F_{C2} \sim v^{2/3}$. Similar results were obtained by Raphael and de Gennes (1989).

12.3 Dependence of dynamic contact angle and capillary pressure on capillary number

Stokes *et al.* (1990) studied the velocity dependence of $\cos\theta_D - \cos\theta_S$ by a method which, in a sense, belongs to the class of contact angle measurement methods, except that it appears to be more sensitive and its details are also very different from those of the methods discussed above. They measured the capillary pressure P_c across the interface as a function of the contact-line velocity v. P_c can be measured very accurately by superimposing a small-amplitude oscillatory flow on a larger steady-state flow and measuring the response. Stokes *et al.* (1990) generated the oscillatory flow by a plunger that was coupled to the fluid through a latex membrane, and was driven at frequency ω by an audio speaker. The steady-state flow was varied by raising a reservoir of the advancing fluid. Between the reservoir and the sample a narrow-bore and long tube was inserted to guarantee that the flow rate was constant. The interface between a mineral oil and a glycerol-methanol mixture was measured in a 1-mm diameter 30-cm long Pyrex tube. Two Omega pressure transducers were used to measure both the pressure and velocity. The AC (alternating current) output of the transducers was amplified, and the harmonic content measured with several amplifiers. When the glycerol mixture was advanced, $\theta_A = 65°$ and $\theta_R = 45°$ were measured, which indicated that the surface of the tube was somehow disordered. These experiments indicated that for a tube of radius R

$$\cos\theta_D - \cos\theta_A = \frac{R}{2\sigma_{fs}}(P_c - P) = aCa^x \ , \tag{12.2}$$

where P is the total pressure, $a \simeq 3.1 \pm 1$, and $x \simeq 0.4 \pm 0.05$. Equation (2) implies that $P_c \sim v^x$. For *homogeneous* surfaces the velocity dependence of θ_D has been studied by several authors. For example, Cox (1985) found that for θ_D measured a distance r from the contact line one has

$$g_\theta(\theta_D, M) = g_\theta(\theta_0, M) + \left[\ln\left(\frac{r}{\ell_\theta}\right) + B_1\right]Ca \ , \tag{12.3}$$

where g_θ is a simple analytic function, M the viscosity ratio of the two fluids (see Chapter 11), θ_0 the actual contact angle at lengths smaller than ℓ_θ, and B_1 a constant that depends on the model. Molecular dynamics simulations of Koplik *et al.* (1988a) confirmed this equation if one takes $\theta_0 = \theta_S$. Equation (3) implies $x = 1$ for an intermediately-wet fluid on a homogeneous surface, which is different from Eq. (2). This difference may be attributed to the roughness or heterogeneity of the surface, an effect which is apparently strong enough to change the exponent x from unity to about 0.4. This indicates the significance of surface roughness and its effect on contact angle (see also below). On the other hand, Rillaerts and Joos (1980) were able to correlate data from several different systems by plotting $\cos\theta_D - \cos\theta_S$ versus $Ca^{1/2}$, which is close to Eq. (2), and Hoffman (1975) proposed somewhat more complex correlations which can describe many different sets of data for all $0 \leq \theta_D \leq 180°$. For example, at low values of Ca he obtained $\theta_D \sim Ca^{1/3}$ (which is known as Tanner's law). Weitz *et al.* (1987) also measured the velocity dependence of the capillary pressure (and hence θ_D) between two fluids in *a porous medium* and proposed the correlation

$$P_c \simeq \frac{\sigma_{fs}}{r_t}(B_2Ca^x - 1) \ , \tag{12.4}$$

where r_t is a typical radius of pore throats of their porous medium. Their measurements indicated that $x \simeq 0.5 \pm 0.1$, and $B_2 \simeq 300$, which are compatible with the results of Stokes *et al.* (1990). de Gennes (1988) used scaling arguments and proposed that

$$P_c \simeq \frac{\sigma_{fs}}{r_t}Ca^{2/3}T_p^{5/3} \ , \tag{12.5}$$

where T_p is a tortuosity factor. Although this result is almost consistent with Eq. (4), it does not seem to agree with the result of Stokes *et al.* There is, however, one major difference between the experiments of Stokes *et al.* (1990) and Weitz *et al.* (1987): the former experiments were done in a *tube*, whereas the latter ones were performed in *a porous medium*. Whether this can explain the discrepancies between these experiments and de Gennes' prediction, Eq. (5), is not yet clear.

12.4 Fluids on fractal surfaces: Hypodiffusion and hyperdiffusion

In Chapter 5 we discussed fractal properties of pore surfaces. Fractality of pore surfaces can have interesting implications for fluid distributions at low wetting-phase saturations. This was discussed by de Gennes (1985), Melrose (1988), Davis (1989), Davis *et al.* (1990) and Toledo *et al.* (1990). In this section we briefly discuss their results and their implications.

We assume that one of the fluids strongly wets the system, so that even at very low saturations the wetting phase remains hydraulically connected through thin films. The capillary pressure is given by the Young-Laplace equation

$$P_c = 2\mathcal{H}\sigma_{ff} \quad , \tag{12.6}$$

where $\mathcal{H}$ is the mean curvature of the interface between the two phases, and σ_{ff} the interfacial tension between the two fluids. Equation (6) is valid when both phases are present in large amounts. However, if the wetting phase is present only in the form of thin films, then one has to use the augmented Young-Laplace equation

$$P_c = 2\mathcal{H}\sigma_{ff} + \Pi(h) \quad , \tag{12.7}$$

where h is the thickness of the film, and $\Pi(h)$ is called the disjoining pressure. At saturations below the percolation threshold, the wetting phase exists as thin films or *pendular structures* at intergranular contacts, or in nooks and crannies provided by the pore surface features or overhangs. Thus at a given capillary pressure, the liquid volume is proportional to r^{3-D_f}, and since $r \sim 1/P_c$, we must have

$$S_w \sim P_c^{D_f-3} \quad , \tag{12.8}$$

where D_f is the fractal dimension of the pore space (see Chapter 5).

Davis (1989) analyzed Melrose's (1988) data and found that they are well-described by Eq. (8) with $D_f \simeq 2.55$; see Fig. 12.2. This is in the range of fractal dimensions reported by Katz, Thompson, Krohn and others discussed in Chapter 5. If the capillary contribution, $2\mathcal{H}\sigma_{ff}$, is small compared with the disjoining pressure, then $P_c \simeq \Pi(h) \sim h^{-m}$, and therefore

$$S_w \sim P_c^{-1/m} \quad . \tag{12.9}$$

Similar relations can be developed for the hydraulic conductivity K_w of the wetting phase. Thus, if only thin films are present then, since $K_w \sim h^3$, we find

$$K_w \sim P_c^{-3/m} \sim S_w^3 \quad , \tag{12.10}$$

and if only the pendular structures are present, distributed fractally, then

$$K_w \sim S_w^{3/[m(3-D_f)]} \tag{12.11}$$

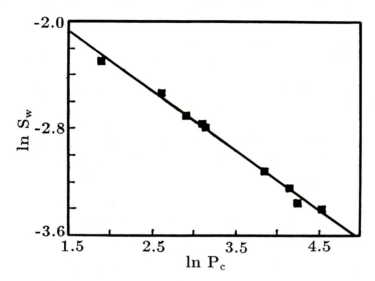

Figure 12.2: Fractal analysis of Melrose's data for the wetting-phase saturation S_w versus the capillary pressure P_c (after Davis, 1989).

If a porous medium with a low wetting-phase saturation is immersed in a reservoir of a wetting fluid, it will spontaneously imbibe the wetting phase. The saturation of the wetting phase will obey a convective-diffusion equation (CDE) in which the dispersion coefficient D_{Lc} is given by

$$D_{Lc} = -\frac{K_w}{\eta_w} \frac{dP_c}{dS_w} \quad , \tag{12.12}$$

where η_w is the viscosity of the wetting phase. D_{Lc} is usually called the capillary dispersion coefficient. Depending on the wetting-phase saturation, one can have three distinct regimes. If $D_{Lc} \to 0$ as $S_w \to 0$, the invading front of the wetting phase will disperse less than that in a diffusive front. This is called *hypodiffusion*. If D_{Lc} approaches a constant as $S_w \to 0$, then we have a diffusive dispersal of the front. Finally, if $D_{Lc} \to \infty$ as $S_w \to 0$, then the front will disperse faster than that in a diffusive front, and is called *hyperdiffusion*. From Eqs. (9)-(12) we obtain

$$D_{Lc} \sim S_w^\alpha \quad , \tag{12.13}$$

$$\alpha = \frac{3 - m(4 - D_f)}{m(3 - D_f)} \quad . \tag{12.14}$$

Therefore, if $m < 3/(4 - D_f)$, we will have capillary hypodiffusion, if $m > 3/(4 - D_f)$, we will be in the regime of capillary hyperdiffusion, while $m = 3/(4 - D_f)$ will give rise to a capillary diffusive front.

Bacri *et al.* (1985) carried out experiments in which water imbibed in a pre-wet sandstone (i.e., a porous medium at its IWS), and observed hyperdiffusion.

Bacri *et al.* (1990b) studied the same problem in a porous medium made of glass beads, which are totally wetted by water, and with polymethylmethacrylate (PMMA) beads which are partially wetted by oil and water. In the porous medium with glass beads, they observed hyperdiffusion again, whereas PMMA beads did not allow this to happen. They also studied the same phenomenon in a porous medium with a mixture of glass and PMMA beads, and observed hyperdiffusion even when the fraction of glass beads was small (but not smaller than the critical volume fraction for percolation). Toledo *et al.* (1990) analyzed the data of Nimmo and Akstin (1988) using the above scaling laws. Nimmo and Akstin (1988) reported measurements of P_c and K_w at low saturations in the presence of air in several compacted samples of Oakley sands, which is a clayey soil. On the other hand, Viani *et al.* (1983) provided data on disjoining pressure of clayey soils, which indicate that they are well-described by $\Pi(h) \sim h^{-1/2}$, implying that $m = 1/2$. Using this value of m, Toledo *et al.* (1990) showed that the above scaling laws hold for 9 different samples with a fractal dimension D_f between 2.35 and 2.67, with an average of about 2.5. Davis *et al.* (1990) also analyzed Ward and Morrow's (1987) data on capillary pressure-saturation curves in the presence of air in several low permeability sandstones. Two distinct regimes were found. One was in the lowest saturation region where $D_f \simeq 2$. The other was in a higher saturation region for which $2.6 \leq D_f \leq 2.9$, which is consistent with the range of fractal dimensions discussed in Chapter 5. Finally, Novy *et al.* (1989) developed a network model of two-phase flow in porous media to test the above scaling laws, and found qualitative agreement between their simulations and the scaling laws.

The wettability of a medium has a strong effect on its transport as well as thermodynamic properties such as the capillary pressure curves. There is a strong correlation between the shape of the capillary pressure curves of a medium and its transport properties. Thus we first discuss the effect of wettability on capillary pressure curves.

12.5 Effect of wettability on capillary pressure

We already discussed in Chapter 5 capillary pressure curves and how they are used for extracting information about the pore size distribution of the porous medium. In this section, we discuss the effect of wettability on capillary pressure curves. Let us first introduce the terminology that is frequently used in the oil industry. In *drainage* a non-wetting phase displaces a wetting phase from a porous medium, while during *imbibition* a wetting phase displaces a non-wetting phase.

With sintered porous teflon and clean fluid pairs, Morrow (1970, 1976), Mc-Caffery and Benion (1974), and Morrow and McCaffery (1978) studied two-phase relative permeabilities (discussed below) and capillary pressure curves in porous media that are uniformly wettable. They took, as an independent measure of wettability, the intrinsic contact angle θ_E made by the fluid pairs on smooth teflon surfaces, and studied two processes. The first is *primary displacement*, i.e., the reduction of the saturation of a reference phase from 100% to the residual saturation (RS) by injection of a non-reference phase. The second is *secondary displacement* that follows primary displacement, i.e., reduction of the non-reference phase saturation to the RS by injection of the reference phase. On the basis of the capillary pressure (and relative permeability) behavior in the sequence of primary and secondary displacements, they identified three regimes of wettability: (i) *wetted*, in which primary displacement is drainage and secondary displacement is imbibition; (ii) *intermediately wetted*, in which primary and secondary displacements are *both* drainage, and (iii) *non-wetted*, in which primary displacement is imbibition and secondary displacement is drainage. The second case is very interesting: because primary displacement is drainage, one intuitively expects secondary displacement to be imbibition. However, in the intermediate wettability regime the operative contact angle appears to give the more strongly wetting characteristics to the phase being displaced, whether it be the reference phase or the non-reference phase.

Figure 12.3(b) shows the capillary pressure curves for a typical wetted regime measured by Killins *et al.* (1953). The measurements were done on a water-wet Berea sandstone. The primary process, denoted by 1 on the curve, is drainage in which oil displaced water and was terminated at A, where IWS was reached. This is followed by process 2, a spontaneous imbibition of water into the core, up to point B at which $P_c = 0$. Beyond B, the water has to be forced into the medium (curve 3), characterized by a negative P_c, until point C is reached at which the ROS is reached. Note the two typical knees, one at the beginning of drainage, and the other at B. This figure should be compared with Fig. 12.3(a) which is typical of capillary pressure curves in non-wetted porous media. The curves were measured in an oil-wet Berea sandstone treated with Drifilm (to render it strongly oil-wet). Note that in both processes 1 (spontaneous imbibition of oil) and 2 (drainage of oil by water) the capillary pressure curve takes on negative values. Moreover, the imbibition curve rises steeply, and the drainage curve proceeds slowly, except near the original starting point A. Finally, Fig. 12.3(c) shows schematic capillary pressure curves for typical intermediately-wet systems.

In order to compare capillary pressure curves measured on different cores from the same porous medium, and to take into account the effect of permeability and porosity of each core, the curves are usually replotted in terms of the Leverett J-function, Eq. (5.10). As long as the wettability of all cores is the same, the $\cos\theta$ term of Eq. (5.10) is a simple numerical factor, but if different fluids are used with cores of the same reservoir, then this term becomes important. Figure 12.4 shows the effect of contact angle on capillary pressure, as measured by Morrow (1970). As the contact angle increases, the return curve of the capillary pressure (to the left of the dashed curve) becomes steeper, which is in agreement

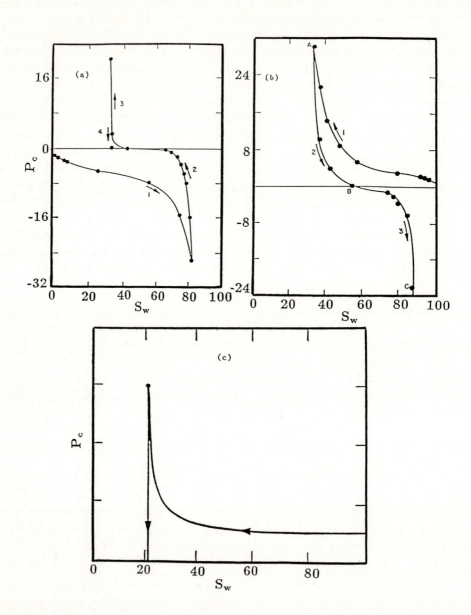

Figure 12.3: (a) Typical capillary pressure P_c (in cm of Hg) versus saturation S_w (in percent) curves for a non- wetted regime, (b) for a wetted porous medium (after Killins *et al.*, 1953), and (c) for an intermediately-wet system. Curves 1 represent drainage, 2 represent spontaneous imbibition, 3 denote forced imbibition, and curve 4 is secondary drainage.

with Fig. 12.3(c), since increasing θ implies the tendency of the system towards intermediate wettability. In fact, in the last curve on the left (shown by ■) the advancing contact angle θ_A is about 77° which, as discussed above, is well within the range of intermediate wettability.

There is yet another wettability regime that is of great practical importance. This is the so-called *mixed wettability* regime (Owens and Archer, 1971; Treiber *et al.*, 1972; Salathiel, 1973; Craig, 1977), in which some pores are wetted by one fluid, while others are wetted by the other fluid. There is evidence that over the course of geological times oil displaces brine from a portion of the pore space. Depending on the nature and composition of the oil, those portions of the pore wall that are separated from the crude oil by only a thin film or a tiny pendular structure of brine may become oil-wet owing to diffusion and adsorption or stronger chemical interactions with constituents of the oil phase. One widely cited interaction is the deposition by the crude of polar organic surfactants upon the surface of the rock (Benner *et al.*, 1943; Salathiel, 1973; Melrose, 1982). In any event, it appears that migration, accumulation and deposition processes can generate a distribution of surface wetting preferences, the small pores that have not been invaded by oil remaining water-wet, while the larger pores that have been occupied by oil becoming more or less oil-wet. Figure 12.5 shows capillary pressure curves for a system in which a fraction Y of the pores was invaded by oil and became oil-wet. $Y = 0$ corresponds to a totally water-wet system, while $Y = 1$ is representative of a totally oil-wet porous medium. These various wettability regimes also affect transport properties of porous media, which are discussed below. More complete discussions of the effect of wettability on capillary pressure curves are given by Melrose (1965, 1968), Heiba (1985), and Anderson (1987a).

12.6 Immiscible displacement processes

We now discuss displacement of one fluid by another immiscible fluid. This process is controlled and affected by a variety of factors that were mentioned at the beginning of this Chapter. We already discussed the effect of wettability and contact angles on capillary pressure, and shall discuss their effect on transport properties of the porous medium in two-phase flow later. Among the remaining factors, the capillary number Ca and the mobility ratio M have the greatest importance. Depending on how the displacement process proceeds, many different regimes may arise. A very careful discussion and classification of imbibition processes and how to distinguish between them was given by Payatakes and Dias (1984). We give here a summary of their classification and discussion and expand on them if appropriate.

(i) *Spontaneous imbibition*, which we already mentioned in the context of capillary pressure curves discussed in Chapter 5.

(ii) *Constant influx, constant Ca imbibition*, which occurs if a pressure drop

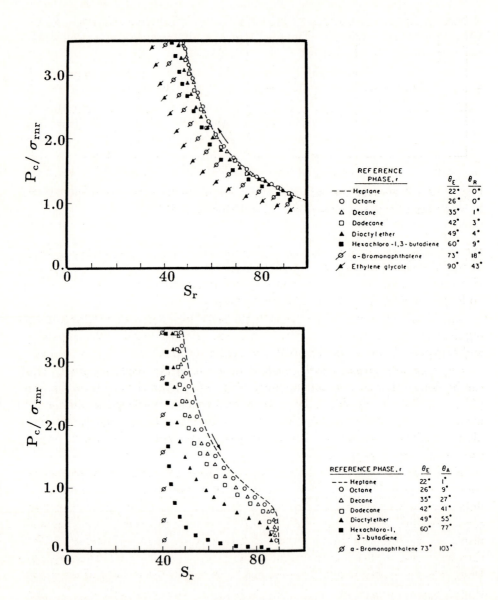

Figure 12.4: Effect of contact angle on capillary pressure P_c, divided by the surface tension between the reference and non-reference fluids, as a function of the saturation S_r of the reference fluid (in percent). P_c/σ_{rnr} is in cm^{-1} (after Morrow, 1970).

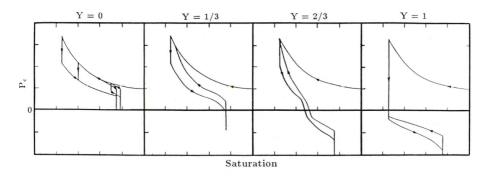

Figure 12.5: Capillary pressure curves for porous media with mixed wettability. Y is the fraction of the oil-wet pores (after Heiba, 1985).

ΔP is applied to the medium, and if it is adjusted as the invading fluid expells more fluids from the medium. If $M \leq 1$, then we must maintain $\Delta P \geq 0$, and vice versa. If for $M \leq 1$ the applied pressure dominates capillary forces, we will no longer be dealing with an imbibition process.

(iii) *Quasi-static imbibition*, which happens when the flow rate of the displacing fluid is vanishingly small. In this case, the interface between the displacing and displaced fluid advances in only *one* pore at a time.

(iv) *Dynamic invasion with constant flow rate of the displacing fluid*. This can be done either at favorable mobility ratio ($M \leq 1$) or at an unfavorable one ($M > 1$). To achieve this, a large pressure drop ΔP is applied to the porous medium, which can be so large that it would dominate capillary forces. The interface advances in *many* pores at any given time. For $M > 1$ viscous fingers can develop (see Chapter 11).

Note that the reverse of at least some of these imbibition processes can be considered as drainage processes. At the end of these processes, the displaced fluid exists only in the form of isolated blobs or clusters of finite sizes that cannot be displaced by any of the above processes. In order to mobilize and displace such blobs, the capillary number has to be significantly increased, which then gives rise to three other displacement processes which are (Payatakes and Dias, 1984) quasi-static and dynamic displacements of blobs, both of which are time-dependent phenomena, and steady-state dynamic displacement. The last process can be carried out if the displacing and displaced fluids are simultaneously injected into the porous medium. After some time, a dynamic equilibrium is reached and the macroscopic flow rate becomes constant. This problem was investiaged by Siddiqui (1989) using computer simulations, and has been discussed, from an experimental point of view, by Craig (1971). Payatakes (1982) has given an excellent discussion of blob mobilization and displacement in porous media. Therefore we do not discuss the last three processes in great detail.

Before describing various theoretical and experimental studies of immiscible displacement in porous media, let us remind the reader that, unlike miscible displacement processes discussed in Chapter 11, immiscible displacements in Hele-Shaw cells have no direct relevance to those in a porous medium. This can be seen easily by realizing that the interface in a Hele-Shaw cell is sharp and continuous, whereas it is diffused in a porous medium due to its disordered structure. The interested reader can consult Bensimon *et al.* (1986), Saffman (1986), and Homsy (1987) regarding two-phase flows in a Hele-Shaw cell. Let us now discuss various imbibition processes and their properties.

12.6.1 Spontaneous imbibition

The driving force for spontaneous imbibition is capillary suction, and because of this the smallest pore bodies which are next to the interface are always invaded. At any given time step, many pore-level interfaces advance in the porous medium. Usually, the displacement takes place at small but finite capillary numbers. Moreover, the value of Ca, which depends on the extent of the process, does not remain a constant, but varies over a range of values. This was demonstrated nicely in the experiments of Legait and Jacquin (1982) who studied spontaneous imbibition of water in sandstones containing oil at $M = 225$. They reported that the rate of oil displacement increased strongly with time. When the same experiment was carried out at $M = 1$, the same phenomenon was observed, albeit in a much weaker manner.

When relatively large values of Ca are created, a transition zone develops in which there is a high saturation gradient. As the interface advances in the medium, two separate regions develop. One is in front of the transition zone in which the saturation of the displaced fluid is high. The other moves behind the transition zone in which the saturation of the displacing fluid is high. This region expands as the interface advances. The transition zone remains essentially the same throughout the displacement, except when the interface nears the end of the porous medium. For this reason, this region is called the *stabilized zone.* Although Bail (1956) argued that under conditions used in regular imbibition in oil reservoirs, the length of this zone is not very large, its dynamics are interesting because they affect the efficiency of the displacement, and it is also in this region where oil blobs are formed by disconnection of the displaced fluid by the displacing fluid.

12.6.2 Quasi-static imbibition

One main difference between this process and spontaneous imbibition is that in this process at any given step only one pore is invaded by the displacing fluid. This can be done by adjusting the backpressure so that the *narrowest* pore throat is invaded, while the interface at other larger throats remains essentially motionless. Since even the largest pore throats are smaller than the pore body to which they are connected, once a pore body is invaded, all of the throats that are connected to it are also invaded. As soon as the interface enters such throats, the smallest pore body that is connected to them is invaded, and so on. Thus at any given step of the displacement the smallest pore body that is accessible from the surface through a continuous path of the displacing fluid is invaded.

When the displacing fluid forms a sample-spanning cluster of invaded pore bodies (and the associated pore throats), a *breakthrough* is achieved. Just before the breakthrough the displaced fluid is mostly connected. However, as the displacement proceeds small blobs of the displaced fluid are formed which get trapped. At the end of this process, one may end up with a large number of isolated blobs whose volume fraction is significant. The value of this volume fraction depends on the morphology of the pore space. In unconsolidated media it varies between 0.14 and 0.2 (Chatzis *et al.*, 1983), whereas in consolidated porous media it is anywhere between 0.4 and 0.8 (Wardlaw and Cassan, 1978).

Experimental data of Raimondi and Torcaso (1964), Egbogah and Dawe (1980), and Chatzis *et al.* (1983) indicate that the size distribution of the blobs, when expressed in terms of the number of pore bodies they occupy (remember that most of the porosity and volume of a porous medium are provided by its pore bodies), obey the following power law

$$n_s \sim s^{-\tau} \; , \tag{12.15}$$

where n_s is the number of blobs of size s. Equation (15) reminds us immediately of Eq. (3.29), the scaling law for the number of finite percolation clusters of s sites. Indeed, as we discuss below, an appropriate percolation model can be devised to model such a quasi-static displacement. Egbogah and Dawe (1980) found that the size of the blobs varied between 1 and 10 grain volumes, but most of them were around $s = 1$.

12.6.3 Imbibition at constant flow rates

This is very similar to spontaneous imbibition except that in this case we need to adjust a backpressure in order to keep the flow rate constant. There has been some controversy regarding the role of M in the displacement processes. There are some older papers (for example, Geffen *et al.*, 1951; Donaldson *et al.*, 1966) in which it was claimed that for fluids with identical wettability characteristics M does not have any significant effect. More recent works do not agree with this. Le Febvre du Prey (1973) investigated systematically the effect of M on relative permeabilities (i.e., the permeability to a fluid phase divided by the permeability of the medium; see below) to two-phase flows, by studying displacements in sintered porous media in which Ca varied between 10^{-7} and

5×10^{-3}. He found that the higher the viscosity of one of the fluids, the lower the relative permeability of the other fluid, and this effect was found to be even more important than the wettability effect. This effect is presumably due to the fact that the high viscosity of the fluid gives rise to a film of the fluid residing on the pore walls, which denies pore volume to the other fluid and decreases its relative permeability.

Abrams (1975) found that the ROS in short porous media correlates well with the group $M^{-0.4}Ca$, where $10^{-7} \leq Ca \leq 10^{-2}$. Egbogah and Dawe (1980) found that for $M \leq 1$ the blob size distribution became much broader, took on bimodal and even trimodal shapes, and the average blob size increased dramatically.

12.6.4 Dynamic invasion at constant flow rates

The driving force for this process is an applied pressure drop, since the role of capillary forces is of secondary importance. If $M > 1$, we shall have an unstable displacement which is discussed later. Therefore, for now, we consider only the $M < 1$ case. There is a small transition zone in this process in which the saturations of both phases change with time. If the capillary pressure is negligible compared with the applied pressure, we shall have several advancing interfaces in as many pores at any given stage of the displacement. Because the driving force is the applied pressure, the microscopic interfaces choose the *largest* accessible pore throats (to minimize the resistance). Thus the structure of the sample-spanning cluster of the displacing fluid resembles that in a drainage process. However, this does not necessarily mean that smaller throats will not be selected: local pressures are also important and can cause the invasion of smaller throats by the advancing fluid. Since in the transition zone the saturation of the phases changes with time, the value of Ca cannot remain constant, even though the flow rate can be kept constant.

As in the previous cases, the advancing fluid creates isolated blobs of the displaced fluid. Whether these blobs become stranded or not depends on many factors. Ng and Payatakes (1980) and Payatakes *et al.* (1980) argued that the stranding of the blobs depends on Ca, the length of the blob in the direction of the macroscopic flow, and the sizes of the pore bodies and pore throats in which the blobs reside. If a very large blob is created, initially it is mobile, but later on it breaks into several smaller blobs, and the breaking process continues until they are small enough to be stranded.

12.6.5 Trapping of blobs

If the displaced fluid is incompressible, then at the end of both imbibition and dynamic invasion one obtains many isolated blobs or clusters of the displaced

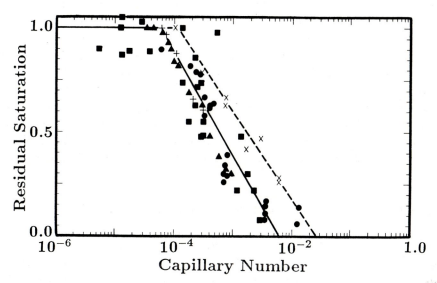

Figure 12.6: Capillary desaturation curves, compiled with the available data for wetting fluids (dashed lines) and non-wetting fluids (solid line). The porous medium is Berea core.

fluid, whose displacement is the main goal of oil recovery processes. The most common way of correlating experimental data for the trapped oil is in terms of a relationship between the residual wetting phase saturation S_{rw} or that of the non-wetting phase saturation S_{rnw} and a local capillary number Ca_l. This relationship is usually called the capillary desaturation curve (CDC). If Ca_l is small, then both S_{rw} and S_{rnw} are roughly constant. As Ca_l increases, one reaches a typical knee in the curve, beyond which the residual saturations decrease. Most waterflooding processes, those in which water is injected into an oil reservoir to displace the remaining oil, are well onto the plateau of a CDC where, as a general rule, S_{rw} is smaller than S_{rnw}. In most cases the two CDCs are normalized by their corresponding plateau values. Figure 12.6 shows a CDC constructed on the basis of experimental data obtained by several authors. Observe that the CDC for the wetting phase is to the right of the CDC for the non-wetting phase, which is not only typical, but also indicates the significance of wetting properties of porous media. Moreover, as can be seen, at some value of Ca_l the two curves intersect, corresponding to some intermediate wetting condition. It also has been observed that the pore size distribution of the pore space can have a strong influence on the CDC. For a pore space with a broad pore size distribution, e.g., a carbonate rock (see Chapters 4 and 5), the horizontal part of the CDC is very small or non-existent, whereas for a typical sandstone one observes a significant horizontal part, such as that shown in Fig. 12.6 .

12.6.6 Displacement of blobs: Choke-off and pinch-off

Mobilization and displacement of trapped oil blobs require relatively high capillary numbers, say $Ca > 10^{-4}$. However, the value of Ca depends on several factors, including the shape and size of the blobs, the morphology of the porous medium, especially around the regions where the blobs reside, and the contact angle. For $Ca > Ca_c$, where Ca_c is a critical value of Ca, the blobs start to move. If $Ca - Ca_c$ is small, then we obtain what is called quasi-static displacement of the blobs. During this process one blob moves downstream, while one or two may move upstream. A blob may get re-entrapped if it arrives at a pore body where all throats that are connected to it are too small for the blob's movement, in which case one needs an even higher Ca to move such blobs.

A moving blob is almost certain to break into smaller blobs by one of the following mechanisms. In *pinch-off* the velocity of the moving blob becomes small for a long enough time that the blob collapses into several smaller blobs. In *dynamic breakup* (Payatakes, 1982) a blob advances in two or more pore throats simultaneously, which can easily happen if the coordination number of the pore space is large enough. For this to happen, the value of Ca has to be large enough that even if two pore throats connected to the same pore body have different effective sizes, there can still be enough force to move the blob into both pores. If the blob completely evacuates the pore body, it breaks into two or more smaller blobs, depending on how many pore throats it enters.

There is yet another mechanism for blob breakup which is usually called *choke-off* or *snap-off*, and was first discussed by Pickell *et al.* (1966). Roof (1970), Mohanty *et al.* (1980, 1987), and Arriola *et al.* (1983) have discussed this phenomenon in detail. Choke-off signals the breakup of a small drop from the leading tip of a non-wetting thread that tries to pass through a narrow constriction. Roof (1970) conducted several experiments and demonstrated this phenomenon nicely. He considered choke-off in a toroidal pore throat, assumed that there is a thin lubrication film on the surface, and showed that choke-off occurs if the curvature of the interface at the throat is larger than the curvature of the tip of the thread. If the throat is non-axisymmetric, then Roof showed that the same phenomenon happens except that it takes place faster than the toroidal case, because the film has easier access to the point of rupture as a result of the fact that some parts of the cross section are not filled by the non-wetting film. He also showed that before choke-off occurs, the tip of the non-wetting fluid thread has to travel a distance of several pore-throat diameters beyond the original throat. Therefore choke-off is important if the ratio of the pore body and pore throat diameters is large, and this was demonstrated by Wardlaw (1982) and Li and Wardlaw (1986a,b). From their detailed experimental observations,

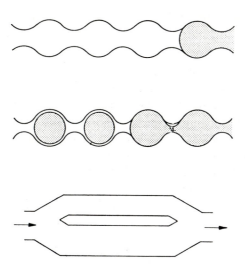

Figure 12.7: Snap-off geometries (top two) and the pore doublet model (bottom). Dotted areas represent the oil phase, while the white area is the water phase (after Chatzis *et al.*, 1983).

Chatzis *et al.* (1983) determined that roughly 80% of trapping of the non-wetting phase occurs by snap-off geometries, while the remaning 20% happens in pore doublets, or a combination of both pore geometries; see Fig. 12.7. Hammond (1983) made a careful study of this phenomenon by solving the problem of slow adjustment of lubricated threads and drops in axisymmetric, straight capillaries, and constricted tubes. He showed that for choke-off it is not sufficient for the lubrication film to be unstable. One must also have a sufficient amount of wetting liquid in the film near the incipient neck to form a bridge across the tube because, while the neck is being formed, the thread on either side bulges and isolates the local wetting fluid from the rest. He estimated that for rupture to occur, the thickness ℓ_w of the wetting fluid, normalized by the tube radius, must be larger than $\pi/6$. He also estimated that the time scale for the growth of a perturbation large enough to cause the rupture of a thread in a constriction is at least of the order $\ell_w^{-2} D\eta_1/\sigma_{fs}$, where D is the diameter of the tube.

A simple criterion for snap-off can be derived using these observations. We assume that the non-wetting phase flows through a single path of variable cross-sectional area. The sides of the flow path are lubricated by the wetting phase, so that it is possible to define a local capillary pressure everywhere along the path, which may however vary from point to point. For example, if the path is narrow, then the local capillary pressure is large. Depending on the potential gradient and the geometry of the path, the capillary pressure across the path can be larger than the potential gradient across the same segment of the path. If so, the external force would not be large enough to force the non-wetting phase to enter the next pore segment, in which case the non-wetting phase snaps off into globules that reside in the pore bodies along the path. If this description is reasonable, then the condition for mobilizing the trapped globules is given by

$$\Delta\Phi + g\ell_g\Delta\rho\sin\alpha_g \geq \Delta P_c \ , \tag{12.16}$$

where $\Delta\Phi$ and ΔP_c are the wetting phase potential and capillary pressure changes across the golbule, ℓ_g is the size of the globule, $\Delta\rho$ is the difference between the densities of the two phases, and α_g is the angle between the globule's major axis and the horizontal axis. Based on inequality (16) and the work of Melrose and Brandner (1974), Stegemeier (1974, 1976) derived the following condition for mobilization of a trapped globule

$$Ca_l = \frac{K|\nabla P_w|}{\sigma_{wnw}} \geq \frac{\phi}{c}\left[J\cos\theta_A - \frac{\cos\theta_R}{\sqrt{2T_p}}\right]^2 \ , \tag{12.17}$$

where Ca_l is a local capillary number, ∇P_w is the pressure gradient in the wetting phase, σ_{wnw} is the interfacial tension between the wetting and non-wetting phases, ϕ is the porosity of the medium, J is the Leverett J-function [see Eq. (5.10)], θ_A and θ_R are the advancing and receding contact angles, T_p is the tortuosity factor of the porous medium, and c is a constant whose value is about 20. Observe that condition (17) expresses the condition for mobilization of non-wetting phases globule in terms of measurable, or easily estimated, quantities. Moreover, it allows the effect of contact angles to be included in the condition, thus making it more realistic.

Finally, just as blobs breakup into several smaller blobs, they can also coalesce. This happens when two interfaces that belong to two different blobs pass through the same throat and are pressed against each other for a long enough time. Constantinides and Payatakes (1991) used a network model to investigate the likelihood of collision and coalescence of blobs in a porous medium. Their results indicated that the wetting characteristics are more important than Ca, M, or θ, and that the probability of coalescence, given a collision, decreases as θ increases. They found this probability to vary between 0.03 to 0.15. Thus the breakup and coalescence phenomena give rise to a series of complex dynamical processes, in which the displaced fluid can break, but form a large cluster again at a later time. Lenormand and Zarcone (1985b) presented nice experimental realizations of these phenomena in a micromodel. These are fascinating phenomena for a review of which the reader can consult Payatakes (1982).

12.7 Models of two-phase flow and displacement

Similar to all the phenomena discussed so far, there are two classes of models of two-phase flows and displacements in porous media. One of them relies on continuum equations, averaged over suitably defined representative volume. This is the classical engineering approach whose major elements and achievements were already discussed in the previous Chapters. The literature on this class of model is enormous, and except for its most important aspects, there is no possibility of discussing it in this Chapter. We refer the reader to Collins (1961), Bear (1972), Scheidegger (1974), and Marle (1981) for comprehensive discussions of this class models. The models in the second class are discrete or statistical. Over the past fifteen years the literature on this class of models has also grown dramatically, and it would be difficult to discuss everything that has been done.

12.7.1 Continuum equations and relative permeabilities

Whitaker (1986b) studied flow of two immiscible fluids in porous media. Starting with the continuity and Stokes' equations for each phase, and using the appropriate boundary and initial conditions, he derived the following equations for the average flow velocity and saturations (volume fractions) of the phases β and γ

$$\mathbf{v}_\beta = -\frac{\mathbf{K}_\beta}{\eta_\beta} \cdot \left(\boldsymbol{\nabla} \langle P_\beta \rangle^\beta - \rho_\beta \mathbf{g} \right) + \mathbf{K}_{\beta\gamma} \cdot \mathbf{v}_\gamma \quad , \tag{12.18}$$

$$\phi \frac{\partial S_\beta}{\partial t} + \boldsymbol{\nabla} \cdot \mathbf{v}_\beta = 0 \quad , \tag{12.19}$$

$$\mathbf{v}_\gamma = -\frac{\mathbf{K}_\gamma}{\eta_\gamma} \cdot \left(\boldsymbol{\nabla} \langle P_\gamma \rangle^\gamma - \rho_\gamma \mathbf{g} \right) + \mathbf{K}_{\gamma\beta} \cdot \mathbf{v}_\beta \quad , \tag{12.20}$$

$$\phi \frac{\partial S_\gamma}{\partial t} + \boldsymbol{\nabla} \cdot \mathbf{v}_\gamma = 0 \quad , \tag{12.21}$$

where all notations are as before. Note that Eqs. (18) and (20) contain two terms, the first of which is the usual Darcy's law, written for each phase, while the second one is a cross term that couples the two phases. Equations (18) and (20) were first proposed by Raats and Klute (1968) and de Gennes (1983b) based on physical arguments, although somewhat similar equations had been conjectured by Rose (1972). In analogy with the thermodynamics of irreversible processes, one may assume that $\mathbf{K}_{\beta\gamma} = \mathbf{K}_{\gamma\beta}$. These equations are valid if the capillary number $Ca \ll 1$, and if moving contact lines discussed above do not have a significant effect. Generally speaking, the coupling terms in Eqs. (18) and (20) are not significant unless $\eta_\beta \simeq \eta_\gamma$ (in which case a thin film of one phase coats the walls of a pore whose bulk volume is filled with the other phase). Some of the most convincing evidence for insignificance of the cross terms of Eqs. (18) and (20) was provided by Yadav *et al.* (1987). They experimented with a wetting and a non-wetting fluid and showed that the permeability of both phases in drainage, measured when the opposite phase was solidified *in situ*, was the same as that typically measured for Berea sandstones. However, Kalaydjian and Legait (1987), and Goode and Ramakrishnan (1993) showed that such coupling terms might be important in certain cases even if η_β and η_γ are not close to each other.

The *phase* permeabilities $\mathbf{K}_\beta$ and $\mathbf{K}_\gamma$ are supposed to be known, but in practice one calculates them by the use of the relations

$$\mathbf{K}_\beta = \mathbf{K} k_{r\beta} \quad , \tag{12.22}$$

with a similar equation for the γ-phase, where $k_{r\beta}$ is called the *relative permeability* (RP) to the β-phase, a concept that has been used for many decades in the petroleum industry. A major problem in multiphase flows in porous media is the prediction of the RPs. Unlike the absolute permeability K, $k_{r\beta}$ has been found to depend on many parameters, including saturation and saturation histories of the fluids (Johnson *et al.*, 1959; Naar *et al.*, 1962), pore space morphology (Morgan and Gordon, 1970), the wetting characteristics of the fluids (Owens and Archer, 1971; McCaffery and Benion, 1974), sometimes on the viscosity ratio (Odeh, 1959; Le Febvre du Prey, 1973), and the capillary number Ca (Leverett, 1939; Taber, 1969). Moreover, forty years ago it was recognized (Richardson *et al.*, 1952) that the RP to a phase typically becomes small or altogether negligible when its saturation is less than a critical value which is distinctly larger than zero. This is of course the signature of a percolation problem. Thus various percolation models have been developed for modelling two-phase flows in porous media which are discussed below. The reason $k_{r\beta}$ apparently depends on the saturation history of a phase, i.e., the way that saturation has been reached, is that there are presumably multiple shapes that satisfy the Stokes' equation, which is made non-linear by the free interfaces. This dependence naturally gives rise to *hysteresis* in RPs which is discussed shortly.

12.7.2 Measurement of relative permeability

There are many methods of measuring RPs (see Anderson, 1987b). One method that is routinely used is as follows. The porous medium is initially filled with the wetting phase, and both wetting and non-wetting fluids are injected into the medium at a constant flow rate. When the steady-state has been reached, the pressure drop across the medium is recorded. From the knowledge of the flow rate and pressure drop the phase permeabilities K_β and K_γ are calculated, using Eqs. (18) and (20) (and ignoring the cross terms). The phase saturations can also be determined by several methods, the simplest of which is by weighing the sample before and after the injection. Since the absolute permeability of the medium is already known (see Chapter 8), the RP to the wetting phase at this particular value of saturation can be calculated. In the next stage, the injected amount of non-wetting fluid is increased, and the procedure is repeated. In this

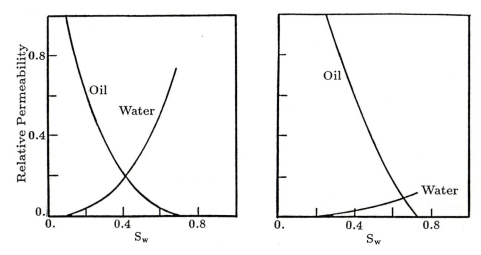

Figure 12.8: Typical relative permeability curves for an oil-wet (left) and a water-wet (right) porous medium (after Craig, 1971).

way, the RP to the wetting phase is obtained. By reversing the procedure one can obtain the RP to the non-wetting phase, and so on. The reader should consult Anderson (1987b) and Heaviside (1991) for more details on measurement of RPs.

12.7.3 The effect of wettability on relative permeability

We have already discussed the effect of wettability on capillary pressure. It is clear that wettability and contact angles should also affect the RPs, and this is indeed the case. Figure 12.8 shows typical oil-water RP curves for a strongly water-wet system, and for a strongly oil-wet porous medium. While the difference between the two RP curves for the oil phase is not very large, there is a dramatic difference between the RPs to the water phase. Note also the existence of a finite saturation (i.e., a percolation threshold) at which the RP vanishes. Normally, if a system is strongly water-wet, there is little or no hysteresis in the RPs to the water phase. This can be seen clearly in the experimental data of Morrow and McCaffery (1978). They measured the RPs in a teflon core with nitrogen as the non-wetting fluid and heptane ($\theta = 20°$) and dodecame ($\theta = 42°$) as the wetting phase. They found that there is no hysteresis in the wetting-phase RP. Figure 12.9 shows the data of McCaffery and Benion (1974) and Morrow and McCaffery (1978) for the three different wettability regimes discussed above, namely, wetted, intermediately-wetted, and non-wetted cases. In this figure the reference phase (with saturation S_r) refers to the displaced phase, while the non-reference phase refers to the displacing phase. The differences between the three cases are rather large. For example, in the non-wetting case, the contact angle

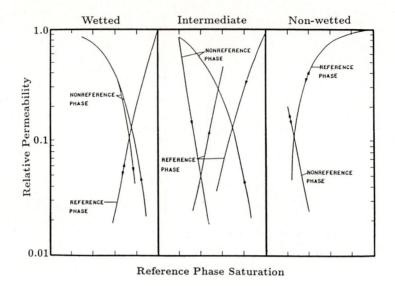

Figure 12.9: Relative permeability versus the reference phase saturation S_r for various wettability regimes. The contact angles for the wetted and non-wetted cases were 49 and 130 degrees, respectively (after McCaffery and Benion, 1974, and Morrow and McCaffery, 1978).

for the non-reference phase was more than 130°, whereas in the wetted case it was at most 49°. It is obvious that wettability strongly affects the trends in RPs, and one major theoretical challenge is to predict such trends.

12.7.4 Fractional flows and the Buckley-Leverett equation

We consider a one-dimensional displacement and ignore the cross terms in Eqs. (18) and (20). We also assume that there are only two phases in the system. If the displacement is along the z direction, then Eqs. (18)-(21) become

$$v_w = -\frac{K k_{rw}}{\eta_w}\left(\frac{\partial P_w}{\partial z} - \rho_w g\right) \ , \tag{12.23}$$

$$v_{nw} = -\frac{K k_{rnw}}{\eta_{nw}}\left(\frac{\partial P_{nw}}{\partial z} - \rho_{nw}\right) \ , \tag{12.24}$$

$$\phi\frac{\partial S_w}{\partial t} + \frac{\partial v_w}{\partial z} = 0 \ , \tag{12.25}$$

$$\phi\frac{\partial S_{nw}}{\partial t} + \frac{\partial v_{nw}}{\partial z} = 0 \ . \tag{12.26}$$

Moreover, we also have $S_w + S_{nw} = 1$ and $P_{nw} - P_w = P_c$. We now define new variables f_w and f_{nw}, called *fractional flows*, by

$$f_w = \frac{v_w}{v} \ , \tag{12.27}$$

$$f_{nw} = \frac{v_{nw}}{v} = 1 - f_w \ , \tag{12.28}$$

where $v = v_w + v_{nw}$. It is not difficult to show that

$$f_w = \frac{\dfrac{\eta_{nw}}{k_{rnw}} + \dfrac{K(\rho_w - \rho_{nw})g}{v}}{\dfrac{\eta_w}{k_{rw}} + \dfrac{\eta_{nw}}{k_{rnw}}} + \frac{K}{v} \frac{1}{\dfrac{\eta_w}{k_{rw}} + \dfrac{\eta_{nw}}{k_{rnw}}} \frac{\partial P_c}{\partial z} \ . \tag{12.29}$$

If the capillary pressure gradient is negligible, $\partial P_c / \partial z \simeq 0$, which is the case if the flow rate or the fluid velocity is large, and if the density difference $\rho_w - \rho_{nw}$ is also small (or the gravity is not important), we obtain

$$f_w = \left(1 + \frac{K_{nw}}{K_w} \frac{\eta_w}{\eta_{nw}}\right)^{-1} = \left(1 + \frac{k_{rnw}}{k_{rw}} \frac{\eta_w}{\eta_{nw}}\right)^{-1} \ . \tag{12.30}$$

If we now use the obvious relation, $\partial f_w / \partial z = (df_w/dS_w)(\partial S_w/\partial z)$, then the continuity equation for the wetting phase, Eq. (25), becomes

$$\left(\frac{v}{\phi} \frac{df_w}{dS_w}\right) \frac{\partial S_w}{\partial z} = -\frac{\partial S_w}{\partial t} \ , \tag{12.31}$$

with a similar equation for the non-wetting phase. This is a non-linear equation (since df_w/dS_w depends on S_w), and therefore in general does not have an explicit analytical solution. On the other hand, we can write

$$\frac{dS_w}{dt} = \frac{\partial S_w}{\partial z} \frac{dz}{dt} + \frac{\partial S_w}{\partial t} \ , \tag{12.32}$$

so that if $z = z(t)$ is selected to coincide with a surface of fixed S_w, we have $dS_w/dt = 0$, and

$$\left(\frac{dz}{dt}\right)_{S_w} = -\frac{\partial S_w}{\partial t} \left(\frac{\partial S_w}{\partial z}\right)^{-1} \ , \tag{12.33}$$

which, together with Eq. (31) yield

$$\left(\frac{dz}{dt}\right)_{S_w} = \frac{v}{\phi} \frac{df_w}{dS_w} \ . \tag{12.34}$$

This is the celebrated Buckley-Leverett equation (Buckley and Leverett, 1942), widely used in the petroleum industry. We should emphasize that the Buckley-Leverett equation is valid when the gradient of the capillary pressure and the

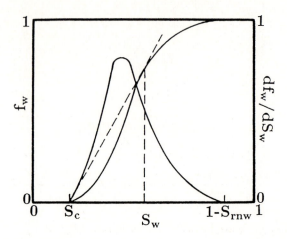

Figure 12.10: Wetting phase fractional flow f_w and its derivative versus the wetting phase saturation S_w, constructed based on the relative permeabilities shown on the left side of Fig. 12.8.

density difference between the two fluids (or gravity) are negligible. Moreover, we should keep in mind that, because f_w depends on the phase permeabilities or relative permeabilities, it also depends on the phase saturation S_w and hence $S_{nw} = 1 - S_w$. Integrating the Buckley-Leverett equation with respect to time, we obtain

$$z(t) - z(0) = \frac{v(t) - v(0)}{\phi} \frac{df_w}{dS_w} . \tag{12.35}$$

Consider now, as an example, the relative permeability curves shown in Fig. 12.8. From these curves we can construct the fractional flow curves, if the ratio of the viscosities of the two fluids is known. Consider the simplest case in which the viscosity ratio is one. Then, given the relative permeability curves, the corresponding fractional flow curve and its derivative df_w/dS_w can be constructed and are shown in Fig. 12.10. Suppose that the initial saturation profile consists of all saturations $S_c < S_w < 1$ that exist at $z = 0$ and $t = 0$, where S_c is the connate, or irreducible, wetting phase saturation discussed above. If we use Eq. (35) to construct the saturation profile at time t, we obtain what is shown in Fig. 12.11. This figure shows that one may obtain multiple values of the saturation, an unphysical situation. To remove this multiplicity, we write a volumetric balance for the wetting phase

$$Q(t) = \int_0^{z_c} A\phi(S_w - S_c)dz , \tag{12.36}$$

where A is the cross-sectional area of the system. Here z_c is a cutoff point beyond which $S_w = S_c$. This cutoff introduces a discontinuity in the saturation profile. If we let S_{cw} to be the upper saturation at the discontinuity, use $Q(0) = 0$,

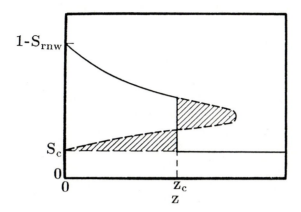

Figure 12.11: The Buckley-Leverett prediction of the distribution of the wetting phase, indicating the discontinuity.

$z(0) = 0$, and $f_w = 0$ at $S_w = 1 - S_{rnw}$, and carry out integration by parts, we obtain

$$\frac{df_w(S_{cw})}{dS_w} = \frac{f_w(S_{cw})}{S_{cw} - S_c} \ . \qquad (12.37)$$

This equation indicates that the multiplicity in the saturation profile can be removed, if we remove all saturations below S_{cw}. This particular saturation is evaluated as the saturation at which a straight line passing through the point $f_w = 0$ and $S_w = S_c$ is tangent to the curve $f_w(S_w)$. The existence of such multiplicities makes analysis of immiscible flows difficult. Moreover, use of Eq. (29) in (25) leads to a second-order non-linear equation for S_w for the general problem of flow of two immiscible fluids in a porous medium that can be solved mostly by numerical simulations. Marle (1981) provides a nice qualitative discussion of this problem, to whom the interested reader is referred.

12.8 Discrete models of capillary-controlled two-phase flow

In this section we discuss statistical and network models of two-phase flow and displacement in porous media. Some of these are based on percolation concepts and their variants and, strictly speaking, are applicable only when Ca is very small. There are other models that are more general and are presumably valid even when the capillary number Ca is not too small (in which case both capillary and viscous forces are relevant). The limit $Ca \to \infty$ is of course the case of miscible displacements already discussed in Chapter 11.

12.8.1 Random-percolation models

At the outset we should point out that a fundamental assumption in *all* percolation models of two-phase flow is that the (bond or site) occupation probability p is proportional to the capillary pressure. Without such an assumption, it would be difficult to make a one-to-one correspondence between a percolation model and the two-phase flow problem. Although in some percolation models such as invasion percolation (see below) the occupation probability is not explicitly defined, an analog of it can be calculated.

The first random percolation model of two-phase flows in porous media was suggested by Larson (1977), with the details given in Larson *et al.* (1977, 1981a,b). In Larson *et al.* (1981a) the authors proposed a model for drainage. The porous medium was represented as a cubic network of bonds and sites with distributed sizes. It was assumed that a bond next to the interface is penetrated by the displacing fluid if the capillary pressure at that point exceeds a critical value, which implies that the radius of the bond has to exceed a critical radius r_{min}. This is the same radius that is defined by Eq. (5.28), which implies that during drainage the *largest* pore throats are invaded by the non-wetting fluid (which is similar to dynamic invasion discussed above). All bonds that are connected to the (non-wettting) displacing fluid by a path of pores or bonds, whose effective radii are larger than r_{min}, are considered *accessible*, the accessibility being defined in the sense of percolation discussed in Chapter 3. It was also assumed that all accessible bonds, whose radii are at least as large as r_{min}, are filled with the non-wetting fluid. This is of course not true, since an interface which starts at one external face of a porous medium has to travel along a certain path before it reaches an accessible and potentially eligible bond. Larson *et al.* (1981a) also assumed that the displaced fluid is compressible, so that even if a blob of it is surrounded by the displacing fluid, it can still be invaded. This does not, however, result in a serious error, as we discuss below.

In their next paper, Larson *et al.* (1981b) proposed a percolation model of imbibition in order to calculate the residual non-wetting phase saturation S_{rnw} and its dependence on Ca (that is a CDC discussed above). To do this, they modelled the creation of isolated blobs of the non-wetting phase by a random site percolation (see Chapter ·3). At the site percolation threshold of the network, they calculated the fraction $\hat{g}(s)$ of the active sites that are in clusters of length s in the direction of flow and argued that this represents the desired blob size distribution. To calculate S_{rnw}, they assumed that once a blob is mobilized, it is permanently displaced which, as discussed above, is not always the case, because a blob can get trapped again, can join another blob and create a larger one, and so on.

The fundamental assumption behind Larson *et al.*'s work is that pore-level events are controlled by capillary forces. It is possible to use simple scaling arguments to estimate the values of Ca for which this assumption is valid. The capillary pressure across the interface is proportional to

$$P_c \sim \frac{\sigma_{wnw} \cos \theta}{\ell_g} \quad , \tag{12.38}$$

where ℓ_g is a typical grain size. On the other hand, the viscous pressure drop is proportional to

$$P_{vis} \sim \frac{\eta_w v \ell_g}{K} \quad . \tag{12.39}$$

Therefore

$$\frac{P_{vis}}{P_c} \sim \frac{Ca}{K_d} \quad , \tag{12.40}$$

where $K_d = K/\ell_g^2$ is a dimensionless permeability which is small (of order 10^{-3} or smaller), because K, the permeability of the porous medium, is controlled by the narrowest throats in the medium. It follows that for capillary-controlled displacements, one must have $Ca \ll 1$, and in practice one has $Ca \sim 10^{-5} - 10^{-6}$. Experimental data (Le Febvre du Prey, 1973; Amaefule and Handy, 1982; Chatzis and Morrow, 1984) seem to support this since, as discussed above (see Fig. 12.7), they indicate that S_{rnw} is constant for $Ca < Ca_c$, where Ca_c is the critical value of Ca for capillary-controlled displacement, and S_{rnw} decreases only when $Ca > Ca_c$. Larson *et al.* (1981b) compiled a wide variety of experimental data and compared them with their predictions.

Heiba *et al.* (1982, 1983, 1984, 1992) further developed these ideas and used them for calculating RPs for all regimes of wettability discussed above. Heiba *et al.* (1982, 1992) distinguished between bonds (pore throats) that are *allowed* to a phase, and those that are actually occupied by the phase. Then, given a pore size distribution of the pore space, they calculated the pore size distribution of the allowed pores and the occupied pores. Consider, for example, a displacement process in which one fluid is strongly wetting, while the other one is completely non-wetting. Then, according to the percolation model of Heiba *et al.* (1982, 1992), during primary drainage the pore size distribution of the pores occupied by the displacing (non-wetting) phase is given by Eq. (5.29), since the *largest* throats are occupied by the non-wetting fluid, and during imbibition the pore size distribution of the pores occupied by the displacing (wetting) phase is given by Eq. (5.32), because the *smallest* pores are occupied by the wetting phase. One can, in a similar fashion, derive expressions for the pore size distribution of the pores occupied by the displacing and displaced fluids during secondary imbibition and drainage. Once such pore size distributions are determined, calculating the permeability of each fluid phase (and therefore its RP) reduces to a problem of percolation conductivity, because when we calculate the permeability of a given phase, the conductance (or effective radii) of the bonds occupied by the other phase can be set to zero, since the two phases are immiscible. Therefore any of the

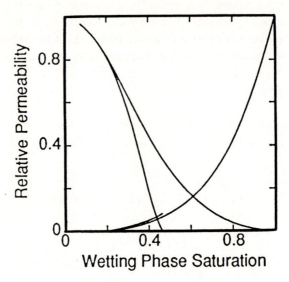

Figure 12.12: Two-phase relative permeabilities, as predicted by the random percolation model of Heiba *et al.* (1982, 1992), using a simple-cubic network.

methods discussed in Chapter 8 can be used for calculating the RPs to the phases. These ideas were first developed by Heiba *et al.* (1982) and were implemented by them and by Sahimi *et al.* (1986a) on a variety of networks. Figure 12.12 shows the results obtained with a cubic network. A comparison between this figure and Fig. 12.8 shows that all qualitative, and many quantitative, aspects of the experimental data are reproduced by the model. Note that, as discussed in Chapter 5, drainage is better described by a bond percolation process, whereas imbibition is more complex (see below).

In two subsequent papers, Heiba *et al.* extended their model to the case in which the porous medium is intermediately-wet, or has mixed wettability characteristics (Heiba *et al.*, 1983), and to the case where there are *three immiscible phases* in the medium (for example, oil, water, and gas). Consider the case of an intermediately-wetted medium. As discussed above, in such a case both primary and secondary displacements can be considered as a drainage process. Therefore the formulae developed by Heiba *et al.* (1982, 1992) for drainage can be easily extended and modified for this case. Heiba *et al.* (1983) showed that their model can predict all relevant features of RPs and capillary pressure for intermediately-wetted porous media (see Fig. 12.10). Ramakrishnan and Wasan (1984) used similar ideas and developed expressions for the RPs, and also considered the effect of Ca on them. Just as the residual saturations S_r depend on Ca (in fact $S_r \to 0$ as $Ca \to \infty$, which is the limit of miscible displacements), the RPs also depend on Ca. Normally, if Ca is small RPs do not show a great sensitivity to Ca. Evidence for this is provided by the experimental data of Amaefule and Handy (1982). However, as Ca increases the RP curves lose their curvature, and in the limit $Ca \to \infty$ they become straight lines. Ramakrishnan and Wasan (1984) developed formulae that can take this effect into account. Levine and Cuthiell (1986) used an effective-medium approximation (see Chapter 8) and a percolation model similar to that of Heiba *et al.* to calculate the RPs to two-phase flows in porous systems.

12.8.2 Random site-correlated bond percolation models

Chatzis and Dullien (1982) used a network model in which the sites represented the pore bodies to which random radii were assigned, and the bonds represented the pore throats whose effective radii were correlated with those of the sites. Using this model, Chatzis and Dullien (1982, 1985), Diaz *et al.* (1987), and Kantzas and Chatzis (1988) simulated RP and capillary pressure curves for sandstones. On the other hand, Wardlaw *et al.* (1987) determined experimentally the correlations between the pore bodies and pore throats sizes, and found that there are little, if any, such correlations in Berea sandstones, but that there may be some correlations between pore bodies and pore throats of the Indiana limestone. Li *et al.* (1986), Constantinides and Payatakes (1989), and Maier and Laidlaw (1990, 1991b) also proposed network models in which the sizes of the pore bodies and pore throats were correlated. In spite of the fact that the correlated model is much more detailed than a simple random bond model, the predicted RPs by this model are not different in any significant way from those of random percolation model discussed above.

12.8.3 Invasion percolation

Invasion percolation was first proposed by Lenormand and Bories (1980), Chandler *et al.* (1982), and Wilkinson and Willemsen (1983). In this model the network is initially filled with a fluid called the *defender*. To each site of the network is assigned a random number uniformly distributed in [0,1]. Then the displacing fluid (the *invader*) is injected into the medium which displaces the defender at each time step by choosing the site next to the interface that has the smallest random number. If we interpret the random numbers as the resistance that the sites offer to the invading fluid, then choosing the site with the smallest random number is equivalent to selecting a pore with the largest size, and hence this model simulates a drainage process. A slightly more tedius procedure can be used for working with bonds instead of sites. Two versions of the model have been developed. In one model the defender is incompressible, and therefore if its blobs are surrounded by the invader they become trapped. This was studied by Chandler *et al.* and Wilkinson and Willemsen. In the second model, trapping is ignored (i.e., the displacing fluid displaces an infinitely-compressible defender). This was studied by Wilkinson and Barsony (1984). Note that invasion percolation represents a dynamical growth process, as opposed to random percolation which is static.

There is a close connection between invasion percolation without trapping and random percolation. This connection was first found by Wilkinson and Barsony (1984) by Monte Carlo simulations. To see this connection, define an *acceptance profile* $a_n(r)$ such that $a_n(r)dr$ is the probability that the random number r selected at the nth step of the invasion is in the interval $[r, r + dr]$. Then, as $n \to \infty$ one has

$$a_\infty(r) = \frac{1}{p_c} \quad r < p_c \;, \tag{12.41}$$

and $a_\infty = 0$ for $r > p_c$, where p_c is the percolation threshold. Monte Carlo simulation of Wilkinson and Barsony (1984) and theoretical work of Chayes *et al.* (1986) support Eq. (41). In practice a cumulative acceptance profile $b_n(r)$ is computed by

$$b_n(r) = (n + \langle n_b \rangle_n - 1)^{-1} \sum_{m=1}^{n} a_m(r) \;, \tag{12.42}$$

where $\langle n_b \rangle_n$ is the average number of sites on the boundary immediately before the site is selected. In the limit $n \to \infty$ one has

$$b_n(r) = 1 \quad r < p_c \;, \tag{12.43}$$

and $b_n(r) = 0$ for $r > p_c$. Wilkinson and Barsony (1984) also defined two other quantities

$$B_1(n) = \int_0^{p_c} [1 - b_n(r)]dr \;, \quad B_2(n) = \int_{p_c}^{1} b_n(r)dr \;, \tag{12.44}$$

and postulated that $B_1(n) \sim B_2(n) \sim n^{-1/\Delta_p}$, where $\Delta_p = \beta_p + \gamma_p = \nu_p D_p$ is called the gap exponent of percolation (see Chapter 3), β_p, γ_p and ν_p are the critical exponents of percolation, and D_p is the fractal dimension of the sample-spanning percolation cluster (see Chapter 3). They hypothesized that Δ_p is the same for invasion and random percolation, and this was supported by their Monte Carlo simulations. An exact solution of the problem on the Bethe lattice (Nickel and Wilkinson, 1983) also confirmed this. Therefore invasion percolation without trapping seems to be in the universality class of random percolation.

For invasion percolation with trapping, Monte Carlo simulations of Wilkinson and Willemsen (1983) and those of others indicated that the fractal dimension D_p of the sample-spanning cluster is $D_p(d = 2) \simeq 1.82$, somewhat smaller than $D_p(d = 2) = 91/48 \simeq 1.896$ for random percolation (see Table 3.3), whereas for $d = 3$ no significant difference between invasion and random percolation models was observed. Therefore the effect of trapping seems to be negligible in three dimensions. However, it is not yet established rigorously that invasion percolation with trapping does not belong to the universality class of random percolation.

From a conceptual point of view, invasion percolation is definitely a more appropriate model of immiscible displacements than the random percolation models discussed above. The most obvious reason for this is the fact that there is a well-defined interface that starts from one side of the system and displaces the defender in a systematic and realistic way. Thus, the concepts of history, and the sequence of invading pores according to a physical rule are naturally built into the model.

Let us summarize some *experimental* evidence in support of invasion percolation model of two-phase flows in porous media. Lenormand and Zarcone (1985a) displaced oil (the wetting fluid) by air (the non-wetting fluid) in a large and transparent two-dimensional etched network and obtained $D_p \simeq 1.82$, consistent with two-dimensional computer simulations of invasion percolation with trapping. Jacquin (1985) and Shaw (1987) also performed experiments that gave strong support to the validity of invasion percolation. For example, Shaw (1987) showed that if a porous medium, filled with water, is dried by hot air, the dried pores (i.e., those filled with air) form a percolation cluster with the same fractal dimension as that of invasion percolation. Stokes *et al.* (1986) used a cell packed with unconsolidated glass beads, an essentially three-dimensional pore system. The wetting fluid was water or a water-glycerol mixture, while the non-wetting fluid was oil. When oil displaced water (drainage), the resulting patterns were consistent with an invasion percolation description of the process. Chen and Wada (1986) used a technique in which one uses index matching of the fluids to the porous matrix to "look" inside the porous medium. Their observations were consistent with an invasion percolation model. Chen and Koplik (1985) used small two-dimensional etched networks, with oil and air as the wetting and non-wetting fluids, respectively, and found that their drainage patterns were consistent with the assumptions and results of invasion percolation. Finally, Lenormand and Zarcone (1985b) used two-dimensional etched networks and a variety of wetting and non-wetting fluids (oil, different water-sucrose solutions, air), and showed that their drainage experiments are all completely consistent with an invasion percolation description of this phenomenon. Therefore although all these porous media were man-made, and we still do not have any experimental evidence from two-phase flow in a natural porous medium, there is little doubt that invasion percolation is an appropriate description of capillary-controlled two-phase flow in porous media, especially in the case of drainage.

A large number of authors have used an invasion percolation algorithm or its variants to simulate two-phase flow in porous media. Some of them (Lin and Slattery, 1982; Mohanty and Salter, 1982; Katz *et al.*, 1988; Roux and Wilkinson, 1988; Blunt and King, 1990, 1991; Jerauld and Salter, 1990) were concerned mainly with calculating the RPs and conductivities of invasion percolation clusters. Some of these authors did not mention invasion percolation, although their model was similar to invasion percolation. Others concerned themselves with fundamental properties of invasion percolation clusters. For example, Wilkinson (1986) and Sahimi and Imdakm (1988) derived the scaling laws that the capillary pressure, the RPs, and dispersion coefficients obey near the residual saturations (see below). Furuberg *et al.* (1988) studied the probability $P_i(r,t)$ (where $r = |\mathbf{r}|$) that a site, a distance r from the injection point, is invaded at time t. They found that a dynamic scaling governs $P_i(r,t)$

$$P_i(r,t) \sim r^{-1} f(r^{D_p}/t) \quad , \tag{12.45}$$

where $f(u)$ is a scaling function with the *unusual* property that $f(u) \sim u^{\zeta_1}(u \ll 1)$, *and* $f(u) \sim u^{-\zeta_2}(u \gg 1)$, i.e., $f(u)$ vanishes at *both* ends. This dynamic scaling implies that the most probable point at which the advancement of the interface between the two fluids takes place is at $r \sim t^{1/D_p}$. The reason for this unusual limiting behavior of $f(u)$ is that at time t, most of the region within the distance r has already been invaded, and new sites close to the interface that can be invaded are rare. Roux and Guyon (1989) argued that the exponents ζ_1 and ζ_2 are given by, $\zeta_1 = 1$, and $\zeta_2 = \tau_p + \sigma_p - D_h/D_p - 1$, where τ_p, σ_p, and D_p are the usual percolation exponents and fractal dimension (see Chapter 3), and D_h is the fractal dimension of the *hull* (or external surface) of percolation clusters, where $D_h(d=2) = 1 + 1/\nu_p = 7/4$, and $D_h(d=3) = D_p \simeq 2.52$.

Laidlaw *et al.* (1988) considered two different algorithms for invasion percolation. One of them was the usual one defined above, while in the other one the displacing fluid invaded *all* accessible sites of less than a given size. They found that while the fractions of invading fluid in the two cases are different (which is expected), their scaling properties are the same. Meakin (1991) studied invasion percolation on substrates with multifractal distribution of bond threshold probabilities (see Chapters 3 and 6). He found that the spatial correlations do not change the fractal properties of invasion clusters. However, certain differences may appear in the scaling of the hulls of the clusters. Maier and Laidlaw (1991a) investigated the existence of dimensional invariants (such as B_c defined for random percolation in Chapter 3; see Tables 3.1 and 3.2) in invasion percolation. Finally, Bakurov *et al.* (1990) proposed a dynamical percolation model of oil displacement, and developed a quasi-quantum-mechanical formulation for it. Their model is closely related to invasion percolation.

12.8.4 Random percolation with trapping

Random percolation with trapping was developed by Sahimi (1985) and Sahimi and Tsotsis (1985) to model catalytic pore plugging of porous media. In this problem the pores of a porous medium plug as the result of a chemical reaction and deposition of the solid products on the surface of the pores. Large

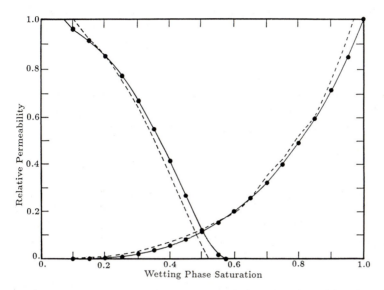

Figure 12.13: Comparison of the relative permeabilities predicted by the random percolation model (solid curves) and invasion percolation (dashed curves), with the experimental data (circles).

(macro-) pores take a long time to be plugged, and if they are surrounded by small (micro-) pores that quickly plug, they are trapped and cannot be reached by the reactants. Accurate computer simulations of Dias and Wilkinson (1986), who proposed the same model for two-phase flow problems in porous media, indicated that most properties of random percolation with trapping in *both* two and three dimensions are the same as those of random percolation discussed in Chapter 3. However, this may not be the case if, as in the problem considered by Sahimi and Tsotsis, the pore sizes are broadly distributed.

More rigorously, a trapping transition for random percolation can be defined as follows (Pokorny *et al.*, 1990). One starts with a random percolation model and at any fraction p of open or occupied bonds removes from the network all the bonds which are part of the infinite cluster. A trap is then defined as the connected component of what remains of the network. For $p \simeq 1$, all traps are of finite size. As p approaches p_c, a trapping transition occurs at p_t below which there is an infinite trap. Aizeman and Grimmett (quoted by Pokorny *et al.*, 1990) proved that $p_t > 1 - p_c$. Pokorny *et al.* (1990) showed that $p_t \simeq 0.52$ for the square network, and that trapping transition is described by the critical exponents of random percolation. Thus, this kind of trapping does not change the fundamental properties of percolation.

We now can compare invasion and random percolation models of two-phase flows. Permeability and the RPs are controlled by narrow pore throats and, moreover, the trapping of one phase by another seems to be insignificant in three dimensions. On the other hand, the random percolation model of Heiba *et al.* (1982, 1992) provides analytical expressions for the pore size distributions of the pores occupied by each phase during imbibition and drainage. These expressions greatly facilitate calculation of RPs, since many methods of calculating the permeability of a porous medium, discussed in Chapter 8, can be used with such expressions for estimating the RPs. Moreover, simulation of invasion percolation in a large three-dimensional network is costly and time consuming. Therefore, a practical question is: how do the predictions of invasion and random percolation models for the RPs compare with the experimental data? Siddiqui (1989) compared the predictions of the two models, obtained with a simple-cubic network, with the experimental data of Talash (1976) for Berea sandstones; see Fig. 12.13. It is clear that the predictions of the random percolation model are at least as accurate as those of invasion percolation. We emphasize that this comparison is only for predicting the RPs, an important problem for simulation of two-phase flows. However, for other quantities of interest invasion percolation is probably a more realistic model than the random-percolation model.

12.9 Crossover from fractal to compact displacement

Although we discussed the RP curves for *both* imbibition and drainage in terms of a percolation model, there are certain qualitative differences between the two that need to be discussed. A clue to these differences is already evident in the RP curves. The RP to the non-wetting phase during primary imbibition by a strongly wetting fluid vanishes only at $S_{rnw} = 0$, i.e., the non-wetting phase is completely expelled from the medium and the wetting phase fills the system; see Fig. 12.12. This indicates that imbibition is an essentially compact displacement. However, during drainage by a completely non-wetting fluid, the RP to the wetting phase vanishes at a finite value of S_{rw}, i.e., the non-wetting phase does not fill the porous medium, and a fractal percolation cluster may form. This was already predicted by the percolation model of Heiba *et al.* (1982, 1992), and was also nicely demonstrated by Lenormand and Zarcone (1984) who used a two-dimensional etched network, injected mercury into the system (drainage), and then withdrew it (imbibition). The cluster formed during imbibition was totally compact and filled the etched network.

A definitive study of this problem was made by Cieplak and Robbins (1988, 1990). In this study the porous medium was represented by a two-dimensional array of disks with random radii, where the underlying lattice was either a triangular or a square network. The limit of low Ca was considered, and the displacement dynamics were modelled as a stepwise process where each unstable section of the interface moved to the next stable or nearly stable configuration. Their simulations showed that there are three basic types of instability and the corresponding mechanisms of the advancement of the interface. (i) *Burst*, which

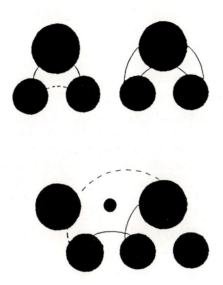

Figure 12.14: Three kinds of growth and instability that occur during an immiscible displacement in a porous medium: Burst (top left), touch (top right), and overlap (bottom) (after Cieplak and Robbins, 1990).

happens when, at a given capillary pressure P_c, no stable arc connects two disks, and therefore the interface simply jumps forward to connect to the nearest disk. (ii) *Touch*, which happens when an arc that connects two disks, intersects another disk at a wrong contact angle θ. In this case, the interface connects to this third disk. (iii) *Overlap*, which happens when two nearby arcs overlap. There is no need for the disk to which both arcs are connected, and it can be removed from the interface. Figure 12.14 illustrates these three mechanisms.

To simulate advancement of the interface P_c is fixed and the stable arcs are found. If instabilities are found, local changes are made to remove them. Then, P_c is increased by a small amount (to simulate a capillary-controlled invasion), the interface is advanced, possible instabilities are removed again, and so on. As in invasion percolation with trapping, if the invading fluid surrounds a blob of the displaced fluid, the blob is kept intact for the rest of the simulation. If all disks have the same radius, the resulting patterns are very regular and faceted which also preserve the symmetry of the lattice. This is in agreement with the experiments of Ben-Jacob *et al.* (1985). However, when the radii of the disks are randomly distributed, then the behavior of the system depends on the contact angle θ. To quantify the effect of θ, we define an interface width w. When θ is near 180°, i.e., one has a drainage process, then the phenomenon is an invasion percolation and w is of the order of pore size. However, Cieplak and Robbins (1988, 1990) showed that as θ decreases the cluster of invading fluid becomes more compact and w increases; see Fig. 12.15. Finally, at a critical contact

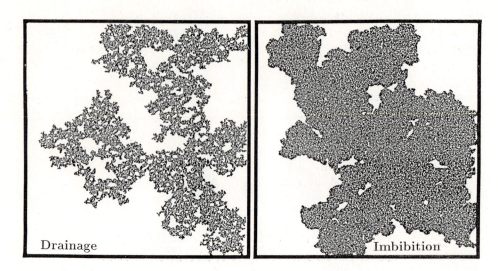

Figure 12.15: Displacement patterns for $\theta = 179°$ (left), and $\theta = 58°$ (right) (after Cieplak and Robbins, 1990).

angle θ_c, w diverges according to a power law

$$w \sim (\theta - \theta_c)^{-\nu_\theta} \quad , \tag{12.46}$$

where $\nu_\theta \simeq 2.3$ in their two-dimensional simulations. The critical angle θ_c was found to depend on the porosity ϕ of the system, for example, $\theta_c \simeq 29°$ for $\phi = 0.322$, and $\theta_c \simeq 69°$ for $\phi = 0.73$. The exponent ν_θ was found to be universal. The compactness of the cluster for $\theta < \theta_c$ is consistent with the imbibition picture discussed above.

The divergence of w at θ_c is clearly due to the transition from fractal to compact displacement. For large θ, the interface advances mainly by burst, similar to invasion percolations, and its pattern is independent of θ. However, as $\theta \to \theta_c$, the overlap and touch phenomena become more important, and the interface is unstable for almost any configuration of the local geometry. Thus the growth pattern of the interface changes, and hence w diverges.

12.10 Pinning of an interface: Dynamic scaling of rough surfaces

Although the available experimental data, the percolation model of Heiba *et al.* (1982, 1992) for the RPs, two-dimensional experiments of Lenormand and Zarcone (1984), and simulation of Cieplak and Robbins (1988, 1990) all indicated a major difference between drainage and imbibition, we still have not discussed the nature of the interface during imbibition. The cluster of the invading fluid during imbibition is compact, but capillary forces lead to random local pinning of the interface which results in its roughening. This rough interface has a self-affine (anisotropic) fractal structure (see Chapter 3), which was demonstrated by the experiments of Rubio *et al.* (1989) and Horváth *et al.* (1991a). The self-affinity of such rough interfaces was first suggested by Cieplak and Robbins (1988), but it was not quantified.

Rubio *et al.* (1989) performed their experiments in a thin (essentially two-dimensional) porous medium made of tightly packed clean glass beads of various diameters. Water was injected into the porous medium to displace the air in the system. The motion of the interface was recorded and digitized with high resolution. The experiments of Horváth *et al.* (1991a) were very similar (see below). But before embarking on an analysis of the results of Rubio *et al.* (1989) and Horváth *et al.* (1991a), let us review briefly the dynamics of rough surfaces and interfaces.

The roughness of the interface is characterized by the width $w(L)$ defined as, $w(L) = \langle [h(x) - \langle h \rangle_L]^2 \rangle^{1/2}$, where h is the height of the interface at position x, and $\langle h \rangle_L$ is its average over a horizontal segment of length L. According to the scaling theory of Family and Vicsek (1985) for growing rough surfaces, one has the following scaling form at time t

$$h(x) - \langle h \rangle_L \sim t^\beta f(x/t^{\beta/\alpha}) \quad , \qquad (12.47)$$

where α and β are two critical exponents that satisfy the following scaling relation

$$\alpha + \frac{\alpha}{\beta} = 2 \quad , \qquad (12.48)$$

and the scaling function $f(u)$ has the properties that $|f(u)| < c$ for $u \gg 1$, and $f(u) \sim L^\alpha f(Lu)$ for $u \ll 1$, where c is a constant. It is then easy to see that

$$w(L,t) \sim t^\beta g\left(t/L^{\alpha/\beta}\right) \quad , \qquad (12.49)$$

where $g(u)$ is another scaling function, and therefore

$$w(L,\infty) \sim L^\alpha \quad . \qquad (12.50)$$

Note that $w(L,t)$ is a measure of the correlation length along the direction of growth. A variety of surface growth models and the resulting dynamical scaling can be described by the stochastic differential equation proposed by Kardar, Parisi, and Zhang (KPZ)(1986)

$$\frac{\partial h}{\partial t} = \sigma \nabla_T^2 h + \frac{v}{2} |\nabla h|^2 + \mathcal{N}(\mathbf{r}, t) \quad , \qquad (12.51)$$

where σ is the surface tension, v is the growth velocity perpendicular to the interface, and $\mathcal{N}$ is a noise term. Kardar *et al.* (1986) considered the case in which the noise was assumed to be Gaussian with the correlation

$$\langle \mathcal{N}(\mathbf{r}, t) \mathcal{N}(\mathbf{r}', t) \rangle = 2A\delta(\mathbf{r} - \mathbf{r}', t - t') \quad , \qquad (12.52)$$

where A is the strength of the noise. For this model it has been proposed that (Kim and Kosterlitz, 1989; Hentschel and Family, 1991)

$$\alpha = \frac{2}{d+2} , \tag{12.53}$$

$$\beta = \frac{1}{d+1} , \tag{12.54}$$

for a d-dimensional system.

Another stochastic equation was proposed by Koplik and Levine (1985)

$$\frac{\partial h}{\partial t} = \sigma \nabla_T^2 h + v + A\mathcal{N}(\mathbf{r}, h) , \tag{12.55}$$

a linear equation whose noise term is more complex than that of the KPZ equation. For this model, the numerical work of Kessler *et al.* (1991) indicated that $\alpha(d = 2) \simeq 0.75$. It is now easy to see why a pinning transition occurs by considering Eq. (55) in zero transverse dimension

$$\frac{\partial h}{\partial t} = v + A\mathcal{N}(h) . \tag{12.56}$$

If $v > A\mathcal{N}_{max}$, where $\mathcal{N}_{max}$ is the maximum value of $\mathcal{N}$, then $\partial h/\partial t > 0$, and the interface always moves with a velocity fluctuating around v. If, however, v $< A\mathcal{N}_{max}$, the interface will eventually arrive at a point where v $+A\mathcal{N} = 0$, and get pinned down. Therefore for a fixed v there has to be a pinning transition at some finite value of A. Indeed, Stokes *et al.* (1988) performed fluid displacement experiments in random packs of monodisperse glass beads in pyrex tubes and measured the capillary pressures at which such a pinning transition takes place. The reader is referred to Family and Vicsek (1991) for a variety of models and experiments in random media and the accompanying surface growth phenomena.

Now that we have equipped ourselves with this description of rough interfaces, let us now go back to the experiments of Rubio *et al.* (1989) and Horváth *et al.* (1991a). Rubio *et al.* (1989) found that $\alpha \simeq 0.73$, significantly different from $\alpha = 1/2$, predicted by Eq. (52), but consistent with the result of Kessler *et al.* (1991). Horváth *et al.* (1990) reanalyzed Rubio *et al.*'s data and obtained, $\alpha \simeq 0.91$, larger than all other values. Horváth *et al.* (1991a) conducted their own experiments in a Hele-Shaw-like cell, packed randomly and homogeneously with glass beads, and displaced the air in the system with glycerol-water mixture, and obtained, $\alpha \simeq 0.81$, and $\beta \simeq 0.65$. Although this value of α is close to Rubio *et al.*'s, as analyzed by Horváth *et al.* (1990), and although these α and β satisfy scaling relation (48), they are significantly different from the predictions of Eqs. (53) and (54), but their α is consistent with Kessler *et al.*'s result. Martys *et al.* (1991) employed the model of Cieplak and Robbins (1988) discussed above and showed that below θ_c one has, $\alpha \simeq 0.81$, in perfect agreement with Horváth *et al.*'s (1991a) result. We point out that if the interface formed during imbibition were self-similar (as in percolation models) instead of being self-affine, we should obtain $\alpha = 1$. Thus, α can serve to distinguish between a self-affine and a self-similar interface.

How can we explain these beautiful results? Several explanations have been proposed, but as of the time of writing this book this question has not found a definitive answer. For example, Zhang (1990a) proposed a modification of the KPZ model in which the distribution of the noise amplitude is of power-law form

$$\mathcal{P}(A) \sim A^{-(\mu_n+1)} , \tag{12.57}$$

which is interesting, since such a distribution implies long-range correlations in the noise (similar to fractional Brownian motion discussed in Chapter 3). Horváth *et al.* (1991b) showed that the above experimental data can be fitted to this model if $\mu_n \simeq 2.7$. Havlin *et al.* (1991) used an analogy between these phenomena and Lévy flights [i.e., random walks in which the walker takes steps whose length is distributed according to a distribution similar to (57)] to show that

$$\alpha = \frac{3}{\mu_n + 1} , \tag{12.58}$$

$$\beta = \frac{3}{2\mu_n - 1} . \tag{12.59}$$

so that with $\mu_n \simeq 2.7$ we obtain $\alpha \simeq 0.81$, in perfect agreement with the data. Equations (58) and (59) had been proposed by Zhang (1990b) and Krug (1991) as *lower bounds* to the true values of α and β. Of course, we still do not know why the noise amplitude should have a power-law distribution, or if it does, why this particular value of μ_n should fit the data. It may have to do with the geometry and the pore size distribution of the porous media used in these studies.

Another model was proposed by Tang and Leschhorn (1992) and Buldyrev *et al.* (1992). The latter authors also carried out an interesting experiment in which a paper was clipped to a ring, and was dipped into a basin filled with suspensions of ink or coffee. The fluid invaded the paper and formed a rough interface between the wet and dry regions. The roughness exponent was found to be $\alpha \simeq 0.63$, completely different from the above data. Tang and Leschhorn, and Buldyrev *et al.* argued that this phenomenon is related to *directed percolation*. In this process (for a reviews and references see Kinzel, 1983, and Duarte *et al.*, 1992a), the bonds of the network in a given direction are directed and diode-like, in the sense that they allow transport in only one direction; if the direction of the macroscopic potential is reversed, no global transport would take place. This induces a global anisotropy such that one needs *two* correlation lengths for characterizing the network. One is ξ_L, the longitudinal correlation length in the direction of macroscopic potential, while the other is ξ_T, in the transverse direction. The percolation thresholds p_{cd} of directed networks are much larger than those of random percolation. Near p_{cd}

$$\xi_L \sim (p - p_{cd})^{-\nu_L} , \tag{12.60}$$

$$\xi_T \sim (p - p_{cd})^{-\nu_T} \ . \qquad (12.61)$$

One has, $\nu_L \simeq 1.734$ and 1.27, and $\nu_T \simeq 1.1$ and 0.735, for two and three dimensions, respectively. Of course, because of its anisotropy, the sample-spanning cluster formed in directed percolation is self-affine.

Tang and Leschhorn (1992) and Buldyrev *et al.* (1992) argued that $w(L) \sim \xi_T$ and $L \sim \xi_L$, and therefore

$$w(L) \sim L^{\nu_T/\nu_L} \ , \qquad (12.62)$$

i.e., $\alpha = \nu_T/\nu_T \simeq 0.63$ for $d = 2$, which agrees with the experiments of Buldyrev *et al.* (1992), and the simulations of Tang and Leschhorn (1992). On the other hand, if the interface were isotropic and self-similar, we would have $\nu_L = \nu_T$ and $\alpha = 1$, as pointed above. Barabási *et al.* (1992) extended these experiments to three dimensions, and obtained $\alpha \simeq 0.5$. However, we should point out that the roughness of the interface in a natural porous medium such as sandstone is far more complex than whatever that has been considered in all of these experiments, and depite their elegance it remains to be seen whether these models and experiments are directly relevant to imbibition in natural porous media.

A final, and perhaps the most plausible, explanation was proposed by Nolle *et al.* (1993). They calculated the local fluid velocities at the interface, and showed that they obey a power-law distribution similar to Eq. (57) with $\mu_n \simeq 2.7$, in complete agreement with the result of Horváth *et al.* (1991b) discussed above. Thus, the most plausible explanation for these experimental results may be a power-law distribution of the local fluid velocities at the interface, which is caused by the disordered morphology of the porous medium. However, why this distribution should be of power-law type is not clear yet.

12.11 Finite-size effects and Devil's staircase

Most of our theoretical discussion so far has been limited to systems that are essentially of infinite extent. If the system is of finite size, the dependence of macroscopic properties on the size L of the system can be investigated using finite-size scaling discussed in Chapter 3. But we have not investigated the effect of the size of a porous medium on its capillary pressure and the RP curves. Thompson *et al.* (1987b) measured the electrical resistance of a porous medium during mercury injection (drainage), and showed that the resistance decreases (the permeability increases) during the injection process in steps on Devil's staircase; this is shown in Fig. 12.16. The steps were irreversible in that small hysteresis loops did not retrace the steps, and they were not reproduced on successive injections. When the number $N_{\Delta R}$ of resistance steps larger than ΔR was plotted versus ΔR, a power-law relation was found

$$N_{\Delta R} \sim (\Delta R)^{\lambda_R} \ . \qquad (12.63)$$

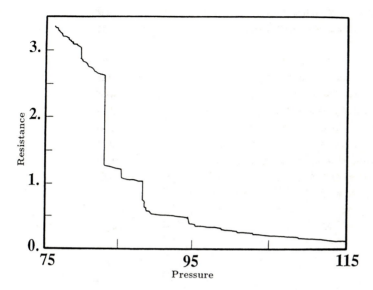

Figure 12.16: Resistance of a sandstone versus the injection pressure during mercury porosimetry (after Thompson *et al.*, 1987).

λ_R was found to vary between 0.57 and 0.81. It presumably depends on the strength of the competition between capillary and gravitational forces: $\lambda_R \simeq 0.57$ signifies the limit of no gravitational forces, while $\lambda_R \simeq 0.81$ presumably represents the limit in which gravitational forces are prominent. Based on the stepwise decrease of the resistance and the apparent first-order (discontinuous) phase transition (see Fig. 12.16), Thompson *et al.* (1987b) concluded that mercury injection should not be modelled by percolation which usually represents a second-order phase transition. A second-order phase transition, such as percolation, is characterized by a continuous vanishing or divergence of a physical quantity, such as the permeability or conductivity, as a critical point, such as the percolation threshold, is approached.

However, simulation of this process by Katz *et al.* (1988), Roux and Wilkinson (1988) and Sahimi and Imdakm (1988), and related simulation of Batrouni *et al.* (1988) showed that such a stepwise decrease in the resistance can be predicted by a (random or invasion) percolation process. The reason for this stepwise decrease in the sample resistance is that, in a *finite sample* penetration of any pore by mercury causes a *finite* change in the resistance (or, in order to cause a finite change in the resistance, the capillary pressure should also change by a finite amount), but as the sample size increases, the size of the step changes decreases such that for a very large sample the steps would vanish and the resistance curves become continuous and smooth. Siddiqui (1989) simulated this process using invasion and random percolation models which showed clearly the effect of sample size on the resistance curve. Using a percolation model, Roux and Wilkinson (1988) showed that for a three-dimensional porous medium of size L

$$N_{\Delta R} \sim L^{3(\mu - \nu_p)/(\mu + 3\nu_p)}(\Delta R)^{-3\nu_p/(\mu + 3\nu_p)} \quad , \tag{12.64}$$

where μ is the critical exponent of the conductivity of the porous medium (see Chapter 3). Thus, $\lambda_R = 3\nu_p/(\mu + 3\nu_p) \simeq 0.57$, which agrees well with the experimental result in the absence of gravity. Thus, while sample size effects are important, the stepwise decrease in the resistance of the sample during mercury injection is still consistent with a percolation description of this process discussed in section Chapter 5, and with that of two-phase flows discussed in this Chapter.

12.12 Displacement under the influence of gravity: Gradient percolation

So far, we have neglected the effect of gravity on an immiscible displacement. However, this effect cannot be neglected for three-dimensional porous media. The hydrostatic component of pressure adds to the applied pressure, and this creates a vertical gradient in the effective injection pressure. Because of this gradient, the fraction of accessible pores decreases with the height of the system. This effect was not taken into account in the percolation models described above. However, a modification of the invasion percolation by Wilkinson (1984), and by Sapoval *et al.* (1985) and Gouyet *et al.* (1988), succeeded in taking into accounts such effects. But before discussing these models, let us briefly describe a few experimental studies regarding the effect of gravity.

Clément *et al* (1987) and Hulin *et al.* (1988b) used the following procedure to study gravitational effects. They injected Wood's metal, which is a low-melting point liquid alloy, into the bottom of a vertical and evacuated crushed-glass column. The experiments were carried out at low values of the capillary number Ca by controlling the flow velocity v. After the front reached a given height, they stopped the injection and let the liquid solidify. The horizontal sections of the front corresponding to various heights were then analyzed, and the correlation function $C(r)$ [see Eq. (5.68)] of the metal distribution in the horizontal planes was determined to see whether a fractal structure was formed. Another series of experiments were carried out by Birovljev *et al.* (1991) in a *two-dimensional* porous medium. They used transparent two-dimensional models consisting of a monolayer of 1-mm glass beads placed at random and sandwiched between two plates. The system was filled with a glycerol-water mixture, which was displaced by air invading the system at one end.

The competition between gravity and capillary forces is usually quantified by the Bond number Bo which is defined as

$$Bo = \frac{\Delta \rho g \ell_g^2}{\sigma_{wnw}} \tag{12.65}$$

where $\Delta\rho$ is the density difference between the two fluids, g the gravity, and ℓ_g the typical size of the grain. Wilkinson (1984) showed that in an immiscible displacement under gravity, the correlation length ξ_g does *not* diverge (unlike random and invasion percolation which have a diverging correlation length ξ_p), but it can reach a maximum value which is given by

$$\xi_g \sim Bo^{-\nu_p/(1+\nu_p)} , \tag{12.66}$$

so that $\xi_g \sim Bo^{-0.47}$, and $\xi_g \sim Bo^{-4/7}$ in three and two dimensions, respectively. In three dimensions, there is a transition region where both fluid phases can span the system (percolate), and the width w of this region is given by

$$w \sim Bo^{-1} . \tag{12.67}$$

Similar results were obtained by Sapoval *et al.* (1985) and Gouyet *et al.* (1988) in the context of *gradient percolation*, which is a model in which a gradient G_r for the occupation probability p is imposed on one direction of the network (such a model had in fact been considered earlier by Trugman, 1983, who called this a *graded percolation*). They used arguments similar to Wilkinson's to show that

$$\xi_g \sim G_r^{-\nu_p/(1+\nu_p)} , \tag{12.68}$$

which is completely similar to Eq. (66), in which Bo has been replaced with G_r. The three-dimensional experiments of Hulin *et al.* (1988b), and the two-dimensional experiments of Birovljev *et al.* (1991) were completely consistent with these results. For example, Birovljev *et al.* (1991) obtained $\xi_g \sim Bo^{-0.57}$, where the exponent 0.57 agrees with the prediction, $\nu_p/(1+\nu_p) = 4/7 \simeq 0.57$.

Wilkinson (1984) also derived an important result regarding the effect of gravity on the ROS. He showed by a scaling argument that the difference $S_{ro}-S_{ro}^0$, where S_{ro} is the ROS for $Bo \neq 0$ and S_{ro}^0 is the corresponding value when $Bo = 0$, is given by

$$S_{ro} - S_{ro}^0 \sim Bo^{\lambda_B} , \tag{12.69}$$

where $\lambda_B = (1 + \beta_p)/(1 + \nu_p)$, ν_p and β_p being the usual percolation exponents (see Chapter 3), so that $S_{ro} - S_{ro}^0 \sim Bo^{0.74}$, for a three-dimensional porous medium. Wilkinson (1984) also proposed a simple model for simulating invasion percolation under the influence of gravity, in which a bias, linearly proportional to the height of the interface, is added to the random numbers that are assigned to the sites in the usual invasion percolation. The simulation proceeds as in the usual invasion percolation by invading those sites in the vicinity of the interface that have the smallest numbers.

12.14 Dispersion in two-phase flows

An important problem in enhanced oil recovery is dispersion in multiphase flows through a porous medium (see, for example, Thomas *et al.*, 1963; Shelton and Schneider, 1975; Salter and Mohanty, 1982; Delshad *et al.*, 1985). This

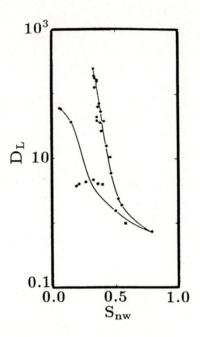

Figure 12.17: Longitudinal dispersion coefficient D_L as a function of the non-wetting phase saturation S_{nw} during drainage (circles), imbibition (squares), and secondary drainage (diamonds). D_L has been normalized by its value at $S_{nw} = 0$ (after Salter and Mohanty, 1982).

phenomenon is also relevant to ground water movement in soils that are partially saturated with air (see, for example, Gardner and Brooks, 1957; Biggar and Nielsen, 1960, 1962; Krupp and Elrick, 1968; Gaudet *et al.*, 1977; De Smedt and Wierenga, 1979). In Chapter 9 we discussed dispersion in single-phase flow through a porous medium. Although there are certain similarities between dispersion in single- and two-phase flows, there are also significant differences between the two. Similar to capillary pressure and relative permeabilities, dispersion coefficients and dispersivities in two-phase flow depend on the saturation and the way it has been reached. In other words, dispersion coefficients during imbibition and drainage are different. If neither fluid phase saturates the porous medium (i.e., the saturation is less than 1), then dispersion in a fluid phase, in the presence of another immiscible phase, is similar to dispersion in a sample-spanning percolation cluster discussed in Chapter 9. In this analogy, the fluid phase in which dispersion is occuring plays the role of the sample-spanning cluster of open or occupied pores, and the second phase is similar to the cluster of closed or unoccupied pores. This analogy breaks down if there is significant interphase mass transfer that brings the solute particles from one fluid phase to another. But we ignore this possibility here, since it seems that interphase mass transfer is not very important in many cases.

Thus, reduction of the saturation of a fluid phase (which is similar to reduction of the fraction of occupied bonds in a percolation network) should result in larger dispersivities and dispersion coefficients, and indeed this has been observed in several experiments. Figure 12.17 presents longitudinal dispersion coefficient D_L as a function of the non-wetting phase saturation during both drainage and imbibition. In this figure D_L has been renormalized by its value when the wetting phase saturates the system (i.e., at $S_{nw} = 0$). Note that during drainage as S_{nw} decreases D_L appears to rise without bound. However, in imbibition D_L first increaes as S_w decreases, reaching a maximum, and then decreases as S_w is decreased further. This is attributed to the existence of thin films of the wetting phase that coat the pore surfaces, as a result of which the wetting phase retains its macroscopic connectivity. Thus, although initial reduction of S_w results in a corresponding reduction in the connectivity of the wetting phase and thus an increase in D_L, at low values of S_w there is a network of thin films of the wetting phase through which solute particles can be transported in the fluid phase, as a result of which D_L decreases again.

Another important difference between dispersion in one- and two-phase flows is in the velocity-dependence of D_L and D_T during drainage. As long as the saturation of a fluid phase during drainage is not too close to its residual value, dispersion coefficients pertaining to that phase have a fluid velocity-dependence similar to one-phase flow discussed in Chapter 9, e.g., $D_L/D_m \sim Pe \ln Pe$, where D_m is the molecular diffusivity and Pe is the Péclet number. However, as the phase saturation approaches its residual value (its percolation threshold), the volume fraction of the dead-end pores in which the fluid is residing increases. If the time that the solute particles spend in the dead-end pores is significant, we have a crossover to a new velocity-dependence of D_L given by (see Chapter 9), $D_L/D_m \sim Pe^2$. However, for this scaling relation to be observed, the phase saturation has to be very close to its residual value. These are unique features of dispersion in two-phase flows through a porous medium.

Sahimi *et al.* (1982, 1983a, 1986a) extended their network model of dispersion in single-phase flow, that was discussed in Chapter 9, to two-phase flows. The main difference between simulation of dispersion in one- and two-phase flows is in the distribution of the two immiscible phases. First, one has to fix the fluid saturations, and distribute the two phases according to one of the models discussed above (for example, random or invasion percolation). Next, one determines the flow fields throughout the two fluid phases (which is similar to calculating the flow field in a sample-spanning percolation cluster). Finally, once the flow fields have been determined, simulation of dispersion in either phase during imbibition or drainage is carried out by exactly the same method described in Chapter 9. Figure 12.18 presents the results of such a simulation, which has striking

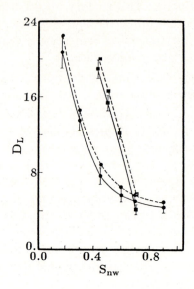

Figure 12.18: Simulation results for the longitudinal dispersion coefficient D_L versus the non-wetting phase saturation S_{nw} during drainage (circles) and imbibition (squares). Dashed curves indicate the simulation results that include diffusion into the dead-end pores (after Sahimi *et al.*, 1986a).

similarity with the experimental data shown in Fig. 12.17.

12.13 A phase diagram for displacement processes

Lenormand (1989) studied the crossovers between three regimes of displacements, namely, capillary-controlled displacements (represented by percolation models), unstable viscous displacements (represented by diffusion-limited aggregation (DLA) models and their generalizations) studied in Chapter 11, and stable viscous displacements (represented by anti-DLAs; see Chapter 11). If L is the linear size of a porous medium, then the boundaries of a percolation-type displacement scales as

$$Ca \sim L^{-(\mu+\nu_p+1)/\nu_p} \tag{12.70}$$

towards the stable viscous displacements, and as

$$Ca \sim L^{-(\nu_p+1)/\nu_p} \ , \tag{12.71}$$

towards the unstable regime. Unstable viscous displacements can occur for

$$Ca \sim L^{-1} \ , \tag{12.72}$$

which extends towards percolation-type displacements as L increases. Stable displacements do not depend on the size of the system. These considerations

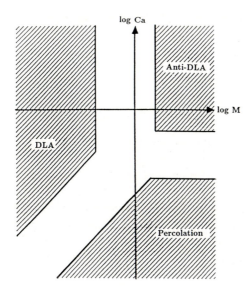

Figure 12.19: Phase diagram for three types of displacement as a function of the capillary number Ca and the viscosity ratio M (after Lenormand, 1989).

lead to the phase diagram shown in Fig. 12.19. Fernández *et al.* (1991) also studied the crossover from invasion percolation to a DLA-type displacement. They found that on length scales much smaller (larger) than a crossover length scale L_{co}, invasion percolation (DLA) patterns are obtained. Moreover, they argued that

$$L_{co} \sim \left(\frac{\delta P_c}{Ca} \right)^{2/(2+D_i)} \, , \tag{12.73}$$

where δP_c is a measure of the spatial variations of P_c and D_i is the interface fractal dimension on *small length scales*, which they found to be $D_i \simeq 1.3$ in two dimensions.

12.15 Scaling laws for relative permeability and dispersion coefficients

One can derive scaling laws for capillary pressures, the RPs and dispersion coefficients near the residual saturations, to see whether the experimental data agree with such scaling laws. Wilkinson (1986) derived such scaling laws for capillary pressure and the RPs, while Sahimi and Imdakm (1988) did the same for the dispersion coefficients. The main problem in deriving such scaling laws is relating the saturations S_w and S_{nw} of the wetting and non-wetting phases (and their residual values S_{rw} and S_{rnw}) to the occupation probability p and the percolation threshold of the system.

During drainage we are concerned with the point where the displacing non-wetting fluid first percolates. Since the trapping of the wetting phase is not important, S_{nw} is proportional to L^{D_p}/L^3, if $L \ll \xi_p$, where ξ_p is the correlation length of percolation and L is the linear size of the system. For $L \gg \xi_p$, we replace L with ξ_p, so that, $S_{nw} \sim \xi_p^{-\beta_p/\nu_p}$, and therefore

$$S_{nw} \sim (p - p_c)^{\beta_p} \ . \tag{12.74}$$

On the other hand, during imbibition we have

$$S_{nw} - S_{rnw} \sim \int_{p_c}^{p} X^A dp \sim (p - p_c)^{1+\beta_p} \ , \tag{12.75}$$

where X^A is the accessible fraction of percolation defined in Chapter 3. Given these two equations, it is not difficult to derive the scaling laws for the RPs and dispersion coefficients. Thus the RP k_{rnw} to the non-wetting phase during drainage obeys $k_{rnw} \sim (p - p_c)^e$, where e is the critical exponent of the permeability (see Chapter 3). We assume that e is not the same as μ, the critical exponent of the conductivity of the system, although in most cases $e = \mu$. In view of (74)

$$k_{rnw} \sim (S_{nw})^{e/\beta_p} \ , \tag{12.76}$$

during drainage, and

$$k_{rnw} \sim (S_{nw} - S_{rnw})^{e/(1+\beta_p)} \ , \tag{12.77}$$

during imbibition. Similar results can be derived for the dispersion coefficients. Thus if near the residual saturations holdup dispersion is the dominant mechanism (see Chapter 9), then $D_L \sim (p - p_c)^{-2\nu_p + 2e - \mu - \beta_p}$. Therefore in drainage

$$D_L \sim (S_{nw})^{(2e-\mu-\beta_p-2\nu_p)/\beta_p} \ , \tag{12.78}$$

and in imbibition

$$D_L \sim (S_{nw} - S_{rnw})^{(2e-\mu-\beta_p-2\nu_p)/(1+\beta_p)} \ . \tag{12.79}$$

These results appear to agree with both simulations and experimental data. For example, Eqs. (78) and (79) imply that $D_L \sim S_{nw}^{-0.41}$, and $D_L \sim (S_{nw} - S_{rnw})^{-0.12}$, for drainage and imbibition, respectively. These are in agreement with experimental data of Delshad *et al.* (1985), which indicate very weak divergence of D_L near the residual saturations.

12.16 Network models of immiscible displacements at finite capillary numbers

So far we have discussed capillary-controlled displacements, and in such processes viscous forces do not play any role. However, in practice, especially in oil recovery processes, it is often true that for a given displacement such as water-flooding of an oil reservoir, the capillary number Ca is relatively large so that viscous forces become important. In this section, we discuss network models of such processes.

The first of such models was apparently developed by Singhal and Somerton (1977), followed by the works of Mohanty *et al.* (1980) and Payatakes *et al.* (1980). In particular, Mohanty *et al.* (1980) used a square network of pore bodies and pore throats with distributed sizes, modelled displacement of a non-wetting fluid by a wetting one, investigated the effect of pore body and pore throat size distributions, and simulated both low and relatively high Ca regimes.

Detailed, and to some extent quantitative, models of these phenomena were developed by Koplik and Lasseter (1984, 1985), Dias and Payatakes (1986a,b), Leclerc and Neal (1988), and Lenormand *et al.* (1988). In Koplik and Lasseter's work, the pore space is modelled by a two-dimensional, but non-planar, network of cylindrical pore throats and spherical pore bodies with distributed effective sizes. The local coordination number of the network was random. In general, the equations that have to be solved are those for the pressure field throughout the network, and those for saturations of the two phases. In a pore throat, the pressure drop ΔP is given by

$$\Delta P = -\frac{1}{g_{p1}}Q + P_c - \frac{1}{g_{p2}}Q \ , \tag{12.80}$$

where Q is the flow rate, and g_{pi} is the single-phase flow conductance of fluid region i. The basic assumption behind equation (80) is that away from the interface the flow field in each fluid region is unaffected by the other fluid. Equation (80) gives rise to a non-linear problem, if it is assumed that the radius of the meniscus between the two fluids, and thus the capillary pressure P_c change in some way as the meniscus passes from a pore body or pore throat into the contiguous pore throat or pore body. The non-linear problem can be converted into a constrained linear one if one assumes that the meniscus stops at the interface during this passing period. The flow in the pore throat is, therefore, zero until the constraints are violated and the meniscus either moves forward into the pore body or back into the pore throat. If only one fluid is present in a pore body or pore throat, then P_c is of course dropped from Eq. (80). Using Eq. (80) and the fact that for each pore body one has the mass conservation law, $\sum_i Q_i = 0$, one obtains a set of equations for the pressure at the *center* of each pore body which can be solved by a number of numerical methods. Then in a time step Δt, a meniscus m with velocity v_m moves a distance $v_m \Delta t$. This gives a new fluid distribution, and the process is repeated. Koplik and Lasseter assumed that the two fluids have the same viscosity ($M = 1$), and did detailed computations to calculate the rate of change of saturations from the fluid fluxes and a knowledge of which fluids are crossing the pore-throat boundaries. As already discussed above, at relatively high Ca the viscosity ratio M is expected to have a significant effect, which Koplik and Lasseter's model did not capture. Their simulations were restricted to very small networks (10×10).

The model of Dias and Payatakes (1986a,b), is in some sense, more sophisticated than that of Koplik and Lasseter, and at the same time it is simple enough to allow computations with larger networks. They used a square network of pores having converging-diverging segments with a sinusoidal profile. This model was first used by Payatakes *et al.* (1980) for simulating blob mobilization and dynamics that were discussed above. For single-phase flow through a pore the solution due to Tilton and Payatakes (1984) was used according to which

$$Q = \frac{\pi c_0 d_p^3}{4\eta(-\Delta P_1)}\Delta P_{cd} \quad , \tag{12.81}$$

where c_0 is a constant, d_p the smallest diameter of the pore (at the minimum of the sinusoidal profile), ΔP_1 a dimensionless pressure drop along the pore (which is a function of d_p) when the flow is creeping and the Reynolds number is unity, and ΔP_{cd} the pressure drop along the converging-diverging pore. For two-phase flow in the pore the solution of the problem due to Sheffield and Metzner (1976) was used. For the capillary pressure across the interface the Washburn approximation, Eq. (5.9), was used. Various mechanisms of imbibition, similar to those discussed above, were then simulated. In their second paper, Dias and Payatakes (1986b) simulated mobilization of oil blobs using physical mechanisms that were discussed above. The calculated quantities included the ROS and the distribution of the blobs. They found that the ROS decreases with decreasing M, even for very small values of Ca. Moreover, for $M < 1$ the ROS decreases as Ca does (if $Ca > 10^{-7}$), while for $M > 1$ the ROS increases slightly with Ca in the range $10^{-7} \leq Ca \leq 5 \times 10^{-5}$, but for still higher values of Ca the ROS decreases rapidly as Ca increases. Finally, they found that a waterflood at finite values of Ca gives rise to blob populations in which most blobs occupy only one pore body, whereas as $Ca \to \infty$ larger blobs are also formed. These findings are all in qualitative agreement with the experimental data which we already discussed above.

The final model is due to Lenormand *et al.* (1988) which is completely similar to that of Leclerc and Neal (1988), that was mentioned in our discussion of miscible displacements in Chapter 11. More details of the work of Leclerc and Neal (1988) can be found in Kiriakidis *et al.* (1991). Blunt and King (1990, 1991) also used a similar model. Whereas both Koplik and Lasseter, and Dias and Payatakes replaced the actual non-linear problem by a sequence of linear problems, Lenormand *et al.* (1988) solved the actual non-linear problem. In their model the porous medium is represented by a network of bonds with distributed effetive radii. Consider a pore between nodes i and j with radius r_{ij} for which the flow rate Q_{ij} is given by

$$Q_{ij} = \frac{\pi r_{ij}^4}{8\ell\eta_{ij}}(P_i - P_j - P_{cij})^+ \quad , \tag{12.82}$$

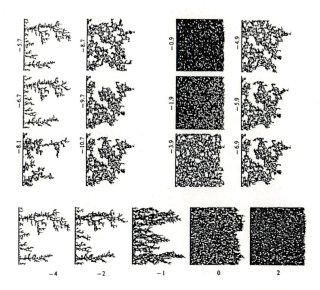

Figure 12.20: Displacement patterns obtained with the model of Lenormand *et al.* Numbers in the horizontal and vertical directions refer to log M and log Ca, respectively (after Lenormand *et al.*, 1988).

where ℓ is the length of the pore, P_i and P_j the pressures at i and j, and P_{cij} the capillary pressure in the pore. The mixture viscosity η_{ij} was taken to be $\eta_{ij} = 0.5[\eta_2(\alpha_i+\alpha_j)+\eta_1(2-\alpha_i-\alpha_j)]$, where α_i is the fraction of the pore occupied by fluid i. In Eq. (82) + means that $Q_{ij} = 0$ as long as $P_i - P_j < P_{cij}$. Because of this constraint, Eq. (82) is actually a non-linear relation between the flow rate, and the nodal and capillary pressures. Because the actual non-linear equations are solved, one can use any value of the capillary number Ca and mobility ratio M. Figure 12.20 shows the displacement patterns obtained with various values of Ca and M. Only the displacement of the wetting phase by a non-wetting phase was considered. Thus very low values of Ca correspond to invasion percolation, while very large values of Ca represent miscible displacements. These results are also in excellent agreement with the experiments of Lenormand *et al.* (1983) in two-dimensional etched networks. Lenormand *et al.* (1988) were also able to use relatively large networks (100×100) which is a distinct advantage over the models Koplik and Lasseter, and Dias and Payatakes.

12.17 Stability of immiscible displacements in porous media

We now discuss the stability of immiscible displacements. But we first provide a qualitative discussion of the subject, and then follow it up with a more quantitative analysis.

Many years ago, van Meurs (1957) used displacing and displaced fluids of the same refraction index and studied immiscible displacements in porous media. Chouke *et al.* (1959), Perkins and Johnston (1969), White *et al.* (1976), Peters and Flock (1981), Paterson *et al.* (1982, 1984a,b), Måløy *et al.* (1985, 1987), Stokes *et al.* (1986), and Frette *et al.* (1990) provided more experimental results and insight. From these observations it appears that immiscible fingering can take place over many length scales, up to a macroscopic one, and therefore one may even use a characteristic length scale for characterizing what is seen. The main implications of most of these experimental works can be summarized as follows. (i) If the invading fluid wets the porous medium, fingering can be characterized by some macroscopic length scale, such as the width of the fingers, whereas if it does not, fingering is limited to pore scales, in which case shielding dominates spreading (see Chapter 11) (recall that a displacement with a wetting fluid is compact). The characteristic macroscopic length scale decreases as Ca increases. These were confirmed by the careful experiments of Stokes *et al.* (1986), who also showed that when macroscopic fingering takes place, the width w of the finger scales with the permeability and Ca as $w/\sqrt{K} \sim 1/\sqrt{Ca}$. (ii) If the invading fluid is non-wetting, then one obtains fingers that form a percolation cluster. This was already discussed above. (iii) Finally, there is a transition zone just behind the interface where both phases are flowing. We should mention that Måløy *et al.* (1985, 1987) and Oxaal (1991) claimed that their viscous fingering patterns obtained with *immiscible* fluids are similar to DLA clusters which, as discussed in Chapter 11, are strictly applicable to miscible fluids in the limit $M = \infty$. Therefore it is not clear at all why there should be any similarities between the two phenomena.

Chouke *et al.* (1959) were the first to attempt a theoretical analysis of stability of two-phase flows in a porous medium. They ignored the transition zone and assumed that one fluid completely displaces another fluid. As the boundary condition at the interface they used Eq. (11.46), but replaced the microscopic surface tension by an *effective* surface tension for the *macroscopic* system. Although there is no theoretical justification for doing this, such a macroscopic surface tension did provide a reasonable description of some experimental works. Other authors (Outmans, 1962; Rachford, 1964; Hagoort, 1974; Peters and Flock, 1981; Huang *et al.*, 1984; Jerauld *et al.*, 1984a,c; King *et al.*, 1984; Yortsos and Huang, 1986; Chikhliwala and Yortsos, 1988; Chikhliwala *et al.*, 1988; Yortsos and Hickernell, 1989) analyzed stability of immiscible displacements.

We consider the results that are obtained from displacements at a constant velocity v of a non-wetting fluid by the injection of a wetting fluid. The system is initially at a uniform saturation S_{iw}. Unidirectional flow is described by the equation (see above)

$$\phi \frac{\partial S_w}{\partial t} + v \frac{\partial f_w}{\partial S_w} \cdot \frac{\partial S_w}{\partial z} = -\frac{\partial}{\partial z} \left(f_w \frac{K k_{rnw}}{\eta_w} \frac{dP_c}{dS_w} \cdot \frac{\partial S_w}{\partial z} \right), \qquad (12.83)$$

with the initial and boundary conditions

$$S_w = S_{iw} \qquad t = 0 \ , \tag{12.84}$$

$$S_w \to S_{iw} \qquad \text{as } z \to \infty \ , \tag{12.85}$$

$$f_w \left(1 + \frac{K k_{rnw}}{v \eta_{nw}} \cdot \frac{dP_c}{dS_w} \frac{\partial S_w}{\partial z}\right) = f_0 \qquad z = 0 \ , \tag{12.86}$$

where f_w is the fractional flow of the wetting phase discussed above. If f_w has an upward convex segment, which is usually the case during imbibition, or during drainage of not strongly wetting phases, then the base state can be taken as the steady-state solution of Eq. (83). Various upstream conditions, denoted by $-\infty$, are graphically shown in Fig. 12.21. We will use dimensionless notations to simplify the discussion; that is, a reduced saturation

$$S_{rd} = \frac{S_w - S_w^\infty}{S_w^{-\infty} - S_w^\infty} \tag{12.87}$$

and normalized mobilities (see Chapter 11),

$$\lambda_w(S_{rd}) = \frac{k_{rw}}{k_{rw}^{-\infty}} \ , \tag{12.88}$$

$$\lambda_{nw}(S_{rd}) = \frac{k_{rnw}}{k_{rnw}^{-\infty}} \ , \tag{12.89}$$

$$M = \frac{k_{rw}^{-\infty}}{k_{rnw}^{-\infty}} \frac{\eta_{nw}}{\eta_w} \ , \tag{12.90}$$

$$\lambda_t = M \lambda_w + \lambda_{nw} \ , \tag{12.91}$$

$$\lambda_c = \sqrt{\frac{K}{\phi}} \frac{1}{\sigma_{wnw} \cos \theta} \frac{dP_c}{dS_w} < 0 \ . \tag{12.92}$$

Length scales will be measured by $L = \sqrt{K/\phi}(S_w^{-\infty} - S_w^\infty)k_{rw}^{-\infty}/Ca$, which is the length over which the viscous and capillary forces are balanced. In this new notation we have

$$\lambda_{nw} \lambda_w \lambda_c \frac{dS_{rd}}{d\xi} = \lambda_t(f S_{rd} + f_w^\infty) - M \lambda_w \ , \tag{12.93}$$

$$\lambda_{nw} \frac{dP_c}{d\xi} = f S_{rd} + f_w^\infty - 1 \ , \tag{12.94}$$

where $f = f_w^{-\infty} - f_w^\infty$, and $\xi = z/L$. We note that if $k_{rw}(S_w = S_{iw}) > 0$, i.e., if the saturation is mobile, then the downstream decay of the base state is exponential.

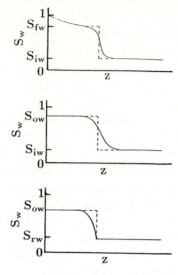

Figure 12.21: Various upstream conditions for wetting-phase saturation S_w versus the axial distance z, used in the stability analysis of immiscible flows.

In the opposite case we have

$$\frac{dS_{rd}}{d\xi} \sim -S_{rd}^{-m} \quad , \tag{12.95}$$

where $0 < m < 1$ for secondary displacements, and $m > 1$ for primary displacements. Equation (95) is a manifestation of hypodiffusion discussed above.

To examine the stability of the system, the above equations are written in a moving coordinate and the stability of the resulting equations are studied using the general method discussed in Chapter 11. The following results are obtained.

(i) The rate of growth ω is bounded from above for all wave numbers α by the expression

$$\omega < -A_1\alpha^2 + A_2A_3\alpha \quad , \tag{12.96}$$

where $A_1 = \min(-\lambda_{nw}\lambda_w\lambda_c/\lambda_t) > 0$, $A_2 = \max(df_w/dS_{rd}) > 0$, and $A_3 = \lambda_t(1)/\lambda_t(0) - 1 > 0$, where 1 and 0 correspond to upstream and downstream values of S_{rd}, respectively; see Eq. (87). Therefore, ω lies below the parabola on the right side of (96) and a cutoff wave number $\alpha_{co} \leq A_2A_3/A_1$, while the maximum growth rate ω_m does not exceed $(A_1A_3)^2/(4A_1)$. These general results confirm what is expected, that long-wave instability is driven by mobility contrast (A_3), while short-wave stabilization is a result of capillarity (A_1).

(ii) One can develop large wavelength asymptotics, $\omega = \omega_1\alpha + \omega_2\alpha^2 + \cdots$, where

$$\omega_1 = f\frac{\lambda_t(1) - \lambda_t(0)}{\lambda_t(1) + \lambda_t(0)} \quad , \tag{12.97}$$

which is a generalization of the Saffman-Taylor condition, $\omega = \alpha(M-1)/(M+1)$, for Hele-Shaw cells mentioned in Chapter 11. ω_2 is not zero but varies according to the upstream decay of the base state. For example, if the upstream decay is algebraic (see the top portion of Fig. 12.21), then one has $\omega = \omega_1\alpha + \omega_2^*\alpha^2 \ln\alpha + \omega_3\alpha^2 + \cdots$, where

$$\omega_2^* = \frac{4\lambda_t(0)d\lambda_t/dS_{rd}|_{S_{rd}=1}}{k[\lambda_t(0) + \lambda_t(1)]} \quad , \tag{12.98}$$

where k is a measure of the algebraic decay. Since $d\lambda_t/dS_d|_{S_d=1}$ is positive for unstable displacements, its existence indicates a stabilizing effect. Interestingly, this is due to mobility effects alone which adds to the process stability. Chikhliwala and Yortsos (1988) obtained the numerical solutions of the above equations to demonstrate the adequacy of the above expansions.

The geometries of the interface and the medium can have an important effect. For miscible displacements Tan and Homsy (1987) suggested an algebraic (rather than exponential) dependence for the time evolution of the disturbances. This suggestion was applied by Yortsos (1987b) to immiscible fluids in radial displacements. In this case, the base state profiles are sole functions of the similarity variable $\xi_s = r^2/t$. Perturbations are then sought in the following forms

$$S_{rd} = \bar{S}(\xi_s) + t^\delta s(\xi_s)e^{i\alpha\theta_a} \quad , \tag{12.99}$$

$$P = \bar{P}(\xi_s) + t^\delta p(\xi_s)e^{i\alpha\theta_a} \quad , \tag{12.100}$$

where $\bar{P}$ and P are dimensionless capillary pressures, θ_a is the azimuthal angle, and the bars indicate the mean (unperturbed) values (see Chapter 11). If $\delta > 0$, then the interface is unstable. At low values of α one has the asymptotic expansion

$$\delta = -1 + \delta_1\alpha + \delta_2\alpha^2 + \cdots \tag{12.101}$$

where δ_1 is equivalent to the Saffman-Taylor term, i.e., $\omega = \alpha(M-1)/(M+1)$, and δ_2 is inversely proportional to a capillary number Ca_m. Large Ca_m leads to instability, but if Ca_m is not too large, even $M > 1$ may not lead to instability. For $Ca_m < Ca_c$ the displacement is *stable*, where Ca_c is a critical value of Ca_m given by $Ca_c = 2f/\omega_m$, and ω_m is a parameter independent of Ca. Finally, it should be pointed out that, as Yortsos (1987a) showed, immiscible displacements are equivalent to miscible displacements *with equilibrium adsorption*. This analogy relates (S, f_w) to the flowing and adsorbed concentrations (C_f, C_a).

12.18 Two-phase flow in megascopic and stratified rock

So far, our discussion has been restricted to macroscopic porous media. But similar to every phenomenon discussed so far, two-phase flows in megascopic porous media have also been studied both experimentally and by numerical simulations. Let us first describe a few key experimental papers, and then discuss the theoretical studies.

Ogandzanjanc (1960) was perhaps the first to study experimentally flow in a stratified porous medium. He used an unconsolidated porous medium and showed that there is significant *crossflow* between the strata. He observed that initially the flow velocity is higher in the more permeable layer and that flow in each layer was similar to a single-phase flow system. However, because of the crossflow the distance between the interfaces in the two layers stabilizes. Since this early work, there have been several other experimental studies, including those of Novosad *et al.* (1984), Sorbie *et al.* (1987), Ahmed *et al.* (1988), and Bertin *et al.* (1990). The last authors studied waterflooding in a system of two strata, where one stratum was made of Aerolith-10, an artificial sintered porous medium, and the other was Berea sandstone. The two strata had the same thickness. The results indicated the strong effect of heterogeneities within each stratum and the contrast between them on the performance of the waterflood and the volume fraction of the recovered oil. Bertin *et al.* (1990) found that even small scale heterogeneities within the sandstone layer could strongly affect the waterflood process.

Most of the theoretical studies of two-phase flow in stratified porous media have two goals. The first is to examine the crossflow between the strata, its significance, and its variations with time as the displacement proceeds. The second goal is to determine its properties and those of the displacement process. For example, Douglass *et al.* (1959) studied imbibition in a layered system and used an averaged form of Darcy's law to describe the two-phase flow problem. Goddin *et al.* (1966) studied waterflood processes in stratified media and concluded that crossflows can be caused by both capillary and viscous forces, while Yokoyama and Lake (1981) undertook an extensive study of the effect of capillary forces on crossflows. Kyte and Berry (1975) used large scale numerical simulation to study immiscible displacements in stratified media, while Coats *et al.* (1971) used a hydrostatic distribution of various fluid phases in the vertical direction to obtain large-scale capillary pressure and relative permeability curves for their stratified medium. With this introduction, let us now describe two-phase flow in stratified and heterogeneous porous media more quantitatively.

12.18.1 Continuum models and large-scale averaging

Perhaps the simplest heterogeneous system to consider is a stratified medium of several layers with no crossflow between them. Each stratum is characterized by an effective permeability. Dykstra and Parsons (1950) considered two-phase flow in such a medium and, assuming that flow in each stratum was piston like (constant velocity), they derived an expression for the amount of recovered oil just at the breakthrough point, i.e., at the point where the interface in one of the strata reaches the outlet of the system. Reznik *et al.* (1984) generalized this model and derived expressions that can be used for calculating the amount of oil recovered at any stage of the process, and the RPs at that stage. If we do calculate these, we find that the shape of the resulting RP curves are not similar to those of macroscopic porous media discussed above.

The literature on two-phase flow in stratified media with communicating layers is relatively extensive. Starting with Goddin *et al.* (1966), many authors (Coats *et al.*, 1967, 1971; Martin, 1968; Hearn, 1971; Jacks *et al.*, 1973; Kyte and Berry, 1975; Killough and Foster, 1979; Yokoyama and Lake, 1981; Kortekaas, 1983; Wright and Dawe, 1983; Ypma, 1983; Bertin *et al.*, 1990) have studied two-phase flows in stratified media using numerical simulations. To give the reader some idea about how such calculations are carried out, we consider a two-dimensional system with only two layers. Suppose that the two fluids are water and oil, and that the fluids and the rock are incompressible. If we write a material balance for the water phase, we obtain

$$\frac{\partial}{\partial x}\left[\frac{K_{xw}}{\eta_w}\left(\frac{\partial P_0}{\partial x} - \frac{\partial P_c}{\partial x}\right)\right] + \frac{\partial}{\partial y}\left[\frac{K_{yw}}{\eta_w}\left(\frac{\partial P_0}{\partial y} - \frac{\partial P_c}{\partial y}\right)\right] = \phi\frac{\partial S_w}{\partial t} \quad, \qquad (12.102)$$

where P_0 is the pressure in the oil phase, and K_{xw}, K_{yw} are the water phase permeabilities in the x- and y-directions, respectively (recall that a stratified medium is anisotropic). If functional forms for K_{xw}, K_{yw} and P_c are assumed, then Eq. (102) can be solved numerically for various boundary conditions at the interface between the two layers, and injection conditions. If the layers are homogeneous, then the results of the previous sections can be used for K_{xw}, K_{yw} and P_c *within* each layer. The crossflows between various layers can be of various natures. One may have systems with only viscous crossflows, or with viscous and capillary crossflows, etc. If we rewrite Eq. (102) in a dimensionless form, then *direction-dependent* capillary numbers appear. The two capillary numbers Ca_x and Ca_y would be related to one another by

$$Ca_y = Ca_x R_L^2 \quad, \qquad (12.103)$$

$$R_L^2 = \left(\frac{L}{H}\right)^2 \left(\frac{K_{yref}}{K_{xref}}\right)^{1/2} \quad, \qquad (12.104)$$

where L and H are the length and thickness of the medium, and K_{xref} and K_{yref} are the permeabilities to a reference phase (which can be taken to be either oil or water). In effect R_L is some kind of aspect ratio which has been corrected by the permeability anisotropy.

Normally, the results of such simulations are averaged over the vertical direction. If this is done, then one ends up with quantities that are referred to as *psuedo functions* in the petroleum engineering literature. For example, one can use psuedo functions for RPs and another psuedo function for the capillary pressure. These psuedo functions are in fact nothing but what Quintard and Whitaker (1988) refer to as *large-scale* RPs and capillary pressure, i.e., the RPs and capillary pressure for a megascopic porous medium (which, in general, may or may not be anisotropic or stratified). Of course, from a scientific point of view, *large-scale functions* are much more appealing than pseudo functions, since as Quintard and Whitaker (1988) pointed out, the word *psuedo* suggests that these functions are something less than what they purport to be, whereas an analysis such as that of Quintard and Whitaker (see below) shows that such functions can be deduced from a rigorous analysis for *any* megascopic porous medium.

Quintard and Whitaker (1988) developed a large-scale averaging technique for two-phase flow in megascopic porous media. Starting from Eqs. (18)-(21) as the *locally-averaged* equations, they developed large-scale averaged continuity and momentum equations for each fluid phase. In their equations, they allowed for the possibility that a portion of a fluid phase may be trapped by another phase (see our discussion of trapping phenomenon given above). The technique they used is along the same lines as those discussed in Chapter 9 for dispersion in megascopic porous media. In order to make the theory tractable, they assumed that the system is in local mechanical equilibrium, which means that the *local* fluid distribution is determined by capillary pressure-saturation relations, and is not limited by the solution of an evolutionary transport equation. In two later papers, the authors studied the effect of large-scale spatial and temporal gradients (Quintard and Whitaker, 1990a), and investigated two-phase flow in a heterogeneous and stratified medium under quasi-static and dynamic conditions (Quintard and Whitaker, 1990b), using the theory developed in Quintard and Whitaker (1988). Quasi-static condition in the present context means that, the local capillary pressure, everywhere in the averaging volume, is set equal to the large-scale P_c evaluated at the centroid of the averaging volume, and that the large-scale P_c is given by the difference between the large-scale pressures in the two fluid phases. As such, the large-scale P_c is assumed to be independent of such complex factors as transient, gravitational and flow effects. If there is significant departure from the quasi-static conditions, then one is in the dynamic regime. They found that even at relatively low flow rates, dynamic effects may be important. Bertin *et al.* (1990) used this theory and compared its predictions with their experimental data, and found only qualitative agreement.

12.19 Two-phase flow in fractured rock

Although two-phase flow in fractured rocks is important to enhanced oil and gas recovery, isolation of radioactive waste, exploitation of geothermal fields for generating energy, and recovery of coalbed methane, very little is known about the laws governing such flows. The conventional approach has been based on the assumption that Darcy's law is applicable to both fluid phases, and that the cross terms of Eqs. (18) and (20) can be neglected. Moreover, it is usually assumed that the RP to each phase is equal to its saturation. This assumption is supported to some extent by the experimental work of Romm (1966), in which oil and water were confined to different regions of a *smooth* fracture by controlling the wettability of the fracture surface. Pruess *et al.* (1983) analyzed some field data from geothermal reservoirs and provided further support for this work. However, theoretical analysis of Pruess and Tsang (1990) for *rough* fractures, and computer simulations of Mukhopadhyay and Sahimi (1992), who used a network of interconnected discrete fractures to study two-phase flow and heat transfer in a geothermal field, indicated that this assumption may be in serious error. In the petroleum engineering literature, the double-porosity model that was discussed in Chapter 10 has been used for simulating two-phase flows in rocks with fractures and pores. However, given that two-phase flows depend crucially on the connectivity of the fluid phases, and that double-porosity models completely ignore this important effect, it is clear that this approach can not be useful. This field of research remains largely undeveloped.

12.20 Conclusions

Multiphase phase flows and displacements in porous media involve a complex set of facinating phenomena, some of which are well-understood, while many of them are not understood yet. Percolation and fractal phenomena have provided us with much deeper understanding of some of these phenomena, while many multiphase flows can not be described by percolation. Network models and simulation are very useful tools of studying these processes. However, multiphase flows in fractured rock and in megascopic porous media have been understood yet. These fields of research are wide open.

Chapter 13

Flow and Transport in Unconsolidated Porous Media

13.0 Introduction

In this Chapter we study and discuss single-phase flow and transport in unconsolidated porous media. We focus on low-Reynolds-number flow, where Darcy's law is applicable. Unconsolidated porous media usually are modelled by random or regular packings of particles, and have been used by chemical engineers for a long time for investigating various flow phenomena in porous media (for a review see de Santos *et al.*, 1991). Thus, we first discuss morphological properties of such packings. We then study flow and transport (diffusion, conduction, and dispersion) in such packings. As usual, we attempt to contrast classical models of flow and transport, which are based on the continuum equations, with more the modern and recent approaches that are based on discrete models of the pore space and large scale computer simulations.

13.1 Morphology of random and regular packings of particles

Random or regular packings of particles, also called granular arrays, appear in many problems of scientific and industrial importance, ranging from soil mechanics to materials science, agronomy, etc. From a fundamental point of view, they are highly interesting because they constitute a particular class of heterogeneous systems with the co-existence of two phases: The *particles* or the grains form generally a disordered-amorphous-like-phase, and the *pore space* consists of the voids between the grains. In spite of their geometrical complementarity, the two phases behave quite differently. For example, the mechanical properties of the packing, e.g., its elastic moduli, are controlled mostly by the *real* contacts between the particles, or contacts that are so close that they can become real under applied stress, and do not depend on the large pockets between particles. On the other hand, in flow through a packing of particles, the fluid flows mostly in the large pores connected by narrower tube-like structures, but not in the small gaps between the two particles. Some of the most important properties of packings of particles are, (i) the shape and distribution of contacts which control transport properties of the system; (ii) the mechanical rigidity of the packing; and (iii) the non-linearities that may arise due to the asymmetry of the contact behavior between two particles.

The problem of packing of spheres (in three dimensions) or cylinders (in two dimensions) in the most compact way is a very old subject which has motivated numerous studies. This problem has been studied by two distinct groups. Engineers have been interested in the properties of packings of particles and their modelling, because they are useful models of unconsolidated porous media, powders, fluidized beds, and many more systems. Condensed matter physicists have also studied these systems for over 30 years (Bernal, 1964). The first goal of their studies was to understand the structure of disordered materials, such as those encountered in systems with hard-sphere interactions (for example, dense liquids, or metallic glasses). Many different classes of packings have been constructed and studied experimentally or by numerical simulations. Different binding procedures or algorithms have been used, which can lead to different packing fractions and mean coordination numbers. For packing of particles, the mean coordination number is the mean number of *real* contacts that a particle has with other particles. Many numerical simulations of *dense* random packings use a particle stability criterion which imposes on each new particle added to the packing at least three contacts with its neighbors in three dimensions (or two in two dimensions), and consequently a coordination number of six in three dimensions (four in two dimensions). These values are obtained if the particles fall under gravity one after another with any reorganization of the packing after the fall of the particles. There are always some correlations in the structure of a packing of particles, because construction of the packing always starts with a seed particle, which imposes some geometrical correlations at short distances. However, the effect of such correlations is usually weak and negligible at large distances. However, there are also some other geometrical correlations that can influence the effective properties of a packing at large distances. For example, weak orientational correlations initiated by a wall of the system can greatly modify the spatial correlations in the transmission of stress through the packing. This amplification of a weak geometrical wall effect is also of importance in the pore (void) space of the packing.

More recently, it has been realized that the problem of packing particles falls within a class of problems referred to as *complex* or *non-polynomial-complete* (Willie and Bennick, 1985). This means that any algorithm which aims at finding the densest packing in a given volume requires a computation time which increases faster than any power-law of the number of particles (spheres in three dimensions or cylinders in two dimensions). The problem of packing of monodisperse discs in two dimensions appears to be very different, at first sight, from the three-dimensional case, because a triangular network (at sites of which the discs are located) satisfies both criteria of providing the *local* largest packing fraction and a *long-range order*. This correspondence between local and long-range criteria can progressively disappear if we modify the boundary conditions in such a way as to make it incompatible with local ordered packing. The simplest way of doing this is introducing some impurities into the system. For example, we can insert some particles in the system whose radii are different from the rest. Even a few particles of different sizes can destroy the translational order, and create great heterogeneities in the real contacts between the particle.

Of particular interest is the determination of the coordination number Z and other characteristic properties of a packing which control the effective transport properties of the particle phase, and hence those of the pore phase. The geometry of a packing can be studied by image analysis treatment of random cuts of the structure (Jernot and Lantuejoul, 1989). The geometry of the packing is frozen by injecting a polymerizable resin which saturates the pore space. Then a plane cut is made across the sample. A number of exact correspondences can be made between averages in two-dimensional planes and three-dimensional characteristic properties such as porosity of the packing, or its specific surface area. The coordination number can be obtained from the distribution of closest approaches of the discs obtained from an intersection by a random plane. Pomeau and Serra (1985) showed that this inversion problem is exactly solvable by making use of a relation between a two-dimensional average measured by an image-analysis technique and the three-dimensional result. It is, however, important to realize that the exact estimate of an average number of real contacts (and thus the coordination number) is somewhat artificial. A mechanical contact may not be counted for a conduction experiment, because of insufficient pressure or area of contact to break an axide layer or to give an appreciable flux in the tunnel junction.

Any regular packing of spheres (or any other particle shape) is *anisotropic*, because intersecting the packing with parallel planes results in different area porosities in the sections. A packing of spheres is stable if each sphere is in contact with at least four other spheres in such a way that at least four of the contact points are not in the same hemisphere. The loosest form of a packing of spheres has a porosity of about 0.88. A cubic packing of spheres (one in which the spheres reside at the nodes of a simple-cubic lattice) has the smallest coordination number among all the regular packings; its coordination number is six. An FCC packing of spheres (one in which the spheres reside at the nodes of a face-centered cubic lattice) has the highest coordination number; for this packing $Z = 12$. In contrast with regular packings, there is no known exact value of the coordination number of a random packing of spheres. Haughey and Beveridge (1969) have classified various forms of random packing of spheres, which are as follows.

(i) *Very loose random packing*, whose porosity is about 0.44, and can be constructed by sedimentation of spheres.

(ii) *Loose random packing*, whose porosity is about 0.4, and can be constructed by dropping the spheres into a container as a loose mass.

(iii) *Poured random packing*, whose porosity is between 0.375 to 0.391, and can constructed by pouring spheres into a container.

(iv) *Close random packing*, whose minimum porosity ranges from 0.36 to 0.375, and can constructed by vigorously shaking a random packing of spheres. Ridgway and Tarbuck (1967) have suggested the following empirical relation between the porosity ϕ of a random packing of spheres and its (mean) coordination number $\langle Z \rangle$

$$\phi = 0.0043\langle Z \rangle^2 - 0.1193\langle Z \rangle + 1.072 \ . \tag{13.1}$$

Note that in a random packing of particles the local porosity varies from region to region, so that when we speak of the porosity of a random packing, what we really mean is its average porosity, with the averaging taken over a large volume of the system. Note also that for any kind of packing, the porosity of the system is *independent* of the size of the spheres.

How do we characterize a packing of particles? In Chapter 6 we discussed some continuum models of porous media, which essentially were packings of particles which could or could not overlap with each other. Several expressions were given for the porosity and specific surface area of such models. At a more fundamental level a packing of particles is characterized by several n-point probability functions, some of which are as follows. The n-point probability function $S_n(\mathbf{x}^n)$ gives the probability of simultaneously finding n points with positions $\mathbf{x}^n \equiv \{\mathbf{x}_1, \cdots, \mathbf{x}_n\}$ in one of the two phases (i.e., the particle or pore phase). The point/q-particle distribution function $G_n(\mathbf{x}_1; \mathbf{r}^q)$ $(n = q+1)$ gives the correlations associated with finding a point at $\mathbf{x}_1$ in one of the phases and any q particles with positions $\mathbf{r}^q$. The surface-surface correlation function $F_{ss}(\mathbf{x}_1, \mathbf{x}_2)$ gives the correlations associated with finding a point at $\mathbf{x}_1$ on the two-phase interface and another point at $\mathbf{x}_2$ on the interface. Finally, the nearest-neighbor distribution function $H(r)$ gives the probability of finding a nearest neighbor at a distance r from a sphere located at the origin. $H(r)$ is a particularly useful quantity as it provides other properties of fundamental interest. For example, the exclusion probability $E(r)$, the probability of finding no particle centers within a sphere of radius r surrounding a particle at the origin, is given by

$$E(r) = 1 - \int_0^\infty H(r)dr \ . \tag{13.2}$$

The mean nearest-neighbor distance ℓ_m is given by

$$\ell_m = \int_0^\infty r H(r)dr \ . \tag{13.3}$$

One can also calculate the random-close packing density for hard spheres by determining the density at which ℓ_m is equal to the diameter of the spheres. Figure 13.1 shows $H(r)$ for a random packing of spheres at two different particle (sphere) volume fractions ϕ_p. It can be shown that all morphological properties of packings of particles can be obtained from these probability functions and their moments. In fact, the results for the porosity and specific surface area of these models that were discussed in Chapter 6 have been obtained through such functions. However, except for the first two moments, experimental measurements of the moments of these functions are not easy, or have not been very accurate. In Chapter 5, where we discussed experimental measurements of the correlation

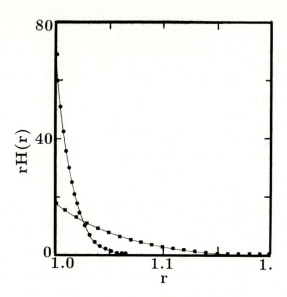

Figure 13.1: Nearest-neighbor distribution function $H(r)$ for three-dimensional hard spheres of diameter r at sphere volume fraction $\phi_p = 0.3$ (squares) and $\phi_p = 0.5$ (circles). Solid lines are theoretical predictions, while symbols represent simulation results (after Torquato, 1991).

function $C(\mathbf{r})$, we described how the first few moments of this function can be measured. Torquato (1991) has given an extensive discussion of these functions, to whose paper the interested reader is referred.

13.2 Flow and conduction in unconsolidated porous media

Calculating flow and transport properties of unconsolidated porous media, such as a packing of particles, is an old problem. For example, the problem of calculating the effective (thermal or electrical) conductivity of a regular array of spherical particles of radius R and conductivity g_s embedded in a homogeneous matrix of unit conductivity goes back to Lord Rayleigh (1892). Over the past *century* this problem has enjoyed continued attention from a large number of researchers, including Runge (1925), Meredith and Tobias (1960), Levine (1966), Jeffrey (1973), Batchelor and O'Brien (1977), Zuzovski and Brenner (1977), McPhedran and McKenzie (1978), McPhedran *et al.* (1978), McKenzie *et al.* (1978), O'Brien (1979) and Perrins *et al.* (1979). The problem that these authors considered is not exactly what we are interested in in this Chapter, because we are interested in the electrical conductivity of a fluid-saturated and unconsolidated porous medium in which the conductivity of the particles is much smaller than that of the fluid (for example, brine). Despite this difference, we discuss this problem briefly, as it is relevant to many porous media problems, but our main focus is on single- and two-phase flows and dispersion in unconsolidated porous media.

13.2.1 Exact results

As already pointed out in Chapter 8, before discussing exact results, we should define precisely what we mean by exact. There are currently no exact results for the permeability and conductivity of unconsolidated porous media with an *arbitrary* microstructure. Therefore, when we refer to exact results or upper and lower bounds, we mean those results that have been obtained for a *specific morphology* of the system.

Most of the available exact results for the permeability and conductivity of packings of particles are for spatially-periodic arrays of spherical or cylindrical particles of radius R whose centers are placed at the nodes of a regular lattice such as the simple-cubic lattice. We assume that the volume fraction of the particles is ϕ_p. Hasimoto (1959) was the first to treat slow fluid flow through a dilute ($\phi_p << 1$) cubic array of spheres. He derived the periodic fundamental solution to the Stokes' equation (see below). He then expanded the velocity profile in terms of the fundamental solution and obtained an expression for the permeability K of the array. Sangani and Acrivos (1982) modified and extended Hasimoto's work, and obtained expressions for the permeability of all three types of cubic lattices of spheres, namely, the simple-cubic, the body-centered and the face-centered cubic lattices. We discuss their approach in some details, since it represents a highly versatile and valuable method for such problems. The equation to be solved is the Stokes' equation, coupled with the usual continuity equation. If v_i is the fluid velocity component in the x_i-direction, then

$$\eta \nabla^2 v_i = \frac{\partial P}{\partial x_i} \; , \tag{13.4}$$

$$\frac{\partial v_i}{\partial x_i} = 0 \; , \tag{13.5}$$

where P is the pressure, and η is the viscosity of the fluid. Since the lattice is periodic and symmetric, the velocity components should satisfy the following equations (Sørensen and Stewart, 1974)

$$v_1(x_1, x_2, x_3) = v_1(-x_1, x_2, x_3) = v_1(x_1, -x_2, x_3) = v_1(x_1, x_2, -x_3) \; , \tag{13.6}$$

$$v_2(x_1, x_2, x_3) = -v_2(-x_1, x_2, x_3) = -v_2(x_1, -x_2, x_3) = -v_2(x_1, x_2, -x_3) \; , \tag{13.7}$$

$$v_3(x_1, x_2, x_3) = -v_3(-x_1, x_2, x_3) = v_3(x_1, -x_2, x_3) = -v_3(x_1, x_2, -x_3) \; , \tag{13.8}$$

$$v_2(x_1, x_2, x_3) = v_3(x_1, x_3, x_2) \; , \tag{13.9}$$

$$v_1(x_1, x_2, x_3) = v_1(x_1, x_3, x_2) \ , \tag{13.10}$$

$$\mathbf{v}(\mathbf{r} + \mathbf{r}_n) = \mathbf{v}(\mathbf{r}) \ , \tag{13.11}$$

$$\mathbf{v} = \mathbf{0} \quad \text{on the particle} \ , \tag{13.12}$$

where the last equation is the usual no-slip boundary condition on a solid surface, and $\mathbf{r}_n$ is a vector denoting the center of a sphere. The variables are now made dimensionless, all distances by a length h (for example, the distance between the centers of two neighboring particles in the lattice), the velocity components with the mean flow velocity v_m, and the pressure with $\eta v_m / h$. Thus, unless specified, from now on all quantities are dimensionless.

The periodic fundamental solution of the problem is obtained by solving

$$\nabla^2 v_i = \frac{\partial P}{\partial x_i} + \delta_{i1} \sum_n \delta(\mathbf{r} - \mathbf{r}_n) \ , \tag{13.13}$$

$$\frac{\partial v_i}{\partial x_i} = 0 \ , \tag{13.14}$$

where $\delta(\mathbf{r} - \mathbf{r}_n)$ is the Dirac delta function. Hasimoto (1959) showed that

$$v_i = v_m \delta_{i1} - \frac{1}{4\pi} \left(S_1 \delta_{i1} - \frac{\partial^2 S_2}{\partial x_1 \partial x_i} \right) \ , \tag{13.15}$$

$$\frac{\partial P}{\partial x_i} = -\frac{\delta_{i1}}{V_c} + \frac{\partial^2 S_1}{\partial x_1 \partial x_i} \ , \tag{13.16}$$

where V_c is the dimensionless volume of the basic unit cell of the lattice. Thus, $V_c = 1, 1/2$ and $1/4$ for the simple-cubic, BCC and FCC lattices, respectively. The functions S_1 and S_2 are the solutions of the following equations

$$\nabla^2 S_2 = S_1 \ , \tag{13.17}$$

$$\nabla^2 S_1 = 4\pi \left[\frac{1}{V_c} - \sum_n \delta(\mathbf{r} - \mathbf{r}_n) \right] \ , \tag{13.18}$$

and are given by

$$S_1 = \frac{1}{\pi V_c} \sum_{\mathbf{k}_n \neq 0} \frac{\exp[-2\pi i(\mathbf{k}_n \cdot \mathbf{r})]}{k_n^2} \ , \tag{13.19}$$

$$S_2 = -\frac{1}{4\pi^3 V_c} \sum_{\mathbf{k}_n \neq 0} \frac{\exp[-2\pi i(\mathbf{k}_n \cdot \mathbf{r})]}{k_n^4} \ , \tag{13.20}$$

where $i = \sqrt{-1}$. Here $\mathbf{k}_n$ are vectors in the reciprocal lattices

$$\mathbf{k}_n = \sum_{i=1}^{3} n_i \mathbf{b}_i \ , \tag{13.21}$$

where b_is are the unit vectors in the reciprocal lattices,

$$\mathbf{k}_n \cdot \mathbf{a}_i = n_i \quad i = 1,2,3 \ , \tag{13.22}$$

and a_is are the unit vectors in the original lattices. For the simple-cubic lattice the a_is and b_is are identical, and thus, $\mathbf{a}_1 = \mathbf{b}_1 = (1,0,0)$, $\mathbf{a}_2 = \mathbf{b}_2 = (0,1,0)$, and $\mathbf{a}_3 = \mathbf{b}_3 = (0,0,1)$. For the BCC lattice they are given by

$$\mathbf{a}_1 = \frac{1}{2}(1,1,-1) \quad \mathbf{b}_1 = (1,1,0) \ , \tag{13.23}$$

$$\mathbf{a}_2 = \frac{1}{2}(-1,1,1) \quad \mathbf{b}_2 = (0,1,1) \ , \tag{13.24}$$

$$\mathbf{a}_3 = \frac{1}{2}(1,-1,1) \quad \mathbf{b}_3 = (1,0,1) \ , \tag{13.25}$$

while for the FCC lattice we have $(\mathbf{a}_i)_{FCC} = \frac{1}{2}(\mathbf{b}_i)_{BCC}$ and $(\mathbf{b}_i)_{FCC} = 2(\mathbf{a}_i)_{BCC}$.

To obtain the complete solution for $\mathbf{v}$ and P, Sangani and Acrivos (1982) added to the fundamental solution the derivatives of $\mathbf{v}$ and S_1 multiplied by some unknown coefficients. From this sum those terms that do not obey the Stokes' and continuity equations and the above symmetry conditions were subtracted. Then the complete solution has the following form

$$v_1 = v_m - \frac{1}{4\pi}\left[G_1\left(S_1 - \frac{\partial^2 S_2}{\partial x_1^2}\right) + G_2\frac{\partial^2 S_1}{\partial x_1^2} - G_3\left(\frac{\partial^4}{\partial x_2^4} - 6\frac{\partial^4}{\partial x_2^2\partial x_3^2} + \frac{\partial^4}{\partial x_3^4}\right)S_1\right] \ , \tag{13.26}$$

$$v_2 = \frac{1}{4\pi}\left[G_1\frac{\partial^2 S_2}{\partial x_1\partial x_2} - G_2\frac{\partial^2 S_1}{\partial x_1\partial x_2} - G_3\frac{\partial}{\partial x_1}\left(\frac{\partial^3}{\partial x_2^3} - 3\frac{\partial^3}{\partial x_3^2\partial x_2}\right)S_1\right] \ , \tag{13.27}$$

$$v_3 = \frac{1}{4\pi}\left[G_1\frac{\partial^2 S_2}{\partial x_1\partial x_2} - G_2\frac{\partial^2 S_1}{\partial x_1\partial x_3} - G_3\frac{\partial}{\partial x_1}\left(\frac{\partial^3}{\partial x_3^3} - 3\frac{\partial^3}{\partial x_2^2\partial x_3}\right)S_1\right] \ , \tag{13.28}$$

$$\frac{\partial P}{\partial x_i} = -\frac{6\pi RK}{hV_cK_s}\delta_{i1} + \frac{1}{4\pi}G_1\frac{\partial^2 S_1}{\partial x_1\partial x_i} \ . \tag{13.29}$$

Here G_1 is an operator given by

$$G_1 = \sum_{M=0}^{\infty}\sum_{m=0}^{m\leq M/2} A_{nm}\left\{\frac{\partial^{2n}}{\partial x_1^{2n}}\left[\left(\frac{\partial}{\partial \xi_1}\right)^{4m} + \left(\frac{\partial}{\partial \xi_2}\right)^{4m}\right]\right\} \ , \tag{13.30}$$

with similar expressions for G_2 and G_3 (but with coefficients B_{nm} and C_{nm}), where $\xi_1 = x_2 + ix_3$, $\xi_2 = x_2 - ix_3$ and $M = n + 2m$. Here K_s is the Stokes' permeability $K_s = 2R^2/(9\phi_p)$. The coefficients A_{nm}, B_{nm} and C_{nm} are unknown and are determined by using the no-slip boundary condition, Eq. (12). The permeability and the average flow velocity are then given by (Sangani and Acrivos, 1982)

$$K = K_s\frac{hA_{00}}{3\pi R} \ , \tag{13.31}$$

$$v_m = 1 + \frac{2B_{00}}{V_c} \ . \tag{13.32}$$

Therefore, calculating A_{00} and B_{00} suffices for determining K and v_m. The final expression for the permeability of a simple-cubic array of spheres is

$$\frac{K_s}{K} = 1 - 1.7601\phi_p^{1/3} + \phi_p - 1.5593\phi_p^2 + 3.9799\phi_p^{8/3} - 3.0734\phi_p^{10/3} + O(\phi_p^{11/3}) \ . \tag{13.33}$$

Sangani and Acrivos (1982) also presented the numerical coefficients in the above expansion for the BCC and FCC lattices. These expansions are convergent for $0 < \phi_p/\phi_{max} < 0.85$, where ϕ_{max} is the maximum volume fraction of the spheres for a given packing and, $\phi_{max} = \pi/6$, $\sqrt{3}\pi/8$ and $\sqrt{2}\pi/6$ for the simple-cubic, BCC and FCC lattices, respectively.

The case of a square or hexagonal array of parallel circular cylinders can be treated by the same method. Assume that the mean flow is in the x_1-direction, and that the axes of the cylinders are parallel to the x_3-direction. In this case, $v_3 = 0$, and the general solutions for v_1 and v_2 are given by Eqs. (26) and (27), provided that we set $G_3 \equiv 0$. Then the operator G_1 has the following form

$$G_1 = \sum_{n=0}^{\infty} A_n \frac{\partial^{2n}}{\partial x_1^{2n}} \ , \tag{13.34}$$

with a similar expression for G_2 (but with coefficients B_n). The coefficients A_n and B_n are calculated by applying the no-slip boundary condition. The final results are (Sangani and Acrivos, 1982)

$$\frac{K_s}{K} = -\frac{1}{2}\ln\phi_p - 0.738 + \phi_p - 0.887\phi_p^2 + 2.039\phi_p^3 + O(\phi_p^4) \ , \tag{13.35}$$

for the square array, and

$$\frac{K_s}{K} = -\frac{1}{2}\ln\phi_p - 0.745 + \phi_p - \frac{1}{4}\phi_p^2 + O(\phi_p^4) \ , \tag{13.36}$$

for the hexagonal array, where now ϕ_p should be interpreted as the area fraction of the particles.

Using a method similar to the above, Sangani and Acrivos (1983) calculated the thermal conductivity G of a regular packing of spheres of conductivity g_s embedded in a matrix of unit conductivity. Their final result is given by

$$G = 1 - 3\phi_p \left[-\frac{1}{L_1} + \phi_p + c_1 \frac{L_2\phi_p^{10/3}(1 + c_2 L_3\phi_p^{11/3})}{1 - c_3 L_2\phi_p^{7/3}} + \sum_{n=4}^{6} c_n L_{n-1}\phi_p^{(4n-2)/3} \right]^{-1} , \tag{13.37}$$

where higher order terms in the expansion are of order of $O(\phi_p^{25/3})$,

$$L_n = \frac{g_s - 1}{g_s + \frac{2n}{2n-1}} \ , \tag{13.38}$$

and $c=\{c_i\}$ $(i = 1, \cdots, 6)$ for the three cubic lattices are

$$c = (1.3047, 0.2305, 0.4054, 0.07231, 0.1526, 0.0105) \ ,$$

for the simple-cubic lattice,

$$c = (0.129, -0.41286, 0.76421, 0.2569, 0.0113, 0.00562) \ ,$$

for the BCC lattice, and

$$c = (0.07529, 0.69657, -0.741, 0.04195, 0.0231, 9.14 \times 10^{-7}) \ ,$$

for the FCC lattice. These equations are highly accurate and, together with the results of Zick and Homsy (1982) discussed below, are considered as the definitive results for this class of models.

Zick and Homsy (1982) also considered the same problem and obtained the solutions for the cubic family of lattices, but used a different method than that of Sangani and Acrivos. Instead of trying to solve the Stokes' equation directly, they reduced the set of partial differential equations to a set of Fredholm's integral equations of the first kind. Similar to Sangani and Acrivos (1982), Zick and Homsy (1982) used Hasimoto's fundamental solutions. The kernel of the integrals in Zick and Homsy's method is a three-dimensional Fourier series which is difficult to evaluate. However, their method has its advantages. The unknown in their method is the surface stress vector, and therefore the domain of the problem is the two-dimensional surface of a sphere, as opposed to the three-dimensional domain of the original problem in terms of the velocity and pressure. Thus, the number of unknowns is three, the components of the surface stress vector, one less than the number of unknowns in the original problem which are the three velocity components and the pressure. Zick and Homsy (1982) used a Galerkin method in which the unknown, say, $f(x)$, is expressed in terms of a linear combination of some basis functions $\psi_i(\mathbf{x})$

$$f(\mathbf{x}) = \sum_{i=1}^{\infty} a_i \psi_i(\mathbf{x}) \ . \tag{13.39}$$

The unknowns are now the a_is. In practice, the above infinite sum is truncated after a few terms. One possible set of basis functions is $\psi_i(\mathbf{x}) = x^p y^q z^s$, where p, q, and s are integer or zero. For the present problem the coefficient a_1 is non-zero only if p, q, and s are even, a_2 is non-zero only if p and q are odd and s is even, a_3 is non-zero if p and s are odd and q is even, and so on. Such properties greatly facilitate the determination of a_is, and were exploited by Zick and Homsy (1982). Their numerical solutions for various cubic lattices agreed with the expansions of Sangani and Acrivos (1982).

The above results are in some sense exact because, as can be seen, one expresses the quantities of interest in terms of expansions and infinite series, every term of which can, in principle, be calculated. But we should keep in mind that such methods are generally applicable to regular arrays of spheres or cylinders. For a random packing of particles, the permeability and conductivity of the system have to be determined by different techniques, some of which are purely numerical. Even for regular arrays of spheres full numerical simulations have been carried out (see, for example, Sørensen and Stewart, 1974; see also below). Childress (1972), Howells (1974), and Hinch (1977) considered flow through a random array of spheres and obtained an asymptotic expression for the permeability of the system for low values of the particle volume fraction ϕ_p. Hinch's (1977) results can be summarized by the following equation

$$\frac{K_s}{K} = 1 + \frac{3}{\sqrt{2}}\sqrt{\phi_p} + \frac{135}{64}\phi_p \log \phi_p + 16.456\phi_p \ldots \quad . \tag{13.40}$$

Actually, Hinch's result contained a few numerical errors which were corrected by Kim and Russel (1985), who also derived a few higher order terms of the above expansion.

13.2.2 Experimental data, empirical correlations and numerical simulation

There are several empirical or semi-empirical correlations for representing pressure drop-flow velocity (or flow rates) relations in packings of particles. One of the best-known of such correlations is the *Blake-Kozeny* equation

$$v_s = \frac{1}{c_k}\frac{\Delta P}{L}\frac{d_p^2}{36\eta}\frac{\phi^3}{(1-\phi)^2} \quad , \tag{13.41}$$

where v_s is the superficial velocity of the fluid, which is related to the mean flow velocity by $v_s = v_m\phi$. Here c_k is called the Kozeny constant whose value is about 5, and d_p is an equivalent particle diameter defined by

$$d_p = \frac{6}{\Xi} \quad , \tag{13.42}$$

where Ξ is the specific surface area of the particles (see Chapter 5). Equation (41) is used at low Reynolds numbers, where the inertial effects are not important. If such effects are important, then the *Blake-Plummer* equation is combined with the Blake-Kozeny equation, where the former equation is given by

$$v_s^2 = \frac{4}{7}\frac{\Delta P}{L}\frac{d_p}{\rho}\frac{\phi^3}{1-\phi} \quad , \tag{13.43}$$

and ρ is the density of the fluid. If Eqs. (41) and (43) are simply added together, one obtains the *Ergun* equation

$$\frac{\Delta P}{L} = \frac{36c_k\eta}{d_p^2}\frac{(1-\phi)^2}{\phi^3}v_s + \frac{7\rho}{4d_p^2}\frac{1-\phi}{\phi^3}v_s^2 \quad , \tag{13.44}$$

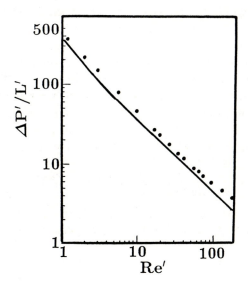

Figure 13.2: Comparison of experimental (circles) and calculated (curve) pressure drops for flow through an in-line bundle of tubes (after Eidsath *et al.*, 1983).

which interpolates between Eqs. (41) and (43).

Macdonald *et al.* (1979) reviewed the pressure drop-flow rate data for a large number of unconsolidated porous media and suggested that, if the particles are smooth, the data can be represented by the following equation

$$\frac{\Delta P'}{L'} = \frac{180}{Re'} + 1.8 \; , \tag{13.45}$$

where

$$\Delta P' = \frac{\Delta P}{\rho v_s^2} \; , \tag{13.46}$$

$$L' = \frac{L}{d_p} \frac{1 - \phi}{\phi} \; , \tag{13.47}$$

$$Re' = \frac{\rho v_s d_p}{\eta} \frac{\phi}{1 - \phi} \; . \tag{13.48}$$

Typical experimental data that follow Eq. (45) are shown in Fig. 13.2. Another popular empirical correlation is the *Kozeny-Carman* equation

$$\frac{K_s}{K} = \frac{10\phi_p}{(1 - \phi_p)^3} \; . \tag{13.49}$$

Recall that $1 - \phi_p = \phi$ is the porosity of the system, if the particles do not overlap. The predictions of Eq. (49) fall within 15% of the results for at least one of the three types of periodic packings discussed above if $\phi_p > 0.5$, and in the case of a random packing of spheres it is relatively close to the lower bound. These bounds were discussed in Chapter 8, and thus are not considered here.

Numerical simulation of flow through packed beds can be done in several different ways. One of most heavily-used methods is a Galerkin finite-element method. The main problem in the numerical simulation of flow through a packing of particles is the curved boundaries at the fluid-solid interface, and finite-element methods provide a flexible and accurate technique for handling such complex boundaries. For flow through packed beds, various versions of such methods have been used, among others, by Sørensen and Stewart (1974), Zick and Homsy (1982), and Eidsath *et al.* (1983). Another method is to map the packing onto an equivalent network of bonds and sites, with appropriate assigments of volumes of the sites and the bonds. If this mapping can be done accurately, then network simulation techniques that were discussed in Chapter 8 can be used for calculating the permeability and other properties of the system. This method was used by Bryant *et al.* (1993a,b). Finally, one can take advantage of the approximate (but accurate) relation between the permeability and conductivity of a porous medium, proposed by Johnson *et al.* (1986) and discussed in Chapter 8; see Eq. (8.141). Since this relation has been shown to be accurate for packings of particles, one can first calculate the electrical conductivity of the system by a random walk method (see Chapter 8), which can be done efficiently and without any need for solving any equation explicitly, and then use the relation between the permeability and electrical conductivity to estimate K.

13.3 Dispersion in unconsolidated porous media

As already discussed in Chapter 9, dispersion in flow through consolidated porous media is similar to that in unconsolidated ones, and therefore all theories of dispersion discussed there are equally applicable to what we discuss in this Chapter. Over the years there have been many experimental studies of dispersion in flow through a packed bed. Some of the oldest works that this author is aware of are those of Ebach and White (1958), Carberry and Bretton (1958), Blackwell *et al.* (1959), Grane and Gardner (1961), Blackwell (1962), Harleman and Rumer (1963), Pfannkuch (1963), Edwards and Richardson (1968), Hassinger and von Rosenberg (1968), and Gunn and Pryce (1969). Most of these authors measured both the longitudinal and transverse dispersion coefficients in a variety of packed beds, using a wide variety of fluids, and over several orders of magnitude in the Péclet number Pe. Their results for both dispersion coefficients are completely similar to those shown in Figs. 9.5 and 9.6. On the other hand, Han *et al.* (1985) investigated the conditions in a packed bed under which one may expect constant (independent of time) dispersion coefficients. If we define a particle

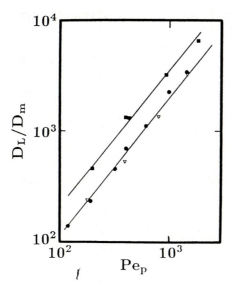

Figure 13.3: The effect of particle size distribution on the longitudinal dispersion coefficient D_L in packed beds. D_m is the molecular diffusivity and Pe_p is the Péclet number [see Eq. (50)] (after Han *et al.*, 1985).

Péclet number by

$$Pe_p = \frac{v_s d_p}{D_m} \frac{\phi}{1 - \phi} \ , \tag{13.50}$$

where D_m is the molecular diffusivity, then Han *et al.* (1985) showed that constant dispersion coefficients are achieved when a dimensionless time t_d

$$t_d = \frac{1}{Pe_p} \frac{L}{d_p} \frac{1 - \phi_p}{\phi_p} \gg 0.3 \ , \tag{13.51}$$

where L is the length of the bed. Han *et al.* (1985) also studied the effect of a particle size distribution on the dispersion coefficients. Figure 13.3 shows their results for D_L for three types of bed. One is a uniform bed in which all the spherical particles have the same size, while the other two contain a range of particle sizes, but with the same average particle size as the uniform system. The porosities of all the three systems are also the same. As this figure indicates, the broader particle size distribution gives rise to larger values of D_L, which is expected, since a broader distribution generates more tortuous transport paths, and thus a larger D_L (see the discussion in Chapter 9). On the other hand, the results of Han *et al.* (1985) indicated no appreciable effect of the particle size distribution on the transverse dispersion coefficient.

Unlike the problem of flow and conduction through packed beds discussed above, theoretical investigation of dispersion through packed beds has not received a lot of attention. This is partly because theoretical studies of dispersion in *random* packings of particles is too complex and, as far as dispersion phenomena are concerned, regular or periodic arrays of particles do not provide realistic models of unconsolidated porous media. The main reason for this is that, as discussed in Chapter 9, dispersion phenomena are sensitive to the microstructure of the porous medium, and a spatially-periodic array of particles does not allow any disorder or heterogeneities in the system. Brenner (1980), Brenner and Adler (1982), Carbonell and Whitaker (1983), Eidsath *et al.* (1983), Koch *et al.* (1989), and Salles *et al.* (1993) investigated dispersion in spatially-periodic systems. For example, Koch *et al.* (1989) showed that at high values of the Péclet number, the mechanical dispersion that is caused by a stochastic velocity field (see the discussion in Chapter 9) is absent, since flow in a spatially-periodic system is completely deterministic, and at high values of Pe molecular diffusion that can generate some microscopic stochasticity in the solute paths is not important. Instead, both D_L and D_T were found to depend *quadratically* on the Péclet number, in contradiction with the experimental data for dispersion in *disordered* porous media (including *random* packings of particles) that indicate a much weaker dependence of the dispersion coefficients on the Péclet number, e.g., $D_L \sim Pe^{1.2}$. Eidsath *et al.* (1983) carried out a numerical simulation of dispersion in a square array of parallel cylinders, in which the flow was parallel to the axis of the cylinders. Using a finite-element method, they first solved for the flow field, and then solved the convective-diffusion equation. From their numerical simulations, Eidsath *et al.* found that $D_L \sim Pe^{1.7}$, which is closer to the experimental data. However, from a theoretical view, their result cannot be explained, because at high Pe one *has* to obtain the quadratic dependence of D_L on Pe. Moreover, at very low values of Pe (i.e., when the flow field essentially has vanished) the results should be compatible with the conduction calculations discussed above. But the results of Eidsath *et al.* were also not compatible with this limit. It is not known whether these discrepancies are due to numerical inaccuracies, to not having achieved the high-Pe limit, or reflect some other form of error.

13.4 Two-phase flow in unconsolidated porous media

Two-phase flows through packed beds have been studied for many decades, as they are relevant to many phenomena of industrial importance. It is not possible to discuss everything that has been done, and thus we restrict our attention to the key points and concepts. There are many articles that review various aspects of two-phase flows through packed beds. Among them are those of Satterfield (1975), Charpentier (1981) and Herskowitz and Smith (1983). The most recent review article that we are aware of is the excellent paper of de Santos *et al.* (1991), which we follow closely and summarize its discussions.

Two-phase flows in packed beds operate on the principle of having sample-spanning fluid phases, despite having free surfaces between the two phases. As discussed by de Santos *et al.*, the outcome of any two-phase flow phenomenon in a packed bed is the product of four factors.

(i) The first factor is the mass-transfer coefficient (see Chapter 9) which measures the rate of transfer of a substance by convection and diffusion. This depends on the interfacial area between the two phases, usually a liquid and a gas, and on the departure from equilibrium between the two phases. Direct contact between the two phases can facilitate mass transfer, but this is not always useful since it can lead to dispersing of one fluid in the other and formation of drops and bubbles, mists, and foams in the flowing fluids. Direct contact between the two phases is enhanced by local convection in the direction of diffusion. If this is to happen, one needs near the phase boundary an expandable and contractible interface, and an accelerating and decelerating surface velocity. This requires varying cross sections of gas flow. However, this may lead to the local separation of the gas flow which makes the process inefficient.

(ii) The second important factor is the interfacial area per unit volume of the contacting particles.

(iii) The third factor is the residence time of the fluids in the packed bed. For larger superficial velocities, the packed bed has to be longer in order to achieve a certain residence time for a fluid phase, which in turn requires a higher gas pressure drop, and thus higher pumping cost.

(iv) The final important factor is the departure from mass-transfer equilibrium between the two fluid phases. This departure can be measured in terms of the difference between local concentrations in the two fluid phases. This concentration difference is often called the *driving force* in the chemical engineering literature. Perhaps the best way of maintaining the driving force at a certain level is by *countercurrent* flow of the two phases (see below). However, as pointed out by de Santos *et al.*, this also causes a larger shear stress at the interface between the two fluids, and thus a larger cost of pumping the fluids into the packed bed. In some cases the driving force is maintained by a chemical reaction within a fluid phase, after a substance has been transferred from the other phase.

Similar to two-phase flows in pipes, there are several flow regimes in a packed bed. These have been discussed in detail by de Santos *et al.* (1991). A summary of their discussion is as follows.

13.4.1 Countercurrent flows

If the flow rates of the two phases are low, then both fluid phases are sample-spanning, also called *trickling*. However, if the liquid-phase flow rate is larger than a certain critical value, then the amount of liquid that stays in the bed, and the pressure drop that is needed to drive the gas phase, both increase sharply. If the liquid-phase flow rate is high enough, the liquid accumulates at the top of

Figure 13.4: Rivulet (left) and film (right) flows in cocurrent downflows (after Christensen *et al.*, 1986).

the bed. This phenomenon is called *flooding*. Under these conditions the liquid can even be carried away by the gas phase. This can also happen if the liquid flow rate is fixed, and the gas flow rate is increased. Before flooding occurs, the liquid starts to accumulate in a range called *loading*. The loading phenomenon might seem attractive because it increases the mass-transfer coefficient, but it can also be counterproductive because it is close to flooding which is not desirable, since flooding limits the flow rates at which the two phases can pass through the packed bed.

Various mechanisms have been proposed for flooding, with apparently no general agreement or consensus between them. For example, Shearer and Davidson (1965) proposed that a standing wave of flowing liquid is raised by the effect of counterflowing gas, and if the flow rate is high enough, the wave growth and the counterflowing gas reinforce each other to the extent that they block flow passages between the particles and induce flooding. On the other hand, Hutton *et al.* (1974) suggested that the onset of flooding is linked to the instability of liquid films that flow down a solid wall, and this instability is accentuated by the shear stress induced by the counterflowing gas. Factors that increase liquid holdup are both the liquid and gas flow rates, the liquid viscosity, and the porosity of the packed bed. Factors that decrease liquid holdup are liquid density, surface tension between the liquid and the gas, larger particles in the packing, and particles that are not wetted very well by the liquid. Moreover, as the liquid holdup increases, so does the pressure difference needed to drive the gas. There have been many attempts to correlate this pressure difference with various factors, such as the flow rates, fluid properties, and packing characteristics (see Ellman *et al.*, 1988, for a review). Most of such correlations are empirical or, at best, semi-empirical, and while some of them do provide reasonable predictions for some of the quantities of interest, they often lack rigorous theoretical foundation.

13.4.2 Cocurrent downflows

In this type of flow there are at least four regimes (Herskowitz and Smith, 1983) which are as follows.

(i) *Trickling*, which happens when the liquid and gas flow rates are low enough. Then the liquid flows as films over the solid particles, and the gas occupies and flows in the bulk of the pore space. An important consideration in this regime is the way the liquid is distributed in the system. If the liquid flow rate is small enough, it can even flow through the localized rivulets, even if it does not completely wet the solid particles. Figure 13.4 shows the difference between film and rivulet flows. Rivulet flow is undesirable because it leads to poor contact between the fluids and also between the fluids and the solid particles, and therefore a smaller mass-transfer coefficient. It can also be caused by grooves, by incomplete wetting, and by capillary action. Kan and Greenfield (1979), Christensen *et al.* (1986), and Levec *et al.* (1988) observed hysteresis loops in the trickling flow through a packed bed of particles with small diameters (<1.8 mm). More specifically, the pressure drop and liquid holdup for specified gas and liquid flow rates were found to depend on the manner by which they had been reached. This is of course similar to two-phase flows in consolidated porous media discussed in Chapter 12. Figure 13.5 shows a typical hysteresis loop. The upper branch of the loop corresponds to film flow, in which all the particles are wetted by the liquid films. As discussed in Chapters 5 and 12, the origin of such hysteresis loops is linked to the distribution of the liquid phase throughout the system. Smaller particles enhance the influence of capillarity, and thus the likelihood of hysteresis. Transverse dispersion can counter this effect.

(ii) *Pulsing*, which happens at intermediate values of the fluids flow rates. Under this condition liquid-rich and gas-rich slugs alternate as they travel down the bed, giving rise to pulsing discharge. The origin of pulsing flow and its characteristics have been studied by several research groups. Pulsing flow is characterized by large fluctuations in pressure drop and liquid holdup. According to Satterfield (1975), several industrial reactors do in fact operate near the trickling-to-pulsing transition.

(iii) *Bubbling*, which occurs at high liquid and low gas flow rates. Under these conditions the liquid controls the pore space and spans the system, while the gas phase is carried along in the form of small bubbles.

(iv) *Spray*, which happens at conditions opposite to those in bubbling. In this case, the gas phase spans the system, while the liquid phase is carried along in the form of dispersed droplets.

13.4.3 Cocurrent upflows

In this type of operation, both fluids enter the system at the bottom of the

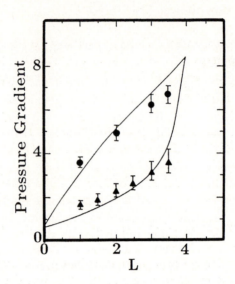

Figure 13.5: Comparison of the predicted pressure gradient (in kPa/m, shown with the symbols) with the experimental data. The gas flow rate is 756 kg/m^{2}hr, and L is in 10^4kg/m^{2}hr (after Chu and Ng, 1989).

packed bed. This can be a very useful operation if the ratio of gas to liquid flow rates is small, or if one needs a large residence time for the liquid. Turpin and Huntington (1967) identified three flow regimes for cocurrent upflows which are pulsing, bubbling and spray, very similar to cocurrent downflows discussed above.

By now it should be clear to the reader that two-phase flow of gases and liquids in packed beds involves a set of complex and fascinating phenomena. Moreover, as discussed in Chapter 12, flow of two immiscible fluids in any porous medium involves competition between various forces, and as usual this competition is expressed in terms of dimensionless numbers. The relevant dimensionless numbers are the Reynolds number Re, the capillary number Ca, and the Bond number Bo. Table 13.1, adopted from de Santos *et al.* (1991), presents the ranges of these number for two-phase flow operations in packed beds, and compares them with those for consolidated porous media, typically encountered in oil recovery operations. As can be seen, the ranges of the dimensionless numbers that are used in the two systems are vastly different, and one cannot yield much insight into the other. This is one reason for considering separately flow phenomena in unconsolidated porous media. It also indicates the rich variety of phenomena that one may observe in such porous media, which explains why flow in unconsolidated porous media has been of great interest for a long time.

Table 13.1

Values of the Reynolds number Re, the capillary number Ca, and the Bond number Bo that are typically encountered in two-phase flow through packed beds, and their comparison with those for consolidated porous media that are used in oil recovery processes.

Porous media	Re	Ca	Bo
Packed beds	$10^{-2} - 10^3$	$10^{-1} - 10$	$10^{-1} - 10$
Consolidated	$10^{-9} - 10^{-2}$	$10^{-7} - 10^{-3}$	$10^{-9} - 10^{-2}$

13.5 Modelling two-phase flows in unconsolidated porous media

As with all flow phenomena in porous media that we have discussed so far in this book, two-phase flows in unconsolidated porous media have been modelled both by continuum and discrete or network models. What follows is a brief description of each.

13.5.1 Continuum models

Continuum models of two-phase flow in packed beds are to some extent similar to those for two-phase flows in consolidated porous media discussed in Chapter 12. We discuss a typical continuum model developed by Grosser *et al.* (1988) and Dankworth and Sundaresan (1989). There are many other continuum models in the literature, for which the interested reader can consult these two papers. The model we discuss consists of the volume-averaged equations of motion for each fluid phase (see Chapters 8, 9, and 12), together with the appropriate constitutive equations. As discussed in Chapter 12, continuum models are valid if both fluid phases are sample-spanning and are not near macroscopic disconnection (percolation threshold), and the linear size of the system (for example, the height of the packed bed) is much larger than the length scale at which the system can be considered as homogeneous.

We assume that the volume fractions of the gas and liquid phases are ϕ_g and ϕ_l, respectively. Thus we must have $\phi_g + \phi_l = \phi$. The continuity and momentum equations for the fluid phases are (see Chapter 12, Section 12.7.1)

$$\frac{\partial \phi_i}{\partial t} + \boldsymbol{\nabla} \cdot (\phi_i \mathbf{v}_i) = 0 \ , \tag{13.52}$$

$$\rho_i \phi_i \left(\frac{\partial \mathbf{v}_i}{\partial t} + \mathbf{v}_i \cdot \boldsymbol{\nabla} \mathbf{v}_i \right) = -\phi_i \boldsymbol{\nabla} P_i + \phi_i \rho_i \mathbf{g} + \mathbf{F}_{di} + \phi_i \boldsymbol{\nabla} \cdot \boldsymbol{\tau}_i + \boldsymbol{\nabla} \cdot \mathbf{R}_i \ , \tag{13.53}$$

where i denotes either the liquid or the gas phase, $\mathbf{F}_{di}$ is the drag force exerted on phase i, $\boldsymbol{\tau}_i$ is the viscous stress tensor, and $\mathbf{R}_i$ is a pseudo-turbulence stress. Normally, $\nabla \cdot \boldsymbol{\tau}_i$ and $\mathbf{R}_i$ are small and can be safely ignored. The next step is to provide constitutive equations for the drag force $\mathbf{F}_{di}$. Many empirical or semi-empirical correlations have been proposed in the past that relate $\mathbf{F}_{di}$ to the flow velocity, and the packed bed characteristics. In fact, the main difference between various continuum models of two-phase flows in packed beds is in the type of correlations that are used for $\mathbf{F}_{di}$. For example, Sáez and Carbonell (1985) proposed the following Ergun-type equations

$$\mathbf{F}_{dg} = \left[\frac{A\eta_g(1-\phi)^2\phi^{1.8}}{d_p^2\phi_g^{2.8}} + \frac{B\rho_g(1-\phi)\phi^{1.8}}{d_p\phi_g^{1.8}}|\mathbf{v}_g| \right]\mathbf{v}_g \,, \tag{13.54}$$

$$\mathbf{F}_{dl} = -\left(\frac{\phi - s_l}{\phi_l - s_l}\right)^{2.43}\left[\frac{A\eta_l(1-\phi)^2\phi_l^2}{d_p^2\phi^3} + \frac{B\rho_l(1-\phi)\phi_l^3}{d_p\phi^3}|\mathbf{v}_l| \right]\mathbf{v}_l \,. \tag{13.55}$$

Here A and B are constants, and s_l is called the *static holdup* which is the volume fraction of the liquid in the packed bed in the absence of any flow. There are several empirical equations for s_l. For example, it can be estimated from

$$s_l = \frac{1}{20 + 0.9Bo} \,, \tag{13.56}$$

where Bo is the Bond number which, for a packed bed, is defined as

$$Bo = \frac{\rho_l g d_p^2 \phi^3}{\sigma_{lg}(1-\phi)^2} \,, \tag{13.57}$$

where σ_{lg} is the surface tension between the two fluids. Different correlations for the drag force were also given by Hutton and Leung (1974). The final step for formulating the continuum model is to provide a correlation, or else experimental data, for the capillary pressure P_c between the two phases. For example, Reed *et al.* (1987) measured P_c in beds of up to 1-mm-diameter particles, and found that the Leverett J-function (see Chapter 5) describes the capillary pressure well.

In practice, the above set of equations would be too difficult to solve analytically or even numerically, and thus one has to make many simplifying assumptions. For example, it is typically assumed that the macroscopic flow is one dimensional, and that the packed bed is long enough that the domain of the problem can be considered as infinite. As in any two-phase flow problem, one also has to consider the possibility of instability phenomena (see Chapters 11 and 12). Using this type of continuum models, it has been shown that liquid holdup and pressure gradients below the flooding point can be approximated by a uniform state where all temporal and spatial derivatives in the macroscopic equations vanish. Such a model also predicts the existence of two uniform solutions, a low holdup state which is commonly observed below the flooding point, and an upper holdup state which is, however, not accessible to a steady flow.

Thus, the maximum flooding conditions can be defined as those beyond which no uniform state solution exists. Sáez and Carbonell (1985) and Levec *et al.* (1986) showed that Eqs. (54) and (55) can provide satisfactory correlations for pressure drop and liquid holdup in the cocurrent downflow regime, and Dimenstein and Ng (1986) and Sundaresan (1987) were able to describe the size, velocity, and holdup of pulses in cocurrent downflows by use of the type of continuum equations described above. Thus, for flow in packed beds continuum models have been relatively successful. One reason for this is that in both rivulet and film flows one deals with a percolation system that is far from the percolation threshold (see Fig. 13.4), and thus continuum models are expected to be relatively accurate.

13.5.2 Network models

Two-phase flows in packed beds are to some extent similar to flow of two immiscible fluids, e.g., oil and water, in consolidated porous media discussed in Chapter 12. Therefore, the application of network models and percolation theory to modelling such phenomena is natural. The first of such applications was made by Crine *et al.* (1980a,b). Their model was very simple: one starts with a network in which a randomly-selected fraction p of the bonds are active (they allow fluid flow), and the rest are inactive. Crine *et al.* argued that p corresponds to irrigation rate or the liquid flow rate. Thus, if p is small and below the percolation threshold p_c, then the pattern of the clusters of active bonds corresponds to rivulet flow, while for $p > p_c$ the pattern is similar to the film flow. As discussed above, the idea is that in film flow a sample-spanning cluster of the wetted particles is formed, whereas in rivulet flow, the liquid flow rate is too small to wet all the particles. Figure 13.4 shows nicely the difference between these two flow patterns. To make these patterns look more realistic, Crine *et al.* (1980a,b) converted the random-decorated lattice of active and inactive bonds to a sort of continuum percolation (see Chapter 3) by drawing circles around the active sites of the lattice, with a radius equal to half of the length of a lattice bond. Then they drew contours of the overlapping circles, and "colored" them, depending on how many active bonds (per site of the lattice) are connected to the same site. Ahtchi-Ali and Pedersen (1986) used a similar method, except that instead of decorating their lattice randomly, they started at one face of the lattice, and by moving along a particular direction, selected the active bonds randomly at the interface between the wetted and dry parts of the lattice. This algorithm, which is somewhat similar to invasion percolation discussed in Chapter 12, is more realistic than that of Crine *et al.*, because it simulates in some sense the flow of a liquid that enters the bed at one face of the system, and leaves it at the opposite face.

Although such percolation algorithms did represent significant advancement over many of the classical models that completely ignored the effect of the morphology of the packed bed, they did not contain enough microscopic physics of the phenomenon to be predictive. More refined and quantitative models were proposed by Zimmermann *et al.* (1987), Chu and Ng (1989), and Melli and Scriven (1991). In the model of Zimmermann *et al.* (1987), a two-dimensional network was used to represent a packing of spherical particles. The size of the network was very small (8×22), and its coordination number was 4. Note that, since a bicontinuous structure (a two-phase system in which *both* phases are sample-spanning) cannot exist in two dimensions, one can simulate only the flow of one of the phases (liquid or vapor), if a two-dimensional network is used. For this reason Zimmermann *et al.* (1987) assumed that the gas-phase flow rate was zero or so small that it did not affect the flow of the liquid. The bonds were assumed to have the same effective sizes, and thus the effect of the pore size distribution of the packing was ignored.

One now needs some rules for how the particles are wetted, and how the liquid, once it has arrived at a node, splits into the outgoing bonds. Zimmermann *et al.* (1987) assumed that, when the sum of the flow rates reaching a node (the contact point between two spheres) was less than a critical value, one has partial wetting of the spheres. Otherwise, complete wetting was assumed. At completely wetted spheres each outflow was assumed to be half the total flow rate arriving there. At partially wetted spheres the left and right outflows (in the two-dimensional network) were assumed to be equal to their inflow counterparts. Using this simple model Zimmermann *et al.* (1987) studied the effect of various factors on the flow of the liquid down the packed bed.

As a more realistic model, Chu and Ng (1989) used a three-dimensional simple-cubic lattice, in which each bond represented a pore. The packed bed was a regular cubic packing of spheres of equal sizes, but it was tilted in such a way that if viewed from above, each sphere rests on top of three other spheres spaced evenly apart. Thus, the cubic network was also tilted. The networks used were very small (the largest was $8 \times 8 \times 10$). Since all the spheres had the same size, the effective sizes of all bonds have to be the same. However, to make the model more realistic, the effective radii of the pores were selected from a distribution function that mimicked the size distribution of a random three-dimensional packing of spherical particles. Because of the three-dimensional character of the network, it was possible to simulate the flow of both phases. Thus, two bicontinuous flow regimes were studied. Since the upper branch of the hysteresis loop (Fig. 13.5 corresponds to liquid film flow over the pore walls and the flow of the gas in the remaining pore space, one can model the phenomenon as an *annular flow* in which a liquid film of a given thickness flows over the walls of the pores (bonds of the network), and the gas flows in the bulk in the middle of the pores. Assuming that the inclination angle is β and that the flow is in the z-direction, and writing down the momentum equation for each phase, we obtain

$$- S_l \frac{dP}{dz} - \tau_{wl}\ell_l + \tau_i\ell_i + \rho_l S_l g \sin \beta = 0 \ , \tag{13.58}$$

$$- S_g \frac{dP}{dz} - \tau_i\ell_i + \rho_g S_g g \sin \beta = 0 \ . \tag{13.59}$$

207zHere S_l and S_g are, respectively, the cross-sectional areas for the liquid and the gas, ℓ_l and ℓ_i are the liquid-wall and the interfacial lengths, and τ_{wl} and τ_i are the shear stresses at the wall and at the liquid-gas interface, respectively. One needs correlations that relate τ_{wl} and τ_i to the flow velocities and the fluid properties. As in the case of continuum models discussed above, such correlations usually are empirical or semi-empirical. However, for laminar flows in tubes these correlations can be derived from the momentum equations. Chu and Ng (1989) used the following well-known equations

$$\tau_{wl} = 2\frac{\nu_l\ell_l\rho_l}{S_l} v_l \ , \tag{13.60}$$

$$\tau_i = 2\frac{\nu_g\ell_i\rho_g}{S_i} v_g \ , \tag{13.61}$$

where ν_l and ν_g are the kinematic viscosities of the liquid and gas phases, respectively. These equations can easily be derived by using the momentum equation (Chapter 2) for tubes (Bird *et al.*, 1960).

Thus, the procedure to calculate the upper branch of the hysteresis loop is as follows. For every bond of the network one uses Eqs. (58)-(61). At every node of the network one has to have conservation of mass, which means that the algebraic sum of all flow rates reaching the node has to be zero. If we eliminate $\tau_i\ell_i$ between Eqs. (58) and (59) and use Eq. (60) in the resulting equation, we obtain a single equation for Q_g, the gas flow rate, in terms of dP/dz. Thus, if we write a mass balance for the gas phase at every node of the network, we obtain a set of linear equations governing the nodal pressures, from the solution of which we can determine all quantities of interest. It should be clear to the reader that this model is very similar to those discussed in Chapters 8-12 for flow and dispersion in consolidated porous media and fractured rock.

The lower branch of the hysteresis loop corresponds to *segregated* flow. In this case, a pore is either filled with liquid or gas alone, or if both phases co-exist in a pore, they are segregated in the sense that a fraction of the bulk of a pore is filled with the liquid and the rest is filled by the gas. Thus, the same procedure as above, with some modest modifications, can be used for modelling the lower branch of the hysteresis loop. Figure 13.5 shows the results of such calculations and compares them with the experimental data of Christensen *et al.* (1986) mentioned above. Given the small size of the networks used in the computations, and the many simplifying assumptions made in the model, the agreement between the simulation results and the data is very good. This agreement demonstrates once again the power of network simulation for modelling flow phenomena in porous media.

Melli and Scriven (1991) used a variation of the model of Chu and Ng (1989), in which most of the volume of the network was assigned to the sites rather than the bonds. Four flow regimes were also considered which were annular (see above), bubbling, flooded (in which bonds were filled with only liquid), and *bridged*. The last flow regime occurs in pores with constrictions, i.e., with slowly varying diverging-converging sections (which actually exist in flow passages in packed beds). In this situation, at high enough liquid flow rates the standing wave below the constriction grows and bridges over so that only liquid flows, gas pressure builds moderately and the bridge is pushed down and broken. This can be repeated again, giving rise to *cycling*. Melli and Scriven (1991) used small (10×24) two-dimensional networks and *no* Monte Carlo sampling, because their computations proved to be too intensive. Thus, only qualitative results were obtained.

13.6 Conclusions

Flow processes in unconsolidated porous media involve a set of complex and fascinating phenomena that are in many ways more complex than those encountered in consolidated porous media. Exact results have been obtained for certain models of unconsolidated porous media that provide useful estimates for their flow and transport properties. Continuum and network models have both been relatively successful in modelling various phenomena in such systems, although only the latter models can provide deep insight into the microscopic mechanisms of flow processes in packed beds.

Chapter 14

Advances in Computational Methods

14.0 Introduction

We are approaching the end of this book. After reading the previous Chapters, we should have a reasonable understanding of various phenomena associated with flow phenomena in porous media and fractured rock. We know which problems have been more or less solved, and which remain unsolved. However, there is still one issue that needs to be addressed in this book. So far, we have assumed that once a model for a certain phenomenon is developed, the necessary computations can, *in principle*, be carried out. However, *in principle* is quite different from *in practice*. Many of the models that we have described in this book need enormous computations, which may make them impractical. In general, simulation of multiphase, multicomponent flows in natural porous media, often called *reservoir simulation* in the petroleum industry, requires enormous amounts of computer time. Even the three-dimensional percolation and network models described in previous Chapters, which are relatively simple models of the actual phenomena, need a large amount of computer time. In fact, some of such models have not even been studied in three dimensions. The reservoir simulators that are normally used in the petroleum industry are usually one of two types. The first type is the so-called *black oil* simulators in which it is assumed that the fluids (usually oil, gas, and water) are homogeneous, and that gas can dissolve in the oil, or vice versa, in any proportion. This avoids the problem of computing the detailed phase diagrams of the system which requires an accurate thermodynamic equation of state. The second type of simulator is more complex; they are usually called *compositional simulators*. Such simulators represent the oil as a mixture of several hydrocarborns and perform detailed phase equilibria calculations in order to determine the distribution of components between the liquid and vapor phases in the reservoir. Both simulators use a huge amount of computer time, and for this reason devising efficient numerical methods for solving the transport equations has always been an active area of research.

We do not intend to discuss all numerical methods for reservoir simulation, since this subject deserves to have a separate *book* by itself. There are already many books on the subject (see, for example, Peaceman, 1977). There are also a few review papers that discuss this subject. For example, the interested reader can consult Cheshire and Pollard (1988), Christie (1988) and Shiles *et al.* (1990) for reviews of this subject. However, we discuss briefly the finite-difference (FD) approximation to the transport equations that is routinely used in reservoir simulations, and its advantages and disadvantages. We then discuss two new methods of simulating flow in porous media and fractured rock. These new methods are based on lattice-gas automata (LGA), also called cellular automata (CA), and lattice-Boltzmann automata (LBA), which we believe can dramatically change the way people think about simulation of flow in porous media and reservoir simulation.

14.1 Finite-difference methods

Simulation of multiphase flows in porous media involves solving a parabolic pressure equation and a hyperbolic system of conservation laws (see Chapter 2). Most reservoir simulators are based on an FD approximation to these equations. The reservoir is divided into blocks, and the flow field is computed between the blocks. At first it may seem that a standard five-point (in two dimensions) or seven-point (in three dimensions) FD approximation may suffice for solving the governing equations. Indeed, such an approximation does work for smooth solutions in which there are no steep gradients. However, in large-scale simulations that are intended for field-scale systems, the heterogeneities can give rise to large gradients in the quantities of interest, especially around wellbores or in the vicinity of fractures. Given that, from a practical point of view, these are in fact the most interesting areas, the standard FD approximation may result in unphysical solutions. Using a standard FD approximation has both advantages and disadvantages. The main advantage is that it is straightforward to set up the discretized transport equations that are to be solved. Moreover, the resulting matrix of the coefficients is sparse and banded. A main disadvantage is the development of high gradient regions that make the equations very stiff. The solution may not be accurate enough, especially if large blocks are used. One may also have *numerical dispersion* (Lantz, 1971), which can mimic *physical dispersion* discussed in Chapter 9. While development of numerical dispersion can be advantageous, if physical dispersion is actually present, it can also be a disadvantage, in the sense of producing solutions that do not actually mimic the true situation. They may also lead to spurious oscillations or overshoots in the neighborhood of high-gradient areas which result in unstable solutions and unphysical results.

One possible way of handling such problems is using finer grids where high gradients have developed. Thus, one has to use a *non-uniform* grid, such that in the areas where the solutions are expected to be smooth, the grid is not too refined, whereas in the high-gradient areas it is very refined. A non-uniform grid, together with higher-order FD approximations that use nearest- and next-nearest-neighbors, turn out to be very accurate. For example, the quantity $\partial C/\partial x$, where C is a concentration, at grid point i is approximated by (Liu *et al.*, 1994)

$$\left(\frac{\partial C}{\partial x}\right)_i \simeq a_{1i}C_{i-2} + a_{2i}C_{i-1} - (a_{1i} + a_{2i} + a_{3i})C_i + a_{3i}C_{i+1} , \qquad (14.1)$$

so that evaluating the first derivative at grid point i requires knowledge of C at the nearest and next-nearest-neighbors of i. Here

$$a_{1i} = \frac{d_{i-1}d_{i+1}}{d_{i-2}(d_{i-2}+d_{i+1})(d_{i-2}-d_{i-1})} \ ,$$

$$a_{2i} = -\frac{d_{i-2}d_{i+1}}{d_{i-1}(d_{i-2}-d_{i-1})(d_{i-1}+d_{i+1})} \ ,$$

$$a_{3i} = \frac{d_{i-2}d_{i-1}}{d_{i+1}(d_{i-1}+d_{i+1})(d_{i-2}+d_{i+1})} \ ,$$

where

$$d_{i-2} = \Delta x_{i-1} + \frac{\Delta x_{i-2}+\Delta x_i}{2} \ ,$$

$$d_{i-1} = \frac{\Delta x_{i-1}+\Delta x_i}{2} \ ,$$

$$d_{i+1} = \frac{\Delta x_i+\Delta x_{i+1}}{2} \ .$$

In a similar way, $\partial^2 C/\partial x^2$ is approximated by

$$\left(\frac{\partial^2 C}{\partial x^2}\right)_i \simeq b_{1i}C_{i-2} + b_{2i}C_{i-1} - (b_{1i}+b_{2i}+b_{3i})C_i + b_{3i}C_{i+1} \ , \qquad (14.2)$$

where

$$b_{1i} = 2\frac{d_{i+1}-d_{i-1}}{d_{i-2}(d_{i-2}+d_{i+1})(d_{i-2}-d_{i-1})} \ ,$$

$$b_{2i} = 2\frac{d_{i-2}-d_{i+1}}{d_{i-1}(d_{i-2}-d_{i-1})(d_{i-1}+d_{i+1})} \ ,$$

$$b_{3i} = 2\frac{d_{i-2}+d_{i-1}}{d_{i+1}(d_{i-1}+d_{i+1})(d_{i-2}+d_{i+1})} \ .$$

Note that the above approximations result in a matrix of the coefficients that is denser and has a larger bandwith than the matrix that one obtains with the standard five-point or seven-point FD approximation. By a similar method, the derivatives in the other directions can be approximated by such higher-order FD formulae.

Once the FD equations are set up, they are usually solved by an iterative method. Non-linear equations are solved using the Newton-Raphson method. Direct methods of solving a system of equations, such as Gaussian elimination, are *never* used (although such methods are, in principle, much more accurate than iterative methods), simply because the number of equations is so large that no computer memory can fit the enormous matrix of the coefficients, even though this matrix is usually sparse and banded. Solving these equations by an iterative method means that there has to be a tradeoff between the desired accuracy of the solution and the computer time that one has to consume in order to achieve that accuracy. A very refined discretization also implies a much larger number of equations to be solved. Various finite-element methods are better suited for computations for systems with complex boundaries or high gradients, but the matrix of coefficients in finite-element methods is usually dense, not sparse and banded as in finite-difference methods.

All iterative techniques have to solve a large set of linear equations which has the following form

$$\mathbf{AX} = \mathbf{b} \ , \tag{14.3}$$

where $\mathbf{A}$ is the matrix of the coefficients, the Jacobian matrix which arises when the non-linear equations are solved by the Newton-Raphson method, $\mathbf{X}$ is the vector of the unknowns (the difference between the current iterative solutions and those of the last iteration), and $\mathbf{b}$ is the current right hand side. If the reservoir has irregular boundaries, then at any given time step the grid may contain many cells (in two dimensions) or blocks (in three dimensions) which are inactive. Such cells or blocks are similar to the dead-end bonds of a percolation network. A simulation strategy can be developed in which one has a look-up table of active blocks or cells, so that computations are carried out only for such cells or blocks. This increases the efficiency of the simulation.

The way in which the blocks are numbered can also be important. A natural numbering scheme is one in which the blocks are numbered first in the x-direction, then in the y-direction, and finally in the z-direction. The advantage of this numbering scheme is that calculation of the residual vector $\mathbf{R} = \mathbf{b} - \mathbf{AX}$ (which is zero when the true solution is reached) is easily vectorized for use in a vector computer such as the Cray Y-MP. However, this may not be the most efficient way of numbering the blocks, because if the grid contains a large number of inactive cells or blocks, then a lot of computer time would be wasted on such blocks. With a look-up table for the active cells or blocks, the natural numbering scheme is not used. However, this does not mean that the alternative numbering scheme is not vectorizable, and in fact it has been shown (Cheshire and Pollard, 1988) that vectorizable alternative schemes can be developed.

There are several other aspects of reservoir simulation that need to be considered carefully, before a full simulation is attempted. For example, any iterative technique requires an initial guess for the unknowns. After any iteration, one needs a new guess to continue the iteration, and using the current iteration solution, or a combination of the current and the last iteration solutions which is routinely done in over-relaxation methods, is not the optimal way of proceeding. Conjugate gradient methods are normally used for accelerating the rate of convergence of iterative techniques. However, these methods are most efficient if matrix $\mathbf{A}$ is symmetrical. Unfortunately, matrices that arise in reservoir simulation are not usually symmetrical, and thus conjugate gradient are not as efficient as they can be. An alternative method is the orthomin algorithm developed by Vinsome (1976). In this method, after each iteration $\mathbf{R}^2$ is minimized by explicitly orthogonalizing each new residual to *all* previous residuals. Conjugate gradient methods do this automatically, if **A** is symmetrical. This orthogonalization is vectorizable and consumes only a small fraction of the total computation time.

Another important issue is the required computer memory for the simulation. For large-scale simulation of a multiphase, multicomponent mixture, one needs a large amount of computer memory. Some computers have virtual memory facility, while some, such as the Cray computers, do not, and therefore all the computations have to be carried out with real memory. Given the large size of a realistic simulation, the memory that is provided by a vector computer may not be enough. In this case one keeps the data on a disk and reads them into the real memory of the computer in small blocks, as needed, to carry out the computations. Therefore, memory management is also an important issue to consider, when large-scale reservoir simulations are to be attempted.

14.2 Lattice-gas simulation of fluid flow

Broadwell (1964) was probably the first that developed an automata-type model for fluid flow problems. In his model velocity was a discrete variable, but space and time were both continuous. Later, Hardy, de Pazzis and Pomeau (HPP) (1973,1976) developed a completely discrete model for fluid flow on the square lattice which could mimic several features of real flow problems. But because a square lattice is not isotropic, the HPP model in the continuum limit did not reduce to the Navier-Stokes equation. As a result, the model did not receive much attention.

Lattice gas or CA models (see Wolfram, 1986a; Doolen, 1991; Boon, 1992; Benzi *et al.*, 1992; for reviews) are large lattices where each site can be in one of several discrete states. We first describe the models qualitatively, and then provide a more concrete description of them. The lattice is populated by particles, and the variables describing the state of the system are Boolean indicating the presence or absence of the particles in the bonds of the lattice. The evolution of the system is governed by a set of collision rules that tell us how the particles move in the lattice, and how they are scattered once they collide with each other (see below). Both time and space are discrete, and usually connections between sites are between nearest neighbors only, ideal conditions for high-speed simulations on vector or parallel computers. The LGA approximation of the Navier-Stokes equation in two dimensions is based on particles of unit mass either resting on the site of a lattice or moving with unit velocity on one of the six bonds emanating from each lattice site. Frisch, Hasslacher and Pomeau (FHP) (1986, 1987), using the HPP model, showed that in order that the discrete equations reduce to the usual Navier-Stokes equations, two-dimensional simulations have to be done on a triangular lattice.

On the triangular lattice the particles either have a unit velocity along a bond

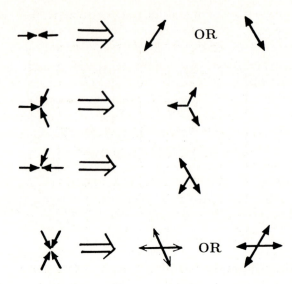

Figure 14.1: Collision rules for particle motion on a triangular lattice during lattice gas simulation of fluid flow.

of the lattice, or are at rest at a site. Up to six particles may reside at any site on the triangular lattice. However, multiple occupancy of a site is subject to the constraint that no two particles at a site can have the same velocity. Figure 14.1 shows how the particles move on this lattice. If particles hit each other, they are scattered according to the laws of momentum conservation. For example, one particle hitting another one that is at rest may be scattered such that its direction changes by 60°, and that the previously resting particle moves in a direction inclined at −60° with respect to the direction of the incoming particle. However, this two-particle collision alone leads to anisotropy. One needs three-particle collisions in order to remove the anisotropy (see Fig. 14.1). Thus two- and three-particle collisions constitute the minimum set of collision rules necessary to obtain an isotropic behavior for the LGA. Usually up to four particle collisions are employed (5 and 6 particle collisions almost never happen). If a particle hits a solid wall, it is reflected by 180° to simulate the no-slip boundary condition.

The extension of LGA to three dimensions is more complex since no regular three-dimensional lattice is isotropic, and thus in the continuum limit one has spurious terms, in addition to those in the Navier-Stokes equation, which are caused by the anisotropy of the lattice. However, there are now several methods of circumventing this difficulty. For example, one can use a three-dimensional topologically-random lattice, such as the Voronoi lattice which is macroscopically isotropic (Sahimi, 1989, unpublished). Alternatively, one can use (d'Humieres *et al.*, 1986) the FHP model on a *four-dimensional* face-centered hypercubic (FCHC) lattice. The nodes (x_1, x_2, x_3, x_4) of the lattice satisfy the condition that

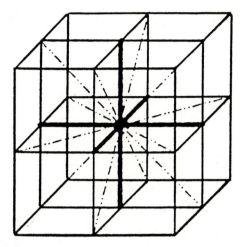

Figure 14.2: Three-dimensional projection of the elementary cell of the FCHC lattice. Two particles along the thick black lines can propagate with velocities v = ±1. Along the dotted lines the particle velocity is zero.

$x_1 + x_2 + x_3 + x_4$ is even, where x_is are integer numbers. For this lattice, which has a coordination number of 24, all pairwise symmetric fourth-order tensors are isotropic, and therefore one can simulate the Navier-Stokes equation on such a lattice. We may then make the observation that any solution of the four-dimensional equation which does not depend on the fourth dimension is a solution of the three-dimensional equation. This suggests the use of a FCHC lattice which wraps around periodically in the fourth direction. One actually uses a lattice which is only one lattice unit long in the fourth dimension, and therefore has an effectively three-dimensional structure. Figure 14.2 shows the three-dimensional projection of an elementary cell of the FCHC lattice. The disadvantage of this model is that, although the fourth dimension is very thin, the discrete velocities still have components in *all* directions; therefore the model is bit intensive (24 or 25 bits per site as compared with 6 in two dimensions). A third approach (d'Humieres *et al.*, 1986) is to use a three-dimensional cubic lattice in which the particles move with speeds 0, 1 and $\sqrt{2}$ (instead of 0 and 1 in two dimensions). This model uses only 19 bits per site. Note that having a large number of discrete particle velocities is not necessarily a disadvantage, because it allows more collisions between the particles, and therefore higher Reynolds number flows. Despite the discrete nature of LGA, these models are capable of exhibiting rich macroscopic complexity such as turbulence (d'Humieres and Lallemand, 1986).

To make the discussion more concrete, and also to lay the foundation for our discussion of LBA, let us consider the LGA on the triangular lattice and describe it in terms of the particle velocities. In this model, the particles have momenta selected from the vector

$$\mathbf{M}_i = \left[\cos \frac{\pi(i-1)}{3}, \sin \frac{\pi(i-1)}{3} \right] , \quad i = 1, \cdots, 6 . \qquad (14.4)$$

The microscopic densities corresponding to the number of particles and momentum conservation at position $\mathbf{x}$ at time t are, respectively

$$n(\mathbf{x}, t) = \sum_i n_i(\mathbf{x}, t) , \qquad (14.5)$$

and

$$M_l(\mathbf{x}, t) = \sum_i M_{i,l} n_i(\mathbf{x}, t) , \qquad (14.6)$$

where $n_i(\mathbf{x}, t)$ is the Boolean variable that indicates the presence $(n_i = 1)$ or absence $(n_i = 0)$ of a particle moving with momentum $\mathbf{M}_i$ at site $\mathbf{x}$ and time t. We have used $i, j, k, l, \cdots$ to label Cartesian coordinates. An equation for the evolution of the system can then be written, if we assume that the particles first move in the direction of their velocity vector and then collide with other particles. Under these conditions the evolution of the system is described by

$$n_i(\mathbf{x} + \mathbf{v}_i, t + 1) = n_i(\mathbf{x}, t) + \mathcal{B}_i[n_i(\mathbf{x}, t)] , \qquad (14.7)$$

where $\mathcal{B}_i$ is the microscopic collision operator, which is the Boolean algebra expression that corresponds to the set of collisions described above, and $\mathbf{v}_i$ is the particle velocity vector.

Let us emphasize once again that, in order for LGA (or LBA to be described below) to represent a physical phenomenon, the lattice used in the simulation must have sufficient symmetry, mass, momentum and energy must be conserved, and local equilibrium must exist and depend only upon the conserved quantities.

14.3 Lattice-Boltzmann simulation of fluid flow

Lattice-gas automata are not yet free of some drawbacks. For example, they suffer from statistical noise. This results in some physical phenomena that lead, among other things, to the divergence of viscosity in two dimensions. Moreover, they have a velocity-dependent pressure, which is unphysical. Their applicability is also limited to low flow velocities.

Various models have been proposed to circumvent the difficulties that LGA encounter (McNamara and Zanetti, 1988; Higuera *et al.*, 1989; Qian *et al.*, 1992). McNamara and Zanetti proposed a LBA model for simulating flow problems that retains the advantages that LGA have, but does not have some of their drawbacks. In their LBA $n_i(\mathbf{x}, t)$ in Eqs. (5) and (7) is a *continuous* variable between 0 and 1, and replaces the Boolean operations in $\mathcal{B}_i$ for the collisions with the appropriate arithmetic operations. Thus, instead of working with n_i, which is subject to large fluctuations and noise, one works with its mean value $N_i = \langle n_i \rangle$ which is no longer a Boolean variable. This implies that instead of following each particle in detail, one only considers the average behavior of a collection of them. Thus the evolution of the system is described by

$$N_i(\mathbf{x} + \mathbf{v}_i, t + 1) = N_i(\mathbf{x}, t) + \mathcal{B}_i[N_i(\mathbf{x}, t)] \; . \tag{14.8}$$

This scheme removes the statistical noise that LGA usually suffer. However, it can be shown that this scheme is not practical for flow simulations in three dimensions, because in practice the expression for $\mathcal{B}_i$ is a polynomial in N_i that has as many terms as there are collisions. Since there can be up to 25 different kinds of collisions in three dimensions, the usefulness of the method is lost. Higuera *et al.* (1989) proposed a simpler method in which $\mathcal{B}_i$ is a linear function. However, both methods still contain a velocity-dependent pressure.

To keep advantages of LGA and LBA and make the method more realistic and efficient, Qian *et al.* (1992) proposed a relaxation method such that the evolution of the system is described by

$$N_i(\mathbf{x} + \mathbf{v}_i, t + 1) = (1 - \omega)N_i(\mathbf{x}, t) + \omega N_{ip}(\mathbf{x}, t) \; , \tag{14.9}$$

where ω is a relaxation parameter, and N_{ip} is the predicted value of N_i at time t, given by

$$N_{ip}(\mathbf{x}, t) = e_p \rho \left[1 + \frac{\mathbf{v}_{i\alpha} \mathbf{v}_\alpha}{v_s^2} + \frac{\mathbf{v}_\alpha \mathbf{v}_\beta}{2v_s^2} \left(\frac{\mathbf{v}_{i\alpha} \mathbf{v}_{i\beta}}{v_s^2} - \delta_{\alpha\beta} \right) \right] \; , \tag{14.10}$$

where α and β denote Cartesian coordinates (with implied summation for repeated indices), ρ is the fluid density, v_s is the speed of sound, $\mathbf{v}_\alpha$ is the particle velocity in the α-direction, p is the square of the particle's velocity, and e_p is the corresponding equilibrium distribution for $\mathbf{v} = 0$. We recall that the particle velocities are 0, 1, $\sqrt{2}$, and $\sqrt{3}$, and therefore $p = 0, 1, 2$, and 3. The e_p's are calculated such that the fourth-order velocity tensor is isotropic. It can then be shown that the pressure is also independent of velocity. For a family of models with s particle velocities on a d-dimensional cubic lattice, denoted by dCs, the values of e_p are given in the following Table.

Table 14.1

Values of e_p for a model on a d-dimensional cubic lattice with s particle velocities.

Model	e_0	e_1	e_2	e_3
2C9	4/9	1/9	1/36	0
3C15	2/9	1/9	0	1/72
3C19	1/3	1/18	1/36	0
4C25	1/3	1/36	0	0

14.4 LGA and LBA simulation of single-phase flow in porous media

So far we have discussed the general principles of LGA and LBA. Let us now discuss how they are used for simulating flow in porous media. However, we should point out that these methods can also be used for simulating flow through a fracture network. All one has to do is to generate the fracture network configuration. Once this is done, simulation of flow through the network is similar to that through porous media (Knackstedt and Sahimi, 1995).

To simulate flow through porous media with the LG or LB model, one first generates the configuration of the pore space. For example, one can distribute obstacles randomly or in a prescribed fashion in a triangular or the FCHC lattice. For example, Rothman (1988) used rectangular obstacles and variations of them in his two-dimensional simulations, while Brosa and Stauffer (1989, 1991), Duarte and Brosa (1990), Kohring (1991a,b,c), Sahimi and Stauffer (1991), and Knackstedt *et al.* (1993) used circular and overlapping or non-overlapping obstacles. One main advantage of using LGA or LBA is that *any* configuration of the pore space can be used. Thus even the exact digital image of a natural porous medium can be used in the simulations.

Once the desired pore space configuration is generated, the simulation can be started. Consider, for example, a LG simulation of fluid flow in porous media. At the beginning of the simulation, one constructs a transition table that tells us how one of the present states (determined by the velocity of the incoming particles) is transformed into the next state, determined by the outgoing velocities. The table contains all of the possible states (for example, in two dimensins there are 2^8 possible states). One then starts the simulation using the two-, three- and four-body collisions at a site which conserve momentum and energy. The rules are such that no new particle is created, and in one time step all particles on the lattice move first to different lattice sites and then undergo transitions at the new sites. One iteration in the simulation consists of updating all lattice sites according to the transition table. The number of required iterations for reaching a steady-state solution depends on the porosity of the medium. If the system is far from its percolation threshold (i.e., if its porosity is large), about 10^3 iterations suffice. However, near the percolation threshold, many more iterations may be required.

An issue of particular importance is the mean free path of the particles and its relation to the size of the obstacles. As Rothman (1988) showed, in order for the LGA results to approach the continuum limit, the mean size of the void area in the lattice must be at least twice the mean free path of the particles, which is about 9 lattice bonds. If this condition is not obeyed in the simulations, the result may not represent a truely macroscopic continuum. Thus, simulations such as those of Balasubramanian *et al.* (1987), in which a fraction of the sites are blocked at random in order to create random obstacles, may not correspond to true flow in an actual porous medium, because their model does not obey the constraint imposed by the mean free path of the particles. This implies that a simple percolation algorithm in which a fraction of the sites of a lattice are blocked at random cannot be used for generating a random porous medium, if it is to be used with a LGA or LBA for simulating flow in porous media.

The LGA and LBA that we described have been used by various authors to investigate flow phenomena in porous media. Rothman (1988) studied single-phase flow in porous media. Succi *et al.* (1989) studied the same problem in three dimensions. Duarte *et al.* (1992b) and Knackstedt *et al.* (1993) used the LGA to study dynamic permeability of a porous medium (see Chapter 8). This is a notoriously difficult problem, because one has to obtain the frequency-dependent permeability by Fourier-transforming the numerical results. As discussed above, LGA results are subject to large noise and fluctuations, and therefore their Fourier transform is very difficult to obtain. Despite this, Knackstedt *et al.* (1993) showed that, if a large number of realizations are used, the LGA results are reliable. In particular, they could reproduce certain exact results for the dynamic permeability of channel-like systems. This indicates the reliability of LGA for simulating flow problems. Brosa and Stauffer (1989,1991), Kohring (1991a,b,c), and Sahimi and Stauffer (1991) looked at flow in two-dimensional porous media with various obstacle shapes and arrangements (random versus regular and periodic), and paid particular attention to the efficiency of the simulation. For example, Kohring (1991a) achieved a speed of 233 million site updates per second (233 MUPS) on a single processor of the Cray Y-MP/832, and 1690 MUPS when running in multiprocessor batch mode. This is the fastest algorithm currently available. Vollmar and Duarte (1992) studied flow through a porous membrane, and investigated the effect of various boundary conditions. Many other hydrodynamical problems in non-porous systems have been treated with this method; see Wolfram (1986b), Kadanoff *et al.* (1989), Zanetti (1989), Doolen (1991), Boon (1992), and Benzi *et al.* (1992) for reviews and details.

We should point out that because of the ease with which flow in complex geometries of a pore space can be simulated with the LG and LB methods, one can use them to estimate the permeability of a three-dimensional porous medium with a given geometry and compare the results with various rigorous results and bounds for the permeability of the same system (see Chapter 8). Cancelliere *et al.* (1990) used a LBA to undertake such a study for the Swiss Cheese or randomly-distributed penetrable sphere model discussed in Chapters 6 and 13. They found that at high solid fractions the calculated permeability is within one order of magnitude of the upper bound due to Weissberg and Prager (1970) and Berryman and Milton (1985). Moreover, one can use these models and random walk methods (see Chapter 8) to calculate the permeability K and the electrical conductivity G of the same model pore space, and thus test the validity of the relations between K and G proposed by Katz and Thompson (1986) and Johnson *et al.* (1986) which were discussed in Chapter 8. It is this author's opinion that LBA offer the most promising method for simulating complex problems of flow, dispersion and displacement processes in natural porous media.

14.5 LG and LB simulation of two-phase flow in porous media

Rothman and Keller (1988), Rothman and Zaleski (1989), Somers and Rems (1989, 1991), and Chen *et al.* (1991a,b,c) generalized the above model to study flow of two immiscible fluids in two-dimensinal porous media. Rem and Somers (1989) also studied the same problem in three dimensions. Let us consider the method of Rothman and co-workers on a triangular lattice. We follow the description given by Rothman (1990). Two types of particles, say red and blue, reside on the lattice. The velocities at a given site are numbered from 0 to 6 with $v_0 = 0$, and $v_{j+1} = [\cos(2\pi j/6), \sin(2\pi j/6)]$ [see Eq. (4)]. Two Boolean variables $r_i(\mathbf{x})$ and $b_i(\mathbf{x})$ are introduced that indicate the presence or absence of a red or blue particle with velocity v_i at lattice site $\mathbf{x}$. The system evolves in two steps. In the first step particles are propagated to the nearest-neighbor sites where they may collide with other particles. The number of both red and blue particles is conserved. To simulate the surface tension between the two fluids, the configuration resulting from a collision depends on the configuration at the nearest-neighbor sites on the lattice. Those collisions which carry red particles in the direction where the neighbors contain a relative majority of red particles are favored, and likewise for blue particles.

Two vectors are now introduced. The first one is the *color flux* given by

$$\mathbf{q} = \sum_i \mathbf{v}_i [r_i(\mathbf{x}) - b_i(\mathbf{x})] \quad , \tag{14.11}$$

while the second is the *color field* defined by

$$\mathbf{f} = \sum_i \mathbf{v}_i \sum_j [r_j(\mathbf{x} + \mathbf{v}_i) - b_j(\mathbf{x} + \mathbf{v}_i)] \quad . \tag{14.12}$$

The particle configuration that results from a collision is that configuration which maximizes $\mathbf{f} \cdot \mathbf{q}$, such that the number of red and blue particles, and also the total number of particles are conserved. This rule ensures that the fluxes of the red and blue particles in the direction of their respective mass gradient is maximized. Rothman and Keller (1988) showed that Darcy's law is valid within the regions of the red and blue particles, and that when the particles are at rest, the pressure drop across the interface between the red and blue regions obeys the Young-Laplace equation, Eq. (5.3). To simulate the effect of wettability, the sites of the lattice that belong to the solid matrix and are at the interface between the solid and the fluid are colored red or blue, so that the vector $\mathbf{f}$ calculated in the vicinity would be biased to point in the direction of the solid site. Thus red (blue) particles would preferentially spread on the red (blue) solid sites. If the "density" of coloring on the solid sites is varied, then various regimes of wettability discussed in Chapter 12 can be simulated. Chen *et al.* (1991a,b,c) used somewhat different rules for simulating two-phase flows which they claimed to be more realistic and also more efficient.

Rothman (1990) used this method to study flow in two-dimensional model, similar to the etched network of Lenormand *et al.* (1983). Good agreement between his simulations and Lenormand *et al.*'s data was found. We should mention that Appert and Zaleski (1990) proposed a LBA for liquid-vapor phase transition of the same fluid species. Thus, their model may be used to simulate two-phase flow of liquids and vapors in a porous medium, an important problem in thermal oil recovery processes. Gunstensen and Rothman (1991) extended the model further and developed a model for a three-phase fluid flow problems (for example, oil-water-gas mixtures).

14.6 Conclusions

Lattice-gas and lattice-Boltzmann automata offer efficient and simple methods of simulating flow phenomena in complex geometries, such as porous media and fractured rock. They are ideal tools for vector computers and parallel computations, which are believed to be the way massive computations of the future will be carried out. As such models become more sophisticated and realistic, more traditional numerical methods of simulating fluid flow in porous media and fractured rock, such as the finite-difference methods, lose their competitive edge, and will be phased out in the future.

References

Ababou, R. and L.W. Gelhar, in *Dynamics of Fluids in Heirarchical Porous Media*, J. H. Cushman, ed. (Academic, San Diego, 1990), p. 394.

Abbasi, M. H., J. W. Evans and I. S. Abramson, *AIChE J.* **29**, 617 (1983).

Abdassah, D. and I. Ershaghi, *SPE Form. Eval.* **1**, 113 (1986).

Abrams, A., *Soc. Pet. Eng. J.* **15**, 437 (1975).

Acuna, J. A. and Y. C. Yortsos, *Water Resour. Res.* (1994).

Adamson, A. W., *Physical Chemistry of Surfaces*, 5th edition (Wiley, New York, 1990).

Adler, P. M., *Porous Media: Geometry and Transport* (Butterworth, Boston, 1992).

Adler, P. M., *Int. J. Multiphase Flow* **11**, 91 (1985a); **11**, 213 (1985b); **11**, 241 (1985c); **11**, 853 (1985d).

Adler, P. M. and H. Brenner, *Physicochem. Hydro.* **5**, 245 (1984a); **5**, 269 (1985b); **5**, 287 (1985c).

Aharony, A. and A. B. Harris, *J. Stat. Phys.* **54**, 1091 (1989).

Ahmed, G. and J. A. Blackman, *J. Phys. C* **12**, 837 (1979).

Ahmed, G., L. M. Castanier and W. E. Brigham, *SPE Reservoir Eng.* **3**, 45 (1988).

Ahtchi-Ali, B. and H. Pedersen, *Ind. Eng. Chem. Fund.* **25**, 108 (1986).

Akanni, K. A., J. W. Evans and I. S. Abramson, *Chem. Eng. Sci.* **42**, 1945 (1987).

Aki, K., in *Earthquake Prediction: An Introductional Review*, D. W. Simpson and P. G. Richards, eds. (American Geophysical Union, Washington, D. C., 1981), p. 566.

Alben, R., G. S. Cargill, III and J. Wenzel, *Phys. Rev. B* **13**, 835 (1976).

Alexander, S. and R. Orbach, *J. Physique Lett.* **43**, 625 (1982).

Alexandroff, P., *Elementary Concepts of Topology* (Dover, New York, 1961).

Alexandrowicz, Z., *Phys. Lett.* **80A**, 284 (1980).

Amaefule, J. O. and L. L. Handy, *Soc. Pet. Eng. J.* **22**, 371 (1982).

Ambegaokar, V., B. I. Halperin and J. S. Langer, *Phys. Rev. B* **4**, 2612 (1971).

Amott, E., *Trans. AIME* **216**, 156 (1959).

Anderson, W. G., *J. Pet. Tech.* **38**, 1246 (1986); **39**, 1283 (1987a); **39**, 1453 (1987b).

Andersson, J. and B. Dverstorp, *Water Resour. Res.* **23**, 1876 (1987).

Andersson, T. B. and R. Jackson, *Ind. Eng. Chem. Fund.* **6**, 527 (1967).

Andrews, D. J., *J. Geophys. Res.* **85**, 3867 (1980).

Androutsopoulos, G. P. and R. Mann, *Chem. Eng. Sci.* **34**, 1203 (1979).

Appert, C. and S. Zaleski, *Phys. Rev. Lett.* **64**, 1 (1990).

Araktingi, V. G. and F. M. Orr, Jr., Society of Petroleum Engineers Paper 18095 (1988).

Arbabi, S. and M. Sahimi, *J. Phys. A* **23**, 2211 (1990a).

Arbabi, S. and M. Sahimi, *Phys. Rev. B* **41**, 772 (1990b).

Archie, G. E., *Trans. AIME* **146**, 54 (1942).

Aris, R., *Proc. R. Soc. Lond.* **A235**, 67 (1956).

Aris, R., *Chem. Eng. Sci.* **11**, 194 (1959).

Aris, R., *Vectors, Tensors, and the Basic Equations of Fluid Mechanics* (Prentice-Hall, Englewood Cliffs, 1962).

Aronovitz, J. A. and D. R. Nelson, *Phys. Rev. A* **30**, 1948 (1984).

Arriola, A., G. P. Wilhite and D. W. Green, *Soc. Pet. Eng. J.* **23**, 99 (1983).

Arya, A., T. A. Hewett, R. G. Larson and L. W. Lake, *SPE Reservoir Eng.* **3**, 139 (1988).

Auriault, J.-L., L. Borne and R. Chambon, *J. Accoust. Soc. Am.* **77**, 1641 (1985).

Avellaneda, M. and S. Torquato, *Phys. Fluids A* **3**, 2529 (1991).

Aviles, C. A., C. H. Scholz and J. Boatwrigth, *J. Geophys. Res.* **92**, 331 (1987).

Avnir, D., ed., *The Fractal Approach to Heterogeneous Chemistry* (Wiley, New York, 1990).

Avnir, D., D. Farin and P. Pfeifer, *J. Chem. Phys.* **79**, 3566 (1983).

Avnir, D., D. Farin and P. Pfeifer, *J. Colloid Interface Sci.* **103**, 112 (1985).

Bachmat, Y., *Water Resour. Res.* **5**, 139 (1969).

Bachmat, Y., *Israel J. Tech.* **10**, 391 (1972).

Bacri, J.-C., C. Leygnac and D. Salin, *J. Physique Lett.* **45**, L767 (1984); **46**, L467 (1985).

Bacri, J.-C., N. Rakotamalala and D. Salin, *Phys. Rev. Lett.* **58**, 2035 (1987).

Bacri, J.-C., N. Rakotamalala and D. Salin, *Phys. Fluids* **A2**, 674 (1990a).

Bacri, J.-C., M. Rosen and D. Salin, *Europhys. Lett.* **11**, 127 (1990b).

Bacri, J.-C., D. Salin and R. Wouméni, *Phys. Rev. Lett.* **67**, 2005 (1991).

Baecher, G. B. and H. H. Einstein, *Proceedings of the 19th U. S. Symposium on Rock Mechanics* (1978), p. 1.

Baecher, G. B. and N. A. Lanney, *Proceedings of the 19th U. S. Symposium on Rock Mechanics* (1978), p. 56.

Baecher, G. B., N. A. Lanney and H. H. Einstein, *Proceedings of the 18th U. S. Symposium on Rock Mech.* (1977), p. 5c1.

Bail, P. T., *Producers Monthly* **21**, 20 (1956).

Bakr, A. A., L. W. Gelhar, A. L. Gutjahr and J. R. MacMillan, *Water Resour. Res.* **14**, 263 (1978).

Baker, L. E., *Soc. Pet. Eng. J.* **17**, 219 (1977).

Bakurov, V. G., V. I. Gusev, A. F. Izmailov and A. R. Kessel, *J. Phys. A* **23**, 2507 (1990).

Balasubramanian, K., F. Hayot and W. F. Saam, *Phys. Rev. A* **36**, 2248 (1987).

Balberg, I. and N. Bienbaum, *Phys. Rev. B* **28**, 3799 (1983).

Balberg, I. and N. Bienbaum, *Phys. Rev. A* **31**, 1222 (1985).

Balberg, I., N. Bienbaum and C. H. Anderson, *Phys. Rev. Lett.* **51**, 1605 (1983).

Balberg, I., N. Bienbaum and N. Wagner, *Phys. Rev. Lett.* **52**, 1465 (1984).

Bale, H. D. and P. W. Schmidt, *Phys. Rev. Lett.* **53**, 596 (1984).

Ball, P. C. and R. Evans, *Langmuir* **5**, 714 (1989).

Banavar, J. R., M. Cieplak and D. L. Johnson, *Phys. Rev. B* **37**, 7975 (1988).

Banavar, J. R. and D. L. Johnson, *Phys. Rev. B* **35**, 7283 (1987).

Banavar, J. R., M. Lipsicas and J. E. Willemsen, *Phys. Rev. B* **32**, 6066 (1985).

Banavar, J. R. and L. M. Schwartz, *Phys. Rev. Lett.* **58**, 1411 (1987).

Barabási, A.-L., S. V. Buldyrev, S. Havlin, G. Huber, H. E. Stanley and T. Vicsek, in *Proceedings of the Les Houches Workshop*, R. Jullien, J. Kertész, P. Meakin and D. E. Wolf, eds. (Nova Science, New York, 1992).

Barenblatt, G. E. and I. P. Zheltov, *Dokl. Akad. Nauk. (USSR)* **132**, 545 (1960).

Barenblatt, G. E., I. P. Zheltov and I. N. Kochina, *J. Applied Mathematics* (USSR) **24**, 1286 (1960).

Barnsley, M. F., *Fractals Everywhere* (Academic Press, Boston, 1988).

Barnsley, M. F. and S. Demko, *Proc. Roy. Soc. Lond.* **A399**, 243 (1985).

Barret, L. K. and C. S. Yust, *Metall. (Berlin)* **3**, 1 (1970).

Barrett, E. P., L. G. Joyner and P. P. Halenda, *J. Amer. Chem. Soc.* **73**, 373 (1951).

Bartell, F. E. and F. L. Miller, *Ind. Eng. Chem.* **20**, 738 (1928).

Barton, C. C., in *Fractals and Their Use in the Earth Sciences*, C. C. Barton and P. R. LaPointe, eds. (Geological Society of America, 1992).

Barton, C. C. and P. A. Hsieh, *Physical and Hydrological-Flow Properties of Fractures*, Guidebook T385, (American Geophysical Union, Las Vegas, Nevada, 1989).

Barton, C. C. and E. Larsen, in *Proceedings of the International Symposium on Fundamentals of Rock Joints*, O. Stephenson, ed. (Bjorkliden, Sweden, 1985), p. 77.

Barton, C. C., T. A. Schutter, W. R. Page and J. K. Samuel, *Trans. Amer. Geophys. Union* **68**, 1295 (1987).

Batchelor, G. K., *An Introduction to Fluid Dynamics* (Cambridge University Press, Cambridge, 1967).

Batchelor, G. K. and R. W. O'Brien, *Proc. Roy. Soc. London* **A355**, 313 (1977).

Batrouni, G., B. Kahng and S. Redner, *J. Phys. A* **21**, L23 (1988).

Batycky, J. P., F. G. McCaffery, P. K. Hodgins and D. B. Fisher, *Soc. Pet. Eng. J.* **21**, 296 (1981).

Bear, J., *Dynamics of Fluids in Porous Media* (Elsevier, New York, 1972).

Beavers, G. S. and D. D. Joseph, *J. Fluid Mech.* **30**, 197 (1967).

Ben-Jacob, E., R. Godbey, N. D. Goldenfeld, J. Koplik, H. Levine, T. Mueller and L. M. Sander, *Phys. Rev. Lett.* **55**, 1315 (1985).

Ben-Jacob, E., G. Deutscher, P. Garik, N. D. Goldenfeld and Y. Lareath, *Phys. Rev. Lett.* **57**, 1903 (1986).

Benham, A. L. and R. W. Olson, *Soc. Pet. Eng. J.* **3**, 138 (1963).

Benner, F. C., W. W. Riches and F. E. Bartell, in *Fundamental Research on Occurence and Recovery of Petroleum* (API, New York, 1943), p. 74.

Bensimon, D., L. P. Kadanoff, S. Liang, B. I. Shraiman and C. Tang, *Rev. Mod. Phys.* **58**, 977 (1986).

Benzi, R., S. Succi and M. Vergassola, *Phys. Rep.* **222**, 145 (1992).

Beran, M. J., *Statistical Continuum Theories* (Interscience, New York, 1968).

Bergman, D. J., *Phys. Rev. Lett.* **44**, 1285 (1980).

Berman, D., B. G. Orr, H. M. Jaeger and A. M. Goldman, *Phys. Rev. B* **33**, 4301 (1986).

Bernal, J. D., *Nature* **183**, 141 (1959); **185**, 68 (1960).

Bernal, J. D., *Proc. Roy. Soc. London* **A280**, 299 (1964).

Bernard, R. A. and R. H. Wilhelm, *Chem. Eng. Prog.* **46**, 233 (1950).

Bernasconi, J., *Phys. Rev. B* **9**, 4575 (1974); **18**, 2185 (1978).

Bernasconi, J. and H. J. Wiesmann, *Phys. Rev.* **B13**, 1131 (1976).

Berryman, J. G. and S. C. Blair, *J. Appl. Phys.* **60**, 1430 (1986).

Berryman, J. G. and G. W. Milton, *J. Chem. Phys.* **83**, 745 (1985).

Bertin, H., M. Quintard, Ph. V. Corpel and S. Whitaker, *Trans. Porous Media* **5**, 543 (1990).

Bhattacharya, R. N. and V. K. Gupta, *Water Resour. Res.* **19**, 338 (1983).

Biggar, J. W. and D. R. Nielsen, *J. Geophys. Res.* **65**, 2887 (1960).

Biggar, J. W. and D. R. Nielsen, *Soil Sci. Soc. Amer. Proc.* **26**, 125 (1962).

Billardo, U., G. C. Borgia, V. Bortolotti, P. Fantazzini and E. Mesim, *J. Pet. Sci. Eng.* **5**, 273 (1991).

Billaux, D., Ph.D. Thesis, Ecole des Mines, Paris (1990).

Billaux, D., J. P. Chiles, K. Hestir and J. C. S. Long, *Int. J. Rock. Mech. Min. Sci. Geomech. Abstr.* **26**, 281 (1989).

Bird, R. B., W. E. Stewart and E. N. Lightfoot, *Transport Phenomena* (Wiley, New York, 1960).

Birovljev, A., L. Furuberg, J. Feder, T. Jøssang K. J. Måløy, and A. Aharony, *Phys. Rev. Lett.* **67**, 584 (1991).

Bjerrum, N. and E. Manegold, *Kolloid Z. (USSR)* **43**, 5 (1927).

Blackman, J. A., *J. Phys. C* **9**, 2049 (1976).

Blackwell, R. J., *Soc. Pet. Eng. J.* **2**, 1 (1962).

Blackwell, R. J., J. R. Rayne and W. M. Terry, *Trans. AIME* **216**, 1 (1959).

Blatt, H., G. Middleton and R. Murray, *Origin of Sedimentary Rocks* (Prentice Hall, New York, 1980), Chapter 14.

Blumberg Selinger, R., J. Nittmann and H. E. Stanley, *Phys. Rev. A* **40**, 2590 (1989).

Blunt, M., J. W. Barker, B. Rubbin, M. Mansfield, I. D. Culverwell and M. A. Christie, *SPE Reservoir Eng.* **9**, 73 (1994).

Blunt, M. and P. R. King, *Phys. Rev. A* **37**, 3935 (1988); **42**, 4780 (1990).

Blunt, M. and P. R. King, *Trans. Porous Media* **6**, 407 (1991).

Bonnecaze, R. T. and J. F. Brady, *Proc. R. Soc. Lond.* **A430**, 285 (1990).

Bonnet, J. and R. Lenormand, *Rev. Inst. Franc. Pet.* **42**, 477 (1977).

Boon, J. P., ed., *Proceedings of Advanced Research Workshop on Lattice Gas Automata Theory, Implementation, and Simulation,* J. Stat. Phys. **68**, No. 3/4 (1992).

Bouchaud, J. P., A. George, J. Koplik, A. Provata and S. Redner, *Phys. Rev. Lett.* **64**, 2503 (1990).

Bourdet, D., J. A. Ayoub, T. M. Whittle, Y. M. Pirard and V. Kniazeff, *World Oil*, 77 (October 1983).

Brace, W. F. and A. S. Orange, *J. Geophys. Res.* **73**, 5407 (1968).

Brace, W. F., J. B. Walsh and W. T. Frangos, *J. Geophys. Res.* **73**, 2225 (1968).

Bracewell, R. N., O. Buneman, H. Hao and J. Villasenor, Proc. IEEE **74**, 1282 (1986).

Brandt, W. W., *J. Chem. Phys.* **63**, 5162 (1975).

Brenner, H., *Chem. Eng. Sci.* **17**, 229 (1962).

Brenner, H., *Phil. Trans. R. Soc. Lond.* **A297**, 81 (1980).

Brenner, H. and P. M. Adler, *Phil. Trans. R. Soc. Lond.* **A307**, 149 (1982).

Brigham, W. E., *Soc. Pet. Eng. J.* **14**, 91 (1974).

Brigham, W. E., P. W. Reed and J. N. Dew, *Soc. Pet. Eng. J.* **1**, 1 (1961).

Brinkman, H. C., *Apply. Sci. Res.* **A1**, 27 (1947).

Broadbent, S. R. and J. M. Hammersley, *Proc. Camb. Phil. Soc.* **53**, 629 (1957).

Broadwell, J. E., *Phys. Fluids* **7**, 1243 (1964).

Brosa, U. and D. Stauffer, *J. Stat. Phys.* **57**, 399 (1989); **63**, 405 (1991).

Brown, S. R., *Geophys. Res. Lett.* **14**, 1095 (1987a).

Brown, S. R., *J. Geophys. Res.* **B92**, 1337 (1987b).

Brown, S. R. and C. H. Scholz, *J. Geophys. Res.* **B90**, 12575 (1985).

Brownstein, K. R. and C. E. Tarr, *Phys. Rev. A* **19**, 2446 (1979).

Bruggeman, D. A. G., *Annl. Phys.* **24**, 636 (1935).

Bryant, S. L., P. R. King and D. W. Mellor, *Trans. Porous Media* (1993a).

Bryant, S. L., D. W. Mellor and C. A. Cade, *AIChE J.* **39**, 387 (1993b).

Buckley, J. S., in *Interfacial Phenomena in Petroleum Recovery*, N. R. Morrow, ed. (Marcel Dekker, New York, 1991), p. 157.

Buckley, S. E. and M. C. Leverett, *Trans. AIME* **146**, 107 (1942).

Buka, A., J. Kertész and T. Vicsek, *Nature (London)* **323**, 424 (1986).

Buldyrev, S. V., A.-L. Barabási, F. Caserta, S. Havlin, H. E. Stanley and T. Vicsek, *Phys. Rev. A* **45**, R8313 (1992).

Bunde, A., A. Coniglio, D. C. Hong and H. E. Stanley, *J. Phys. A* **18**, L137 (1985).

Bunde, A. and S. Havlin, (eds.) *Fractals and Disordered Systems* (Springer, Berlin, 1991).

Bussian, A. E., *Geophysics* **48**, 1258 (1983).

Butcher, P. N., *J. Phys. C* **8**, L324 (1975).

Buyevich, Y. A., A. I. Lenov and V. M. Safrai, *J. Fluid Mech.* **37**, 371 (1969).

Cacas, M. C., *et al. Water Resour. Res.* **26**, 479 (1990a); **26**, 491 (1990b).

Caldwell, J. A., in *Proceedings of Symposium on Percolation Through Fissured Rocks* (International Society for Rock Mechanics and International Association of Engineering Geology, Stuttgart, Germany, 1972), p. 115.

Cancelliere, A., C. Chang, E. Foti, D. H. Rothman and S. Succi, *Phys. Fluids A* **2**, 2085 (1990).

Carberry, J. J. and R. H. Bretton, *AIChE J.* **4**, 367 (1958).

Carbonell, R. G. and S. Whitaker, *Chem. Eng. Sci.* **38**, 1795 (1983).

Castillo, E., R. J. Krizeck and G. M. Karadi, *Proceedings of the Symposium on Fundamentals of Transport Phenomena in Porous Media* (Guelph, Canada, 1972), p. 778.

Cates, M. E. and T. A. Witten, *Phys. Rev. Lett.* **56**, 2497 (1986).

Chan, D. Y. C., B. D. Hughes and L. Paterson, *Phys. Rev. A* **34**, 4079 (1986).

Chan, D. Y. C., B. D. Hughes, L. Paterson and C. Sirakoff, *Phys. Rev. A* **38**, 4106 (1988).

Chandler, R., J. Koplik, K. Lerman and J. F. Willemsen, *J. Fluid Mech.* **119**, 249 (1982).

Chang, S.-H. and J. C. Slattery, *Trans. Porous Media* **1**, 179 (1986).

Chapman, A. M. and J. J. L. Higdon, *Phys. Fluids A* **4**, 2099 (1992).

Charlaix, E., *J. Phys. A* **18**, L533 (1986).

Charlaix, E., E. Guyon and N. River, *Solid State Commun.* **50**, 999 (1984).

Charlaix, E., E. Guyon and S. Roux, *Trans. Porous Media* **2**, 31 (1987a).

Charlaix, E., J.-P. Hulin and T. J. Plona, *Phys. Fluids* **30**, 1690 (1987b).

Charlaix, E., J.-P. Hulin, C. Leroy and C. Zarcone, *J. Phys. D* **21**, 1727 (1988a).

Charlaix, E., A. P. Kushnik and J. P. Stokes, *Phys. Rev. Lett.* **61**, 1595 (1988b).

Charpentier, J. C., *Chem. Eng. J.* **11**, 161 (1976).

Chatenever, A. and J. C. Calhoun, *Pet. Trans. AIME* **195**, 149 (1952).

Chatwin, P. C., *J. Fluid Mech.* **80**, 33 (1977).

Chatzis, I. and F. A. L. Dullien, *J. Can. Pet. Tech.* **16**, 97 (1977).

Chatzis, I. and F. A. L. Dullien, *Rev. Inst. Franc. Pet.* **37**, 183 (1982).

Chatzis, I. and F. A. L. Dullien, *Int. Chem. Eng.* **25**, 47 (1985).

Chatzis, I. and N. R. Morrow, *Soc. Pet. Eng. J.* **24**, 555 (1984).

Chatzis, I., N. R. Morrow and H. T. Lim, *Soc. Pet. Eng. J.* **23**, 311 (1983).

Chaudhari, N. M. and A. E. Scheidegger, *Can. J. Phys.* **43**, 1776 (1965).

Chayes, J. T., L. Chayes and C. M. Newman, *Commun. Math. Phys.* **107**, 611 (1986).

Chelidze, T. and Y. Gueguen, *Int. J. Rock Mech. Min. Sci. Geomech. Abstr.* **27**, 223 (1990).

Chen, J.-D. and J. Koplik, *J. Colloid Interface Sci.* **108**, 304 (1985).

Chen, J.-D. and N. Wada, *Exp. Fluids* **4**, 336 (1986).

Chen, J.-D. and D. Wilkinson, *Phys. Rev. Lett.* **55**, 1892 (1985).

Chen S., G. D. Doolen, K. Eggert, D. G. Grunau and E. Y. Loh, *Phys. Rev. A* **43**, 245 (1991b).

Chen S., G. D. Doolen and W. H. Matthaues, *J. Stat. Phys.* **64**, 1133 (1991c).

Chen S., K. Diemer, G. D. Doolen, K. Eggert, S. Gutman and B. J. Travis, *Physica D* **47**, 72 (1991a).

Chen, Z.-X., *Trans. Porous Media* **4**, 147 (1989).

Cheshire, I. M. and R. K. Pollard, in *Mathematics of Oil Production*, S. F. Edwards and P. R. King, eds. (Clarendon, Oxford, 1988), p. 253.

Chiew, Y. C. and E. D. Glandt, *J. Colloid Interface Sci.* **99**, 86 (1984).

Chikhliwala, E. D., A. B. Huang and Y. C. Yortsos, *Trans. Porous Media* **3**, 257 (1988).

Chikhliwala, E. D. and Y. C. Yortsos, *SPE Reservoir Eng.* **3**, 1268 (1988).

Childress, S., *J. Chem. Phys.* **56**, 2527 (1972).

Chouke, R. L., P. van Meurs, and C. van der Poel, *Trans. AIME* **216**, 188 (1959).

Christensen, G., S. J. McGovern and S. Sundaresan, *AIChE J.* **32**, 1677 (1986).

Christie, M. A., in *Mathematics of Oil Production*, S. F. Edwards and P. R. King, eds. (Clarendon, Oxford, 1988), p. 269.

Christie, M. A. and D. J. Bond, *SPE Reservoir Eng.* **2**, 514 (1987).

Chu, C. F. and K. M. Ng, *AIChE J.* **35**, 148 (1989).

Cieplak, M. and M. O. Robbins, *Phys. Rev. Lett.* **60**, 2042 (1988).

Cieplak, M. and M. O. Robbins, *Phys. Rev. B* **41**, 11508 (1990).

Clément, E., C. Baudet, E. Guyon and J.-P. Hulin, *J. Phys. D* **20**, 608 (1987).

Clemo, T. M. and L. Smith, *Trans. Amer. Geophys. Union* **70**, 43 (1989).

Closmann, P. J., *Soc. Pet. Eng. J.* **15**, 385 (1975).

Cloud, W. F., *Oil Weekly* **103**, 26 (1941).

Coats, K. H., J. R. Dempsey and J. H. Henderson, *Soc. Pet. Eng. J.* **11**, 63 (1971).

Coats, K. H., R. L. Nielson, M. H. Terhune and A. G. Weber, *Soc. Pet. Eng. J.* **7**, 377 (1967).

Coats, K. H. and B. D. Smith, *Soc. Pet. Eng. J.* **4**, 73 (1964).

Cohen, M. H. and K. S. Mendelson, *J. Appl. Phys.* **53**, 1127 (1982).

Collins, R. E., *Flow of Fluids Through Porous Media* (Pennwell, Tulsa, 1961), p. 196.

Conner, Wm. C., J. Horowitz and A. M. Lane, *AIChE Symp. Series* **88** (**No. 266**), 29 (1988).

Conner, Wm. C. and A. M. Lane, *J. Catal.* **84**, 217 (1984).

Conner, Wm. C., A. M. Lane and A. J. Hoffman, *J. Colloid Interface Sci.* **600**, 185 (1984).

Constantinides, G. N. and A. C. Payatakes, *Chem. Eng. Commun.* **81**, 55 (1989).

Constantinides, G. N. and A. C. Payatakes, *J. Colloid Interface Sci.* **141**, 486 (1991).

Cox, R. G., *J. Fluid Mech.* **168**, 169 (1985).

Craig, F. F., *The Reservoir Engineering Aspects of Waterflooding* (Society of Petroleum Engineers, Richardson, Texas, 1971), p. 12.

Crine, M., P. Marchot and G. Lhomme, *Chem. Eng. Commun.* **7**, 377 (1980a).

Crine, M., P. Marchot and G. Lhomme, *Chem. Eng. Sci.* **35**, 51 (1980b).

Crossley, P. A., L. M. Schwartz and J. R. Banavar, *Appl. Phys. Lett* **59**, 3553 (1991).

Cruden, D. M., *Int. J. Rock. Mech. Min. Sci. Geomech. Abtr.* **14**, 133 (1977).

Cuiec, L., Society of Petroleum Engineers Paper 13211 (1984).

Cushman, J. H., *Water Resour. Res.* **20**, 1668 (1984); **27**, 643 (1991).

Cussler, E. L., *AIChE J.* **28**, 500 (1982).

Cussler, E. L., J. Kopinsky and J. A. Weimer, *Chem. Eng. Sci.* **38**, 2027 (1983).

Daccord, G., J. Nittmann and H. E. Stanley, *Phys. Rev. Lett.* **56**, 336 (1986).

Dagan, G., 1981, *Water Resour. Res.* **17**, 107 (1981); **18**, 813 (1982a); **18**, 1571 (1982b); **22**, 1205 (1986).

Dagan, G., *Annu. Rev. Fluid Mech.* **19**, 183 (1987).

Dankworth, D. C. and S. Sundaresan, *AIChE J.* **35**, 1282 (1989).

David, C., Y. Gueguen and G. Pampoukis, *J. Geophys. Res.* **95**, 6993 (1990).

Davis, H. T., *Europhys. Lett.* **8**, 629 (1989).

Davis, H. T., R. A. Novy, L. E. Scriven and P. G. Toledo, *J. Phys. Condens. Matter* **2**, SA457 (1990).

Davis, H. T., L. R. Valencourt and C. E. Johnson, *J. Amer. Ceram. Soc.* **58**, 446 (1975).

Davis, J. A. and S. C. Jones, *J. Pet. Tech.* **20**, 1415 (1968).

Day, P. R., *Trans. Am. Geophys. Union* **37**, 595 (1956).

de Arcangelis, L., A. Hansen, H. J. Herrmann and S. Roux, *Phys. Rev. B* **40**, 877 (1989).

de Arcangelis, L., J. Koplik, S. Redner and D. Wilkinson, *Phys. Rev. Lett.* **57**, 996 (1986a).

de Arcangelis, L., S. Redner and A. Coniglio, *Phys. Rev. B* **34**, 4656 (1986b).

de Gennes, P. G., *La Recherche* **7**, 919 (1976).

de Gennes, P. G., *C. R. Acad. Sci.* **295**, 1061 (1982).

de Gennes, P. G., *J. Fluid Mech.* **136**, 189 (1983a).

de Gennes, P. G., *Physicochem. Hydro.* **4**, 175 (1983b).

de Gennes, P. G., in *Physics of Disordered Materials*, M. Daoud, ed. (Plenum, London, 1985), p. 227.

de Gennes, P. G., *Europhys. Lett.* **5**, 689 (1988).

de Gennes, P. G. and E. Guyon, *J. Mechanique* **17**, 403 (1978).

DeGregoria, A. J., *Phys. Fluids* **28**, 2933 (1985); **29**, 3557 (1986).

de Josselin de Jong, G., *Trans. Amer. Geophys. Union* **39**, 67 (1958).

Dean, R. H. and L. L. Lo, *SPE Reservoir Eng.* **2**, 638 (1988).

Deans, H. A., *Soc. Pet. Eng. J.* **3**, 49 (1963).

Debye, P., H. R. Anderson, Jr. and H. Brumberger, *J. Appl. Phys.* **28**, 679 (1957).

Delshad, M., D. J. MacAllister, G. A. Pope and B. A. Rouse, *Soc. Pet. Eng. J.* **25**, 476 (1985).

Derrida, B. and J. Vannimenus, *J. Phys. A* **15**, L557 (1982).

Derrida, B., J. G. Zabolitzky, J. Vannimenus and D. Stauffer, *J. Stat. Phys.* **36**, 31 (1994).

de Santos, J. M., T. R. Melli and L. E. Scriven, *Annu. Rev. Fluid Mech.* **23**, 233 (1991).

De Smedt, F. and P. J. Wierenga, *J. Hydrol.* **41**, 59 (1979).

d'Humieres, D. and P. Lallemand, *Physica* **140A**, 326 (1986).

d'Humieres, D., P. Lallemand and U. Frisch, *Europhys. Lett.* **2**, 291 (1986).

Dias, M. M. and A. C. Payatakes, *J. Fluid Mech.* **164**, 305 (1986a); **164**, 337 (1986b).

Dias, M. M. and D. Wilkinson, *J. Phys. A* **19**, 3131 (1986).

Diaz, C. E., I. Chatzis and F. A. L. Dullien, *Trans. Porous Media* **2**, 215 (1987).

Dienes, J. K., Los Alamos Scientific Laboratory, Report LA-8553-PR, (1980), p. 19.

Dimenstein, D. M. and K. M. Ng, *Chem. Eng. Commun.* **41**, 215 (1986).

Dodd, C. G. and O. G. Keil, *J. Phys. Chem.* **63**, 299 (1959).

Donaldson, E. C., P. B. Lorenz and R. D. Thomas, Society of Petroleum Engineers Paper 1562 (1966).

Donaldson, E. C., R. D. Thomas and P. B. Lorenz, *Soc. Pet. Eng. J.* **9**, 13 (1969).

Doolen, G., ed., *Proceedings of NATO Advanced Research Workshop on Lattice Gas Methods for PDE's: Theory, Application and Hardware*, Physica **D47** (1991).

Dougherty, E. L., *Soc. Pet. Eng. J.* **3**, 155 (1963).

Douglas, J., D. W. Peaceman and H. H. Rachford, *Trans. AIME* **216**, 297 (1959).

Doyen, P. M., *J. Geophys. Res.* **93**, 7729 (1988).

Duarte, J. A. M. S. and U. Brosa, *J. Stat. Phys.* **59**, 501 (1990).

Duarte, J. A. M. S., J. de Carvalho and H. Ruskin, *Physica* **A183**, 411 (1992a).

Duarte, J. A. M. S., M. Sahimi and J. de Carvalho, *J. Physicque II*, **2** (1992b).

Duering, E. and D. J. Bergman, *J. Stat. Phys.* **60**, 363 (1990).

Dullien, F. A. L., *AIChE J.* **21**, 299 (1975).

Dullien, F. A. L., *Porous Media: Fluid Transport and Pore Structure* (Academic Press, New York, 1979).

Dullien, F. A. L. and G. K. Dhawan, *J. Colloid Interface Sci.* **52**, 129 (1975).

Dullien, F. A. L., F. S. Y. Lai and I. F. MacDonald, *J. Colloid Interface Sci.* **109**, 201 (1986).

Dussan V., E. B., *Ann. Rev. Fluid Mech.* **11**, 371 (1979).

Dussan V., E. B. and S. H. Davis, *J. Fluid Mech.* **65**, 71 (1974).

Dverstorp, B. and J. Andersson, *Water Resour. Res.* **25**, 540 (1989).

Dykstra, H. and R. L. Parsons, in *Secondary Recovery of Oil in the United States*, 2nd ed. (API, New York, 1950), p. 160.

Ebach, E. A. and R. R. White, *AIChE J.* **4**, 161 (1958).

Edwards, M. F. and J. F. Richardson, *Chem. Eng. Sci.* **23**, 109 (1968).

Egbogah, E. O. and R. A. Dawe, *Bull. Can. Pet. Tech.* **28**, 200 (1980).

Ehrlich, R., P. J. Brown, J. M. Yarus and D. T. Eppler, in *Advanced Particulate Morphology*, J. K. Beddow and T. P. Meloy, eds. (CRC Press, Boca Raton, Florida, 1980), p. 101.

Eidsath, A., R. G. Carbonell, S. Whitaker and L. R. Herrmann, *Chem. Eng. Sci.* **38**, 1803 (1983).

Elam, W. T., A. R. Kerstein and J. J. Rehr, *Phys. Rev. Lett.* **52**, 1516 (1984).

Ellman, M. J., N. Midoux, A. Laurent and J. C. Charpentier, *Chem. Eng. Sci.* **43**, 2201 (1988).

Emanuel, A. S., G. K. Alameda, R. A. Behrens and T. A. Hewett, *SPE Reservoir Eng.* **3**, 311 (1989).

Endo, H. K., J. C. S. Long, C. R. Wilson and P. A. Witherspoon, *Water Resour. Res.* **20**, 1390 (1984).

Englman, R., Y. Gur and Z. Jaeger, *J. Appl. Mech.* **50**, 707 (1983).

Erdös, P. and S. B. Haley, *Phys. Rev. B* **13**, 1720 (1976).

Evans, J. W., M. H. Abbasi and A. Sarin, *J. Chem. Phys.* **72**, 2967 (1980).

Everett, D. H., *Trans. Faraday Soc.* **50**, 1077 (1954).

Everett, D. H., in *The Solid-Gas Interface*, Vol. II, E. Elison Flood, ed. (Marcel Dekker, New York, 1967), p. 1055.

Everett, D. H. and J. M. Haynes, *J. Colloid Interface Sci.* **38**, 125 (1972).

Family, F., in *Random Walks and Their Applications in the Physical and Biological Sciences*, M. F. Shlesinger and B. J. West, eds. (AIP, New York, 1984), p. 33.

Family, F. and T. Vicsek, *J. Phys. A* **18**, L75 (1985).

Family, F. and T. Vicsek, eds., *Dynamics of Fractal Surfaces* (World-Scientific, Singapore, 1991).

Fara, H. D. and A. E. Scheidegger, *J. Geophys. Res.* **66**, 3279 (1961).

Fatt, I., *Trans. AIME* **207**, 144 (1956).

Fatt, I., *Science* **131**, 158 (1960).

Fayers, F. J., *SPE Reservoir Eng.* **3**, 542 (1988).

Feder, J., *Fractals* (Plenum, New York, 1988).

Feeney, R., S. L. Schmidt, P. Strickholm, J. Chadam and P. Ortoleva, *J. Chem. Phys.* **78**, 1293 (1983).

Feng, S., B. I. Halperin and P. N. Sen, *Phys. Rev. B* **35**, 197 (1987).

Ferer, M., R. A. Geisbrecht, W. N. Sams and D. H. Smith, *Phys. Rev. A* **45**, R6973 (1992).

Fernández, J. F., R. Rangel and J. Rivero, *Phys. Rev. Lett.* **67**, 2958 (1991).

Finlayson, B. A., *Nonlinear Analysis in Chemical Engineering* (McGraw-Hill, New York, 1980).

Finney, J., Ph.D. Thesis, University of London (1968).

Firoozabadi, A. and J. Hauge, *J. Pet. Tech.* **42**, 784 (1990).

Firoozabadi, A., K. Ishimoto and B. Dindoruk, Society of Petroleum Engineers Paper 21798 (1991).

Fisher, M. E., in *Critical Phenomena: Enrico Fermi Summer School*, edited by M. S. Green (Academic Press, New York, 1971), p. 1.

Fisher, M. E. and J. W. Essam, *J. Math. Phys.* **2**, 609 (1961).

Fisher, R. L. and P. D. Lark, *J. Colloid Interface Sci.* **69**, 486 (1979).

Flory, P. J., *J. Amer. Chem. Soc.* **63**, 3083 (1941).

Forchheimer, P., *Z. Ver. Deuts. Ing.* **45**, 1782 (1901).

Fowler, A. D., H. E. Stanley and G. Daccord, *Nature (London)* **341**, 134 (1989).

Frette, V., K. J. Måløy, F. Boger, J. Feder, T. Jøssang and P. Meakin, *Phys. Rev. A* **42**, 3432 (1990).

Fried, J. J. and M. A. Combarnous, *Adv. Hydrosci.* **7**, 169 (1971).

Frisch, U., B. Hasslacher and Y. Pomeau, *Phys. Rev. Lett.* **56**, 1505 (1986).

Frisch, U., D. d'Humieres, B. Hasslacher, P. Lallemand, Y. Pomeau and J.-P. Rivet, *Complex Syst.* **1**, 648 (1987).

Frisch, U. and Parisi, G., in *Turbulence and Predictability in Geophysical Fluid Dynamics and Climate Dynamics*, M. Ghil, R. Benzi and G. Parisi, eds. (North-Holland, Amsterdam, 1985).

Furuberg, L., J. Feder, A. Aharony and T. Jøssang, *Phys. Rev. Lett.* **61**, 2117 (1988).

Gale, J. E., *Proceedings of the 28th U. S. Symposium of Rock Mechanics*, Tucson, Arizona (1987), p. 1213.

Gardner, J. W. and J. G. J. Ypma, Society of Petroleum Engineers Paper 10686 (1982).

Gardner, W. R. and Brooks, R. H., *Soil Sci.* **83**, 195 (1957).

Gaudet, J. P., H. Jégat, G. Vachaud and P. J. Wierenga, *Soil Sci. Soc. Amer. J.* **41**, 665 (1977).

Gawlinski, E. T. and S. Redner, *J. Phys. A* **16**, 1063 (1983).

Gawlinski, E. T. and H.E. Stanley, *J. Phys. A* **14**, L291 (1981).

Gefen, Y., A. Aharony and S. Alexander, *Phys. Rev. Lett.* **50**, 77 (1983).

Geffen, T. M., W. W. Owens, D. R. Parrish and R. A. Morse, *Trans. AIME* **192**, 99 (1951).

Gelhar, L. W. and C. L. Axness, *Water Resour. Res.* **19**, 161 (1983).

Gelhar, L. W., A. L. Gutjahr and R. J. Naff, *Water Resour. Res.* **15**, 1387 (1979).

Gelhar, L. W., C. Welty and K. R. Rehfelt, *Water Resour. Res.* **28**, 1955 (1992).

Gilman, J. and H. Kazemi, *Soc. Pet. Eng. J.* **23**, 675 (1983).

Gilvarry, J. J., *Solid State Commun.* **2**, 9 (1964).

Gingold, D. B. and C. J. Lobb, *Phys. Rev. B* **42**, 8220 (1990).

Giordano, R. M. and S. J. Salter, Society of Petroleum Engineers Paper 13165 (1984).

Giordano, R. M., S. J. Salter and K. K. Mohanty, Society of Petroleum Engineers Paper 14365 (1985).

Gist, G. A., A. H. Thompson, A. J. Katz and R. L. Higgins, *Phys. Fluids* **A2**, 1533 (1990).

Goddin, C. S., F. F. Craig, J. O. Wilkes and M. R. Tek, *J. Pet. Tech.* **18**, 765 (1966).

Golden, J. M., *Water Resour. Res.* **16**, 201 (1980).

Goode, P. A. and T. S. Ramakrishnan, *AIChE J.* **39**, 1124 (1993).

Gouyet, J. F., M. Rosso and B. Sapoval, *Phys. Rev. B* **37**, 1832 (1988).

Grane, F. E. and G. H. F. Gardner, *J. Chem. Eng. Data* **6**, 283 (1961).

Grassberger, P., *Phys. Lett. A* **97**, 277 (1983).

Gray, W. G., *Chem. Eng. Sci.* **30**, 229 (1975).

Gray, W. G. and K. O'Niel, *Water Resour. Res.* **12**, 148 (1976).

Greenberg, R. J. and W. F. Brace, *J. Geophys. Res.* **74**, 2099 (1969).

Greenkorn, R. A., C. R. Johnson and R. E. Haring, *J. Pet. Tech.* **5**, 329 (1965).

Griffiths, A., *Proc. Roy. Soc. Lond.* **23**, 190 (1911).

Griffiths, A. A., *Phil. Trans. Roy. Soc. Lond.* **221**, 163 (1921).

Grisak, G. E. and J. F. Pickens, *Water Resour. Res.* **16**, 719 (1980).

Grosser, K., R. G. Carbonell and S. Sundaresan, *AIChE J.* **34**, 1850 (1988).

Gueguen, Y. and J. K. Dienes, *Math. Geol.* **21**, 1 (1989).

Gunn, D. J. and C. Pryce, *Trans. Inst. Chem. Eng.* **47**, T341 (1969).

Gunstensen, A. and D. H. Rothman, *Physica D* **47**, 47 (1991).

Gutjahr, A. L. and L. W. Gelhar, *Water Resour. Res.* **19**, 161 (1981).

Gutjahr, A. L., L. W. Gelhar, A. A. Bakr and J. R. MacMillan, *Water Resour. Res.* **14**, 953 (1978).

Güven, O., R. W. Falta, F. J. Molz and J. G. Melville, *Water Resour. Res.* **21**, 676 (1985).

Güven, O., F. J. Molz and J. G. Melville, *Water Resour. Res.* **20**, 1337 (1984).

Guyer, R. A., *Phys. Rev. A* **32**, 2324 (1985).

Guyon, E., L. Oger and T.J. Plona, *J. Phys. D* **20**, 1637 (1987).

Haan, S. W. and R. Zwanzig, *J. Phys. A* **10**, 1547 (1977).

Habermann, B., *Trans. AIME* **219**, 264 (1960).

Hagiwara, T., Society of Petroleum Engineers Paper 13100 (1984).

Hagoort, J., *Soc. Pet. Eng. J.* **14**, 63 (1974).

Haji-Sheikh, A. and E. M. Sparrow, *SIAM J. Appl. Math.* **14**, 370 (1966).

Haldorsen, H. H., P. J. Brand and C. J. MacDonald, in *Mathematics in Oil Production*, S. F. Edwards and P. R. King, eds. (Clarendon, Oxford, 1988), p. 109.

Haldorsen, H. H. and E. Damsleth, *J. Pet. Tech.* **42**, 404 (1990).

Haldorsen, H. H. and L. W. Lake, *Soc. Pet. Eng. J.* **24**, 447 (1984).

Hall, P. L., D. F. R. Mildner and R. L. Brost, *J. Geophys. Res.* **B91**, 2183 (1986).

Halperin, B. I., S. Feng and P. N. Sen, *Phys. Rev. Lett.* **54**, 2891 (1985).

Halsey, T. C., P. Meakin and I. Procaccia, *Phys. Rev. Lett.* **56**, 854 (1986).

Hammond, P. S., *J. Fluid Mech.* **137**, 363 (1983).

Han, N.-W., J. Bhakta and R. G. Carbonell, *AIChE J.* **31**, 277 (1985).

Hansen, J. P. and A. T. Skjeltorp, *Phys. Rev. B* **38**, 2635 (1988).

Happel, J. and H. Brenner, *Low Reynolds Number Hydrdynamics* (Nijhoff, San Diego, 1983).

Hardy, J., O. de Pazzis and Y. Pomeau, *Phys. Rev. A* **13**, 1949 (1976).

Hardy, J., Y. Pomeau and O. de Pazzis, *J. Math. Phys.* **14**, 1746 (1973).

Haring, R. E. and R. A. Greenkorn, *AIChE J.* **16**, 477 (1970).

Harleman, D. R. F. and R. R. Rumer, *J. Fluid Mech.* **16**, 385 (1963).

Harris, A. B. and A. Aharony, *Europhys. Lett.* **4**, 1355 (1987).

Harris, C. K., *Trans. Porous Media* **5**, 517 (1990); **9**, 287 (1992).

Hasimoto, H., *J. Fluid Mech.* **5**, 317 (1959).

Hassinger, R. C. and D. U. von Rosenberg, *Soc. Pet. Eng. J.* **8**, 195 (1968).

Hassler, G. L. and E. Brunner, *Trans. AIME* **160**, 114 (1945).

Hatziavramidis, D. T., *SPE Reservoir Eng.* **3**, 631 (1988).

Haughey, D. P. and G. S. G. Beveridge, *Can J. Chem. Eng.* **47**, 130 (1969).

Haus, J. W. and K. W. Kehr, *Phys. Rep.* **150**, 263 (1987).

Havlin, S. and D. Ben-Avraham, *Adv. Phys.* **36**, 695 (1987).

Havlin, S., S. V. Buldyrev, H. E. Stanley and G. H. Weiss, *J. Phys. A* **24**, L925 (1991).

Hayes, J. B., *The Society of Paleontologists and Mineralogists Special Publication No. 26*, (1979), p. 127.

Hearn, C. L., *J. Pet. Tech.* **23**, 805 (1971).

Heaviside, J., in *Interfacial Phenomena in Petroleum Recovery*, N. R. Morrow, ed. (Marcel Dekker, New York, 1991), p. 377.

Heiba, A. A., Ph.D. Thesis, University of Minnesota, Minneapolis (1985).

Heiba, A. A., H. T. Davis and L. E. Scriven, Society of Petroleum Engineers Paper 12172 (1983).

Heiba, A. A., H. T. Davis and L. E. Scriven, Society of Petroleum Engineers Paper 12690 (1984).

Heiba, A. A., M. Sahimi, L. E. Scriven and H. T. Davis, Society of Petroleum Engineeris Paper 11015 (1982).

Heiba, A. A., M. Sahimi, L. E. Scriven and H. T. Davis, *SPE Reservoir Eng.* **7**, 123 (1992).

Hele-Shaw, H. J. S., *Nature* **58**, 34 (1898).

Heller, J.P., *J. Appl. Phys.* **37**, 1566 (1966).

Heller, J.P., in *Proceedings of the Second International Conference on Fundamentals of Transport Phenomena in Porous Media*, D. E. Elrick, ed. (International Association of Hydraulic Research, Guelph, Canada, 1972), p. 1.

Hentschel, H. G. E. and F. Family, *Phys. Rev. Lett.* **66**, 1982 (1991).

Hentschel, H. G. E. and Procaccia, I., *Physica D* **8**, 435 (1983).

Herrmann, H. J. and S. Roux, eds., *Statistical Models for the Fracture of Disordered Media* (North-Holland, Amsterdam, 1990).

Herskowitz, M. and J. M. Smith, *AIChE J.* **29**, 1 (1983).

Hestir, K. and J. C. S. Long, *J. Geophys. Res.* **95**, 21565 (1990).

Hewett, T. A., Society of Petroleum Engineers Paper 15386 (1986).

Hewett, T. A. and R. A. Behrens, *SPE Form. Eval.* **5**, 217 (1990).

Hickernell, F. J. and Y. C. Yortsos, *Stud. Appl. Math.* **74**, 93 (1986).

Higuera, F., S. Succi and R. Benzi, *Europhys. Lett.* **9**, 663 (1989).

Hilfer, R., *Phys. Rev.* **B44**, 628 (1991a); **44**, 60 (1991b).

Hill, S., *Chem. Eng. Sci.* **1**, 247 (1952).

Hinch, E. J., *J. Fluid Mech.* **83**, 695 (1977).

Hoffman, R. L., *J. Colloid Interface Sci.* **50**, 228 (1975).

Homsy, G. M., *Ann. Rev. Fluid Mech.* **19**, 271 (1987).

Hori, M. and F. Yonezawa, *J. Phys. C* **10**, 229 (1977).

Horn, F. J. M., *AIChE J.* **17**, 613 (1971).

Horváth, V. K., F. Family and T. Vicsek, *Phys. Rev. Lett.* **65**, 1388 (1990).

Horváth, V. K., F. Family and T. Vicsek, *J. Phys. A* **24**, L25 (1991a).

Horváth, V. K., F. Family and T. Vicsek, *Phys. Rev. Lett.* **67**, 3207 (1991b).

Horváth, V. K., J. Kertész and T. Vicsek, *Europhys. Lett.* **4**, 1133 (1987a).

Horváth, V. K., T. Vicsek and J. Kertész, *Phys. Rev. A* **35**, 2353 (1987b).

Hoshen, J. and R. Kopelman, *Phys. Rev.* **B24**, 3438 (1976).

Hough, S. E., *Geophys. Res. Lett.* **16**, 673 (1989).

Howells, I. D., *J. Fluid Mech.* **64**, 449 (1974).

Howison, S. D., *SIAM J. Appl. Math.* **46**, 20 (1986).

Huang, A. B., E. D. Chikhliwala and Y. C. Yortsos, Society of Petroleum Engineers Paper 13163 (1984).

Huang, J. and D. L. Turcotte, *J. Geophys. Res.* **B94**, 7491 (1989).

Hudson, J. A. and S. D. Priest, *Int. J. Rock. Mech. Min. Sci. Geomech. Abstr.* **17**, 279 (1983).

Hughes, B. D., *Random Walks, Percolation and Fractals* (Oxford University Press, 1994).

Hughes, B. D. and M. Sahimi, *J. Stat. Phys.* **29**, 781 (1982).

Hughes, B. D. and M. Sahimi, *Phys. Rev. Lett.* **70**, 2581 (1993); *Phys. Rev. E* **48**, 2776 (1993).

Huh, C. and L. E. Scriven, *J. Colloid Interface Sci.* **35**, 85 (1971).

Hulin, J. P., E. Charlaix, T. J. Plona, L. Oger and E. Guyon, *AIChE J.* **34**, 610 (1988a).

Hulin, J. P., E. Clément, C. Baudet, J. F. Gouyet and M. Rosso, *Phys. Rev. Lett.* **61**, 333 (1988b).

Hull, L. and K. Koslow, *Water Resour. Res.* **22**, 1731 (1986).

Hull, L., J. Miller and T. Clemo, *Water Resour. Res.* **23**, 1505 (1987).

Hurst, H. E., R. P. Black and Y. M. Simaika, *Long Term Storage: An Experimental Study* (Constable, London, 1965).

Hutton, B. E. T. and L. S. Leung, *Chem. Eng. Sci.* **29**, 1681 (1974).

Hutton, B. E. T., L. S. Leung, P. Brooks and D. J. Nicklin, *Chem. Eng. Sci.* **29**, 493 (1974).

Imdakm, A. O. and M. Sahimi, *Chem. Eng. Sci.* **46**, 1977 (1991).

Ingle, J. C., Pacific Section, *Soc. Econ. Paleon. Mineralogists Special Publication 159* (1981).

Isaacs, C. M., Society of Petroleum Engineers Paper 12733 (1984).

Jacks, H. H., O. J. E. Smith and C. C. Mattax, *Soc. Pet. Eng. J.* **13**, 175 (1973).

Jacquin, C., *C. R. Acad. Sci. Paris* **B300**, 721 (1985).

Jeffrey, D. J., *Proc. Roy. Soc. London* **A335**, 355 (1973).

Jensen, M. H., L. P. Kadanoff, A. Libchaber, I. Procaccia and J. Stavans, *Phys. Rev. Lett.* **55**, 2798 (1985).

Jerauld, G. R., H. T. Davis and L. E. Scriven, Society of Petroleum Engineers Paper 13164 (1984a).

Jerauld, G. R., J. C. Hatfield, L. E. Scriven and H. T. Davis, *J. Phys. C.* **17**, 1519 (1984b).

Jerauld, G. R., L. C. Nitsche, G. F. Teletzke, H. T. Davis and L. E. Scriven, Society of Petroleum Engineers Paper 12691 (1984c).

Jerauld, G. R. and S. J. Salter, *Trans. Porous Media* **5**, 103 (1990).

Jerauld, G. R., L. E. Scriven and H. T. Davis, *J. Phys. C* **16**, 3429 (1984d).

Jernot, J. P. and C. Lantuejoul, in *Disorder and Mixing*, E. Guyon, J. P. Nadal and Y. Pomeau, eds. (Kluwer, Amsterdam, 1989), p. 327.

Joanny, J. F. and P. G. de Gennes, *J. Chem. Phys.* **81**, 552 (1984).

Joanny, J. F. and M. O. Robbins, *J. Chem. Phys.* **92**, 3206 (1990).

Johnson, D. L., *Phys. Rev. Lett.* **63**, 580 (1989).

Johnson, D. L., J. Koplik and L. M. Schwartz, *Phys. Rev. Lett.* **57**, 2564 (1986).

Johnson, D. L., J. Koplik and R. Dashen, *J. Fluid Mech.* **176**, 379 (1987).

Johnson, E. F., D. P. Gassler and V. D. Naumann, *Trans. AIME* **216**, 270 (1959).

Joy, T. and W. Strieder, *J. Phys. C* **11**, L867 (1978); **12**, L279 (1979).

Kadanoff, L. P., *J. Stat. Phys.* **39**, 267 (1985).

Kadanoff, L., G. McNamara and G. Zanetti, *Phys. Rev. A* **40**, 4527 (1989).

Kalaydjian, F. and B. Legait, *C. R. Acad. Sci. Paris* **304**, 1035 (1987).

Kan, K. M. and P. F. Greenfield, *Ind. Eng. Chem. Process Des. Dev.* **17**, 482 (1978).

Kantzas, A. and I. Chatzis, *Chem. Eng. Commun.* **69**, 191 (1988).

Kardar, M., G. Parisi and Y.-C. Zhang, *Phys. Rev. Lett.* **56**, 889 (1986).

Katz, A. J. and A. H. Thompson, *Phys. Rev. Lett.* **54**, 1325 (1985).

Katz, A. J. and A. H. Thompson, *Phys. Rev. B* **34**, 8179 (1986).

Katz, A. J. and A. H. Thompson, *J. Geophys. Res.* **B92**, 599 (1987).

Katz, A. J., A. H. Thompson and R. A. Rashke, *Phys. Rev. A* **38**, 4901 (1988).

Katz, A. J. and S. A. Trugman, *J. Colloid Interface Sci.* **123**, 8 (1988).

Kazemi, H., *Soc. Pet. Eng. J.* **9**, 451 (1969).

Kazemi, H., L. S. Merrill, K. L. Porterfield and P. R. Zoman, *Soc. Pet. Eng. J.* **16**, 317 (1976).

Kazemi, H., M. S. Seth and G. W. Thomas, *Soc. Pet. Eng. J.* **9**, 463 (1969).

Keller, J. B., in *Nonlinear P.D.E. in Engineering and Applied Sciences*, R. L. Sternberg, A. J. Kalinowski and J. S. Papadakis, eds. (Marcel Dekker, New York, 1980).

Kenyon, W. E., P. I. Day, C. Straley and J. F. Willemsen, *SPE Form. Eval.* **3**, 622 (1988).

Kerstein, A. R., *J. Phys. A* **16**, 3071 (1983).

Kertész, J. and T. Vicsek, *J. Phys. A* **19**, L257 (1986).

Kessler, D., J. Koplik and L. Levine, *Adv. Phys.* **37**, 255 (1988).

Kessler, D. and H. Levine, *Phys. Rev. A* **33**, 2634 (1986).

Kessler, D., H. Levine and Y. Tu, *Phys. Rev. A* **43**, 4551 (1991).

Killough, J. E. and H. P. Foster, *Soc. Pet. Eng. J.* **19**, 279 (1979).

Killins, C. R., R. F. Nielsen and J. C. Calhoun, *Producers Monthly* **18**, 30 (1953).

Kim, I. C. and S. Torquato, *J. Appl. Phys.* **68**, 3892 (1990); **69**, 2280 (1991).

Kim, J. L. and E. L. Cussler, *AIChE J.* **33**, 705 (1987).

Kim, J. M. and J. M. Kosterlitz, *Phys. Rev. Lett.* **62**, 2289 (1989).

Kim, S. and W. B. Russel, *J. Fluid Mech.* **154**, 269 (1985).

King, G. C. P., *Pure Appl. Geophys.* **121**, 761 (1984).

King, M. J., W. B. Lindquist and L. Reyna, Society of Petroleum Engineers Paper 13953 (1984).

King, M. J. and H. Scher, *Phys. Rev. A* **35**, 929 (1987); **41**, 874 (1990).

King, P. R., *J. Phys. A* **20**, 3935 (1987a); **20**, L529 (1987b).

King, P. R., *Trans. Porous Media* **4**, 37 (1989).

Kinzel, W., *Ann. Israel Phys. Soc.* **5**, 425 (1983).

Kiriakidis, D. G., E. Mitsoulis and G. H. Neale, *Can. J. Chem. Eng.* **69**, 557 (1991).

Kirkpatrick, S., *Phys. Rev. Lett.* **27**, 1722 (1971).

Kirkpatrick, S., *Rev. Mod. Phys.* **45**, 574 (1973).

Kirkpatrick, S., in *Ill-Condensed Matter*, R. Balian, R. Maynard and G. Toulouse, eds. (North-Holland, Amsterdam, 1979), p. 323.

Kirkpatrick, S., C. D. Gelatt and M. P. Vecchi, *Science* **220**, 671 (1983).

Klafter, J., G. Zumofen and A. Blumen, *J. Phys.* **A24**, 4835 (1991).

Knackstedt, M. A. and M. Sahimi, *Water Resour. Res.* (1985).

Knackstedt, M. A., M. Sahimi and D. Y. C. Chan, *Phys. Rev. E* **47**, 2593 (1993).

Koch, D. L. and J. F. Brady, *J. Fluid Mech.* **154**, 399 (1985).

Koch, D. L. and J. F. Brady, *Chem. Eng. Sci.* **42**, 1377 (1987).

Koch, D. L. and J. F. Brady, *Phys. Fluids* **31**, 965 (1988).

Koch, D. L., R. G. Cox, H. Brenner and J. F. Brady, *J. Fluid Mech.* **200**, 173 (1989).

Kohring, G. A., *Int. J. Mod. Phys. C* **2**, 755 (1991a).

Kohring, G. A., *J. Physique II* **1**, 593 (1991b).

Kohring, G. A., *J. Stat. Phys.* **63**, 411 (1991c).

Koonce, T. K. and R. J. Blackwell, *Soc. Pet. Eng. J.* **5**, 318 (1965).

Kopinsky, J., R. Aris and E. L. Cussler, *AIChE J.* **34**, 2005 (1988).

Koplik, J., *J. Phys. C* **14**, 4821 (1981).

Koplik, J., *J. Fluid Mech.* **119**, 219 (1982).

Koplik, J., J. R. Banavar and J. F. Willemsen, *Phys. Rev. Lett.* **60**, 1282 (1988a).

Koplik, J. and T. J. Lasseter, *Chem. Eng. Commun.* **26**, 285 (1984)

Koplik, J. and T. J. Lasseter, *Soc. Pet. Eng. J.* **25**, 89 (1985).

Koplik, J. and H. Levine, *Phys. Rev. B* **32**, 280 (1985).

Koplik, J., H. Levine and A. Zee, *Phys. Fluids.* **26**, 2864 (1983).

Koplik, J., C. Lin and M. Vermette, *J. Appl. Phys.* **56**, 3127 (1984).

Koplik, J., S. Redner and D. Wilkinson, *Phys. Rev. A* **37**, 2619 (1988b).

Kortekaas, T. F. M., 1983, Society of Petroleum Engineers Paper 12112 (1983).

Kostek, S., L. M. Schwartz and D. L. Johnson, *Phys. Rev. B* **45**, 186 (1992).

Koval, E. J., *Soc. Pet. Eng. J.* **3**, 145 (1963).

Krug, J., *Phys. Rev. A* **44**, R801 (1991).

Kreyszig, E., *Differential Geometry* (University of Toronto Press, Toronto, 1959).

Krizek, R. J., R. M. Karadi and E. Socias, *Proceedings of the International Society of Rock Mechanics Symposium on Percolation Through Fissured Rock* (Stuttgart, Germany, 1972).

Krohn, C. E., *J. Geophys. Res. B* **93**, 3286 (1988a); **93**, 3297 (1988b).

Krohn, C. E. and A. H. Thompson, *Phys. Rev. B* **33**, 6366 (1986).

Krupp, H. K. and D. E. Elrick, *Water Resour. Res.* **4**, 809 (1964).

Ksenzhek, O. S., *J. Phys. Chem.* **37**, 691 (1963).

Kwiecien, M. J., I. F. Macdonald and F. A. L. Dullien, *J. Microscopy* **159**, 343 (1989).

Kyle, C. R. and R. L. Perrine, *Soc. Pet. Eng. J.* **5**, 189 (1965).

Kyte, J. R. and D. W. Berry, *Soc. Pet. Eng. J.* **15**. 269 (1975).

Lado, F. and S. Torquato, *J. Chem. Phys.* **93**, 5912 (1990).

Lahbabi, A. and H.-C. Chang, *Chem. Eng. Sci.* **40**, 435 (1985).

Laidlaw, W. G., G. R. Hamilton, R. B. Fleweilling and W. G. Wilson, *J. Stat. Phys.* **53**, 713 (1988).

Lake, L. W. and H. B. Carroll, Jr., eds., *Reservoir Characterization* (Academic, New York, 1986).

Lake, L. W. and G. J. Hirasaki, *Soc. Pet. Eng. J.* **21**, 459 (1981).

Landauer, R., *J. Appl. Phys.* **23**, 779 (1952).

Landauer, R., in *Electrical Transport and Optical Properties of Inhomogeneous Media, AIP Conference Proceedings* **40**, J. C. Garland and D. B. Tanner, eds. (1978), p. 2.

Lane, A. M., N. Shah and Wm. C. Conner, *J. Colloid Interface Sci.* **109**, 235 (1986).

Lantz, R. B., *Soc. Pet. Eng. J.* **11**, 315 (1971).

LaPointe, P. R., *Int. J. Rock Mech. Min. Sci. Geomech. Abstr.* **25**, 421 (1988).

Larson, R. E. and J. J. L. Higdon, *J. Fluid Mech.* **178**, 119 (1987).

Larson, R. E. and J. J. L. Higdon, *Phys. Fluids A* **1**, 38 (1989).

Larson, R. G., M.S. Thesis, University of Minnesota, Minneapolis, (1977).

Larson, R. G., *Ind. Eng. Chem. Fund.* **20**, 132 (1981).

Larson, R. G., H. T. Davis and L. E. Scriven, *Chem. Eng. Sci.* **36**, 75 (1981a).

Larson, R. G. and N. R. Morrow, *Powder Tech.* **30**, 123 (1981).

Larson, R. G., L. E. Scriven and H. T. Davis, *Nature* **268**, 409 (1977).

Larson, R. G., L. E. Scriven and H. T. Davis, *Chem. Eng. Sci.* **36**, 57 (1981b).

Lastoskie, C., K. E. Gubbins and N. Quirke, *J. Phys. Chem.* **97**, 4786 (1993).

Latour, L. L., R. L. Kleinberg and A. Sezginer, *J. Colloid Interface Sci.* **150**, 535 (1992).

Leath, P. L., *Phys. Rev. B* **14**, 5046 (1976).

Leclerc, D. F. and G. H. Neale, *J. Phys. A* **21**, 2979 (1988).

Le Doussal, P., *Phys. Rev. B* **39**, 4816 (1989).

Lee, J., A. Coniglio and H. E. Stanley, *Phys. Rev. A* **41**, 4589 (1990).

Lee, S. B. and S. Torquato, *J. Chem. Phys.* **89**, 3258 (1988).

Le Febvre du Prey, E. J., *Soc. Pet. Eng. J.* **13**, 39 (1973).

Legait, B. and C. Jacquin, *C. R. Acad. Sci.* **294**, 487 (1982).

Legaski, M. W. and D. L. Katz, *Soc. Pet. Eng. J.* **7**, 43 (1967).

Lenormand, R., *Proc. Roy. Soc. Lond.* **A423**, 159 (1989).

Lenormand, R., in *Hydrodynamics of Dispersed Media*, J. P. Hulin, A. M. Cazabat, E. Guyon and F. Carmona, eds. (North-Holand, Amsterdam, 1990), p. 287.

Lenormand, R. and S. Bories, *C. R. Acad. Sci. Paris* **B291**, 279 (1980).

Lenormand, R., F. Kalaydjan, M.-T. Bieber and J.-M. Lombard, Society of Petroleum Engineers Paper 20475 (1990).

Lenormand, R., E. Toubol and C. Zarcone, *J. Fluid Mech.* **189**, 165 (1988).

Lenormand, R. and C. Zarcone, in *Kinetics of Aggregation and Gelation*, F. Family and D. P. Landau, eds. (Elsevier, Amsterdam, 1984), p. 177.

Lenormand, R. and C. Zarcone, *Phys. Rev. Lett.* **54**, 2226 (1985a).

Lenormand, R. and C. Zarcone, *Physicochem. Hydro.* **6**, 497 (1985b).

Lenormand, R., C. Zarcone and A. Sarr, *J. Fluid Mech.* **135**, 337 (1983).

Levec, J., K. Grosser and R. G. Carbonell, *AIChE J.* **34**, 1027 (1988).

Levec, J., A. E. Sáez and R. G. Carbonell, *AIChE J.* **32**, 369 (1986).

Leverett, M. C., *Trans. AIME* **132**, 149 (1939); **142**, 159 (1941).

Levine, H., *J. Inst. Math. Applics* **2**, 12 (1966).

Levine, S. and D. L. Cuthiell, *J. Can. Pet. Tech.* **25**, 74 (1986).

Levine, S., P. Reed and G. Shutts, *Powder Tech.* **17**, 163 (1977).

Levinshtein, M., M. S. Shur and E. L. Efros, *Soviet Phys.-JETP* **42**, 1120 (1976).

Li, Y., W. G. Laidlaw and N. C. Wardlaw, *Adv. Colloid Interface Sci.* **26**, 1 (1986).

Li, Y. and N. C. Wardlaw, *J. Colloid Interface Sci.* **109**, 461 (1986a); **109**, 473 (1986b).

Liang, S., *Phys. Rev. A* **33**, 2663 (1986).

Liao, K. H. and A. E. Scheidegger, *Bull. Inter. Assoc. Sci. Hydrol.* **12**, 137 (1969).

Lichtner, P. C., E. H. Oelkers and H. C. Helgeson, *J. Geophys. Res.* **B91**, 7531 (1986).

Liggett, J. A. and D. Medina, in *Groundwater Flow and Quality Modeling*, E. Custodio, *et al.*, eds. (1988), p. 363.

Lin, C. and M. H. Cohen, *J. Appl. Phys.* **53**, 4152 (1982).

Lin, C. and J. Hamasaki, *J. Sediment. Pet.* **53**, 670 (1983).

Lin, C., G. Pirei and M. Vermette, *J. Geophys. Res.* **91**, 2173 (1986).

Lin, C.-Y. and J. C. Slattery, *AIChE J.* **28**, 311 (1982).

Lipsicas, M., J. R. Banavar and J. Willemsen, *Appl. Phys. Lett.* **48**, 1544 (1986).

Liu, H., L. Zhang and N. A. Seaton, *Chem. Eng. Sci.* **47**, 4393 (1992).

Liu, H., L. Zhang and N. A. Seaton, *J. Colloid Interface Sci.* **156**, 285 (1993).

Liu, J., G. A. Pope and K. Sepehrnoori, *Appl. Math. Model.* (1994).

Long, J. C. S. and D. Billaux, *Water Resour. Res.* **23**, 1201 (1987).

Long, J. C. S., P. Gilmour and P. A. Witherspoon, *Water Resour. Res.* **21**, 1105 (1985).

Long, J. C. S., K. Karasaki, A. Davey, J. Peterson, M. Landsfeld, J. Kemeny and S. Martel, *Int. J. Rock Mech. Min. Sci. Geomech. Abstr.* **28**, 121 (1991).

Long, J. C. S., J. S. Remer, C. R. Wilson and P. A. Witherspoon, *Water Resour. Res.* **18**, 645 (1982).

Long, J. C. S. and P. A. Witherspoon, *J. Geophys. Res.* **90**, 3087 (1985).

Lowell, R. P., *Water Resour. Res.* **25**, 774 (1989).

Lucido, G., R. Triolo and E. Caponetti, *Phys. Rev. B* **38**, 9031 (1988).

Lumley, J. L. and H. A. Panofsky, *The Structure of Atmospheric Turbulence* (Wiley, New York, 1974).

MacDonald, I. F., M. S. El-Sayed, K. Mow and F. A. L. Dullien, *Ind. Eng. Chem. Fund.* **18**, 199 (1979).

Madden, T. R., *Geophysics* **41**, 1104 (1976).

Madden, T. R., *J. Geophys. Res.* **88**, 585 (1983).

Mahaffey, J. L., W. M. Rutherford and C. S. Mathews, *Soc. Pet. Eng. J.* **6**, 73 (1966).

Maier, R. and W. G. Laidlaw, *Trans. Porous Media* **5**, 421 (1990).

Maier, R. and W. G. Laidlaw, *J. Stat. Phys.* **62**, 269 (1991a).

Maier, R. and W.G. Laidlaw, *Math. Geol.* **23**, 87 (1991b).

Måløy, K. J., J. Feder and T. Jøssang, *Phys. Rev. Lett.* **55**, 2688 (1985).

Måløy, K. J., J. Feder, T. Jøssang and P. Meakin, *Phys. Rev. A* **36**, 318 (1987).

Mandelbrot, B. B., *J. Fluid Mech.* **62**, 331 (1974).

Mandelbrot, B. B., *The Fractal Geometry of Nature* (W. H. Freeman, San Francisco, 1982).

Mandelbrot, B. B., *Phys. Script.* **32**, 257 (1985).

Mandelbrot, B. B. and J. W. van Ness, *SIAM Rev.* **10**, 422 (1968).

Marle, C. M., *Rev. Inst. Francais du Petrole* **22**, 1471 (1967).

Marle, C. M., *Multiphase Flow in Porous Media* (Gulf Publishing Co., Houston, Texas, 1981).

Marle, C. M., P. Simandoux, J. Pacsirsky and C. Gaulier, *Rev. Inst. Fr. Petrol.* **22**, 272 (1967).

Martin, J. C., *Soc. Pet. Eng. J.* **8**, 370 (1968).

Martys, N., D. P. Bentz and E. J. Garboczi, *Phys. Fluids* **6**, 1434 (1994).

Martys, N., M. Cieplak and M. O. Robbins, *Phys. Rev. Lett.* **66**, 1058 (1991).

Mason, G., *J. Colloid Interface Sci.* **88**, 36 (1982).

Mason, G., *Proc. Roy. Soc. Lond.* **A390**, 47 (1983); **A415**, 453 (1988).

Matheron, G., *Rev. Inst. Fr. Petrole* **22**, 443 (1967).

Matheron, G. and G. de Marsily, *Water Resour. Res.* **16**, 901 (1980).

Mathews, J. L., A. S. Emanuel and K. A. Edwards, *J. Pet. Tech.* **41**, 1136 (1989).

Mattax, C. C. and J. R. Kyte, *Oil Gas J.* **59**, 115 (1961).

Maxwell-Garnett, J. C., *Phil. Trans. R. Soc. Lond.* **203**, 385 (1904).

Mayagoitia, V., M. J. Cruz and F. Rojas, *J. Chem. Soc. Faraday Trans.* **85**, 2071, 2079 (1989).

McCaffery, F. G., *J. Can. Pet. Tech.* **11**, 26 (1972).

McCaffery, F. G. and D. W. Bennion, *J. Can. Pet. Tech.* **13**, 42 (1974).

McCall, K. R., D. L. Johnson and R. A. Guyer, *Phys. Rev. B* **44**, 7344 (1991).

McKenzie, D. R., R. C. McPhedran and G. H. Derrick, *Proc. Roy. Soc. London* **A362**, 211 (1978).

McLachlan, D. S., M. B. Button, S. R. Adams, V. M. Gorringe, J. D. Keen, J. Muoe and E. Wedepohl, *Geophysics* **52**, 194 (1987).

McNamara, G. and G. Zanetti, *Phys. Rev. Lett.* **61**, 2332 (1988).

McPhedran, R. C., G. H. Derrick and D. R. McKenzie, *Proc. Roy. Soc. London* **A362**, 211 (1978).

McPhedran, R. C. and D. R. McKenzie, *Proc. Roy. Soc. London* **A359**, 45 (1978).

Meakin, P., *Phys. Rev. A* **33**, 4199 (1986); **36**, 2833 (1987).

Meakin, P., in *Phase Transitions and Critical Phenomena, Vol. 12*, C. Domb and J. L. Lebowitz, eds. (Academic Press, London, 1988), p. 335.

Meakin, P., *Physica* **A173**, 305 (1991).

Meakin, P., A. Coniglio, H. E. Stanley and T. A. Witten, *Phys. Rev. A* **34**, 3325 (1986).

Meakin, P., F. Family and T. Vicsek, *J. Colloid Interface Sci.* **117**, 394 (1987).

Meakin, P., M. Murat, A. Aharony, J. Feder and T. Jøssang, *Physica* **A155**, 1 (1989).

Melli, T. R. and L. E. Scriven, *Ind. Eng. Chem. Res.* **30**, 951 (1991).

Melrose, J. C., *Soc. Pet. Eng. J.* **5**, 259 (1965).

Melrose, J. C., *Ind. Eng. Chem.* **60**, 53 (1968).

Melrose, J. C., Society of Petroleum Engineers Paper 10971 (1982); Paper 18331 (1988).

Melrose J. C. and C. F. Brandner, *Can. J. Pet. Tech.* **13**, 54 (1974).

Mendelson, K. S., *J. Appl. Phys.* **53**, 6465 (1982).

Mendelson, K. S., *Phys. Rev. B* **34**, 6503 (1986).

Mendelson, K. S. and M. H. Cohen, *Geophysics* **47**, 257 (1982).

Meneveau, C. and K. R. Sreenivasan, *Phys. Rev. Lett.* **59**, 1424 (1987).

458 References

Mercado, A., *Int. Assoc. Sci. Hydrol. Publication 72* (Int. Assoc. Hydrol. Sci., Gentbrugge, Belgium, 1967), p. 23.

Meredith, R. E. and C. W. Tobias, *J. Appl. Phys.* **31**, 1270 (1960).

Meyer, H. I., *J. Appl. Phys.* **24**, 510 (1953).

Mildner, D. F. R., R. Rezvani, P. L. Hall and R. L. Brost, *Appl. Phys. Lett.* **48**, 1314 (1986).

Miller, A. and E. Abrahams, *Phys. Rev.* **120**, 745 (1960).

Milton, G. W., in *Physics and Chemistry of Porous Media, AIP Conference Proceedings* **107**, D. L. Johnson and P. N. Sen, eds. (1984), p. 66.

Mitescu, C. D. and J. Roussenq, *C. R. Acad. Sci. Paris* **283**, 999 (1976).

Mizell, S. A., A. L. Gutjahr and L. W. Gelhar, *Water Resour. Res.* **18**, 1053 (1982).

Mohanty, K. K., Ph.D. Thesis, University of Minnesota, Minneapolis (1981).

Mohanty, K. K., H. T. Davis and L. E. Scriven, Society of Petroleum Engineers Paper 9406 (1980).

Mohanty, K. K., H. T. Davis and L. E. Scriven, *SPE Reservoir Eng.* **2**, 113 (1987).

Mohanty, K. K. and S. J. Salter, Society of Petroleum Engineers Paper 11018 (1982).

Molz, F. J., O. Güven and J. G. Melville, *Ground Water* **21**, 715 (1983).

Moreno, L., Y. W. Tsang, C. F. Tsang, F. V. Hale and I. Neretnieks, *Water Resour. Res.* **24**, 2033 (1988).

Morgan, J. T. and D. T. Gordon, *J. Pet. Tech.* **22**, 1194 (1970).

Morrow, N. R., *Ind. Eng. Chem.* **62**, 32 (1970).

Morrow, N. R., *J. Can. Pet. Tech.* **15**, 49 (1976).

Morrow, N. R., in *Interfacial Phenomena in Petroleum Recovery*, N. R. Morrow, ed. (Marcel Dekker, New York, 1991), p. 1.

Morrow, N. R. and F. G. McCaffery, in *Wetting, Spreading, and Adhesion*, G. F. Padday, ed. (Academic Press, New York, 1978), p. 289.

Mukhopadhyay, S. and M. Sahimi, Society of Petroleum Engineers Paper 24043 (1992).

Mukhopadhyay, S. and M. Sahimi, *Water Resour. Res.* (1995).

Mungan, N., *Soc. Pet. Eng. J.* **6**, 247 (1966).

Murat, M. and A. Aharony, *Phys. Rev. Lett.* **57**, 1875 (1986).

Muskat, M., *The Flow of Homogeneous Fluids Through Porous Media* (McGraw-Hill, New York, 1937).

Naar, J., G. R. Wygal and J. H. Henderson, *Soc. Pet. Eng. J.* **2**, 13 (1962).

Nakamo, Y. and J. W. Evans, *J. Chem. Phys.* **78**, 2568 (1983).

Neimark, A. V., *Colloid J. (USSR)* **46**, 727 (1984a); **46**, 1158 (1984b).

Neretnieks, I., *J. Geophys. Res. B* **85**, 4379 (1980).

Neuman, S. P., *Acta. Mech.* **25**, 153 (1977).

Neuman, S. P., *Water Resour. Res.* **26**, 887 (1990).

Ng, K. M. and A. C. Payatakes, *AIChE J.* **26**, 419 (1980).

Nicholson, D., *Trans. Faraday Soc.* **64**, 3416 (1968).

Nicholson, D. and J.H. Petropoulos, *J. Phys. D* **4**, 181 (1971); **8**, 1430 (1975); **10**, 2423 (1977).

Nickel, B. and D. Wilkinson, *Phys. Rev. Lett.* **51**, 71 (1983).

Niemeyer, L., L. Pietronero and H. J. Weismann, *Phys. Rev. Lett.* **52**, 1033 (1984).

Nimmo, J. R. and K. C. Akstin, *Soil Sci. Soc. Am. J.* **52**, 303 (1988).

Nitsche, L. C. and H. Brenner, *Arch. Rational Mech. Anal.* **107**, 225 (1989).

Nittmann, J., G. Daccord and H. E. Stanley, *Nature* **314**, 141 (1985).

Nobles, M. A. and H. B. Janzen, *Pet. Trans. AIME* **213**, 356 (1958).

Nolen-Hoeksema, R. C. and R. B. Gordon, *Int. J. Rock Mech. Min. Sci. Geomech. Abstr.* **24**, 135 (1987).

Nolle, C. S., B. Koiller, N. Martys and M. O. Robbins, *Phys. Rev. Lett.* **71**, 2074 (1993).

Nolte, D. D., L. J. Pyrak-Nolte and N. G. W. Cook, *Pure Appl. Geophys.* **131**, 111 (1989).

Noorishad, J. and M. Mehran, *Water Resour. Res.* **18**, 588 (1982).

Nordqvist, W., Y. W. Tsang, C. F. Tsang, B. Dverstorp and J. A. Andersson, *Water Resour. Res.* **28**, 1703 (1992).

Novak, C. F., R. S. Schechter and L. W. Lake, *AIChE J.* **35**, 1057 (1989).

Novakowski, K.S., G. V. Evans, D. A. Lever and K. G. Raven, *Water Resour. Res.* **21**, 1165 (1985).

Novosad, J., E. Inescu-Forniciov and K. Mannhardt, paper CIM 843542, presented at the 35th Annual Conference of the Society of CIM, Calgary (1984).

Novy, R. A., P. G. Toledo, H. T. Davis and L. E. Scriven, *Chem. Eng. Sci.* **44**, 1785 (1989).

Nyame, B. K. and J. M. Ilbston, in *7th Int. Symp. Chem. Cement Paste*, Paris, **3:VI** (1980), p. 181.

O'Brien, R. W., *J. Fluid Mech.* **91**, 17 (1979).

Odagaki, T. and M. Lax, *Phys. Rev. B* **24**, 5284 (1981).

Odeh, A. S., *Trans. AIME* **216**, 346 (1959).

Odeh, A. S., *Soc. Pet. Eng. J.* **5**, 60 (1965).

Oger, L., J. P. Troadec, D. Bideau, J. A. Dodds and M. Powell, *Powder Tech.* **46**, 121 (1986).

Ogandzanjanc, V. G., *Iz. Akad. Nauk (USSR)* **20**, 129 (1960).

Okubo, P. G. and K. Aki, *J. Geophys. Res. B* **92**, 345 (1987).

Orford, J. D. and W. B. Whalley, *Sedimentology* **30**, 655 (1983).

O'Shaughnessy, B. and I. Procaccia, *Phys. Rev. Lett.* **54**, 455 (1985).

Outmans, H. D., *Soc. Pet. Eng. J.* **2**, 165 (1962).

Owen, J. E., *Trans. AIME* **195**, 169 (1952).

Owens, W. W. and D. L. Archer, *J. Pet. Tech.* **23**, 873 (1971).

Oxaal, U., *Phys. Rev. A* **44**, 5038 (1991).

Pahl, P. J., *Int. J. Rock. Mech. Min. Sci. Geomech. Abstr.* **18**, 221 (1978).

Pakula, R. J. and R. A. Greenkorn, *AIChE J.* **19**, 1265 (1971).

Palciauskas, V. V. and P. A. Domenico, *Geol. Soc. America Bull.* **87**, 207 (1976).

Pandey, R. B., D. Stauffer, A. Margolina and J. G. Zabolitzky, *J. Stat. Phys.* **34**, 427 (1984).

Park, C.-W., S. Gorrell and G. M. Homsy, *J. Fluid Mech.* **141**, 257 (1984).

Parlar, M. and Y. C. Yortsos, *J. Colloid Interface Sci.* **124**, 162 (1988); **132**, 425 (1989).

Parsons, R. W., *Soc. Pet. Eng. J.* **6**, 126 (1966).

Passioura, J. B., *Soil Sci.* **111**, 339 (1971).

Paterson, L., *J. Fluid Mech.* **113**, 513 (1981).

Paterson, L., *Int. J. Hydrogen Energy* **8**, 53 (1983).

Paterson, L., *Phys. Rev. Lett.* **52**, 1621 (1984).

Paterson, L., *Phys. Fluids* **28**, 26 (1985).

Paterson, L., V. Hornof and G. H. Neale, *Powder Tech.* **33**, 265 (1982).

Paterson, L., V. Hornof and G. H. Neale, *Rev. Inst. Fran. Pet.* **39**, 517 (1984a).

Paterson, L., V. Hornof and G. H. Neale, *Soc. Pet. Eng. J.* **24**, 325 (1984b).

Pathak, P., H. T. Davis and L. E. Scriven, Society of Petroleum Engineers Paper 11016 (1982).

Patsoules, M. G. and J. C. Cripps, *Energy Sources* **7**, 15 (1983).

Payandeh, B., *Riv. Nuovo Cim.* **3**, 1 (1980).

Payatakes, A. C., *Ann. Rev. Fluid Mech.* **14**, 365 (1982).

Payatakes, A. C. and M. M. Dias, *Rev. Chem. Eng.* **2**, 85 (1984).

Payatakes, A. C., K. M. Ng and R. W. Flumerfelt, *AIChE J.* **26**, 430 (1980).

Peaceman, D. W., *Fundamentals of Numerical Reservoir Simulation* (Elsevier, Amsterdam, 1977).

Peaceman, D. W. and H. H. Rachford, Jr., *Soc. Pet. Eng. J.* **2**, 327 (1962).

Perkins, T. K. and O. C. Johnston, *Soc. Pet. Eng. J.* **3**, 70 (1963); **9**, 39 (1969).

Perkins, T. K., O. C. Johnston and R. N. Hoffman, *Soc. Pet. Eng. J.* **5**, 301 (1965).

Perrine, R. L., *Soc. Pet. Eng. J.* **1**, 17 (1961); **3**, 205 (1963).

Perrins, T. W., D. R. McKenzie and R. C. McPhedran, *Proc. Roy. Soc. London* **A369**, 207 (1979).

Peters, E. J. and D. L. Flock, *Soc. Pet. Eng. J.* **21**, 249 (1981).

Peters, E. J., W. H. Broman and J. A. Broman, Society of Petroleum Engineers Paper 13167 (1984).

Pfeifer, P. and D. Avnir, *J. Chem. Phys.* **79**, 3558 (1983).

Pfannkuch, H.-O., *Rev. Inst. Fr. Pet.* **18**, 215 (1963).

Pfeifer, P., D. Avnir and D. Farin, *J. Stat. Phys.* **36**, 699 (1984).

Philip, J. R., *Soil Science* **83**, 345 (1957).

Philip, J. R., *Trans. Porous Media* **1**, 319 (1986).

Pickell, J. J., B. F. Swanson and W. B. Hickman, *Soc. Pet. Eng. J.* **6**, 55 (1966).

Pickens, J. F. and G. E. Grisak, *Water Resour. Res.* **17**, 1191 (1981).

Piggott, A. R. and D. Elsworth, *Water Resour. Res.* **25**, 457 (1989).

Pike, G. E. and C. H. Seager, *Phys. Rev. B* **10**, 1421 (1974).

Pirson, R. S. and S. J. Pirson, Society of Petroleum Engineers Paper 101 (1961).

Pis'men, L. M., *Dokl. Akad. Nauk (USSR)* **207**, 657 (1972).

Pittmann, E. D., in *Physics and Chemistry of Porous Media, AIP Conference Proceedings* **107**, D. L. Johnson and P. N. Sen, eds. (1984), p. 1.

Plumb, O.A. and S. Whitaker, *Water Resour. Res.* **24**, 913(1988a); **24**, 927 (1988b).

Pokorny, M., C. M. Newman and D. Meiron, *J. Phys. A* **23**, 1431 (1990).

Pollard, D. D., *Geophys. Res. Lett.* **3**, 513 (1976).

Pollard, P., *Trans. AIME* **216**, 38 (1959).

Pomeau, Y. and J. Serra, *J. Microsc.* **138**, 179 (1985).

Pomeau, Y. and J. Vannimenus, *J. Colloid Interface Sci.* **104**, 447 (1985).

Porod, G., *Kolloid Zhur.* **124**, 83 (1951).

Poulton, M. M., N. Mojtabai and I. W. Farmer, *Int. J. Rock Mech. Min. Sci. Geomech. Abstr.* **27**, 219 (1990).

Power, W. L., T. E. Tullis, S. R. Brown, G. N. Boitnott and C. H. Scholz, *Geophys. Res. Lett.* **14**, 29 (1987).

Prager, S., *Phys. Fluids* **4**, 1477 (1961).

Priest, S. D. and J. A. Hudson, *Int. J. Rock. Mech. Min. Sci. Geomech. Abstr.* **13**, 135 (1976); **18**, 183 (1981).

Press, W. H., B. P. Flanmnery, S. A. Teukalsky and W. T. Vetterling, *Numerical Recipes* (Cambridge University Press, Cambridge, 1986).

Pruess, K., G. S. Bodvarsson and V. Stefansson, in *Proc. of Sixteenth Stanford Geothermal Workshop*, Stanford University (1983).

Pruess, K. and Y. V. Tsang, *Water Resour. Res.* **26**, 1915 (1990).

Qian, Y. H., D. d'Humières and P. Lallemand, *Europhys. Lett.* **17**, 479 (1992).

Quintard, M. and S. Whitaker, *Trans. Porous Media* **3**, 357 (1988); **5**, 341 (1990a); **5**, 429 (1990b).

Raats, P. A. C. and A. Klute, *Soil Sci. Soc. Am. J.* **32**, 452 (1968).

Rachford, H. H., *Soc. Pet. Eng. J.* **4**, 249 (1964).

Raimondi, P. and M. A. Torcaso, *Soc. Pet. Eng. J.* **4**, 49 (1964).

Ramakrishnan, T. S. and D. T. Wasan, Society of Petroleum Engineers Paper 12693 (1984).

Ramakrishnan, T. S. and D. T. Wasan, *Int. J. Multiphase Flow* **12**, 357 (1986).

Rao, P. S. C., D. E. Ralston, R. E. Jessup and J. M. Davidson, *Soil Sci. Soc. Am. J.* **44**, 1139 (1980).

Raphael, E. and P. G. de Gennes, *J. Chem. Phys.* **90**, 7577 (1989).

Raven, K. G., K. S. Novakowski and P. A. Lapcevic, *Water Resour. Res.* **24**, 2019 (1988).

Rayleigh, R. S., *Phil. Mag.* **34**, 481 (1892).

Reed, A. W., H. Meister and J. Sasmor, *Nucl. Tech.* **78**, 54 (1987).

Reed, R. L. and R. J. Healy, in *Improved Oil Recovery by Surfactant and Polymer Flooding*, D. O. Shah and R. S. Schechter, eds. (Academic, New York, 1977).

Rem, P. and J. Somers, in *Discrete Kinetic Theory, Lattice-Gas Dynamics, and Foundations of Hydrodynamics*, edited by R. Monaco (World Scientific, Singapore, 1989), p. 268.

Reynolds, P. J., H. E. Stanley and W. Klein, *Phys. Rev. B* **21**, 1223 (1980).

Reznik, A. A., R. M. Enick and S. B. Panuelker, *Soc. Pet. Eng. J.* **24**, 643 (1984).

Richardson, J. G., J. G. Kerver, J. A. Hafford and J. Osoba, *Trans. AIME* **195**, 187 (1952).

Ridgway, K. and K. J. Tarbuck, *Brit. Chem. Eng.* **12**, 384 (1967).

Rikvold, P. A. and G. Stell, *J. Colloid Interface Sci.* **108**, 158 (1985).

Rillaerts, E. and P. Joos, *Chem. Eng. Sci.* **35**, 883 (1980).

Rink, M. and J. R. Schopper, *Geophys. Prospect.* **16**, 277 (1968).

Rintoul, M. D. and H. Nakanishi, *J. Phys. A* **25**, L945 (1992).

Ritter, H. L. and L. C. Drake, *Ind. Eng. Chem.* **17**, 782 (1945).

Robbins, M. O. and J. F. Joanny, *Europhys. Lett.* **3**, 729 (1987).

Roberts, J. N. and L. M. Schwartz, *Phys. Rev. B* **31**, 5990(1985).

Robinson, J. W. and J. E. Gale, *Ground Water* **28**, 25 (1990).

Robinson, P. C., Ph.D. Thesis, St. Catherine's College, Oxford University (1984a).

Robinson, P. C., *J. Phys. A* **17**, 2823 (1984b).

Roman, H. E., *J. Stat. Phys.* **58**, 375 (1990).

Romm, E. S., *Fluid Flow in Fractured Rocks* (in Russian) (Nedra Publishing House, Moscow, 1966; English Translation, W. R. Blake, Bartlesville, OK, 1972).

Roof, J. G., *Soc. Pet. Eng. J.* **10**, 85 (1970).

Rose, H. E, *Proc. Inst. Mech. Eng. Appl. Mech.* **153**, 141 (1945).

Rose, W. D., *Illinois State Geol. Surv. Circ.* **237** (1957).

Rose, W. D., in *Proceedings of the Second International Conference on Fundamentals of Transport Phenomena in Porous Media*, D. E. Erlick, ed. (International Association of Hydraulic Research, Guelph, Canada, 1972).

Rose, W. D. and W. A. Bruce, *Trans. AIME* **186**, 127 (1949).

Rose, W. D. and P. A. Witherspoon, *Illinois State Geol. Surv. Circ.* **224** (1956).

Ross, B., *Water Resour. Res.* **22**, 823 (1986).

Rothman, D. H., *Geophysics* **53**, 509 (1988).

Rothman, D. H., *J. Geophys. Res.* **B95**, 8663 (1990).

Rothman, D. H. and J. M. Keller, *J. Stat. Phys.* **52**, 1119 (1988).

Rothman, D. H. and S. Zaleski, *J. Physique* **50**, 2161 (1989).

Roux, J.-N. and D. Wilkinson, *Phys. Rev. A* **37**, 3921 (1988).

Roux, S. and E. Guyon, *J. Phys. A* **22**, 3693 (1989).

Roux, S., C. Mitescu, E. Charlaix and C. Baudet, *J. Phys. A* **19**, L687 (1986).

Rubinstein, J. and S. Torquato, *J. Fluid Mech.* **206**, 25 (1989).

Rubio, M. A., C. A. Edwards, A. Dougherty and J. P. Gollub, *Phys. Rev. Lett.* **63**, 1685 (1989).

Runge, I., *Z. Tech. Phys.* **6**, 61 (1925).

Ryan, D., R. G. Carbonell and S. Whitaker, *Chem. Eng. Sci.* **35**, 10 (1980).

Saeger, R. B., L. E. Scriven and H. T. Davis, *Phys. Rev. A* **44**, 5087 (1991).

Sáez, A. E. and R. G. Carbonell, *AIChE J.* **31**, 52 (1985).

Saffman, P. G., *J. Fluid Mech.* **6**, 321 (1959); **7**, 194 (1960); **173**, 73 (1986).

Saffman, P. G., *Stu. Appl. Math.* **50**, 93 (1971).

Saffman, P. G. and G. I. Taylor, *Proc. Roy. Soc. Lond.* **A245**, 312 (1958).

Sahimi, M., *J. Phys. C* **17**, 3957 (1984).

Sahimi, M., *Phys. Rev. Lett.* **55**, 1698 (1985).

Sahimi, M., *J. Phys. A* **20**, L1293 (1987).

Sahimi, M., *Chem. Eng. Commun.* **64**, 177 (1988a).

Sahimi, M., *Chem. Eng. Sci.* **43**, 2981 (1988b).

Sahimi, M., *Phys. Rev. A* **43**, 5367 (1991).

Sahimi, M., *CHEMTECH* **22**, 687 (1992a).

Sahimi, M., *J. Chem. Phys.* **96**, 4718 (1992b).

Sahimi, M., *Physica A* **186**, 160 (1992c).

Sahimi, M., *AIChE J.* **39**, 369 (1993); **40** (1994a).

Sahimi, M., *Applications of Percolation Theory* (Taylor and Francis, London, 1994b).

Sahimi, M. and S. Arbabi, *Phys. Rev. B* **40**, 4975 (1989).

Sahimi, M. and S. Arbabi, *J. Stat. Phys.* **62**, 453 (1991).

Sahimi, M. and S. Arbabi, *Phys. Rev. Lett.* **68**, 608 (1992).

Sahimi, M. and S. Arbabi, *Phys. Rev. B* **47**, 713 (1993).

Sahimi, M., H. T. Davis and L. E. Scriven, *Chem. Eng. Comm.* **23**, 329 (1983a).

Sahimi, M., H. T. Davis and L. E. Scriven, *Soc. Pet. Eng. J.* **25**, 235 (1985).

Sahimi, M., G. R. Gavalas and T. T. Tsotsis, *Chem. Eng. Sci.* **45**, 1443 (1990).

Sahimi, M. and J. D. Goddard, *Phys. Rev. B* **33**, 7848 (1986).

Sahimi, M., A. A. Heiba, B. D. Hughes, H. T. Davis and L. E. Scriven, Society of Petroleum Engineers Paper 10969 (1982).

Sahimi, M., A. A. Heiba, H. T. Davis and L. E. Scriven, *Chem. Eng. Sci.* **41**, 2123 (1986a).

Sahimi, M., B. D. Hughes, L. E. Scriven and H. T. Davis, *J. Chem. Phys.* **78**, 6849 (1983b).

Sahimi, M., B. D. Hughes, L. E. Scriven and H. T. Davis, *Phys. Rev. B* **28**, 307 (1983c).

Sahimi, M., B. D. Hughes, L. E. Scriven and H. T. Davis, *Chem. Eng. Sci.* **41**, 2103 (1986b).

Sahimi, M. and A. O. Imdakm, *J. Phys. A* **21**, 3833 (1988).

Sahimi, M. and A. O. Imdakm, *Phys. Rev. Lett.* **66**, 1169 (1991).

Sahimi, M. and V. L. Jue, *Phys. Rev. Lett.* **62**, 629 (1989).

Sahimi, M. and M. A. Knackstedt, *J. Physique I* **4**, 11 (1994).

Sahimi, M., M. C. Robertson and C. G. Sammis, *Phys. Rev. Lett.* **70**, 2186 (1993).

Sahimi, M., L. E. Scriven and H. T. Davis, *J. Phys. C* **17**, 1941 (1984).

Sahimi, M. and H. Siddiqui, *J. Phys. A* **20**, L89 (1987).

Sahimi, M. and D. Stauffer, *Chem. Eng. Sci.* **46**, 2225 (1991).

Sahimi, M. and B. N. Taylor, *J. Chem. Phys.* **95**, 6749 (1991).

Sahimi, M. and T. T. Tsotsis, *J. Catal.* **96**, 552 (1985).

Sahimi, M. and Y. C. Yortsos, *Phys. Rev. A* **32**, 3762 (1985).

Salathiel, R. A., *J. Pet. Tech.* **25**, 1216 (1973).

Salles, J., J.-F. Thovert, R. Delannay, L. Prevors, J.-L. Auriault and P. M. Adler, *Phys. Fluids A* **5**, 2348 (1993).

Salter, S. J. and K. K. Mohanty, Society of Petroleum Engineers Paper 11017 (1982).

Sammis, C. G., G. C. P. King and R. Beigel, *Pure Appl. Geophys.* **125**, 777 (1985).

Sangani, A. S. and A. Acrivos, *Int. J. Multiphase Flow* **8**, 193 (1982).

Sangani, A. S. and A. Acrivos, *Proc. Roy. Soc. London A* **386**, 263 (1983).

Sapoval, B., M. Rosso and J. F. Gouyet, *J. Physique Lett.* **46**, L149 (1985).

Sarkar, S. K., *Phys. Rev. A* **31**, 3468 (1985).

Sarkar, S. K. and M. H. Jensen, *Phys. Rev. A* **35**, 1877 (1987).

Satterfield, C. N., *AIChE J.* **21**, 209 (1975).

Schaefer, D. W., J. E. Martin, P. Wiltzuis and D. S. Cannell, *Phys. Rev. Lett.* **52**, 2371 (1984).

Scheidegger, A. E., *J. Appl. Phys.* **25**, 994 (1954).

Scheidegger, A. E., *Bull. Int. Ass. Sci. Hydrol.* **10**, 38 (1965).

Scheidegger, A. E., *The Physics of Flow Through Porous Media*, 3rd ed. (University of Toronto Press, Toronto, 1974).

Scheidegger, A. E. and E. F. Johnson, *Can. J. Phys.* **39**, 326 (1961).

Scher, H. and R. Zallen, *J. Chem. Phys.* **53**, 3759 (1970).

Schmidt, E. J., K. K. Velasco and A. M. Nur, *J. Appl. Phys.* **59**, 2788 (1986).

Schmidt, V. and D. A. McDonald, *Society of Economic Paleontologists and Mineralogists Special Publication No. 26* (1979), p. 209.

Schmittbuhl, J., S. Gentier and S. Roux, *Geophys. Res. Lett.* **20**, 639 (1993).

Scholz, C. H. and B. B. Mandelbrot, eds., *Fractals in Geophysics* (Birkhaüser Verlag, Basel, 1989).

Schowalter, W. R., *AIChE J.* **11**, 99 (1965).

Schwartz, F. W., *Water Resour. Res.* **13**, 743 (1977).

Schwartz, F. W., L. Smith and A. S. Crowe, *Water Resour. Res.* **19**, 1253 (1983).

Schwartz, L. M. and J. R. Banavar, *Phys. Rev. B* **39**, 11965 (1989).

Schwartz, L. M., J. R. Banavar and B. I. Halperin, *Phys. Rev. B* **40**, 9155 (1989a).

Schwartz, L. M. and S. Kimminau, *Geophysics* **52**, 1402 (1987).

Schwartz, L. M., N. Martys, D. P. Bentz, E. J. Garboczi and S. Torquato, *Phys. Rev. E* **48**, 4584 (1993).

Schwartz, L. M., P. N. Sen and D. L. Johnson, *Phys. Rev. B* **40**, 2450 (1989b).

Seaton, N. A., *Chem. Eng. Sci.* **46**, 1895 (1991).

Seaton, N. A., J. P. R. B. Walton and N. Quirke, *Carbon* **27**, 853 (1989).

Seeburger, D. A. and A. Nur, *J. Geophys. Res.* **B89**, 527 (1984).

Sen, A. K. and S. Torquato, *J. Chem. Phys.* **89**, 3799 (1988).

Sen, P. N., *Appl. Phys. Lett.* **39**, 667 (1981).

Sen, P. N., *Geophysics* **49**, 586 (1984).

Sen, P. N., C. Scala and M. H. Cohen, *Geophysics* **46**, 781 (1981).

Sevik, E. M., P. A. Monson and J. M. Ottino, *J. Chem. Phys.* **88**, 1198 (1988).

Shankland, T. J. and H. S. Waff, *J. Geophys. Res.* **79**, 4863 (1974).

Shante, V. K. S., *Phys. Rev. B* **16**, 2597 (1977).

Shante, V. K. S. and S. Kirkpatrick, *Adv. Phys.* **20**, 325 (1971).

Shapiro, A. M. and J. Andersson, *Water Resour. Res.* **19**, 959 (1983).

Shaw, T. M., *Phys. Rev. Lett.* **59**, 1671 (1987).

Shearer, C. J. and J. F. Davidson, *J. Fluid Mech.* **22**, 321 (1965).

Sheffield, R. E. and A. B. Metzner, *AIChE J.* **22**, 736 (1976).

Shelton, J. L. and F. N. Schneider, *Soc. Pet. Eng. J.* **15**, 217 (1975).

Sheng, P., *Phys. Rev. B* **22**, 6364 (1980); **41**, 4507 (1990).

Sheng, P. and M.-Y. Zhou, *Phys. Rev. Lett.* **61**, 1591 (1988).

Sherwood, J. D., *J. Phys. A* **19**, L195 (1986).

Sherwood, J. D. and J. Nittmann, *J. Physique* **47**, 15 (1986).

Shiles, G., G. A. Pope and K. Sepehrnoori, in *Science and Engineering on Supercomputers*, E. J. Pitcher, ed. (Springer, New York, 1990), p. 497.

Shimo, M. and J. C. S. Long, in *Flow and Transport Through Unsaturated Rock*, AGU Geophysics Mono. **43**, 121 (1987).

Shlesinger, M. F., B. J. West and J. Klafter, *Phys. Rev. Lett.* **58**, 1100 (1987).

Shraiman, B. I. and D. Bensimon, *Phys. Rev. A* **30**, 2840 (1984).

Siddiqui, H., Ph.D. Thesis, University of Southern California (1989).

Siddiqui, H. and M. Sahimi, *Chem. Eng. Sci.* **45**, 163 (1990a).

Siddiqui, H. and M. Sahimi, *J. Phys A* **23**, L497 (1990b).

Simon, R. and F.J. Kelsey, *Soc. Pet. Eng. J.* **11**, 99 (1971); **12**, 345 (1972).

Sing, K. S. W., D. H. Everett, R. A. W. Haul, L. Moscou, R. A. Pierotti, J. Rouquèrol and T. Siemieniwska, *Pure Appl. Chem.* **57**, 603 (1985).

Singhal, A. K. and W. H. Somerton, *J. Inst. Franc. Pet.* **32**, 897 (1977).

Sinha, S. K., T. Freltoft and J. Kjems, in *Kinetics of Aggregation and Gelation*, F. Family and D. P. Landau, eds. (Elsevier, New York, 1984), p. 87.

Slattery, J. C., *AIChE J.* **13**, 1066 (1967); **15**, 866 (1969).

Slobod, R. L., A. Chambers and W. L. Prekn, *Trans. AIME* **192**, 127 (1951).

Slobod, R. L. and R. A. Thomas, *Soc. Pet. Eng. J.* **3**, 9 (1963).

Smith, D. M. and D. G. Huizenga, in *Proceedings of the 10th IASTED Symposium in Applied Modelling and Simulation* (Acta Press, Calgary, Canada, 1984), p. 13.

Smith, L. and R. A. Freeze, *Water Resour. Res.* **15**, 1543 (1979).

Smith, L. and F. W. Schwartz, *Water Resour. Res.* **16**, 303 (1980); **17**, 351 (1981a); **17**, 1463 (1981b); **20**, 1241 (1984).

Snow, D. T., *Water Resour. Res.* **5**, 1273 (1969).

Somers, J. A. and P. C. Rem, in *Discrete Kinematic Theory, Lattice Gas Dynamics, and Foundation of Hydrodynamics*, R. Monaco, ed. (World Scientific, Singapore, 1989), p. 268.

Somers, J. A. and P. C. Rem, *Physica D* **47**, 192 (1991).

Sorbie, K. S., R. M. Swat and T. C. Rowe, Society of Petroleum Engineers Paper 16706 (1987).

Sørensen, J. P. and W. E. Stewart, *Chem. Eng. Sci.* **29**, 819 (1974).

Stalkup, F. I., *Miscible Displacement* (Society of Petroleum Engineers, Dallas, 1984).

Stanley, H. E. and P. Meakin, *Nature* **335**, 405 (1988).

Stanley, H. E., P. J. Reynolds, S. Redner and F. Family, in *Real-Space Renormalization*, T. W. Burkhardt and J. M. J. van Leeuwen, eds. (Springer, Berlin, 1982), p. 169.

Stauffer, D., J. Adler and A. Aharony, *J. Phys. A* **27**, L475 (1994).

Stauffer, D. and A. Aharony, 1992, *Introduction to Percolation Theory*, 2nd ed. (Taylor and Francis, London, 1992).

Stauffer, D., A. Coniglio and M. Adam, *Adv. Polymer Sci.* **44**, 103 (1982).

Stegemeier, G. L., Society of Petroleum Engineers Paper 4745 (1974).

Stegemeier, G. L., Paper 13c, 81st National Meeting of the American Institute of Chemical Engineers, Kansas City, Missouri (1976).

Stinchcombe, R. B., *J. Phys. C* **7**, 201 (1974).

Stinchcombe, R. B. and B. P. Watson, *J. Phys. C* **9**, 3221 (1976).

Stockmayer, W. H., *J. Chem. Phys.* **11**, 45 (1943).

Stockton, A. D., R. P. Thomas, R. H. Chapman and H. Dykstra, *J. Pet. Tech.* **36**, 2137 (1984).

Stokes, J. P., M. J. Higgins, A. P. Kushnik, S. Bhattacharya and M. O. Robbins, *Phys. Rev. Lett.* **65**, 1885 (1990).

Stokes, J. P., A. P. Kushnik and M. O. Robbins, *Phys. Rev. Lett.* **60**, 1386 (1988).

Stokes, J. P., D. A. Weitz, J. P. Gollub, A. Dougherty, M. O. Robbins, P. M. Chaikin and H. M. Lindsay, *Phys. Rev. Lett.* **57**, 1718 (1986).

Straley, C., A. Mateson, S. Feng, L. M. Schwartz, W. E. Kenyon and J. R. Banavar, *Appl. Phys. Lett.* **51**, 1146 (1987).

Straley, J. P., *J. Phys. C* **10**, 1903 (1977a); **10**, 3009 (1977b).

Stroud, D., *Phys. Rev. B* **12**, 3368 (1975).

Succi, S., E. Foti and F. Higuera, *Europhys. Lett.* **10**, 433 (1989).

Sudicky, E. A. and J. A. Cherry, *Water Pollut. Res. Can.* **14**, 1 (1979).

Sudicky, E. A., J. A. Cherry and E. O. Frind, *Water Resour. Res.* **21**, 1035 (1985).

Sudicky, E. A. and E. O. Frind, *Water Resour. Res.* **18**, 1634 (1982).

Sundaresan, S., *AIChE J.* **33**, 455 (1987).

Swanson, B. F., *J. Pet. Tech.* **33**, 2498 (1981).

Szep, J., J. Cresti and J. Kertész, *J. Phys. A* **18**, L413 (1985).

Tabeling, P. and A. Libchaber, *Phys. Rev. A* **33**, 794 (1986).

Taber, J. J., *Soc. Pet. Eng. J.* **9**, 3 (1969).

Talash, A. W., Society of Petroleum Engineers Paper 5810 (1976).

Tan, C.-T. and G. M. Homsy, *Phys. Fluids* **29**, 3549 (1986); **30**, 1239 (1987)

Tan, C.-T. and G. M. Homsy, *Phys. Fluids A* **4**, 1099 (1992).

Tang, C., *Phys. Rev. A* **31**, 1977 (1985).

Tang, D. H., E. O. Frind, and E. A. Sudicky, *Water Resour. Res.* **17**, 555 (1981).

Tang, D. H., F. W. Schwartz and L. Smith, *Water Resour. Res.* **18**, 231 (1982).

Tang, L.-H. and H. Leschhorn, *Phys. Rev. A* **45**, R8309 (1992).

Tao, R., M. A. Novotny and K. Kaski, *Phys. Rev. A* **A8**, 1019 (1988).

Tartar, L., Appendix 2 of *Nonhomogeneous Media and Vibration Theory, Lecture Notes in Physics* **127** (Springer, Berlin, 1980).

Taylor, G. I., *Proc. R. Soc. Lond.* **A219**, 186 (1953).

Tchalenko, J. S., *Geol. Soc. Amer. Bull.* **81**, 1625 (1970).

Thomas, G. H., G. R. Countryman and I. Fatt, *Soc. Pet. Eng. J.* **3**, 189 (1963).

Thomas, L. D., T. N. Dixon and R. G. Pierson, *Soc. Pet. Eng. J.* **23**, 42 (1983).

Thompson, A. H., *Ann. Rev. Earth Planet Sci.* **19**, 237 (1991).

Thompson, A. H., A. J. Katz and C. E. Krohn, *Adv. Phys.* **36**, 652 (1987a).

Thompson, A. H., A. J. Katz and R. A. Rashke, *Phys. Rev. Lett.* **58**, 29 (1987b).

Thompson, A. H., S. W. Sinton, S. L. Huff, A. J. Katz, R. A. Raschke and G. A. Gist, *J. Appl. Phys.* **65**, 3259 (1989).

Thorpe, M. F. and P. N. Sen, *J. Acoust. Soc. Am.* **77**, 1674 (1985).

Tilton, J. N. and A. C. Payatakes, *AIChE J.* **30**, 1016 (1984).

Timur, A., *Log Anal.* **9**, 8 (1968).

Timur, A., *J. Pet. Tech.* **21**, 775 (1969).

Todd, M. R. and W. J. Longstaff, *Soc. Pet. Eng. J.* **12**, 874 (1972).

Todorovic, P. N., *Water Resour. Res.* **6**, 211 (1970).

Todorovic, P. N., in *Proc. Inter. Symp. Stoch. Hydrol.* (Pittsburgh, Pennsylvania, 1971), p. 632.

Todorovic, P. N., *Water Resour. Res.* **11**, 348 (1975).

Todorovic, P. N., *Adv. Water Resour.* **5**, 42 (1982).

Toledo, P. G., R. A. Novy, H. T. Davis and L. E. Scriven, *Soil Sci. Soc. Am. J.* **54**, 673 (1990).

Tomadakis, M. M. and S. V. Sotirchos, *AIChE J.* **37**, 74 (1991).

Tompson, A. F. B. and W. G. Gray, *Water Resour. Res.* **22**, 591 (1986).

Topp, G. C., *Soil Sci. Soc. Amer. Proc.* **35**, 219 (1971).

Torquato, S., *J. Chem. Phys.* **81**, 5079 (1984).

Torquato, S., *Phys. Rev. Lett.* **64**, 2644 (1990).

Torquato, S., *Appl. Mech. Rev.* **44**, 37 (1991).

Torquato, S. and J. D. Beasley, *Phys. Fluids* **30**, 633 (1987).

Torquato, S., J. D. Beasley and Y. C. Chiew, *J. Chem. Phys.* **88**, 6540 (1988).

Torquato, S. and I. Kim, *J. Appl. Phys.* **72**, 2612 (1992).

Torquato, S. and B. Lu, *Phys. Fluids A* **2**, 487 (1990).

Torelli, L., *Pure Appl. Geophys.* **96**, 75 (1972).

Torelli, L. and A. E. Scheidegger, *J. Hydrol.* **15**, 23 (1972).

Treiber, L. E., D. L. Archer and W. W. Owens, *Soc. Pet. Eng. J.* **12**, 531 (1972)

Trugman, S. A., *Phys. Rev. B* **27**, 7539 (1983).

Tryggvason, G. and A. Aref, *J. Fluid Mech.* **154**, 287 (1985).

Tsakiroglou, C. D. and A. C. Payatakes, *J. Colloid Interface Sci.* **137**, 315 (1990).

Tsallis, C., A. Coniglio and S. Redner, *J. Phys. C* **16**, 4339 (1983).

Tsang, C. F., Y. W. Tsang and F. V. Hale, *Water Resour. Res.* **27**, 3095 (1991).

Tsang, Y. W., C. F. Tsang, I. Neretnieks and L. Moreno, *Water Resour. Res.* **24**, 2049 (1988).

Tsang, Y. W. and P. A. Witherspoon, *J. Geophys. Res.* **86**, 9287 (1981).

Turban, L., *J. Phys. C* **11**, 449 (1978).

Turcotte, D. L., *J. Geophys. Res.* **B91**, 1921 (1986).

Turner, G. A., *Chem. Eng. Sci.* **10**, 14 (1959).

Turpin, J. L. and R. L. Huntington, *AIChE J.* **13**, 1196 (1967).

Underwood, E. E, *Quantitative Stereology* (Addison Wesley, New York, 1970).

Van Brakel, J., *Powder Tech.* **11**, 205 (1975).

Van Damme, H., C. Laroche and L. Gatineau, *Revue. Phys. Appl.* **22**, 241 (1987a).

Van Damme, H., C. Laroche L. Gatineau and P. Levitz, *J. Physique* **48**, 1121 (1987b).

Van den Broeck, C., *Physica* **A112**, 343 (1982).

Van den Broeck, C. and R. M. Mazo, *Phys. Rev. Lett.* **51**, 1309 (1983).

Van den Broeck, C. and R. M. Mazo, *J. Chem. Phys.* **81**, 3624 (1984).

van Meurs, P., *Trans. AIME* **210**, 295 (1957).

Viani, B. E., P. F. Low and C. B. Roth, *J. Colloid Interface Sci.* **96**, 229 (1983).

Vicsek, T., *Phys. Rev. Lett.* **53**, 2281 (1984).

Vicsek, T., *Phys. Rev. A* **32**, 3084 (1985).

Vicsek, T., *Fractal Growth Phenomena*, 2nd ed. (World Scientific, Singapore, 1992).

Visscher, W. M. and M. Bolsterli, *Nature* **239**, 504 (1972).

Vollmar, S. and J. A. M. S. Duarte, *J. Physique II* **2**, 1565 (1992).

Vonk, C. G., *J. Appl. Cryst.* **9**, 433 (1976).

Voss, R. F., in *Fundamental Algorithms for Computer Graphics*, R. A. Earnshaw, ed., NATO ASI Series, Vol. F17 (Springer, Berlin, 1985), p. 805.

Wagner, O. R. and R. O. Leach, *J. Pet. Tech.* **1**, 65 (1959).

Wall, G. C. and R. J. C. Brown, *J. Colloid Interface Sci.* **82**, 141 (1981).

Walsh, M. P., S. L. Bryant, R. S. Schechter and L. W. Lake, *AIChE J.* **30**, 317 (1984).

Ward, J. S. and N. R. Morrow, *SPE Form. Eval.* **2**, 345 (1987).

Wardlaw, N. C., *AAPG Bulletin* **60**, 245 (1976).

Wardlaw, N. C., *J. Can. Pet. Tech.* **21**, 21 (1982).

Wardlaw, N. C. and J. P. Cassan, *Bull. Can. Pet. Geol.* **27**, 117 (1978).

Wardlaw, N. C., Y. Li and D. Forbes, *Trans. Porous Media* **2**, 597 (1987).

Warren, J. E. and H. S. Price, *Soc. Pet. Eng. J.* **1**, 153 (1961).

Warren, J. E. and P. J. Root, *Soc. Pet. Eng. J.* **3**, 245 (1963).

Warren, J. E. and F. F. Skiba, *Soc. Pet. Eng. J.* **4**, 215 (1964).

Washburn, E. W., *Proc. Nat. Acad. Sci. USA* **7**, 115 (1921).

Watson, B. P. and P. L. Leath, *Phys. Rev. B* **9**, 4893 (1974).

Webman, I., *Phys. Rev. Lett.* **47**, 1496 (1981).

Weinbrandt, R. M. and I. Fatt, *J. Pet. Technol.* **21**, 543 (1969).

Weissberg, H. L. and S. Prager, *Phys. Fluids* **5**, 1390 (1962); **13**, 2958 (1970).

Weitz, D. A., J. P. Stokes, R. C. Ball and A. P. Kushnik, *Phys. Rev. Lett.* **59**, 2967 (1987).

Wheatcraft, S. W. and S. W. Tyler, *Water Resour. Res.* **24**, 566 (1988).

Whitaker, S., *AIChE J.* **13**, 420 (1967).

Whitaker, S., *Trans. Porous Media* **1**, 3 (1986a); **1**, 105 (1986b).

White, I., P. M. Calumbera and J. R. Philip, *Soil Sci. Soc. Am. J.* **41**, 483 (1976).

Wilke, S., E. Guyon and G. de Marsily, *Math. Geol.* **17**, 17 (1985).

Wilkinson, D., *Phys. Rev. A* **30**, 520 (1984); **34**, 1380 (1986).

Wilkinson, D., *Phys. Fluids* **28**, 1015 (1985).

Wilkinson, D. and M. Barsony, *J. Phys. A* **17**, L129 (1984).

Wilkinson, D. and J. F. Willemsen, *J. Phys. A* **16**, 3365 (1983).

Willemsen, J. F., *Phys. Rev. Lett.* **52**, 2197 (1984).

Willemsen, J. F., *Phys. Rev. A* **31**, 432 (1985).

Williams, C. E. and B. M. Fung, *J. Magn. Res.* **50**, 71 (1982).

Willie, L. T. and J. Bennick, *J. Phys. A* **18**, L113 (1985).

Wilson, C. R. and P. A. Witherspoon, *Water Resour. Res.* **11**, 328 (1975); **12**, 102 (1976).

Wilson, S. D. R., *J. Colloid Interface Sci.* **51**, 532 (1975).

Winterfeld, P. H., L. E. Scriven and H. T. Davis, *J. Phys. C* **14**, 2361 (1981).

Witten, T. A. and L. M. Sander, *Phys. Rev. Lett.* **47**, 1400 (1981); *Phys. Rev. B* **27**, 5686 (1983).

Wolfram, S., ed., *Theory and Applications of Cellular Automata* (World Scientific, Singapore, 1986a).

Wolfram, S., *J. Stat. Phys.* **45**, 471 (1986b).

Wong, P.-Z., J. Howard and J.-S. Lin, *Phys. Rev. Lett.* **57**, 637 (1986).

Wong, P.-Z., J. Koplik and J. P. Tomanic, *Phys. Rev. B* **30**, 6606 (1984).

Wood, J. R. and T. A. Hewett, *Geochim. et Cosmochim. Acta* **46**, 1707 (1982).

Wood, J. R. and R. C. Surdam, *The Society of Economic Paleontologists and Mineralogists Special Publication No. 26* (1979), p. 243.

Wooding, R.A., *ZAMP* **13**, 255 (1962).

Wooding, R. A., *J. Fluid Mech.* **39**, 477 (1969).

Wooding, R. A. and H. J. Morel-Seytoux, *Annu. Rev. Fluid Mech.* **8**, 233 (1976).

Wright, R.-J. and R. A. Dawe, *Rev. Inst. Franc. Pet.* **38**, 455 (1983).

Wyllie, M. R. J. and W. D. Rose, *Pet. Trans. AIME* **189**, 105 (1950).

Xia, W., and M. F. Thorpe, *Phys. Rev. B* **38**, 2650 (1988).

Yadav, G. D., F. A. L. Dullien, I. Chatzis and I. F. MacDonald, *SPE Reservoir Eng.* **2**, 137 (1987).

Yanuka, M., F. A. L. Dullien and D. E. Elrick, *J. Microscopy* **135**, 159 (1984).

Yokoyama, Y. and L. W. Lake, Society of Petroleum Engineers Paper 10109 (1981).

Yonezawa, F. and M. H. Cohen, *J. Appl. Phys.* **54**, 2895 (1983).

Yortsos, Y. C., *AIChE J.* **33**, 1912 (1987a).

Yortsos, Y. C., *Phys. Fluids* **30**, 2928 (1987b).

Yortsos, Y. C. and F. J. Hickernell, *SIAM J. Appl. Math.* **49**, 730 (1989).

Yortsos, Y. C. and A. B. Huang, *SPE Reservoir Eng.* **1**, 378 (1986).

Yortsos, Y. C. and M. Zeybek, *Phys. Fluids* **31**, 3511 (1988).

Young, A. P. and R. B. Stinchcombe, *J. Phys. C* **8**, L535 (1975).

Ypma, J. G. J., Society of Petroleum Engineers Paper 12158 (1983).

Zanetti, G., *Phys. Rev. A* **40**, 1539 (1989).

Zgrablich, G., S. Mendioroz, L. Daza, J. Pajares, V. Mayagoitia, F. Rojas and Wm. C. Conner, *Langmuir* **7**, 779 (1991).

Zhdanov, V. P., V. B. Fenelonov and D. K. Efremov, *J. Colloid Interface Sci.* **120**, 218 (1987).

Zhang, Y.-C., *J. Physique* **51**, 2113 (1990a).

Zhang, Y.-C., *Physica* **A170**, 1 (1990b).

Zheng, L. H. and Y. C. Chiew, *J. Chem. Phys.* **90**, 322 (1989).

Zhou, M.-Y. and P. Sheng, *Phys. Rev. B* **39**, 12027 (1989).

Zick, A. A. and G. M. Homsy, *J. Fluid Mech.* **115**, 13 (1982).

Ziff, R. M., *Phys. Rev. Lett.* **69**, 2670 (1992).

Ziman, J. M., *Models of Disorder* (Cambridge Univesity Press, Cambridge, 1979), p. 154.

Zimmerman, S. P., C. F. Chu and K. M. Ng, *Chem. Eng. Commun.* **50**, 213 (1987).

Zimmerman, W. B. and G. M. Homsy, *Phys. Fluids A* **3**, 1859 (1991).

Zuzovski, M. and H. Brenner, *J. Appl. Math. Phys. (ZAMP)* **28**, 979 (1977).

Index

As a general rule, numbers refer to the pages where the most important use of the words was made.